CRC HANDBOOK SERIES IN NUTRITION AND FOOD

Miloslav Rechcigl, Jr.

Editor-in-Chief

SECTION OUTLINE

SECTION A: **Science of Nutrition and Food**
Nomenclature, Nutrition Literature, Organization of Research, Training and Extension in Nutrition and Food, Nutrition Societies, and Historical Milestones in Nutrition.

SECTION B. **Part 1: Ecological Aspects of Nutrition**
The Living Organisms, Their Distribution and Chemical Constitution, Habitats and Their Ecology, Biotic Associations and Interactions, Cycling of Water in the Biosphere, Ecosystem Productivity, and Factors Affecting Productivity of Animals and Plants.

SECTION B. **Part 2: Physiological Aspects of Nutrition**
Feeding and Digestive Systems in Various Organisms, Food Assimilation Processes in Microorganisms and Plants, Selection of Food in Various Organisms, Factors Affecting Food and Water Intake, Gastrointestinal Tract-Anatomical, Morphological and Functional Aspects, Passage of Ingesta in Various Organisms, Digestive Enzymes, Nutrient Absorption, Transport, and Excretion, Cellular Digestion and Metabolism.

SECTION C: **The Nutrients and Their Metabolism**
Chemistry and Physiology of Nutrients and Growth Regulators, Antinutrients and Antimetabolites, Nutrient Metabolism in Different Organisms, Regulatory Aspects of Nutrient Metabolism, Nutritional Adaptation, Biogenesis of Specific Nutrients, and Nutrient Interrelationships.

SECTION D: **Nutritional Requirements**
Comparative Nutrient Requirements, Qualitative Requirements of Specific Organisms, Tissues, and Cells, Quantitative Requirements (Nutritional Standards) of Selected Organisms, Nutritional Requirements for Specific Processes and Functions, and Requirements and Utilization of Specific Nutrients.

SECTION E: **Nutritional Disorders**
Nutritional Disorders in Specific Organisms, Nutritional Disorders in Specific Tissues, Effect of Specific Nutrient Deficiencies and Toxicities (3 Vol.), and Nutritional Aspects of Disease.

SECTION F: **Food Composition, Digestibility, and Biological Value**
Nutrient Content and Energy Value of Foods and Feeds, Factors Affecting Nutrient Composition of Plants, Factors Affecting Nutrient Composition of Animals, Effect of Processing on Nutrient Content of Foods and Feeds, and Utilization and Biological Value of Food.

SECTION G: **Diets, Culture Media, and Food Supplements**
Diets for Animals, Diets for Invertebrates, Culture Media for Microorganisms and Plants, Culture Media for Cells, Organisms and Embryos, and Nutritional Supplements.

SECTION H: **The State of World Food and Nutrition**
World Population, Natural and Food Resources, Food Production, Food Losses, Food Usage and Consumption, Socioeconomic, Cultural, and Psychological Factors Affecting Nutrition, Geographical Distribution of Nutritional Diseases, Nutrient Needs - Current and Projected, Agricultural Inputs - Current and Projected, Food Aid, and Food Marketing and Distribution.

SECTION I: **Food Safety, Food Spoilage, Food Wastes, Food Preservation, and Food Regulation**
Naturally Occurring Food Toxicants, Food Contaminants, Food Additives, Food Spoilage and Deterioration, Disposal of Food Wastes, Food-Borne Diseases, Detoxification of Foreign Chemicals, Food Sanitation and Preservation, Food Laws, and Nutrition Labeling.

SECTION J: **Production, Utilization, and Nutritive Value of Foods**
Plant and Animal Sources.

SECTION K: **Nutrition and Food Methodology**
Assessment of Nutritional Status of Organisms, and Measuring Nutritive Value of Food.

CRC Handbook Series

in

Nutrition and Food

Miloslav Rechcigl, Jr., Editor-in-Chief

Nutrition Advisor and Director
Interregional Research Staff
Agency for International Development
U.S. Department of State

Section E: Nutritional Disorders

Volume III

Effect of Nutrient Deficiencies
in Man

CRC Press, Inc.
2255 Palm Beach Lakes Boulevard · West Palm Beach, Florida 33409

Library of Congress Cataloging in Publication Data

Main entry under title:

Effect of nutrient deficiencies in man.

(Nutritional disorders ; v. 3)
Bibliography: p.
Includes index.
   1. Deficiency diseases. I. Rechcigl, Miloslav.
II. Series.
RC620.A1N87 vol. 3 [RC623.5] 616.3'9'008s 77-19152
ISBN 0-8493-2728-8   [616.3'9']

© 1978 by CRC Press, Inc.

International Standard Book Number 0-8493-2700-8 (Complete Set)
International Standard Book Number 0-8493-2728-8 (Volume 3)

Library of Congress Card Number 77-19152
Printed in the United States

PUBLISHER'S PREFACE

In 1913, when the First Edition of the *Handbook of Chemistry and Physics* appeared, scientific progress, particularly in chemistry and physics, had produced an extensive literature but its utility was seriously handicapped because it was fragmented and unorganized. The simple but invaluable contribution of the *Handbook of Chemistry and Physics* was to provide a systematic compilation of the most useful and reliable scientific data within the covers of a single volume. Referred to as the "bible," the Handbook soon became a universal and essential reference source for the scientific community. The latest edition represents more than 65 years of continuous service to millions of professional scientists and students throughout the world.

In the years following World War II, scientific information expanded at an explosive rate due to the tremendous growth of research facilities and sophisticated analytical instrumentation. The single-volume Handbook concept, although providing a high level of convenience, was not adequate for the reference requirements of many of the newer scientific disciplines. Due to the sheer quantity of useful and reliable data being generated, it was no longer feasible or desirable to select only that information which could be contained in a single volume and arbitrarily to reject the remainder. **Comprehensiveness** had become as essential as **convenience.**

By the late 1960's, it was apparent that the solution to the problem was the development of the multi-volume Handbook. This answer arose out of necessity during the editorial processing of the *Handbook of Environmental Control.* A hybrid discipline or, to be more precise, an interdisciplinary field such as Environmental Science could be logically structured into major subject areas. This permitted individual volumes to be developed for each major subject. The individual volumes, published either simultaneously or by some predetermined sequence, collectively became a multi-volume Handbook series.

The logic of this new approach was irrefutable and the concept was promptly accepted by both the scientist and science librarian. It became the format of a growing number of CRC Handbook Series in fields such as Materials Science, Laboratory Animal Science, and Marine Science.

Within a few years, however, it was clear that even the multi-volume Handbook concept was not sufficient. It was necessary to create an information structure more compatible with the dynamic character of scientific information, and flexible enough to accommodate continuous but unpredictable growth, regardless of quantity or direction. This became the objective of a "third generation" Handbook concept.

This latest concept utilizes each major subject within an information field as a "Section" rather than the equivalent of a single volume. Each Section, therefore, may include as many volumes as the quantity and quality of available information will justify. The structure achieves permanent flexibility because it can, in effect, expand "vertically" and "horizontally." Any section can continue to grow (vertically) in number of volumes, and new sections can be added (horizontally) as and when required by the information field itself. A key innovation which makes this massive and complex information base almost as convenient to use as a single-volume Handbook is the utilization of computer technology to produce up-dated, cumulative index volumes.

The *Handbook Series in Nutrition and Food* is a notable example of the "sectionalized, multi-volume Handbook series." Currently underway are additional information programs based on the same organizational design. These include information fields such as Energy and Agricultural Science which are of critical importance not only to scientific progress but to the advancement of the total quality of life.

We are confident that the "third generation" CRC Handbook comprises a worthy contribution to both information science and the scientific community. We are equally certain that it does not represent the ultimate reference source. We predict that the most dramatic progress in the management of scientific information remains to be achieved.

B. J. Starkoff
President
CRC Press, Inc.

PREFACE
CRC HANDBOOK SERIES IN NUTRITION AND FOOD

Nutrition means different things to different people, and no other field of endeavor crosses the boundaries of so many different disciplines and abounds with such diverse dimensions. The growth of the field of nutrition, particularly in the last two decades, has been phenomenal, the nutritional data being scattered literally in thousands and thousands of not always accessible periodicals and monographs, many of which, furthermore, are not normally identified with nutrition.

To remedy this situation, we have undertaken an ambitious and monumental task of assembling in one publication all the critical data relevant in the field of nutrition.

The *CRC Handbook Series in Nutrition and Food* is intended to serve as a ready reference source of current information on experimental and applied human, animal, microbial, and plant nutrition presented in concise tabular, graphical, or narrative form and indexed for ease of use. It is hoped that this projected open-ended multivolume set will become for the nutritionist what the *CRC Handbook of Chemistry and Physics* has become for the chemist and physicist.

Apart from supplying specific data, the comprehensive, interdisciplinary, and comparative nature of the *CRC Handbook Series in Nutrition and Food* will provide the user with an easy overview of the state of the art, pinpointing the gaps in nutritional knowledge and providing a basis for further research. In addition, the *Handbook* will enable the researcher to analyze the data in various living systems for commonality or basic differences. On the other hand, an applied scientist or technician will be afforded the opportunity of evaluating a given problem and its solutions from the broadest possible point of view, including the aspects of agronomy, crop science, animal husbandry, aquaculture and fisheries, veterinary medicine, clinical medicine, pathology, parasitology, toxicology, pharmacology, therapeutics, dietetics, food science and technology, physiology, zoology, botany, biochemistry, developmental and cell biology, microbiology, sanitation, pest control, economics, marketing, sociology, anthropology, natural resources, ecology, environmental science, population, law, politics, nutritional and food methodology, and others.

To make more facile use of the *Handbook,* the publication has been divided into sections of one or more volumes each. In this manner the particular sections of the *Handbook* can be continuously updated by publishing additional volumes of new data as they become available.

The Editor wishes to thank the numerous contributors, many of whom have undertaken their assignment in pioneering spirit, and the Advisory Board members for their continuous counsel and cooperation. Last but not least, he wishes to express his sincere appreciation to the members of the CRC editorial and production staffs, particularly President Bernard J. Starkoff, Mr. Robert Datz, Mr. Paul R. Gottehrer, and Ms. Marsha Baker, for their encouragement and support.

We invite comments and criticism regarding format and selection of subject matter, as well as specific suggestions for new data (and additional contributors) which might be included in subsequent editions. We should also appreciate it if the readers would bring to the attention of the Editor any errors or omissions that might appear in the publication.

Miloslav Rechcigl, Jr.
Editor-in-Chief
August 1978

PREFACE
SECTION E: NUTRITIONAL DISORDERS

This section provides systematic and detailed information on all relevant aspects of the relationship between nutrition and disease. One subsection is devoted to natural and foodborne diseases in various taxa of organisms, ranging from single-cellular type to higher organisms. Another subsection concerns itself with the effect of specific nutrient deficiencies and excess, described in terms of gross, morphological, and biochemical alterations. The changes in specific tissues and organs, due to malnutrition, and the nutritional aspects of disease form the theme of the remaining subsections.

Miloslav Rechcigl, Jr.,
Editor
May 1978

MILOSLAV RECHCIGL, JR., EDITOR

Miloslav Rechcigl, Jr. is Nutrition Advisor and Director of the Interregional Research Staff in the Agency for International Development, U.S. Department of State.

He has a B.S. in Biochemistry (1954), a Master of Nutritional Science degree (1955), and a Ph.D. in nutrition, biochemistry, and physiology (1958), all from Cornell University. He was formerly a Research Biochemist in the National Cancer Institute, National Institutes of Health and subsequently served as Special Assistant for Nutrition and Health in the Health Services and Mental Health Administration, U.S. Department of Health, Education, and Welfare.

Dr. Rechcigl is a member of some 30 scientific and professional societies, including being a Fellow of the American Association for the Advancement of Science, Fellow of the Washington Academy of Sciences, Fellow of the American Institute of Chemists, and Fellow of the International College of Applied Nutrition. He holds membership in the Cosmos Club, the Honorary Society of Phi Kappa Pi, and the Society of Sigma Xi, and is recipient of numerous honors, including an honorary membership certificate from the International Social Science Honor Society Delta Tau Kappa. In 1969, he was a delegate to the White House Conference on Food, Nutrition, and Health and in the last two years served as President of the District of Columbia Institute of Chemists and a Councilor of the American Institute of Chemists.

His bibliography extends over 100 publications, including contributions to books, articles in periodicals, and monographs in the fields of nutrition, biochemistry, physiology, pathology, enzymology, and molecular biology. Most recently he authored and edited *World Food Problem: A Selective Bibliography of Reviews* (CRC Press, 1975), *Man, Food, and Nutrition: Strategies and Technological Measures for Alleviating the World Food Problem* (CRC Press, 1973), *Food, Nutrition and Health: A Multidisciplinary Treatise Addressed to the Major Nutrition Problems from a World Wide Perspective* (Karger, 1973), following his earlier pioneering treatise on *Enzyme Synthesis and Degradation in Mammalian Systems* (Karger, 1971), and that on *Microbodies and Related Particles. Morphology, Biochemistry and Physiology* (Academic Press, 1969). Dr. Rechcigl also has initiated and edits a new series on Comparative Animal Nutrition and is Associated Editor of *Nutrition Reports International.*

CONTRIBUTORS
SECTION E: NUTRITIONAL DISORDERS
VOLUME III

Jerry K. Aikawa
School of Medicine
University of Colorado Medical Center
Denver, Colorado

Charles M. Baugh
Department of Biochemistry
College of Medicine
University of South Alabama
Mobile, Alabama

D. P. Burkitt
Department of Morbid Anatomy
St. Thomas' Hospital Medical School
London, England

Angel Cordano
Mead Johnson Research Center
Evansville, Indiana

Richard Cotter
Baxter Travenol Laboratories
Morton Grove, Illinois

John T. Dunn
Department of Internal Medicine
University of Virginia School of Medicine
Charlottesville, Virginia

Joseph C. Edozien
Department of Nutrition
School of Public Health
The University of North Carolina at Chapel
  Hill
Chapel Hill, North Carolina

Virgil F. Fairbanks
Departments of Internal Medicine and
  Laboratory Medicine
Mayo Clinic and Mayo Foundation
Rochester, Minnesota

C. Gopalan
Indian Council of Medical Research
Ansari Nagar
New Delhi, India

K. Michael Hambidge
University of Colorado Medical Center
Denver, Colorado

Harold E. Harrison
Department of Pediatrics
School of Medicine
The Johns Hopkins University
Baltimore, Maryland

A. V. Hoffbrand
Department of Hematology
The Royal Free Hospital
London, England

Ralph T. Holman
The Hormel Institute
University of Minnesota
Austin, Minnesota

John Eager Howard
Emeritus
School of Medicine
The Johns Hopkins University
Baltimore, Maryland

John Kroes
Cardiovascular Research Center
University of California Medical Center
San Francisco, California

R. M. Leach, Jr.
Department of Poultry Science
College of Agriculture
The Pennsylvania State University
University Park, Pennsylvania

Donald McLaren
Department of Physiology
University Medical School
Edinburgh, Scotland

L. Preston Mercer
Department of Biochemistry
Oral Roberts University
Tulsa, Oklahoma

Ralph A. Nelson
School of Medicine
University of South Dakota
Sioux Falls, South Dakota

H. Mitchell Perry, Jr.
Hypertension Program
Washington University School of Medicine
John Cochran Veterans Administration
Hospital
St. Louis, Missouri

G. P. Pineo
Department of Medicine
Faculty of Medicine
McMaster University
St. Joseph's Hospital
Hamilton, Ontario, Canada

Ananda S. Prasad
Veterans Administration Hospital
Allen Park, Michigan
Department of Medicine
Division of Hematology
Wayne State University School of Medicine
Detroit, Michigan

K. S. Jaya Rao
Endocrinology Unit
National Institute of Nutrition
Indian Council of Medical Research
Hyderabad, India

Sheldon P. Rothenberg
Division of Hematology/Oncology
Department of Medicine
New York Medical College
New York, New York

B. G. Shah
Nutrition Research, Food Directorate
Department of Health and Welfare
Ottawa, Ontario, Canada

Marian E. Swenseid
University of California School of Public
Health
Los Angeles, California

William C. Thomas, Jr.
Veterans Administration Hospital
College of Medicine
University of Florida
Gainesville, Florida

David I. Thurnham
Department of Human Nutrition
London School of Hygiene and Tropical
Medicine
London, England

H. C. Trowell
Former Consultant Physician
Makerere Medical School
Mulago Hospital
Kampala, Uganda

Pirkko R. Turkki
College for Human Development
Syracuse University
Syracuse, New York

Barbara A. Underwood
Division of Biological Health
College of Human Development
The Pennsylvania State University
University Park, Pennsylvania

Richard W. Vilter
Department of Internal Medicine
University of Cincinnati Medical Center
Cincinnati, Ohio

Marian Wang
School of Home Economics
University of Georgia
Athens, Georgia

Kunio Yagi
Institute of Biochemistry
University of Nagoya
Nagoya, Japan

To my inspiring teachers at Cornell: Harold H. Williams, John K. Loosli, Richard H. Barnes, the late Clive M. McCay, and the late Leonard A. Maynard.
And to my supportive and beloved family: Eva, Jack, and Karen.

TABLE OF CONTENTS
SECTION E: NUTRITIONAL DISORDERS
VOLUME III

Water-soluble Vitamins

EFFECT OF SPECIFIC NUTRIENT DEFICIENCIES IN MAN: THIAMIN

D. I. Thurnham

ETIOLOGY AND EPIDEMIOLOGY

Thiamin (also called thiamine, aneurin(e), vitamin B_1) is one of the water-soluble B-group vitamins. The vitamin is not stored within the body and man is dependent on a constant supply to replenish that being utilized by tissue functions. The principal sources of thiamin in the diet are either cereals or starchy roots and tubers which may contribute from 60 to 85% of dietary thiamin in a large part of the developing world.[1] However, in the last 50 years animal protein and dairy sources of thiamin have made an increasing contribution in developed countries. Thiamin intake is lowest in Asia and the Far East where rice is the main staple food. The germ and bran fractions of the cereal grain are the richest sources of thiamin; thus, intake is closely related to milling practices. Parboiled and undermilled rice provide more thiamin than highly milled (polished) rice, thus the Chinese and Japanese, who prefer the latter, have traditionally been associated with thiamin deficiency diseases.

The clinical conditions associated with thiamin deficiency are listed in Table 1. The most well-known of them, the disease beriberi, was particularly serious at the end of the 19th and in the early part of the 20th century when there were seasonal epidemics of wet beriberi with many deaths, while the more chronic form, dry beriberi, appeared in older segments of the population. Beriberi was associated with closed or controlled communities where normal eating practices were prevented[2] and the diet was severely restricted to polished rice of uncertain freshness and quality.[3] The view that "thiamin deficiency" and beriberi are not synonymous terms, however, is repeatedly expressed.[4,5] The acuteness of the thiamin deficiency and the interrelationship of thiamin deficiency with deficiencies of other nutrients are considered of very great importance in determining the pathological changes and lesions produced.[6] Platt[5] considered that some degree of protein-energy deficiency almost always accompanied subacute beriberi, reflecting the impoverished diet of the patient.

It has been suggested that severe beriberi more often attacked the more active, stronger, or supposedly slightly better-nourished members of a poor community[3,7] or the male infant who tended to be overfed.[8] It is probable that their slightly greater energy intake from a food deficient in thiamin reduces the thiamin-energy ratio below requirements and precipitates the attack. In many others, nonspecific pyrexia may be the precipitating factor, causing marginal thiamin intake to become deficient. Platt[5] reports that more than half his mild cases were associated with a nonspecific bout of fever and such cases responded less readily to treatment with thiamin.

Endemic beriberi has declined greatly in recent years, which must reflect changes in the diet, its distribution, variety and preparation, etc., social and economic factors, and in

Table 1
FORMS OF THIAMIN DEFICIENCY IN MAN

Wet beriberi
   Subacute beriberi or cardiac beriberi
Acute or fulminant type of beriberi
Dry beriberi – chronic, atrophic type, or polyneuropathy
Infantile beriberi
Wernicke-Korsakoff syndrome

the availability of synthetic thiamin for treatment and prophylaxis. In previously endemic areas, epidemics of subacute beriberi may still occur in institutions, such as prisons, when there is difficulty in the supply of rice, or seasonally among laboring workers subsisting on the last season's rice before the new harvest. Likewise reports of subclinical evidence of thiamin deficiency from biochemical measurements still occur from such areas.[9,10] Symptoms of thiamin deficiency may be a complication of any disease or treatment which restricts the intake of food. Severe vomiting of pregnancy or other causes of recurrent vomiting, very restricted dietary regimes, or alcoholism can result in thiamin deficiency (Table 1).

CLINICAL FEATURES OF BERIBERI

Beriberi is a state in which usually both cardiac and nervous functions are disturbed. It is generally believed, however, that the more serious the nervous lesions, the greater is the muscular pain and weakness and the less likely is the development of acute fulminating beriberi. The heart is saved from extreme insufficiency by the patient being forced to complete rest at an early stage in the attack.[5]

The course of the disease is extremely variable, so no one or two signs can be relied on to reveal all cases of beriberi.[3,7,11] This variability is illustrated by analyses made by Cruickshank of neurological and cardiovascular features displayed by European prisoners of war in Japanese camps.[11]

Endemic beriberi in adults is usually described as either wet or dry according to the presence or absence of edema. Tables 2, 3, 4, and 5 show the clinical features associated with these forms (Figure 1).

INFANTILE BERIBERI

Infantile beriberi was an important disease affecting most of Southeast Asia at the beginning of this century. Infants raised on breast milk deficient in thiamin developed the disease, usually in its acute form (Figure 2) while their mothers are usually only mildly affected.[15]

The onset of the disease may be mild in nature and may not be perceived by the mother, but when the frank condition develops, the course is rapid and frequently ends fatally.[14] The disease affects infants in the first year of life but rarely in the first month.[14] It can take a variety of forms (Figure 2), but it is the cardiac type (acute, fulminating beriberi) which is the most dangerous. This form was responsible for most of the deaths, the greatest mortality occurring between 2 and 5 months.[8,14,16] Symptoms

Table 2
EARLY SYMPTOMS AND SIGNS COMMON
TO BOTH WET AND DRY BERIBERI

An ill-defined malaise associated with heaviness and weakness of the legs (which may be accompanied by the following):
    Some difficulty in walking
    Slight edema of legs or face
    Precordial pain and palpitations
    Mild anorexia
    Tenderness in the calf muscles on pressure
    Paresthesia in the legs or arms (pins and needles, numbness)
    Sluggish tendon jerks (sometimes exaggerated)
    Anesthesia of the skin, especially over the tibia
    Pulse usually full and rate sometimes moderately increased

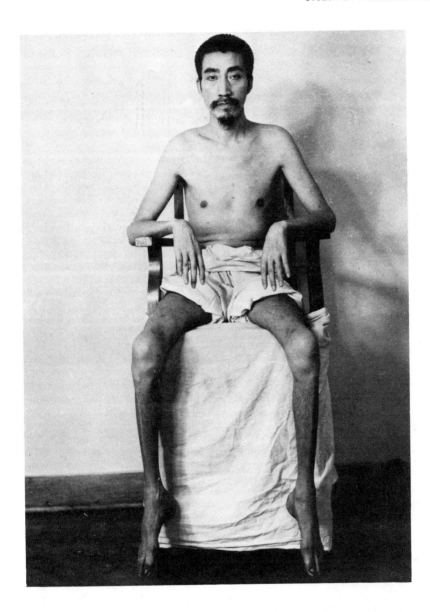

FIGURE 1.   Wrist drop and foot drop in a Chinese patient with dry beriberi. (Courtesy of the late Professor B. S. Platt, formerly head of the Department of Human Nutrition, London School of Hygiene and Tropical Medicine, London, England.)

and signs are in general very similar to adult beriberi, but in infantile beriberi there is frequently greater GI disturbance, anorexia, vomiting, and diarrhea than is found in adults[7] (Table 6). Response to treatment with thiamin is dramatic.

PATHOLOGY OF BERIBERI

Most of the information available on the pathology of beriberi is obtained from accounts written at the beginning of the 20th century, when beriberi was believed to be

Table 3
COMMON CLINICAL FEATURES OF SUBACUTE WET BERIBERI[3,7,11]

Anorexia common but not characteristic

Constipation more common than diarrhea

Aching pain, stiffness, tightness, or cramps in the calf or associated muscles (usually first cause of complaint in male laborers)

  Increasing muscular tenderness and weakness with fatigue pains resembling muscular ischemia

  Pain occurring especially at night before sleep

  Pain on squeezing calves — one of the most useful diagnostic tests of beriberi

  Inability to rise from squatting position without the use of the hands

Diminished reflexes of ankle and knees, though may be exaggerated

  Changes usually bilateral

Hypoesthesia or paresthesia (sensations in the fingers is a frequent first cause of complaint in women)

  Common symptoms: pins and needles, numbness, particularly over the tibia, formication, or itching

Edema of feet and legs — often an early symptom

  Frequently appears first on dorsa of feet and extending up the legs

  May also appear on the back of the hands or as a puffiness of the face

Heart enlarged — tachycardia, bounding pulse

  Raised venous pressure and percussion possibly revealing dilation of right auricle and ventricle

  Heart murmurs, if present, usually systolic

  Downward and outward displacement of the apex beat

  Neck veins possibly distended, showing visible pulsations

  Dyspnea upon exertion

  Palpitations, dizziness, and giddiness

  Extremities possibly cold and pale with peripheral cyanosis; where circulation well maintained, skin warm due to vasodilation

  Electrocardiograms often undisturbed, but QRS complex may show low voltage, and inversion of T waves may indicate disturbed conduction[12]

Nocturia — in ambulatory prisoners of war edema was noted during the day subsiding at night with consequent increased urination

  No albuminuria

Note: Beriberi as described here is the major form, which is typically seasonal in endemic areas.

Table 4
CLINICAL FEATURES OF ACUTE FULMINATING WET BERIBERI[3,7,11]

The whole picture is dominated by indications of insufficiency of heart and blood vessels, which tend to mask all the other features of the subacute form, although these are present and accentuated.

Vomiting is common and usually indicates the onset of acute symptoms.

Often intense thirst, but drinking initiates vomiting.

Pupils dilated: anxious expression on face.

The patients are severely dyspneic, have violent palpitations of the heart, are extremely restless, and experience intense precordial agony. Accessory muscles of respiration only slightly brought into action.

Aphonia frequently present. The patient moans and his cries take on a special character because of the coincident hoarseness produced as a result of paralysis of the laryngeal muscles.

Reflexes of ankle and knee lost or diminished.

Widespread and powerful undulating pulsations visible in the region of the heart, epigastrium, and neck due to a tumultuous heart action.

Facial cyanosis is more marked during inspiration.

Pulse is moderately full, regular, even, frequency of 120—150/min

A wave-like motion may be felt over the heart.

The liver is enlarged and tender and the epigastric region is spontaneously painful.

On percussion, the heart is found to be enlarged both to the left and to the right, but mainly the latter, and the apex beat may reach the axilla.

Raised systolic pressure and low diastolic pressure gives "pistol shot" sound on auscultation over large arteries.

Rapidly increasing edema may extend from legs to trunk and face with associated pericardial, pleural, and other serous effusions.

Oliguria or anuria, but no albuminuria or glycosuria, is present.

Death is accompanied by a systolic pressure falling to 80 or 70 mm, the pulse becomes thinner, and the veins dilate. The rough, whistling respiration deteriorates and rales appear. The patient dies intensely dyspneic, but usually fully conscious.

Table 5
CLINICAL FEATURES OF DRY BERIBERI[3,7,11]

Essentially a chronic condition showing muscular atrophy and polyneuritis and generally occurring in the older adult.

Weak, wasted, and painful musculature makes walking difficult and, at later stages, feeding and dressing almost impossible.

When bedridden and cachectic, very susceptible to infections.

Sensory nerve disturbances are evident from hypoesthesia (impaired sensitivity) in the extremities and progressively over outer aspects of legs, thighs, and forearms. Paresthesia is an early sign and in severe cases, kinesthesia of the foot and knee joints may give rise to Romberg's sign.

Anesthesia may become almost complete.

Motor nerve disturbances likewise begin in the extremities and ascend progressively. Flaccid paralysis of extensor muscles earlier and is more serious than that affecting the flexors and results in "wrist drop" and "foot drop" (see Figure 1).

Achilles tendon reflexes usually hypoactive before the patellar reflex is affected.

Cerebral manifestations are reported uncommon in China and Japan but were found among Caucasian prisoners of war in Japanese camps.[13]

Table 6
CLINICAL SIGNS OF INFANTILE BERIBERI

Chief symptom	%
Hoarseness	83.5
Tachycardia	83.5
Tachypnea	82.7
Vomiting	80.0
Oliguria	64.6
Accentuation of 2nd pulmonary sound	63.4
Paleness	53.8
Slight Fever	51.4
Groaning	49.4
Uneasiness	46.8
Dyspepsia	46.4
Cyanosis	40.0
Blepharoptosis	33.1
Dilatation of Heart	31.6
Tendon Reflex	
Decrease	25.9
Increase	74.1
Emaciation	25.6
Convulsion	17.4
Femoral sound	5.0

From Ota, T., *Nika Zasshi.*, 39, 1736, 1933; quoted by Inouye, K. and Katsura, E., in *Beriberi and Thiamine*, Shimazono, N. and Katsura, E., Eds., Vitamin B Research Committee of Japan, Kyoto University, 1965, 48. With permission.

due to an intoxication or infectious disease; thus, descriptions of pathological changes may have been influenced by these beliefs.[17]

The typically acute beriberi cadaver appears well nourished with frank anasarca and blood flows freely from an incision in subcutaneous veins. The chronic beriberi cadaver is emaciated, with marked muscular atrophy of extremities and no evidence of venous stagnation or fluidity of blood. Most autopsy cases however show a mixture of the above features with anasarca, ascites, plural effusion, pericardial effusion, and petechiae and ecchymoses of pericardium and pleura being common. Table 7 is a summary prepared by Vedder[18] and quoted by Williams.[11] It is essentially very similar to the more detailed descriptions of pathological findings given in the later summaries.[7,11]

WERNICKE-KORSAKOFF SYNDROME

Wernicke's encephalopathy and Korsakoff's psychosis are common neurological disorders which have been recognized since the 1880s. They represent two facets of the same syndrome and arise as a result of moderately severe dietary restriction usually of long duration and usually in association with alcoholism.[19] The distribution of cases is world wide and not geographically limited, as with beriberi.

The association with thiamin deficiency specifically was established in the 1930s.[19,20] However, overt signs of beriberi heart disease are rarely observed in the Wernicke-Korsakoff syndrome, but resting tachycardia and dyspnea on exertion are commonplace. Sudden cardiovascular collapse and death may occasionally follow mild exertion.[19] The more common occurrence of cardiac symptoms than encephalopathic in Oriental beriberi may be related to two factors. Encephalopathy may require a more complete deprivation than is usually accomplished by eating polished rice, while cardiac symptoms may be

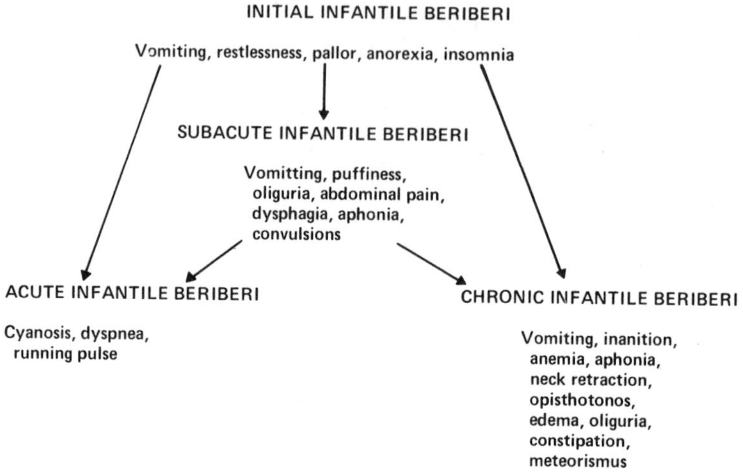

INITIAL INFANTILE BERIBERI

Vomiting, restlessness, pallor, anorexia, insomnia

SUBACUTE INFANTILE BERIBERI

Vomitting, puffiness,
oliguria, abdominal pain,
dysphagia, aphonia,
convulsions

ACUTE INFANTILE BERIBERI

Cyanosis, dyspnea,
running pulse

CHRONIC INFANTILE BERIBERI

Vomiting, inanition,
anemia, aphonia,
neck retraction,
opisthotonos,
edema, oliguria,
constipation,
meteorismus

FIGURE 2.   Course of infantile beriberi. (From Fehily, L., *Caduceus*, 19, 85, 1940. With permission.)

precipitated by tropical conditions where the hot, humid environment promotes congestion of the venous circulation.[21]

Wernicke's encephalopathy may be defined as a neurological disorder of acute onset, characterized by the following features which occur together or in various combinations: nystagmus; abducens and conjugate gaze palsies; ataxia of gait; global confusional state.

Korsakoff's psychosis refers to an abnormal mental state in which memory and learning are affected out of all proportion to other cognitive functions in an otherwise alert and responsive patient.[19] The amnesic condition may not be readily detected in acute Wernicke's disease because of the general mental confusion and apathy but it becomes apparent with improvement of the neurological disorders on treatment. The symptoms of Wernicke's disease all usually respond to treatment with thiamin regardless of the presence of alcohol but the Korsakoff component responds only slowly or not at all.[19] Table 8 is constructed from data of 245 cases of Wernicke-Korsakoff syndrome.[19]

The symptomatology produced by thiamin deficiency probably depends on the degree and acuteness of the deficiency.[21,22] In the more severe deprivations, the brain stem and cranial nerves are affected; in those with less complete deficiencies, only peripheral effects are apparent. Pathological changes, therefore, are as variable as symptoms, but nevertheless certain features are prominent. In Wernicke's encephalopathy, the histological changes are chiefly in the nature of capillary damage in the mammillary bodies, the related walls of the third ventricle, the aquaduct, and in the tegmentum of the medulla. When the condition is severe and acute, there are pinpoint hemorrhages in all these situations, while in more chronic states brown staining, loss of ground substance, endothelial proliferation, and ischemic changes in related nerve cells are seen.

Table 9 indicates gross neuropathological findings in 62 cases of Wernicke-Korsakoff syndrome.[19] The restricted nature of the findings do not correlate very well with the disturbance of function of brain and cerebellum which would be expected from the loss of awareness and ataxia which occurs in the syndrome.

BIOCHEMICAL LESIONS

Some of the earliest investigations on the biochemical defects which are produced by a lack of thiamin were carried out by Sir Rudolph Peters in pigeons. It was observed that lactate levels in vivo were elevated in the brain[23] and some other tissues[24] and that the

Table 7
POSTMORTEM FINDINGS IN 20 CASES OF BERIBERI

Organs affected and findings	No. of cases	% of cases
General		
Anasarca	11	55
Hypercardium	15	75
Hydrothorax	5	25
Ascites	10	50
Hydrops (all serous cavities)	4	20
Anasarca and hydrops (all cavities)	3	15
Heart		
Fatty degeneration (especially right ventricle)	20	100
Massive	7	35
Dilatation		
Right heart	14	70
Both sides	4	20
Hypertrophy		
Left ventricle	5	25
Both ventricles	4	20
Lungs		
Usually hyperemic and edematous		
Liver and kidneys		
General venous congestion with cloudy swelling, fatty degeneration of the cellular elements		
Spleen		
Usually enlarged		
Intestines		
Venous hyperemia of varying extent and intensity, with frequent small hemorrhages in the mucous membrane		
Nervous system		
Brain		
Usually venous hyperemia of the membranes, edema of the pia, and hyperemia of the brain; increased fluid in ventricles; edema and anemia of brain substance		
Cranial nerves		
Vagus degeneration	9	45
Spinal cord (sections)		
Macroscopic: hyperemia of membranes, and serous effusions in peridural and sub-arachnoid spaces	6	30
Microscopic: no abnormalities found	0	0
Atrophy and partial loss of ganglion cells of the anterior horn[a]	1	5
Peripheral nerves		
Severe degeneration and atrophy (most intense in muscular branches, less in nerve trunks)		

[a] In this one case, atrophy and partial loss of ganglion cells of the anterior horn was found, but was thought to be a secondary degeneration, and he considered the softening of the cord described by earlier writers as postmortem changes.

Table 7 (continued)
POSTMORTEM FINDINGS IN 20 CASES OF BERIBERI

Organs affected and findings	No. of cases	% of cases

Findings by osmic acid preparations
   Swelling and constriction of medullary
   sheath; degeneration into droplets;
   invasion of fatty granular cells; com-
   plete absorption of medulla and axis
   cylinder, so that only empty sheath
   of Swann remained		

Findings by cross sections with carmine
  stain
   Loss of fibers
   Degenerated fibers, stained uniformly
   red; no distinction discerned be-
   tween axis cylinder and medulla		

Muscles
  Degeneration of muscle fibers comparable to
  that of nerves		

From E. B. Vedder as cited by Williams, R. R., in *Toward the Conquest of Beriberi,* Harvard University Press, Cambridge, Mass., 1961, 54–55. With permission.

oxygen uptake, in the presence of glucose, of avitaminous brain tissue in vitro was reduced.[25] Subsequently, further work showed that thiamin was part of the pyruvate oxidase system and was active in the form of cocarboxylase (thiamin pyrophosphate or thiamin diphosphate).

In 1935, pyruvate and pyruvate aldehyde (methyl glyoxal) were found to be elevated in the blood of Chinese patients with fulminating beriberi.[26] Both these substances are toxic, but it is debatable to what extent they or the impaired oxidative function of thiamin-deficient tissues are responsible for the histological changes seen.[27] Furthermore, fasting blood levels of pyruvate in the chronic forms of beriberi[5] and Wernicke's disease[28] are normal and although they can be elevated by a loading dose of glucose, pyruvate per se is not considered to be an important pathogenic factor.

Lohmann and Schuster[29] showed that the active form of thiamin in the enzymic decarboxylation of pyruvate was thiamin pyrophosphate. (TPP). The coenzyme TPP is the major form of thiamin in mammalian tissues and takes part in several important enzyme systems of intermediary carbohydrate metabolism. Of those functions which have been most intensively studied in connection with thiamin nutrition, the following are probably the most important:

    1. Embden-Myerhof glycolytic pathway — decarboxylation of pyruvic acid to acetyl coenzyme A
    2. Citric acid cycle — decarboxylation of α-ketoglutamate to succinyl coenzyme A
    3. Hexose monophosphate shunt — transketolase reaction in which a 2-carbon unit from appropriate keto sugars is transferred to a suitable acceptor aldehyde

Studies by Dreyfus and colleagues on the distribution of pyruvate decarboxylase and transketolase in mammalian tissues have resulted in some interesting observations. Transketolase activity was found to be considerably higher in the myelinated structures than in the neuronal masses in the rat,[30] and preliminary studies on human specimens

Table 8
CLINICAL FEATURES OF WERNICKE-KORSAKOFF SYNDROME ON INITIAL EXAMINATION[19]

	No. of cases	% of cases
General medical abnormalities	245	
Tachycardia		51
Disorders of skin and mucous membranes		29
Redness and/or papillary atrophy of the tongue		36
Liver disease		60
Disorders of higher cerebral function	229[a]	
Global confusional state: profound disorientation; apathy; deranged perception; derangement of memory; drowsiness, indifference, inattentiveness		56
Disorder of memory – both retrograde and anterograde amnesia, confabulation		57
Alcohol abstinence syndrome		16
Stupor or coma		5
None		10
Ocular disorders[b]	232	
Nystagmus – almost always horizontal and in over half the cases associated with vertical nystagmus on upward gaze		85
Lateral rectus palsy – bilateral and usually		54
partial (2:1) Conjugate gaze palzy		44
Pupillary abnormality – sluggish reaction to light		19
Others (retinal hemorrhages, ptosis, scotomata)		9
None		4
Ataxia[c]	188[a]	
Gait		87
Legs		20
Arms		12
Speech		87
Polyneuropathy	230[a]	
Limbs only affected usually the legs (all 189 cases) or legs and arms (57 cases) – Common symptoms were weakness, parasthesia, pain, loss of tendon reflexes and of sensation and motor power, but in some a disproportionate effect on motor power took the form of foot drop, wrist drop, or both		82

[a] Number of cases in which it was possible to assess whether the appropriate clinical feature was present or not.

[b] Cruickshank[13] records vomiting and nystagmus as the most prominent early features in ten patients with Wernicke's encephalopathy in Japanese prisoner of war camps. Vomiting and ocular paralysis ceased dramatically on treatment with a small dose of vitamin B_1 (2–6 mg, i.v.).

[c] Cases of severe polyneuritis could not be tested because of muscular weakness.

From Victor, M., Adams, R. D., and Collins, G. H., in *The Wernicke-Korsakoff Syndrome,* Plum, F. and McDowell, F. H., Eds., Blackwell, Oxford, 1971. With permission.

Table 9
GROSS NEUROPATHOLOGICAL CHANGES IN 62 CASES
OF THE WERNICKE-KORSAKOFF SYNDROME

Abnormality	No. of cases	Percent
Opacity of leptomeninges	8	13
Convolutional atrophy	17	27
Decreased brain weight (<1100 g)	3	5
Enlarged lateral ventricles	16	26
Enlarged third ventricles	12	19
Lesion of mammillary bodies	46	74
Walls of third ventricle	15	24
Fornices	3	5
Periaqueductal gray	8	13
Superior colliculus	1	2
Inferior colliculus	3	5
Floor of fourth ventricle	9	15
Hemorrhages diencephalon and/or brain stem	6	10
Atrophy of superior cerebellar vermis[a]	10	36

[a] Based on 28 cases cut in sagittal plane.

From Victor, M., Adams, R. D., and Collins, G. H., in *The Wernicke-Korsakoff Syndrome,* Plum, F., and McDowell, F. H., Eds., Blackwell, Oxford, 1971, 73. With permission.

show similar though smaller differences.[31] In Wernicke's disease, the myelinated fibers of the mammillary bodies characteristically show disproportionately greater damage than the nerve cells in these structures; transketolase function might have a bearing on the pathological changes observed.[32] Furthermore, pyruvate decarboxylase, which tends to show highest in vitro activity in areas rich in neurones, is far less susceptible to thiamin deficiency than is transketolase in the nervous system of the rat.[33] However, organs such as heart, liver, and kidney in the rat suffer a sharp decrease in their ability to decarboxylate pyruvate even at the asymptomatic stage. The importance of these two biochemical lesions in the pathogenesis of beriberi and Wernicke's disease has still not been satisfactorily established,[34] and there is speculation that thiamin may subserve another function in nervous tissue which may be divorced from its coenzyme role.[35]

Approximately 10% of the body's thiamin is in the form of thiamin triphosphate (TTP).[36] Recent work has suggested that in subacute necrotizing encephalomyelopathy (Leigh's disease), there is no TTP in the brain.[37] Leigh's disease is hereditary and pathologically resembles Wernicke's encephalopathy. Additionally, it has been shown that deproteinized extracts of the patients' blood were able to inhibit the enzymic synthesis of TTP using the enzyme prepared from rat brain.[37] TTP may be the neurophysiologically active form of the vitamin which is pathogenically important in chronic beriberi and Wernicke's disease. It has been suggested that TTP is involved in the binding of TPP to apoenzyme moieties[38] or may be involved in ion movements in nervous tissue,[37] but the true function still requires clarification.

REFERENCES

1.   Requirements of vitamin A, thiamine, riboflavine and niacine, *WHO Tech. Rep Ser.*, 362, 1967.
2.   **Burgess, R. C.**, Proceedings of a conference on beriberi, endemic goitre and hypervitaminosis A entitled "Nutritional Disease." I. Epidemiology (section on beriberi), *Fed. Proc. Fed. Am. Soc. Exp. Biol.*, 17 (Suppl.2), 3–8, 1958.
3.   **Platt, B. S.**, in Proceedings of a conference on beriberi, endemic goitre and hypervitaminosis A entitled "Nutritional Disease." II. Clinical features of endemic beriberi, *Fed. Proc. Fed. Am. Soc. Exp. Biol.*, 17 (Suppl.2), 8–20, 1958.
4.   **Follis, R. H., Jr.**, in Proceedings of a conference on beriberi, endemic goitre and hypervitaminosis A entitled "Nutritional Disease." IV. Experimental thiamine deficiency, *Fed. Proc. Fed. Am. Soc. Exp. Biol.*, 17 (Suppl.2), 23–24, 1958.
5.   **Platt, B. S.**, Thiamine deficiency in human beriberi and in Wernicke's encephalopathy, in *Thiamine Deficiency*, (Ciba Found. Study Group, No. 28), Wolstenholme, G. E. W. and O'Connor, M., Eds., Churchill, London, 1967, 135–143.
6.   **Hundley, J. M.**, in Proceedings of a conference on beriberi, endemic goitre and hypervitaminosis A entitled "Nutritional Disease." IV. Discussion of experimental thiamine deficiency, *Fed. Proc. Fed. Am. Soc. Exp. Biol.*, 17 (Suppl.2), 23–31, 1958.
7.   **Inouye, K. and Katsura, E.**, Clinical signs and metabolism of beriberi patients, in *Beriberi and Thiamine*, Shimazono, N., and Katsura, E., Eds., Vitamin B Research Committee of Japan, Kyoto University, 1965, 29–63.
8.   **Fehily, L.**, Infantile beriberi in Hong Kong, *Caduceus*, 19, 78–93, 1940.
9.   **Thanangkul, O. and Whitaker, J. A.**, Childhood thiamine deficiency in Northern Thailand, *Am. J. Clin. Nutr.*, 18, 275–277, 1966.
10. **Chong, Y. H. and Ho, G. S.**, Erythrocyte transketolase activity, *Am. J. Clin. Nutr.*, 23, 261–266, 1970.
11. **Williams, R. R., Ed.**, The pathology and symptomatology of beriberi, in *Towards the Conquest of Beriberi*, Harvard University Press, Cambridge, Mass., 1961, 53–80.
12. **Davidson, S., Passmore, R., Brock, J. F., and Truswell, A. S., Eds.**, in *Human Nutrition and Dietetics*, 6th ed., Churchill, London, 1975, chap. 29, 335–342.
13. **Cruickshank, E. K.**, Wernicke's encephalopathy, *Q. J. Med.*, 19, 327–338, 1950.
14. **Burgess, R. C.**, in Proceedings of a conference on beriberi, endemic goitre and hypervitaminosis A entitled "Nutritional Disease." VI. Special problems concerning beriberi. B. Infantile beriberi, *Fed. Proc. Fed. Am. Soc. Exp. Biol.*, 17 (Suppl.2), 39–46, 1958.
15. **Williams, R. R., Ed.**, Infantile beriberi, in *Towards the Conquest of Beriberi*, Harvard University Press, Cambridge, Mass., 1961, 81–94.
16. **Ackroyd, W. R. and Krishnan, B. G.**, Rice diets and beriberi, *Indian J. Med. Res.*, 29, 551–555, 1941.
17. **Follis, R. M., Jr.**, in Proceedings of a conference on beriberi, endemic goitre and hypervitaminosis A entitled "Nutritional Disease." III. The pathology of endemic beriberi, *Fed. Proc. Fed. Am. Soc. Exp. Biol.*, 17 (Suppl. 2), 20–21, 1958.
18. **Vedder, E. B., Ed.**, in *Beriberi*, Wm. Wood, New York, 1913; quoted by Williams, R. R., Ed., in *Towards the Conquest of Beriberi*, Harvard University Press, Cambridge, Mass, 1961, chap. 4, 54–55.
19. **Victor M., Adams, R. D., and Collins, G. H.**, The clinical findings, in *The Wernicke-Korsakoff Syndrome*, Plum, F. and McDowell, F. H., Eds., Blackwell, Oxford, 1971, 16–34.
20. **Bowman, K. M., Goodhart, R., and Jolliffe, N.**, Observations on the role of vitamin B_1 in the etiology and treatment of Korsakoff Psychosis, *J. Nerv. Ment. Dis.*, 90, 569–575, 1939.
21. **Denny-Brown, D.**, in Proceedings of a conference on beriberi, endemic goitre, and hypervitaminosis A entitled "Nutritional Disease." VI. Special problems concerning beriberi. A. The neurological aspects of thiamine deficiency, *Fed. Proc. Fed. Am. Soc. Exp. Biol.*, 17 (Suppl. 2), 35–39, 1958.
22. **Swank, R. L.**, Avian thiamin deficiency. A correlation of the pathology and clinical behaviour, *J. Exp. Med.*, 71, 683–702, 1940.
23. **Kinnersley, H. W. and Peters, R. A.**, Observations upon carbohydrate metabolism in birds. I. The relation between the lactic acid content of the brain and the symptoms of opisthotonus in rice-fed pigeons, *Biochem. J.*, 23, 1126–1136, 1929.
24. **Fisher, R. B.**, Carbohydrate metabolism in birds. III. The effects of rest and exercise upon the lactic acid content of the organs of normal and rice-fed pigeons, *Biochem. J.*, 25, 1410–1418, 1931.
25. **Gavrilescu, W. and Peters, R. A.**, Biochemical lesions in vitamin B deficiency, *Biochem. J.*, 25, 1397–1409, 1931.

26. **Platt, B. S. and Lu, G. D.,** in *Proc. Physiol. Sect. 3rd Gen. Conf. Chinese Med. Assoc.,* Canton, November 1 to 8, 1935, 16; quoted by Platt, B. S., Vitamins in nutrition: orientations and perspectives, *Br. Med. Bull.,* 12, 78–86, 1956.

27. **Platt, B. S.,** Vitamins in nutrition: orientation and perspectives, *Br. Med. Bull.,* 12, 78–86, 1956.

28. **Bueding, E., Stern, M. H., and Wortis, H.,** Blood pyruvate curves following glucose ingestion in the normal and thiamine-deficient subjects, *J. Biol. Chem.,* 140, 697–703, 1941.

29. **Lohmann, K. and Schuster, P.,** Über die Co-Carboxylase, *Naturwissenschaften,* 25, 26–27, 1937.

30. **Dreyfus, P. M.,** The regional distribution of transketolase in the normal and the thiamine deficient nervous system, *J. Neuropathol. Exp. Neurol.,* 24, 119–129, 1965.

31. **Dreyfus, P. M.,** Transketolase activity in the nervous system, in *Thiamine Deficiency* (*Ciba Found. Study Group,* No. 28), Wolstenholme, G. E. W. and O'Connor, M., Eds., Churchill, London, 1967, 103–111.

32. **Victor, M., Adams, R. D., and Collins, G. H.,** Biochemical aspects of thiamine deficiency, in *The Wernicke-Korsakoff Syndrome,* Plum, F. and McDowell, F. H., Eds., Blackwell, Oxford, 1971, 155–162.

33. **Dreyfus, P. M. and Hauser, G.,** The effect of thiamine deficiency on the pyruvate decarboxylase system of the central nervous system, *Biochim. Biophys. Acta,* 104, 78–84, 1965.

34. **Cooper, J. R. and Pincus, J. H.,** The role of thiamin in nerve conduction, in *Thiamine Deficiency,* Wolstenholme, G. E. W. and O'Connor, M., Eds., Churchill, London, 1967, 112–121.

35. **Von Muralt, A.,** Physiology and clinical use of thiamine: the role of thiamine in neurophysiology, *Ann. N.Y. Acad. Sci.,* 98, 499–507, 1962.

36. **Rindi, G. and de Giuseppe, L.,** A new chromatographic method for the determination of thiamine and its mono-, di-, and tri-phosphates in animal tissues, *Biochem. J.,* 78, 602–610, 1961.

37. **Cooper, J. R., Itokawa, Y., and Pincus, J. H.,** Thiamin triphosphate deficiency in subacute necrotizing encephalomyelopathy, *Science,* 164, 74–75, 1969.

38. **Yusa, T. and Maruo, B.,** Biochemical role of thiamine triphosphoric acid ester, *J. Biochem.* (Tokyo), 60, 735–737, 1966.

EFFECT OF NUTRIENT DEFICIENCIES IN MAN: RIBOFLAVIN

K. Yagi

INTRODUCTION

Deficiency of riboflavin in diet produces ariboflavinosis in man. Ariboflavinosis is characterized by retarded growth of children and infants and the typical symptoms appearing in the borderline area between skin and mucous membrane. These phenomena could be ascribed to the decrease in the level of flavin coenzymes in the tissues and organs of man, caused by a deficiency of riboflavin in the diet. Synthesis of riboflavin by intestinal flora cannot fill the requirement of this vitamin in the human body. Administration of some antibiotics such as chlortetracycline causes ariboflavinosis. The mechanism of action of antibiotics cannot be explained as inhibition of riboflavin synthesis in the intestinal canal. However, it can be explained as the inhibition of flavin coenzyme synthesis from riboflavin and the direct inhibition of flavo-enzymes in vivo.[1] Administration of some hormones such as estrogen also provokes ariboflavinosis,[2] although the mechanism of action is not clear. The effects of antibiotics and hormones should be taken into account in considering dietary deficiency of riboflavin. However, all these can be overcome by simple administration of riboflavin or its adequate derivatives such as riboflavin monophosphate and riboflavin tetrabutyrate, a long-lasting derivative of riboflavin.[3]

RETARDED GROWTH WITH RIBOFLAVIN-DEFICIENT DIET

Riboflavin deficiency always results in retarded growth. Although there have been no reports on growth experiments in man, many studies have been done on animals such as rats, mice, and monkeys. It is believed that retarded growth also occurs in man.

Table 1 illustrates the type and ration of diet that causes ariboflavinosis in rats.[4] This diet contains practically no riboflavin. When rats weighing approximately 50 g were fed the riboflavin-deficient diet shown in Table 1, body weight gain ceased after 5 days of feeding. Thereafter, no growth was observed (see Figure 1). The failure of growth is directly related to the reduction of food intake, caused by riboflavin deficiency.

Plate A1 shows the littermate rats, one of which was fed a riboflavin-deficient diet (Table 1) and the other a normal diet (the diet shown in Table 1 plus 10 µg of riboflavin/day/rat).

LESIONS DUE TO RIBOFLAVIN DEFICIENCY

Animals

In addition to the loss of weight, various lesions caused by riboflavin-deficiency were reported on many experimental animals such as rats, mice, dogs, pigs, chicks, and monkeys. A rat fed a riboflavin-deficient diet (Table 1) is shown in Plate A2.

The characteristic symptoms of riboflavin deficiency observed in the rat are listed in Table 2. Remarkable symptoms are observable in the skin and eye. Neovascularization in the cornea should be emphasized as one of the outstanding ocular changes (Plate A3). Besides the skin and ocular changes, characteristic symptoms were observed in many other animals: ataxia in dogs,[9] diarrhea in pigs,[10] degeneration of nerve tissue in mice,[11] acute paralysis and dystonia in chicks,[12] and seborrheic scaly dermatitis, incoordination of the limbs, and reduction of hemoglobin in monkeys.[13,14] Maternal riboflavin deficiency in a rat causes absorption and congenital malformations in offspring such as

Table 1
COMPOSITION OF
RIBOFLAVIN-DEFICIENT DIET
FOR A RAT

Component	%
Vitamin-free casein	22
Sucrose	64.5
Cornstarch	2
Soybean oil	9
Salts mixture[a]	2.5
Vitamin B supplement[b]	
Fat-soluble vitamin[c]	

[a] NaCl 117 mg, $MgSO_4 \cdot 7H_2O$ 180 mg, $Na_2HPO_4 \cdot 12H_2O$ 235 mg, K_2HPO_4 644 mg, $Ca_3(PO_4)_2$ 365 mg, Ca lactate 879 mg, Fe 80 mg.

[b] Thiamine 5 μg, pyridoxine 5 μg, folic acid 5 μg, pantothenic acid 25 μg, p-aminobenzoic acid 25 μg, myoinositol 625 μg, choline 1.25 mg.

[c] Vitamin A, 10 IU, vitamin D, 1 IU, α-tocopherol, 0.1 mg.

Data from Forker, B. P. and Morgan, A. F., *J. Biol. Chem.*, 209, 303, 1954.

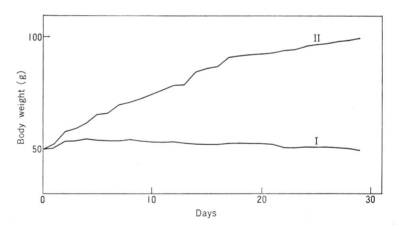

FIGURE 1.   Growth curve of rats fed a riboflavin-deficient diet. I: Riboflavin-deficient rats (*n* = 50); II: controls (*n* = 30). The values of M̄ are given. (From Hasegawa, Y. and Yagi, K., *J. Nutr. Sci. Vitaminol.*, 21, 397, 1975. With permission).

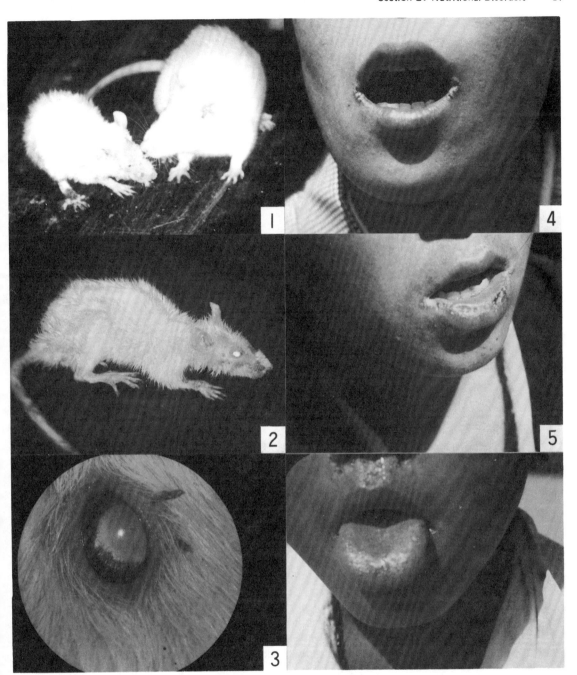

PLATE A.   1. A rat suffering from ariboflavinosis and a normal fat. Littermate rats weighing approximately 50 g were fed a riboflavin-deficient diet and a normal diet for 20 days. 2. A rat with ariboflavinosis. 3. The corneal vascularization of a rat suffering from ariboflavinosis. (From Hasegawa, Y. and Yagi, K., *J. Nutr. Sci. Vitaminol.,* 21, 399, 1975. With permission). 4. Shibi-Gacchakisho (an endemic ariboflavinosis). Angular stomatitis and cheilosis are seen. (Courtesy of Dr. K. Masuda). 5. Shibi-Gacchakisho. The lower lip became dry with a shallow ulceration. Angular stomatitis appeared. (Courtesy of Dr. K. Masuda). 6. Shibi-Gacchakisho. Seborrheic dermatitis around the nose appeared. The tongue became purplish red, and the papillae became swollen and flattened. (Courtesy of Dr. K. Masuda).

Table 2
CHARACTERISTIC SYMPTOMS OF RIBOFLAVIN DEFICIENCY OBSERVED IN A RAT

Changes in appearance[6,7]
    Ragged fur
    Uneven length hair
    Scaly skin
    Alopecia
    Red and swollen lip
    Abnormal filiform papillae of the tongue
Changes in the eye[7,8]
    Inflammation of the conjunctiva
    Blepharitis
    Corneal vascularization
    Corneal opacity

Table 3
RIBOFLAVIN-DEFICIENT RATION FOR HUMAN SUBJECTS

Component	Amount
Cornmeal	9.5 oz
Cow peas	0.48 oz
Lard	1.625 oz
Casein	2.43 oz
Flour	0.75 oz
White bread	3.6 oz
Calcium carbonate	3.0 g
Tomato juice	4.0 oz
Cod liver oil	0.5 oz
Syrup	4.75 oz
Syrup of iron iodide	2 drops
Ascorbic acid	30 mg weekly supplement
Thiamine	3.3 mg weekly supplement

Data from Sebrell, W. H. and Butler, R. E., *Public Health Rep.*, 53, 2282, 1938.

short mandibles, protruding tongues, malpositioned extremities, and various degrees of syndactylism.[15]

Man

The systematic study of experimental riboflavin deficiency in human subjects was first undertaken by Sebrell and Butler in 1938.[16] The dietary ration used (Table 3) contained a relatively small amount of riboflavin. They observed the development of a cheilosis in 10 of 18 women subjects for the period from 94 to 130 days after the beginning of the experiment. The initiation of cheilosis is pallor of the lip in the angles of the mouth, followed by maceration. Within a few days, superficial transverse fissures appeared in the angles of mouth. These lesions were described as resembling those of perlèche. Very little inflammatory reaction occurred, although the lips became abnormally red along the line of closure. In addition to the lesions on the lips, a fine, scaly, slightly greasy desquamation on a mildly erythematous base appeared in the nasolabial folds, the vestibule of the nose and ears, and on the alae nasi. These lesions disappeared after the administration of riboflavin (0.025 mg/kg body weight).

In 1939, Sydenstricker et al.[17] reported on five patients who presented lesions due to riboflavin deficiency. They possessed the typical dermatitis and conjunctivitis which were cured by riboflavin. Ocular manifestations of ariboflavinosis were reported by Sydenstricker et al.[18] in 1940. Using a slit lamp, they observed that the earliest lesions were proliferation and engorgement of the limbic plexus, followed by superficial vascularization of the cornea and the production of interstitial keratitis. These ocular symptoms were cured by administration of riboflavin. These specific lesions of the eye occur in the early stages of the deficiency, permitting easy recognition of ariboflavinosis.

Horwitt et al.[19,20] followed the ariboflavinosis in 15 human subjects with a diet containing 0.5 mg riboflavin in 2200 cal and in 8 subjects with a diet containing 0.55 mg riboflavin, low level of protein (41 g/day), niacin (5.8 mg/day), and vitamin B_{12} (0.001 mg/day). From these studies, the manifestations listed in Table 4 have been established as the common symptoms of the ariboflavinosis. Such findings were reported from Nyasaland, Ceylon, Jamaica, West Africa, Singapore, South India, Malaya, and Japan. In Japan, "Shibi-Gacchakisho" was reported by Masuda[21] as an endemic ariboflavinosis.

Table 4
SYMPTOMS OF ARIBOFLAVINOSIS IN MAN

Changes in skin and mucous membranes
  Angular stomatitis (lesion at the corner of the mouth)
  Cheilosis (lesion on lips, fissuring)
  Seborrheic dermatitis
  Scrotal dermatitis
  Erythema of buccal and palatal mucosa and tongue
Ocular changes
  Conjunctivitis
  Photophobia
  Corneal opacity
  Corneal vascularization
  Keratitis
  Proliferation of the limbic plexus

The changes in the skin and mucous membrane around the mouth and the nose, characteristic of ariboflavinosis are observed in Shibi-Gacchakisho (Plate A4—6). Masuda also pointed out the occurrence of similar lesions around the anus.

DIAGNOSIS AND THERAPY

Level of Flavins in Blood or Plasma

One method of determining the degree of ariboflavinosis is to measure the level of flavins in blood or plasma. Bessey et al.[22] assayed the concentration of free or total riboflavin in red blood cells fluorometrically after prolonged restriction of dietary riboflavin: on restricted diet (0.5 to 0.55 mg/day), the measurement was 10 to 13.1 μg/100 ml red blood cells and on supplemented diet (2.4 to 3.55 mg/day), 20.2 to 27.6 μg/100 ml red blood cells. The plasma flavin level decreases significantly according to the degree of ariboflavinosis as compared with the normal value of 1.7 ± 0.5 μg/100 ml.[23]

The level of riboflavin in blood, plasma, or serum can be determined accurately by the method devised by Yagi et al.[24] represented in Figure 2.

Decrease in the Level of FAD in Tissues

Riboflavin is converted to flavin monophosphate with flavokinase and then to flavin adenine dinucleotide with FAD pyrophosphorylase after absorption. In the tissues, FAD occupies the major part of the flavins. In ariboflavinosis, the level of FAD decreased markedly, as shown in Table 5. The reduction of the amount of coenzyme causes the degradation and disturbs the new synthesis of the respective apoprotein moiety. Yagi's method is recommended for the simultaneous determination of FAD, FMN, and riboflavin in tissue.[25]

Changes in the Activity of Enzyme Containing Flavin as a Coenzyme

In riboflavin-deficient animals, activities of tissue flavo-enzymes are generally reduced.[26,27] Enzymic activity in the liver of a rat with ariboflavinosis displays the following order of reduction: glycolate oxidase, D-amino acid oxidase, glycine oxidase, xanthine oxidase, L-amino acid oxidase, and NADH dehydrogenase.[28] Monoamine oxidase also decreased in both the heart and liver of a rat fed a riboflavin-deficient diet.[29] If it is possible to get specimens by liver biopsy, the examination of the enzyme pattern is valuable for diagnosis. Glutathione reductase, one of the flavo-enzymes in the erythrocyte, is stimulated by FAD in vitro. It is more stimulated in riboflavin-deficient rats than in those administered with riboflavin.[30]

In patients diagnosed as having ariboflavinosis, enzymic activity of glutathione

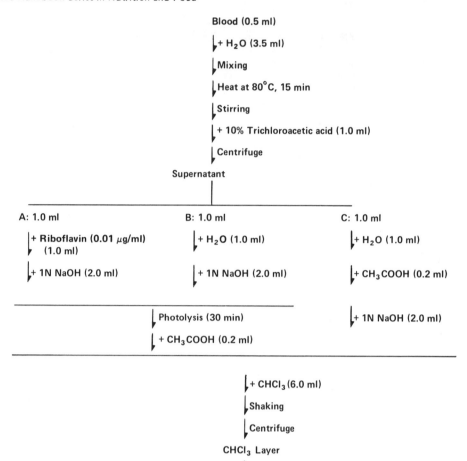

FIGURE 2.   Procedure for the assay of flavins in blood, plasma, or serum. Taking the fluorescence intensity of the tube A, B, and C as f_1, f_2, and f_3, respectively, the total flavin concentration in blood can be expressed in terms of $\mu g/100$ ml of blood: $0.01 \times \dfrac{f_2 - f_3}{f_1 - f_2} \times \dfrac{5.0}{0.5} \times 100 = \dfrac{f_2 - f_3}{f_1 - f_2} \times 10$. This figure was prepared from data obtained in Reference 24.

Table 5

RIBOFLAVIN AND FLAVIN COENZYME LEVELS
IN RIBOFLAVIN-DEFICIENT RAT ORGANS

Diet	Organ	Total flavins ($\mu g/g$)	FAD ($\mu g/g$)	FMN ($\mu g/g$)	Riboflavin ($\mu g/g$)
Control	Liver	32.7	23.7	7.6	1.4
	Kidney	36.7	21.1	14.8	0.8
Deficient	Liver	18.1	10.1	6.8	1.2
	Kidney	25.6	12.5	11.5	1.6

From Yagi, K., Yamamoto, Y., and Kobayashi, M., *J. Vitaminol.*, 14, 275, 1968. With permission.

reductase in blood was lower than in normal subjects. The stimulation of this enzyme by FAD in vitro is more marked in the erythrocyte of ariboflavinosis patients. The glutathione reductase level in blood was in accord with the blood FAD level.[31] From these facts, the evaluation of the erythrocyte glutathione reductase activity seems to be useful as one of the diagnosis of ariboflavinosis.

Urinary Excretion

The amount of urinary excretion is generally proportional to daily intake, and riboflavin is conserved when intake is restricted. Although patients with ariboflavinosis excrete low levels of riboflavin into urine, low urinary excretion of riboflavin is not a definitive sign of ariboflavinosis.

Therapy

Daily intake of riboflavin needed to prevent deficiency is 0.8 to 0.9 mg for the adult. To correct riboflavin deficiency, 5 mg of riboflavin should be administered twice or three times a day. Lesions of riboflavin deficiency disappear rapidly with the administration of riboflavin alone. Riboflavin tetrabutyrate, one of the riboflavin derivatives, is advantageous for clinical use due to its long-lasting effects.[3]

REFERENCES

1. Yagi, K., Yamamoto, Y., and Kobayashi, M., Effect of chlortetracycline to cause deficiency of flavins, *J. Vitaminol.*, 14, 271–277, 1968.
2. Ahmed, F., Bamji, M. S., and Iyengar, L., Effect of oral contraceptive agents on vitamin nutrition status, *Am. J. Clin. Nutr.*, 28, 606–615, 1975.
3. Yagi, K., Okuda, J., and Matsubara, T., Studies on fatty acid esters of flavins. IV. Fate of injected fatty acid esters of riboflavin in rabbits, *J. Vitaminol.*, 10, 275–283, 1964.
4. Forker, B. R. and Morgan, A. F., Effect of adrenocortical hormone on the riboflavin-deficient rat, *J. Biol. Chem.*, 209, 303–311, 1954.
5. Hasegawa, Y. and Yagi, K., Electron microscopic study on the lens of riboflavin-deficient albino rat, *J. Nutr. Sci. Vitaminol.*, 21, 395–401, 1975.
6. Goldberger, J. and Lillie, R. D., A note on an experimental pellagralike condition in the albino rat, *Public Health Rep.*, 41, 1025–1029, 1926.
7. Thatcher, H. S., Sure, B., and Walker, D. J., Avitaminosis. II. Pathologic changes in the albino rat suffering from vitamin G deficiency, *Arch. Pathol.*, 11, 425–433, 1931.
8. Bessey, O. A. and Wolbach, S. B., Vascularization of the cornea of the rat in riboflavin deficiency, with a note on corneal vascularization in vitamin A deficiency, *J. Exp. Med.*, 69, 1–11, 1938.
9. Sebrell, W. H. and Onstott, R. H., Riboflavin deficiency in dogs, *Public Health Rep.*, 53, 83–94, 1938.
10. Wintrobe, M. M., Buschke, W., Follis, R. H., Jr., and Humphreys, S., Riboflavin deficiency in swine, *Bull. Johns Hopkins Hosp.*, 75, 102–114, 1944.
11. Lippincott, S. W. and Morris, H. P., Pathologic changes associated with riboflavin deficiency in the mouse, *J. Natl. Cancer Inst.*, 2, 601–610, 1941.
12. Phillips, P. H. and Engel, R. W., Neuromalacia, its occurrence and histopathology in the chick on low riboflavin diets, *Poult. Sci.*, 17, 444, 1938.
13. Waisman, H. A., Production of riboflavin deficiency in the monkey, *Proc. Soc. Exp. Biol. Med.*, 59, 69–73, 1944.
14. Mann, G. V., Watson, P. L., McNally, A., and Goddard, J., Primate nutrition. II. Riboflavin deficiency in the cebus monkey and its diagnosis, *J. Nutr.*, 47, 225–241, 1952.
15. Warkany, J. and Schraffnberger, E., Congenital malformations induced in rats by maternal nutritional deficiency. VI. The preventive factor, *J. Nutr.*, 27, 477–484, 1944.
16. Sebrell, W. H. and Butler, R. E., Riboflavin deficiency in man, a preliminary note, *Public Health Rep.*, 53, 2282–2284, 1938.
17. Sydenstricker, V. P., Geeslin, L. E., Templeton, C. M., and Weaver, J. W., Riboflavin deficiency in human subjects, *JAMA*, 113, 1697–1700, 1939.

18. Sydenstricker, V. P., Sebrell, W. H., Cleckley, H. M., and Kruse, H. D., The ocular manifestations of ariboflavinosis, *JAMA,* 114, 2437–2445, 1940.

19. Horwitt, M. K., Hills, O. W., Harvey, C. C., Liebert, E., and Steinberg, D. L., Effects of dietary depletion of riboflavin, *J. Nutr.,* 39, 357–373, 1949.

20. Horwitt, M. K., Sampson, G., Hills, O. W., and Steinberg, D. L., Dietary management in a study of riboflavin requirement, *J. Am. Diet. Assoc.,* 25, 591–594, 1949.

21. Masuda, K., "Shibi-Gacchakisho" in Tsugaru area of Aomori Prefecture, *Vitamins,* 3, 189–196, 1950.

22. Bessey, O. A., Horwitt, M. K., and Love, R. H., Dietary deprivation of riboflavin and blood riboflavin levels in man, *J. Nutr.,* 58, 367–383, 1956.

23. Ito, T., Niwa, T., Matsui, E., Ohishi, N., and Yagi, K., Plasma flavin levels of patients receiving long-term hemodialysis, *Clin. Chim. Acta,* 39, 125–129, 1972.

24. Yagi, K., Kikuchi, S., and Kariya, T., Micro-determination of riboflavin in blood, *Vitamins,* 8, 450–451, 1955.

25. Yagi, K., Simultaneous microdetermination of riboflavin, FMN, and FAD in animal tissues, in *Methods in Enzymology,* Vol. 18B, McCormick, D. B., Ed., Academic Press, New York, 1971, 290–296.

26. Decker, L. E. and Byerrum, R. U., The relationship between dietary riboflavin concentration and the tissue concentration of riboflavin-containing coenzymes and enzymes, *J. Nutr.,* 53, 303–315, 1954.

27. Ramakrishnan, S. V. Srinivasan, V., and Nedungadi, T. M. B., Studies on the combined and relative influence of dietary protein and riboflavin in flavoenzymes, *J. Nutr.,* 75, 443–446, 1961.

28. Burch, H. B., Lowry, O. H., Padilla, A. M., and Combs, A. M., Effects of riboflavin deficiency and realimentation on flavin enzymes of tissues, *J. Biol. Chem.,* 223, 29–45, 1956.

29. Leodolter, S. and Genner, M., Monoamine oxidase activity and norepinephrine content of organs from rats fed on a vitamin B_2-deficient diet, *Arch. Int. Pharmacodyn. Ther.,* 190, 393–401, 1971.

30. Glatzle, D., Weber, F., and Wiss, O., Enzymatic test for the detection of a riboflavin deficiency. NADPH-dependent glutathione reductase of red blood cells and its activation by FAD in vitro, *Experientia,* 24, 1122, 1968.

31. Bamji, M. S., Glutathione reductase activity in red blood cells and riboflavin nutritional status in humans, *Clin. Chim. Acta,* 26, 263–269, 1969.

NUTRIENT DEFICIENCIES IN MAN: NIACIN

C. Gopalan and K. S. Jaya Rao

INTRODUCTION

Traditionally, niacin deficiency in man has been equated with pellagra. Recent evidence[1,2] shows that endemic pellagra is probably due to a secondary disturbance of tryptophan-niacin metabolism. Primary nutritional deficiency of niacin is rarely encountered. Sporadic cases may be seen among chronic alcoholics. Spies[3] lists a number of chronic diseases in which associated niacin deficiency may be seen; however, in reality such instances are rare.

Classically, pellagra has been considered to be a disease of the maize-eating population. The disease has practically disappeared from all affluent societies. It is still encountered among the Bantus of South Africa, in the Middle East, in Yugoslavia, and in the maize-eating populations of India. It is also endemic among the poor communities of India where the millet *Sorghum vulgare* is the staple food.

The development of pellagra on maize diets was attributed to the low niacin content of maize. However, Aykroyd and Swaminathan[4] showed that the niacin content of maize is not lower than that of other cereals. Later studies have claimed that niacin in maize is largely in a bound and biologically unavailable form.[5] Maize is also deficient in tryptophan, the amino acid precursor of niacin. Hence, it had come to be accepted that the low tryptophan and niacin content of maize was responsible for its pellagragenic nature. The identification of pellagra in populations subsisting on sorghum, a millet where niacin is neither so low nor biologically unavailable, had raised serious doubts regarding this concept.[1] Now there is compelling evidence indicating that excess leucine and probably an isoleucine-leucine imbalance could be responsible for pellagra in sorghum-eating as well as maize-eating populations.[2,6] Recent studies indicate that concomitant pyridoxine deficiency, either primary or brought about by leucine toxicity, may also have an important role.[7,8]

CLINICAL FEATURES

The main clinical features of pellagra have been popularly classified as the three Ds of pellagra: dermatitis, diarrhea, and dementia. Of the three, dermatitis is the characteristic feature of the disease; the other two may or may not be present.

Skin Changes

The dermal lesions are bilateral, symmetrical, and seen on those parts of the body constantly exposed to sunlight. Therefore, the common sites are the extensor surfaces of the extremities, face, and neck. The lesions are also found, at times, at sites of constant irritation, e.g., regions under the breasts, scrotum, axilla, and perineum. A characteristic feature of these photosensitive lesions is their clear-cut demarcation from the adjoining, unaffected parts. The appearance of the dermatosis varies with the severity of the disease and from patient to patient (Figures 1 to 3).

The lesion starts as an erythematous dermatitis which later gives place to hyperpigmentation. There is dryness and scaling of the skin. Desquamation usually follows, revealing a smooth, shiny, depigmented epidermis. Hyperkeratinization is also very common.

The pigmentation on the face is generally limited to the cheeks and the bridge of the nose and referred to by the descriptive term butterfly pigmentation. The eponym, Casal's

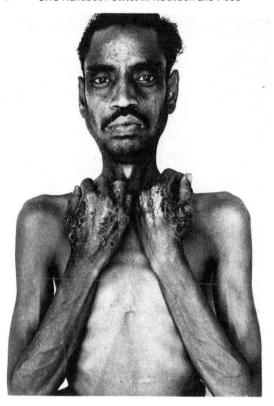

FIGURE 1.   Pellagrous dermatitis: limited to hands.

FIGURE 2.   Extensive dermatitis on extensor surface of forearms and hands.

necklace, describes the lesion extending over the scapuloclavicular area; Casal was the first to give an accurate description of pellagra (Figure 4).

Seborrheic dermatitis of the nasolabial folds is also frequently seen. However, it is considered to be due to an associated riboflavin deficiency rather than to pellagra itself.

Gastrointestinal Changes

Anorexia and a burning or raw sensation of the mouth, stomach, and rectum have been described by various workers. Angular stomatitis, cheilosis, atrophy of the lingual papillae, and glossitis are commonly seen. However, it is not known whether these are manifestations of niacin deficiency or whether they are due to the associated deficiency of riboflavin and other members of the vitamin B-complex group. The tongue and lips are swollen, red, and painful.

Diarrhea, although classified among the classic three Ds of pellagra, is not always present. On the other hand, constipation is a common complaint. Diarrhea, when present, has no characteristic pattern. According to Spies,[3] nearly 60% of the cases have achylia gastrica.

Mental Changes

Striking among the various clinical manifestations of pellagra are the mental changes. These may range from mild symptoms such as insomnia and depression to marked emotional instability and mania. More general symptoms such as headache, irritability, inability to concentrate, and apathy may also be encountered. Insomnia is a very

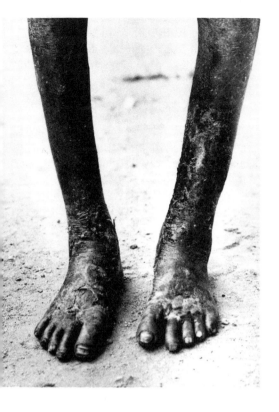

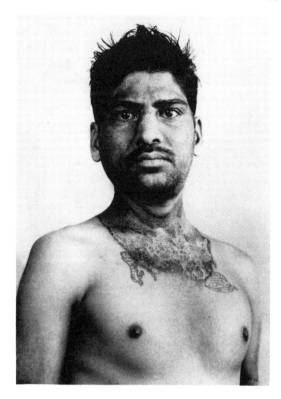

FIGURE 3.   Dermatitis on feet and shins. Clothing worn only up to the knees in this population.

FIGURE 4.   Casal's necklace. The figures on either side at the base are tattoo marks.

common complaint. The more noticeable features are confusion, loss of memory, and disorientation. Various psychoses, including manic-depressive syndromes, severe paranoia, hallucinations, etc., may necessitate confinement in a mental asylum. A good percentage of the inmates of such institutions have been reported to be pellagrins,[9] which bears testimony to the serious nature of the mental changes.

Acute encephalopathies have also been described.[10] Paresthesias, hyperesthesia, loss of vibration and position sense, and exaggerated deep reflexes have all been described.[11,12] Certain of these, for instance paresthesias and polyneuritis, could also be due to a concomitant deficiency of other members of the B-vitamin group.

EPIDEMIOLOGY

This disease afflicts both sexes equally. It is usually seen only in adults. What had earlier been described as infantile pellagra[13] is now identified as kwashiorkor.

Seasonal exacerbations and remissions have been described, the incidence being highest when sunlight is plentiful. In India, where ample sunlight is present throughout the year, the seasonal incidence is related to the food consumption pattern. Sorghum is usually eaten in lean periods, and rice or wheat during the harvest season. The peak incidence of pellagra occurs 3 to 4 months following the lean season.

PATHOLOGICAL CHANGES

Available information regarding histopathological changes is mostly very old. These data were obtained from studies conducted before techniques such as histochemistry and electron microscopy were developed. In the two exhaustive accounts available,[13,14] pellagra and protein malnutrition have not been identified as separate disease entities. Thus, one is left to speculate whether the changes are descriptive of niacin deficiency or of protein-calorie malnutrition.

Skin

There is capillary dilatation in the dermis and hyperkeratosis of the epidermal layer, and the stratum corneum is heavily pigmented. However, the pigment distribution is uneven, and it is believed that this may account for the patchy distribution of the skin pigmentation.[13] There may be edema in the deeper layers.[12]

The sebaceous glands on the face are hypertrophied; the ducts are dilated and filled with lipid material. On the other parts of the body, the glands are atrophied. Sweat glands show no changes.

Gastrointestinal System

The mucous membranes of the lips, tongue, and mouth show capillary dilatation. The mucosa of the tongue, stomach, and intestine show atrophy. However, intestinal absorption has been reported not to be impaired.[15,16] Although Mehta et al.[17] observed malabsorption, the role of concomitant protein-calorie malnutrition was not ruled out. There is glandular degeneration in the stomach and evidence of decreased secretion. Achylia gastrica is frequently seen. Fatty changes in the liver have also been described.

Nervous System

Degeneration of the neurons, peripheral nerves, and the spinal cord has been described. The neurons show an increase in pigment and fat, and the nucleus is pushed to a side.

BIOCHEMICAL CHANGES

As mentioned earlier, recent investigations indicate that pellagra is due to a secondary disturbance of tryptophan-niacin metabolism. Intensive studies have indicated that this is brought about by an excess of dietary leucine. Observations which have led to this concept have been reviewed elsewhere.[6,18]

Urine

The excretion of N'-methyl nicotinamide, its 2-pyridone form, nicotinic acid, kynurenic acid, quinolinic acid, and 5-hydroxy indole acetic acid was found to be low in pellagrins[19,20] (Table 1).

Blood

Plasma tryptophan levels are low in pellagrins (Table 2). No striking alterations are found in the plasma concentrations of other amino acids.[21,22]

The functional forms of niacin in the body are nicotinamide adenine dinucleotide and nicotinamide adenine dinucleotide phosphate. The concentration of the total nucleotides in the erythrocytes of pellagrins is not altered.[23,24] Similar observations were also made in subjects placed on niacin-deficient diets.[25] However, the levels of NAD and NADP individually are low.[26] This discrepancy is explained by the observation that NMN, which is normally negligible, is present in high concentrations in pellagrins (Table 3). Vivian et al.[27] also reported a decrease in NAD levels in subjects placed on niacin-deficient diets. The rate of erythrocyte nucleotide synthesis is also depressed in pellagrins (Table 4). Similar observations were made when L-leucine was orally

Table 1
URINARY EXCRETION OF TRYPTOPHAN METABOLITES IN PELLAGRINS (μmol/24 hr)

	Normals[19]	Pellagrins[19]	Normals[20]	Pellagrins[20]
Nicotinic acid	49.0 ± 6.3	40.0 ± 6.7	7.8 ± 1.4	4.6 ± 2.4
N'-Methyl nicotinamide	44.5 ± 1.6	27.0 ± 2.2	39.0 ± 6.0	5.1 ± 4.9
Quinolinic acid	56.3 ± 14.0	31.1 ± 4.8	37.9 ± 11.3	22.7 ± 12.6
Kynurenine			16.0 ± 4.0	6.8 ± 3.6
Xanthurenic acid			8.0 ± 2.0	5.0 ± 6.9
Tryptophan	887.0 ± 62.7	556.0 ± 48.8		

Table 2
PLASMA AMINO ACIDS IN PELLAGRA (μmol/100 ml)

	Tryptophan	Leucine	Isoleucine	Valine	Lysine	Ref.
Normals	2.8	9.0	5.5	16.5	7.5	22
Pellagrins	2.0	7.8	5.5	13.0	7.0	22
	1.2	13.2	4.8	18.3	17.8	21

Note: Values are adapted from references quoted.

Table 3
FRACTIONATION OF NICOTINAMIDE NUCLEOTIDES IN ERYTHROCYTES[26,28]

	Total concentration (mg/100 ml erythrocytes)	NAD (%)	NADP (%)	NMN (%)
Normal subjects	5.0 ± 0.75	58.0	39.0	5.0
Pellagrins	4.9 ± 0.66	53.4	27.6	18.9
Normal subjects fed 10 g L-leucine daily for 10 days	5.2 ± 0.45	50.0	35.0	16.0

Table 4
CONCENTRATION AND RATE OF SYNTHESIS OF NICOTINAMIDE NUCLEOTIDES IN ERYTHROCYTES: EFFECT OF ORAL ADMINISTRATION OF L-LEUCINE, 10 g/day[23]

	Initial		5 days after leucine	
	Total concentration (mg/100 ml)	Rate of synthesis	Total concentration (mg/100 ml)	Rate of synthesis
Normal subjects	4.7 ± 0.42	11.8 ± 0.90	4.3 ± 0.29	6.9 ± 1.10
Pellagrins	4.9 ± 0.38	7.6 ± 1.80	4.3 ± 0.56	4.7 ± 1.66

administered to normal subjects. Leucine could also inhibit nucleotide synthesis in vitro.[28]

Skin

The pathogenesis of dermatosis has not been adequately investigated. Although altered porphyrin metabolism has been implicated,[13] this has never been proved. Recent studies showed alterations in the collagen content and in the amino acid composition of the dermis.[29] It was also observed that the amino acid composition of the skin of nonpellagrous subjects obtained from an area chronically exposed to sunlight was different from that obtained from a protected area. Such an altered skin is probably more susceptible to niacin deficiency, thus possibly explaining the reason for the photosensitive nature of pellagrous dermatosis. Lowered levels of tryptophan metabolites have been reported in the skins of pellagrins.[30]

Urocanic acid normally constitutes 80% of the UV absorbing substances in the epidermis.[31] Pellagrous skin contains low levels of this compound.[32] The levels of histidine, the precursor amino acid of urocanic acid, are also low in pellagrous skin. Thus, the altered amino acid composition of the skin may be the underlying biochemical lesion in the photosensitive dermatosis of pellagra.

Biochemistry of Mental Changes

Serotonin is believed to play an important role in the normal functioning of mental processes.[33] Pellagrins exhibiting mental depression or emotional instability had low levels of serotonin in their platelets. Pellagrins without mental changes had normal levels of the bioamine.[34] However, Stratigos et al. observed high levels of plasma serotonin in pellagrins.[35]

An excess theta or delta activity of the brain is evident in most pellagrins, particularly those with mental changes.[36] This pattern could also be precipitated by the administration of L-leucine.

BIOCHEMICAL ASSESSMENT OF NIACIN STATUS

Estimation of the urinary excretion of the two major metabolites of niacin, N'-methyl nicotinamide and its 2-pyridone form, is the only reliable method available for estimation of niacin status. The ratio of these two metabolites is considered to be the best criterion for evaluation.[37,38] A ratio of less than 1.0 is considered indicative of niacin deficiency while ratios ranging between 1 and 4 are considered acceptable.[39]

INTERRELATIONSHIP BETWEEN DEFICIENCY OF NIACIN AND OTHER B-COMPLEX VITAMINS

Early in the history of the discovery of vitamins, there arose a question as to whether pellagra is due to niacin deficiency alone or whether there could be multiple vitamin deficiencies involving niacin, riboflavin, and pyridoxine. This confusion probably arose because, in animals as well as humans, some of the signs are common to all three vitamin deficiencies. In 1939, Vilter et al.[40] claimed that riboflavin was beneficial to some pellagrins. However, Sebrell's studies[41] showed that pellagra and ariboflavinosis are two distinct syndromes. It is, however, still believed that riboflavin may have a role in the tryptophan-niacin pathway.[42,43] Henderson et al.[44] demonstrated decreased quinolinic acid excretion in urine of riboflavin-deficient rats. Recently, Verjee[45] also demonstrated disturbances in tryptophan metabolism in baboons rendered riboflavin deficient. Some of the clinical features of pellagra such as nasolabial dyssebacia, angular stomatitis, and glossitis are probably due to an associated riboflavin deficiency.

It is now well established that some of the enzymes in the tryptophan-niacin pathway are pyridoxine dependent.[46] Animals reared on diets deficient in vitamin B_6 have been shown to excrete decreased amounts of N-methyl nicotinamide following a tryptophan load.[47,48] The synthesis of nicotinamide nucleotides in erythrocytes has also been demonstrated to be low.[45,49]

Recent studies show that administration of large doses of vitamin B_6 can prevent some of the changes brought about by feeding leucine. Thus, the increase in the activities of hepatic tryptophan oxygenase and kynureninase observed in rats fed leucine could be prevented by simultaneous administration of vitamin B_6.[7] Similarly, the alterations in urinary excretion of quinolinic acid and 5-hydroxyindole acetic acid brought about in normal human subjects during supplementation with leucine could be prevented by simultaneous administration of vitamin B_6.[8] Preliminary clinical trials showed that pellagrins may also have vitamin B_6 deficiency, and it was also found that the dermatitis of pellagra could be treated with vitamin B_6.[50] The authors postulated that excess leucine may bring about a conditioned deficiency of vitamin B_6. How this might have been brought about has yet to be elucidated.

TREATMENT

Pellagra can be treated by oral administration of 100 to 300 mg of niacinamide daily, in three divided doses. Niacinamide does not possess the unpleasant vasoactive effects of nicotinic acid. Mental changes disappear within 24 to 48 hr and dermal lesions take 3 to 4 weeks for total disappearance. Most cases require concomitant administration of riboflavin and vitamin B_6, the former particularly for the healing of oral lesions. Mental changes of a violent nature can be controlled by intramuscular injections of niacinamide, 150 mg twice daily. A diet high in calories and protein is prescribed for all patients, since most of them are undernourished.

REFERENCES

1. Gopalan, C. and Srikantia, S. G., Leucine and pellagra, *Lancet,* 2, 954–957, 1960.
2. Gopalan, C., Possible role for dietary leucine in the pathogenesis of pellagra, *Lancet,* 1, 197–199, 1969.
3. Spies, T. D., Niacinamide malnutrition and pellagra, in *Clinical Nutrition,* 2nd ed., Jolliffe, N., Ed., Harper & Brothers, New York, 1962, 622–634.
4. Aykroyd, W. R. and Swaminathan, M., The nicotinic-acid content of cereals and pellagra, *Indian J. Med. Res.,* 27, 667–677, 1940.
5. Kodicek, E., The availability of bound nicotinic acid to the rat. 2. The effect of treating maize and other materials with sodium hydroxide, *Br. J. Nutr.,* 14, 13–24, 1960.
6. Gopalan, C., Leucine and pellagra, *Nutr. Rev.,* 26, 323–326, 1968.
7. Rao, S. B., Raghuram, T. C., and Krishnaswamy, K., Role of Vitamin B_6 on leucine-induced metabolic changes, *Nutr. Metab.,* 18, 318–325, 1975.
8. Krishnaswamy, K., Rao, S. B., Raghuram, T. C., and Srikantia, S. G., Effect of vitamin B_6 on leucine-induced changes in human subjects, *Am. J. Clin. Nutr.,* 29, 177–181, 1976.
9. Gopalan, C. and Jaya Rao, K. S., Pellagra and amino acid imbalance, *Vitam. Horm.* (Leipzig), 33, 505–528, 1975.
10. Jolliffe, N., Bowman, K. M., Rosenblum, L. A., and Fein, H. D., Nicotinic acid deficiency encephalopathy, *JAMA,* 114, 307–312, 1940.
11. Davidson, S., Passmoe, R., Brock, J. F., and Truswell, A. S., Pellagra, in *Human Nutrition and Dietetics,* 6th ed., Churchill Livingstone, Edinburgh, 1975, 347–352.
12. Napier, L. E., Pellagra, in *Encyclopaedia of Medical Practice,* 2nd ed., C. V. Mosby, St. Louis, 1953, 489–503.
13. Gillman, J. and Gillman, T., What is pellagra?, in *Perspectives in Human Malnutrition,* Grune & Stratton, New York, 1951, 15–24.

14.   **Follis, R. H., Jr.,** Protein depletion syndromes, in *Deficiency Disease,* Charles C Thomas, Springfield, Ill., 1958, 315—349.

15.   **Halsted, C. H., Sheir, S., Sourial, N., and Patwardhan, V. N.,** Small intestinal structure and absorption in Egypt, *Am. J. Clin. Nutr.,* 22, 744—754, 1969.

16.   **Van Heerden, P. D. R., Grieve, R., and Metz, J.,** Fat absorption in pellagrins with observations on the effect of induced diarrhea, *Trans. R. Soc. Trop. Med. Hyg.,* 60, 241—244, 1966.

17.   **Mehta, S. K., Kaur, S., Avasthi, G., Wig, N. N., and Chhuttanie, P. N.,** Small intestinal deficit in pellagra, *Am. J. Clin. Nutr.,* 25, 545—549, 1972.

18.   **Gopalan, C. and Narasinga Rao, B. S.,** Experimental niacin deficiency, in *Nutritional Pathobiology,* Bajusz, E. and Jasmin, E., Eds., S. Karger, Basel, 1972, 49—80.

19.   **Belavady, B., Srikantia, S. G., and Gopalan, C.,** The effect of the oral administration of leucine on the metabolism of tryptophan, *Biochem. J.,* 87, 652—655, 1963.

20.   **Hankes, L. V., Leklem, J. E., Brown, R. R., and Mekel, R. C. P. M.,** Tryptophan metabolism in patients with pellagra: Problem of vitamin B_6 enzyme activity and feedback control of tryptophan pyrrolase enzyme, *Am. J. Clin. Nutr.,* 24, 730—739, 1971.

21.   **Truswell, A. S., Hansen, J. D. L., and Wannenburg, P.,** Plasma tryptophan and other amino acids in pellagra, *Am. J. Clin. Nutr.,* 21, 1314—1320, 1968.

22.   **Ghafoorunissa and Narasinga Rao, B. S.,** Plasma amino acid pattern in pellagra, *Am. J. Clin. Nutr.,* 28, 325—328, 1975.

23.   **Raghuramulu, N., Srikantia, S. G., Narasinga Rao, B. S., and Gopalan, C.,** Nicotinamide nucleotides in the erythrocytes of patients suffering from pellagra, *Biochem. J.,* 96, 837—839, 1965.

24.   **Axelrod, A. E., Spies, T. D., and Elvehjem, C. A.,** The effect of a nicotinic acid deficiency upon the coenzyme I content of the human erythrocyte and muscle, *J. Biol. Chem.,* 138, 667—676, 1941.

25.   **Klein, J. R., Perlzweig, W. A., and Handler, P.,** Determination of nicotinic acid in blood cells and plasma, *J. Biol. Chem.,* 145, 27—34, 1942.

26.   **Srikantia, S. G., Narasinga Rao, B. S., Raghuramulu, N., and Gopalan, C.,** Pattern of nicotinamide nucleotides in the erythrocytes of pellagrins, *Am. J. Clin. Nutr.,* 21, 1306—1309, 1968.

27.   **Vivian, V. M., Chaloupka, M. M., and Reynolds, M. S.,** Some aspects of tryptophan metabolism in human subjects, *J. Nutr.,* 66, 587—598, 1958.

28.   **Belavady, B., Rao, P. U. S., and Khan, L.,** Effects of leucine and isoleucine on nicotinamide nucleotides of erythrocytes, *Int. J. Vitam. Nutr. Res.,* 43, 442—453, 1973.

29.   **Vasantha, L.,** Collagen content and dermal amino acid pattern in pellagra, *Clin. Chim. Acta,* 27, 543—547, 1970.

30.   **Calandra, P.,** Tryptophan-niacin pathway in the human epidermis, *Acta Vitaminol. Enzymol.,* 28, 189—194, 1974.

31.   **Tabachnick, J.,** Urocanic acid, the major acid-soluble, ultraviolet-absorbing compound in guinea pig epidermis, *Arch. Biochem. Biophys.,* 70, 295—298, 1957.

32.   **Vasantha, L.,** Histidine, urocanic acid and histidine α deaminase in the stratum corneum in pellagrins, *Indian J. Med. Res.,* 58, 1079—1084, 1970.

33.   **Woolley, D. W. and Shaw, E. N.,** Evidence for the participation of serotonin in mental processes, *Ann. N.Y. Acad. Sci.,* 66, 649—667, 1957.

34.   **Krishnaswamy, K. and Ramanamurthy, P. S. V.,** Mental changes and platelet serotonin in pellagrins, *Clin. Chim. Acta,* 27, 301—304, 1970.

35.   **Stratigos, J., Katsambas, A., and Galanopoulou, P.,** Pellagra, serotonin and 5-HIIA, *Br. J. Dermatol.,* 90, 451—452, 1974.

36.   **Srikantia, S. G., Reddy, V., and Krishnaswamy, K.,** Electroencephalographic patterns in pellagra, *Electroencephalogr. Clin. Neurophysiol.,* 25, 386—388, 1968.

37.   Beaton, G. H. and McHenry, E. W., Eds., Academic Press, New York, 1966, 265—315.
      Beaton, G. H. and McHenry, E. W., Eds., Academic Press, New York 1966, 265—315.

38.   **Sauberlich, H. E., Dowdy, R. P., and Skala, J. H.,** Laboratory tests for the assessment of nutritional status, *CRC Crit. Rev. Clin. Lab. Sci.,* 4(3), 284—288, 1973.

39.   Manual for Nutrition Surveys, 2nd ed., Interdepartmental Committee on Nutrition for National Defense, U.S. Government Printing Office, Washington, D.C., 1963.

40.   **Vilter, R. W., Vilter, S. P., and Spies, T. D.,** Relationship between nicotinic acid and a codehydrogenase (cozymase), *JAMA,* 112, 420—422, 1939.

41.   **Sebrell, W. H.,** Public health implications of recent research in pellagra and ariboflavinosis, *J. Home Econ.,* 31, 530—536, 1939.

42.   **Mason, M.,** The metabolism of tryptophan in riboflavin deficient rats, *J. Biol. Chem.,* 201, 513—518, 1953.

43. **Charconnet-Harding, F., Dalgliesh, C. E., and Neuberger, A.,** The relation between riboflavin and tryptophan metabolism, studied in the rat, *Biochem. J.,* 53, 513–521, 1953.

44. **Henderson, L. M., Weinstock, I. M., and Ramasarma, G. B.,** Effect of deficiency of B Vitamins on the metabolism of tryptophan by the rat, *J. Biol. Chem.,* 189, 19–21, 1951.

45. **Verjee, Z. H. M.,** Tryptophan metabolism in baboons: Effect of riboflavin and pyridoxine deficiency, *Int. J. Biochem.,* 2, 711–718, 1971.

46. **Sauberlich, H. E.,** Vitamin B_6 group, IX. Biochemical systems and biochemical detection of deficiency, in *The Vitamins,* Vol. 2, 2nd ed., Sebrell, W. H., Jr. and Harris, R. S., Eds., Academic Press, New York, 1968, 44–80.

47. **Schweigert, B. S. and Pearson, P. B.,** Effect of Vitamin B_6 deficiency on the ability of rats and mice to convert tryptophane to N'-methylnicotinamide and nicotinic acid, *J. Biol. Chem.,* 168, 555–561, 1947.

48. **Rosen, F., Huff, J. W., and Perlzweig, W. A.,** The role of B_6 — deficiency in the tryptophane-niacin relationships in rats, *J. Nutr.,* 33, 561–567, 1947.

49. **Ling, C. T., Hegsted, D. M., and Stare, F. J.,** The effect of pyridoxine deficiency on the tryptophan-niacin transformation in rats, *J. Biol. Chem.,* 174, 803–812, 1948.

50. Annual Report for 1975–1976, National Institute of Nutrition, Hyderabad, India, 1976, 46–48.

EFFECTS OF SPECIFIC NUTRIENT DEFICIENCIES IN MAN: PANTOTHENIC ACID

Ralph A. Nelson

At present, no specific syndrome of pantothenic acid deficiency has been defined for human beings. However, Bean and Hodges and co-workers have evaluated the effect of a pantothenic acid-deficient diet on the clinical and laboratory responses of human volunteers. The difference between these human studies and experimental animal studies is that a sharp picture of the deficient state was not clearly delineated in humans because the deficient state could not be carried to the point of imminent death, as were the animal studies. In human subjects, a high degree of caution and restraint was practiced by these investigators because the clinical state of volunteers usually deteriorated rapidly when they were placed on a pantothenic acid-deficient diet or a diet containing an antagonist to pantothenic acid, such as omega-methyl pantothenic acid.

Giving pantothenic acid to malnourished patients suspected of having pantothenic acid deficiency has, in the main, failed to reverse clinical signs, symptoms, and laboratory values indicative of undernutrition. Some isolated studies have reported reversal of signs and symptoms of undernutrition, but their occurrence is too rare to permit definition of a syndrome specific for pantothenic acid deficiency in human beings.

CONTROLLED STATES OF PANTOTHENIC ACID DEFICIENCY PRODUCED BY FEEDING EITHER A PANTOTHENIC ACID-DEFICIENT DIET OR PANTOTHENIC ACID ANTAGONISTS (OR BOTH) [1-3]

Signs and Symptoms

The general symptoms experienced by volunteers on these study diets were malaise, muscle soreness, easy fatigability, weakness, somnolence, diaphoresis, and blepharitis. Other signs and symptoms were referred to specific organ systems as follows:

1. Emotional — hyperventilation, irritability, insomnia, and depression
2. Neurologic — headache, dizziness, tremor, incoordination, and paresthesias (described as burning on the soles of the feet)
3. Resistance to infection — decreased resistance to infectious disease such as the common cold, tonsillitis, and bronchitis
4. Cardiovascular system — tachycardia and arrhythmias
5. Gastrointestinal system — epigastric burning, nausea, vomiting, and diarrhea (on some occasions, constipation was recorded)
6. Genitourinary system — burning and smarting urination and nocturia

Symptoms most commonly present were persistent and annoying fatigue, headache, and weakness. Some of the subjects demonstrated impaired motor coordination and peculiar gait.

Laboratory Studies

Changes in blood and urine have varied considerably; few consistent abnormal findings have been reported. Those changes that appear to be associated with pantothenic acid deficiency are an increase in erythrocyte sedimentation rate and eosinopenia with an inability to respond to adrenal corticotropic hormones. Such subjects tend to be more sensitive to the hypoglycemic effect of insulin. After the deficient diet is begun, urinary excretion of pantothenic acid rapidly approaches zero. Usually, less than 3 mg of pantothenic acid is excreted per 24 hr while the deficient diet is being consumed. When 4000 mg of pantothenic acid was given as therapy, about 1000 mg was excreted in 24 hr. Giving coenzyme A by mouth did not change the character of the response.

ANTIBODY PRODUCTION IN RESPONSE TO FEEDING A PANTOTHENIC ACID–DEFICIENT DIET [4,5]

Men eating a diet deficient in pantothenic acid were less able to form antibodies against tetanus when compared with control subjects, but there were no differences in ability to produce antibodies against the antigens of typhoid fever and Asian influenza. Volunteers demonstrated hastening in rejection of skin grafts; the length of time for healing of the rejection sites was slightly longer in the pantothenic acid-deficient subjects. In combined pantothenic acid and pyridoxine deficiency produced in human volunteers,[5] antibody production against tetanus and typhoid was impaired, and hypogamma-globulinemia developed in the subjects.

SUSPECTED PANTOTHENIC ACID DEFICIENCY IN HUMAN BEINGS

There have been several states in human beings in which pantothenic acid nutrition has been suspected as being unsatisfactory. However, none of these conditions has clearly been defined as a specific syndrome attributable to pantothenic acid deficiency. Nevertheless, these states of unsatisfactory nutritional status for pantothenic acid should be briefly mentioned.

Burning-feet Syndrome

Burning feet is a syndrome occasionally described for malnourished people in developing countries, but was mostly described in prisoners of World War II who were interned in the Far East. The burning-feet syndrome consists of burning on the soles of the feet, although occasionally it affects the palms of the hands. Hyperhidrosis of the affected areas has been noticed. A paucity of objective signs of this syndrome has been the rule. In some instances, hypertension and exaggerated reflexes have been recorded.[6,7] Some authors have reported good response to therapy with nicotinic acid and have speculated that riboflavin deficiency may contribute to the disorder.[7] One report noted improvement in all ten cases treated daily for 2 to 3 weeks with 20 to 40 mg of pantothenic acid intramuscularly.[6] Presently, however, the association between panto-thenic acid and the burning-feet syndrome is not clear.[8]

Pellagra, Beriberi, and Riboflavin Deficiency

Pantothenic acid levels have been found to be low in the blood of patients with pellagra, beriberi, or riboflavin deficiency. Range of levels recorded was 0.05 to 0.09 μg/ml. These values represented a 23 to 50% decrease in circulating levels of pantothenic acid when compared with normal levels established by the investigators' laboratories.[9]

Pantothenic Acid and Its Relationship to Alcoholic Liver Disease

In a study of 172 alcoholic patients, low levels of pantothenic acid were not commonly found.[10] Decreased blood levels were found in 20% of patients with cirrhosis, 15% with fatty liver, and 12% with normal liver function. The level of pantothenic acid was low in only 2 of 40 patients with neuropathy, compared with low levels of blood thiamine found in all 40 patients. In 60 patients with glossitis, chelitis, and atrophy of lingual papillae, 48 had low levels of niacin and riboflavin but none had low levels of pantothenic acid. None of the patients with Wernicke's encephalopathy, pellagra, or hyperchromic anemia had decreased levels of pantothenic acid. Another report showed that pantothenic acid levels in biopsied liver samples showing fatty metamorphosis were 21 pg/μg dried tissue, compared with normal values of 66 pg/μg dried tissue.[11]

Studies in Human Beings With Inflammatory Bowel Disease

A recent study investigated the apparent relationship between chronic ulcerative and granulomatous colitis and pantothenic acid deficiency. Colonic tissues were obtained at the time of colectomy in 29 patients with inflammatory bowel disease and in 31 patients having colectomy for carcinoma or diverticulitis. Only normal tissues were used from the latter group. It was found that the concentrations of free, bound, and total pantothenic acid in blood and colonic mucosa did not differ between the two groups of patients. Colonic mucosa concentrated free pantothenic acid to about 50 times the level in blood. However, compared with normal gut mucosa, coenzyme A activity was markedly low in mucosa from patients with chronic ulcerative or granulomatous colitis despite the presence of normal amounts of free and bound pantothenic acid. The data suggested a block in the conversion of bound pantothenic acid to coenzyme A in the mucosa in patients with inflammatory bowel disease.[12]

Studies in Women in Whom Pantothenic Acid Status is Considered Unsatisfactory

Pregnant, postpartum, and nonpregnant teenagers tend to eat less pantothenic acid and, thus, are considered to have an unsatisfactory pantothenic acid nutritional status.[13] Intake varied between 3.3 and 4.7 mg/day with recommended dietary allowances in the range of 5 to 10 mg/day. In a study of women 18 to 24 years old in normal health, eating diets varying in content of pantothenic acid, a dietary level of 2.8 mg/day resulted in decreased urinary excretion of pantothenic acid of 3.2 mg/day.[14] When dietary levels were in the range of 7.8 and 12.8 mg/day, urinary excretion was less than intake, namely 4.5 and 5.6 mg/day, respectively. Healthy girls, ages 7 to 9 years, consumed between 2.8 and 5.0 mg/day, but urinary excretion was always less than intake, varying between 1.3 and 2.9 mg/day.[15]

TRIALS WITH PANTOTHENIC ACID AS A THERAPEUTIC AGENT

Baldness and gray hair in humans have been treated with pantothenic acid because of the loss of hair and graying noted in many different experimental animals subjected to pantothenic acid deficiency. There have been no positive therapeutic results of the program. Use of pantothenic acid has been advocated in cases of adynamic ileus. The fact that most deficient states of pantothenic acid in animals and humans are associated with diarrhea helps explain why the use of pantothenic acid in adynamic ileus has been disappointing. Results of the use of pantothenic acid in treatment of peripheral neuritis, ataxia, and other central nervous and peripheral nervous system diseases have been disappointing.

Pantothenic acid has also been advocated for use in humans in times of stress. However, its possible effectiveness is limited because of its prompt excretion in urine when taken in excess, regardless of the rate of injection.

REFERENCES

1. **Bean, W. B. and Hodges, R. E.,** Pantothenic acid deficiency induced in human subjects, *Proc. Soc. Exp. Biol. Med.,* 86, 693–698, 1954.
2. **Hodges, R. E., Ohlson, M. A., and Bean, W. B.,** Pantothenic acid deficiency in man, *J. Clin. Invest.,* 37, 1642–1657, 1958.
3. **Hodges, R. E., Bean, W. B., Ohlson, M. A., and Bleiler, R.,** Human pantothenic acid deficiency produced by omega-methyl pantothenic acid, *J. Clin. Invest.,* 38, 1421–1425, 1959.
4. **Hodges, R. E., Bean, W. B., Ohlson, M. A., and Bleiler, R. E.,** Factors affecting human antibody response. III. Immunologic responses of men deficient in pantothenic acid, *Am. J. Clin. Nutr.,* 11, 85–93, 1962.
5. **Hodges, R. E., Bean, W. B., Ohlson, M. A., and Bleiler, R. E.,** Factors affecting human antibody response. V. Combined deficiencies of pantothenic acid and pyridoxine, *Am. J. Clin. Nutr.,* 11, 187–199, 1962.
6. **Gopalan, C.,** The 'burning-feet' syndrome, *Indian Med. Gaz.,* 81, 23–26, 1946.
7. **Cruickshank, E. K.,** Painful feet in prisoners-of-war in the Far East: review of 500 cases, *Lancet,* 2, 369–372, 1946.
8. **Glusman, M.,** The syndrome of "burning feet" (nutritional melalgia) as a manifestation of nutritional deficiency, *Am. J. Med.,* 3, 211–233, 1947.
9. **Stanberg, S. R., Snell, E. E., and Spies, T. D.,** A note on an assay method for pantothenic acid in human blood (letter to the editor), *J. Biol. Chem.,* 135, 353–354, 1940.
10. **Leevy, C. M., Baker, H., tenHoue, W., Frank, O., and Cherrick, G. R.,** B-complex vitamins in liver disease of the alcoholic, *Am. J. Clin. Nutr.,* 16, 339–346, 1965.
11. **Frank, O., Luisada-Opper, A., Sorrell, M. F., Thomson, A. D., and Baker, H.,** Vitamin deficits in severe alcoholic fatty liver of man calculated from multiple reference units, *Exp. Mol. Pathol.,* 15, 191–197, 1971.
12. **Ellestad-Sayed, J. J., Nelson, R. A., Adson, M. A., Palmer. W. M., and Soule, E. H.,** Pantothenic acid, coenzyme A, and human chronic ulcerative and granulomatous colitis, *Am. J. Clin. Nutr.,* 29, 1333–1338, 1976.
13. **Cohenour, S. H. and Calloway, D. H.,** Blood, urine, and dietary pantothenic acid levels of pregnant teenagers, *Am. J. Clin. Nutr.,* 25, 512–517, 1972.
14. **Fox, H. M. and Linswiler, H. L.,** Pantothenic acid excretion on three levels of intake, *J. Nutr.,* 75, 451–454, 1961.
15. **Pace, J. K., Stier, L. B., Taylor, D. D., and Goodman, P. S.,** Metabolic patterns in preadolescent children. V. Intake and urinary excretion of pantothenic acid and of folic acid, *J. Nutr.,* 74, 345–351, 1961.

EFFECT OF SPECIFIC NUTRIENT
DEFICIENCIES IN MAN: BIOTIN

L. Mercer and C. M. Baugh

HISTORY

Several lines of inquiry led to the discovery and structure elucidation of the vitamin biotin. In 1916, Bateman observed toxic effects in rats receiving diets that contained a high concentration of egg white.[1] Later investigators showed that rats fed diets with raw egg white as the sole protein source developed a syndrome characterized by neuromuscular disorders, severe dermatitis, and loss of hair.[2] The condition was named egg-white injury and could be prevented by cooking the protein or by administering other foods such as yeast and liver.[3] György felt the egg-white injury protecting substance was a vitamin which he called vitamin H (German word Haut meaning skin).[4] In 1936, Kögl isolated a crystalline compound which was a yeast growth factor. The compound came from egg yolk and was named biotin.[5]

In 1940, György demonstrated that biotin and vitamin H were the same substance.[6] The structural formula for biotin was established in 1942 by du Vigneaud,[7] and the following year it was synthesized by Harris.[8]

The antagonist for biotin which caused the observed egg-white injury was determined by Eakin in 1940 to be a protein he called avidin (hungry protein).[9]

STRUCTURE

The physiologically active form of biotin is shown in Figure 1.[10] Of eight possible steroisomers, only *d*-biotin (*cis*-form, β-isomer) serves as a vitamin. Biotin metabolites have been isolated from natural material in the form of biocytin (ε-*N*-Biotinyl-*l*-lysine), the *d*- and *l*-sulfoxides of biotin, bisnorbiotin (2-carbon shorter side chain) and tetranorbiotin (4-carbon shorter side chain).

Avidin, the antagonist of biotin found in egg whites, is a basic glycoprotein, mol wt 60,000, containing 10% carbohydrate and 6% tryptophan.[11] It is a tetramer which binds four molecules of biotin. The biotin-avidin complex is extremely stable with a dissociation constant of 10^{-15} M. The complex is heat resistant, stable over a wide pH range, and not readily dissociated by various proteolytic enzymes or by incubation with liver, kidney muscle, or blood.[12]

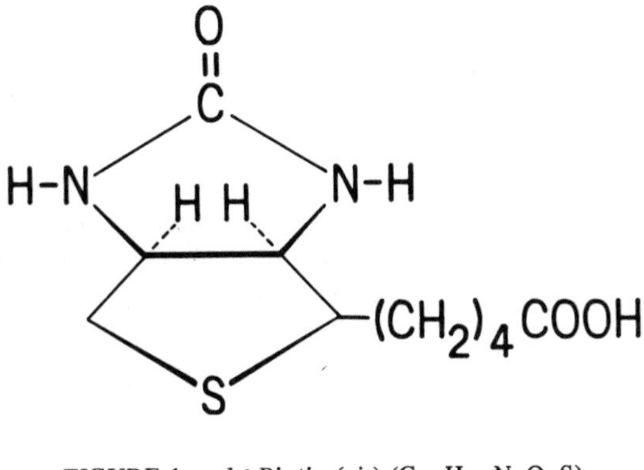

FIGURE 1.   *d*-β-Biotin (*cis*) ($C_{10}H_{16}N_2O_3S$).

SOURCES OF BIOTIN

Biotin is widely distributed in nature and is present in almost all foods. However, the best dietary sources of biotin are organ meats (liver, kidney), egg yolk, legumes, and nuts. All other meats, dairy products, and cereals are considered relatively poor sources.

Biotin, along with other vitamins, is synthesized by intestinal bacteria in man.[13] The availability of vitamins to the host from this source is not known. Mickelsen concluded that healthy adults, but not infants, might obtain some of their biotin from bacterial synthesis.[14]

HUMAN DEFICIENCY

There is no evidence of spontaneous biotin deficiency in humans. This is probably due to the ubiquitous nature of the vitamin. The average daily diet is thought to contain from 150 to 300 μg of biotin. This amount, plus any benefit derived from bacterial synthesis, appears to be adequate to prevent deficiency symptoms from occurring. Most cases describing biotin deficiency found in the literature are based on the antagonistic effect of the egg-white protein avidin. In 1942, Sydenstricker described the clinical manifestations of biotin deficiency induced in four normal volunteers.[15] These volunteers were fed a diet in which 30% of the calories were provided by raw egg white. Beginning in the 3rd to 4th week, all developed a fine, scaly dermatitis. Other symptoms were atrophic glossitis, anorexia, nausea, vomiting, mental depression, pallor, muscle pains, paresthesias, and precordial pain. Examination of the blood showed decreased hemoglobin and increased cholesterol levels. Urinary excretion of biotin decreased to about one tenth of normal levels. Prompt improvement occurred within 3 to 5 days after treatment with an injectable biotin concentrate. The minimal amount required for prompt relief was judged to be 150 μg/day.

Biotin deficiency was reported in a 62-year-old woman with Laennec's cirrhosis. She had, on her physician's advice, ingested six raw eggs daily for 18 months in an effort to regenerate liver tissue. Her symptoms included anorexia, nausea, pallor, vomiting, lassitude, scaly dermatitis, and desquamation of the lips. All symptoms cleared or improved markedly after 2 to 5 days therapy providing a daily i.m. dose of 200 μg biotin.[16]

BIOTIN RESPONSIVE SYNDROMES

Recently, several investigators have reported syndromes which responded to biotin therapy even though biotin deficiency was not demonstrated. Clinical and therapeutic data for 25 infants with generalized seborrhoeic dermatitis were reported by Messaretakis.[17] In 22 of the 25 patients, the skin lesions improved 5 to 8 days after a treatment consisting of 5 mg of i.v. administered biotin. Three patients with bacterial skin infections and staphylococcal pneumonia required a second dose 10 days later. In 15 to 30 days, all skin lesions disappeared, and there was no recurrence in follow-up periods of 4 to 27 months.

In 1970, Gompertz reported a case of biotin responsive propionicacidemia in a child with ketotic hyperglycinemia.[18] An oral load of L-isoleucine was given and caused a rise in plasma propionate due to a decreased propionyl-CoA-carboxylase activity. After treatment with biotin, 5 mg of isoleucine was again administered orally, twice daily for 5 days. Treatment with biotin reduced the resting level of plasma propionate. The rise in propionate level after isoleucine loading was also greatly diminished. Those reactions in Figure 2 marked with a circled 2 appear to be the reactions of importance in this instance of propionicacidemia.

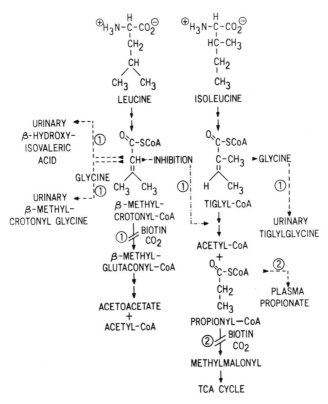

FIGURE 2.   Abnormal metabolism (1) in patient reported by Gompertz.[20] Impairment of β-methylcrotonyl-CoA carboxylase causes a buildup of β-methylcrotonyl-CoA. This compound is either conjugated with glycine or oxidized. It may also inhibit the normal metabolism of tiglic acid causing the formation of tiglylglycine. Block in propionyl-CoA carboxylase (2).[18] This causes a rise in plasma propionate levels as well as a diminution in the rate of isoleucine degradation.

Gompertz reported another patient with similar problems.[19] A liver mitochondrial preparation from this patient was found to have propionyl-CoA-carboxylase activity that was only 9.7% of the controls. Administration of biotin to this patient did not bring about a clinical or biochemical improvement.

In 1971, Gompertz reported a case of biotin-responsive β-methylcrotonylglycinuria.[20] The patient was admitted at 5 months of age to the hospital because of persistent vomiting and erythematous rash. His blood pH was 7.3 and P_{CO_2} 18 mmHg. His urine had a peculiar odor. Gas chromatography of the urine showed β-methylcrotonylglycine, tiglylglycine, and β-hydroxyisovaleric acid, indicating an impairment of β-methylcrotonyl-CoA-carboxylase activity.

An empirical dose of biotin (10 mg/day) was given orally. Hyperventilation and vomiting stopped; the ketosis disappeared, and blood pH became normal within 24 hr. β-Methylcrotonylglycine, tiglylglycine, and β-hydroxyisovaleric acid disappeared totally from the urine and did not reappear. Examination of the reactions (indicated by a circled 1) in Figure 2 suggests that the primary biochemical lesion occurs in the conversion of β-methylcrotonyl-CoA to β-methylglutaconyl-CoA. An accumulation of β-methylcrotonyl-CoA in turn leads to increased urinary levels of β-hydroxyisovaleric acid and β-methylcrotonylglycine. The presence of urinary tiglylglycine is explained as an inhibition in the isoleucine pathway (by β-methylcrotonyl-CoA) of the conversion of tiglyl-CoA to acetyl-CoA. The accumulated tiglyl-CoA is conjugated to glycine prior to urinary excretion.

Scriver has broken down propionicacidemia and β-methylcrotonylglycinuria into biotin-sensitive and biotin-resistant types.[21] These and related syndromes are also discussed by Nyhan.[22] Figure 2 depicts schematically the biotin related metabolism of leucine and isoleucine.

NORMAL LEVELS OF BIOTIN IN HUMAN TISSUES

Several investigators have reported human tissue biotin levels, as shown in Table 1. There is some discrepancy in reported values, probably due to different assay organisms and preassay preparation of samples. There is a very large discrepancy in the values reported by Denko. While more current investigators have consistently found values in the range of 300–500 pg/ml, Denko reported 12,000 pg/ml.[27] This figure cannot be ignored since it is often quoted in reference books. However, the weight of evidence would seem to lean towards the lower numbers, especially when Baker's assay[25] is used. All investigators were in agreement that spontaneous biotin deficiency due to dietary restriction was very rare in the populations studied.

Table 1
BIOTIN LEVELS IN HUMAN TISSUE

Subjects	Preparation	Assay	Source	Biotin (pg/ml)			Ref.
				Low	Mean	High	
7 Healthy young men	HCl Hydrolysis	*Lactobacillus casei*	Blood	7,500	12,300	17,300	27[a]
			Plasma	9,500	12,700	16,600	
12 Normal subjects	Papain	*Ochromonas dancia*	Blood	170	220	279	28
			Serum	213	313	404	
30 Healthy adults	Not specified	*Ochromonas dancia*	Blood	Approximately 300 displayed graphically			26
30 Normal infants	H_2SO_4 Hydrolysis	*Lactobacillus plantarum*	Blood		324 ± 114		23
25 Normal adults					258 ± 74		
68 Healthy gravidas	Papain	*Ochromonas dancia*	Blood	200	590	1,000	29
68 Neonates		(Complete assay in Baker[25])	Cord blood	525	810	1,436	
Mothers at parturition, infants of mothers above,	Papain	*Ochromonas dancia*	Blood	240	420	740	24
				490	820	1,240	
76 nonpregnant healthy females				200	590	1,100	

[a] Figures reported by Denko are much higher than more recent estimates of biotin in blood. However, Denko's figures are often found in handbooks of nutrition, tables, etc. The reasons for the higher values he reports are unknown.

REFERENCES

1. **Bateman, W. G.,** The digestability and utilization of egg proteins, *J. Biol. Chem.,* 26, 263–291, 1916.
2. **Parsons, H. T., Lease, J. G., and Kelly, E.,** Interrelationship between dietary egg white and requirement for protective factor in cure of nutritive disorder due to egg white, *Biochem. J.,* 31, 424–432, 1937.
3. **György, P.,** Curative factor (Vitamin H) for egg white injury with particular reference to its presence in different foodstuffs and in yeast, *J. Biol. Chem.,* 131, 733–744, 1939.
4. **György, P.,** Rachitis und andere Avitaminosen, *Z. aerztl. Fortbild.,* 28, 377, 1931.
5. **Kögl, F. and Tönnis, B.,** Plant-growth substances XX, the bios problem, isolation of crystalline biotin from egg yolk, *Z. Physiol. Chem.,* 242, 43–73, 1936.
6. **György, P., Rose, C. S., Hofman, K., Melville, D. B., and du Vigneaud, V.,** A further note on the identity of Vitamin H with biotin, *Science,* 92, 609, 1940.
7. **du Vigneaud, V., Melville, D. B., Folkers, K., Wolf, D. E., Mozingo, R., Keresztesy, J. C., and Harris, S. A.,** The structure of biotin: a study of desthiobiotin, *J. Biol. Chem.,* 146, 475–485, 1942.
8. **Harris, S. A., Wolf, D. E., Mozingo, R., and Folkers, K.,** Synthetic biotin, *Science,* 97, 447–448, 1943.
9. **Eakin, R. E., Snell, E. E., and Williams, R. J.,** A constituent of raw egg white capable of inactivating biotin *in vitro, J. Biol. Chem.,* 136, 801–802, 1940.
10. **Kutsky, R. J.,** Biotin, in *Handbook of Vitamins and Hormones,* Van Nostrand Reinhold, New York, 1973, 79–86.
11. **Somogyi, J. C.,** Antivitamins, in *Toxicants Occurring Naturally in Foods,* National Academy of Sciences, Washington, D.C., 1973, 254–275.
12. **Lee, H., Wright, L. D., and McCormack, D. B.,** Metabolism, in the rat, of biotin injected intraperitoneally as the avidin-biotin complex, *Proc. Soc. Exp. Biol. Med.,* 142, 439–442, 1973.
13. **Matthews, D. M.,** Absorption of water-soluble vitamins, *Biomembranes,* 43, 847–915, 1974.
14. **Mickelsen, O.,** Intestinal synthesis of vitamins in the non-ruminant, *Vitam. Horm.,* (N.Y.), 14, 1–95, 1956.
15. **Sydenstricker, V. P., Singal, S. A., Briggs, A. P., and De Vaughn, N. M.,** Observations on "egg white injury" in man and its cure with biotin concentrate, *JAMA,* 118, 1199–1200, 1942.
16. **Baugh, C. M., Malone, J. H., Butterworth, C. E., Jr.,** Human biotin deficiency – A case history of biotin deficiency induced by raw egg consumption in a cirrbotic patient, *Am. J. Clin. Nutr.,* 21, 173–181, 1968.
17. **Messaritakis, J., Kattamis, C., Karabula, C., Matsanitos, N.,** Generalized seborrhoeic dermatitis – Clinical and therapeutic data of 25 patients, *Arch. Dis. Child,* 50, 871–874, 1975.
18. **Gompertz, D., Balgobin, L., Baines, N. D., and Hull, D.,** Biotin responsive propionicacidemia, *Lancet,* 1, 244–245, 1970.
19. **Gompertz, D., Storrs, C. N., Barr, D. C. K., Peters, T. J., and Hughes, E. A.,** Localization of enzymic defect in propionicacidaemia, *Lancet,* 1, 1140–1143, 1970.
20. **Gompertz, D., Draffan, G. H., Watts, J. L., and Hull, D.,** Biotin responsive β-methylcrotonyl-glycinuria, *Lancet,* 2, 22–24, 1971.
21. **Daum, R. S., Scriver, C. R., Mamer, O. A., Devlin, E., Lamm, P., and Goldman, H.,** An inherited disorder of isoleucine catabolism causing accumulation of methylacetoacetate and methyl-hydroxybutyrate and intermittent metabolic acidosis, *Pediatr. Res.,* 7, 149–160, 1973.
22. **Nyhan, W. L.,** Propionicacidemia and the ketotic hyperglycinemia syndrome, in *Heritable Disorders of Amino Acid Metabolism: Patterns of Clinical Expression and Genetic Variation,* John Wiley & Sons, New York, 1974, 37–60.
23. **Bhagavan, H. N. and Coursin, D. B.,** Biotin content of blood in normal infants and adults, *Am. J. Clin. Nutr.,* 20, 8, 903–906, 1967.
24. **Baker, H., Orca, E., Thompson, A. D., Langer, A., Munices, E. D., De Angelio, B., and Kaminetzky, H. A.,** Vitamin profile of 174 mothers and newborns at parturition, *Am. J. Clin. Nutr.,* 28, 56–65, 1975.
25. **Baker, H. and Frank, O.,** in *Clinical Vitaminology: Methods and Interpretations,* Interscience, New York, 1968, 136.
26. **Leevy, C. M., Cardi, L., Oscar, F., Gellene, R., and Baker, H.,** Incidence and significance of hypovitaminemia in a randomly selected municipal hospital population, *Am. J. Clin. Nutr.,* 17, 259–271, 1965.
27. **Denko, C. W., Grundy, W. E., and Porter, J. W.,** Blood levels in normal adults on a restricted dietary intake of β-complex vitamins and tryptophan, *Arch. Biochem.,* 13, 481–484, 1947.

28.   **Baker, H., Frank, O., Matovitch, V. B., Pasker, I., Aaronson, S., Hutner, S. H., and Sobatka, H.,** A new assay method for biotin in blood serum, urine and tissues, *Anal. Biochem.,* 3, 31—39, 1962.

29.   **Kaminetzky, H. A., Baker, H., Frank, O., and Langer, A.,** The effects of intravenously administered water-soluble vitamins during labor in normovitaminemic and hypovitaminemic gravidas on maternal and neonatal blood vitamin levels at delivery, *Am. J. Obstet. Gynecol.,* 120, 697—703, 1974.

EFFECT OF NUTRIENT DEFICIENCIES IN MAN: CHOLINE

P. R. Turkki

Choline has proven to be one of the most difficult dietary constituents to study, both in regard to its functions in the body and its role as an essential nutrient. Neither the existence of choline deficiency nor dietary need for choline in man has been conclusively demonstrated. However, in at least 12 species of animals, lack of dietary choline results in a common defect — fatty infiltration of the liver — even though considerable differences exist in regard to how easily this condition is produced and what other symptoms, if any, are present.[1]

There are several factors which make it practically impossible to obtain conclusive evidence that choline is or is not required in the diet of man. First, choline is ubiquitous in both animal and plant cells and, consequently, widely distributed in foods. Thus, even the most nutritionally inadequate diets provide considerable amounts of choline. The choline content of various food products, as compiled by Griffith and Nyc, is shown in Tables 1 through 4.[2]

Second, choline can by synthesized in the body by addition of methyl groups from S-adenosylmethionine to aminoethanol or phosphatidylaminoethanol. Thus, the dietary level of methionine and other methyl donors (betaine) and precursors (B_{12} and folacin) influences the need for choline in the diet. The adult daily intake of choline from mixed diets has been estimated to range from 400 to 900 mg.[3]

Third, the functions of choline, both as a lipotropic factor and as a labile methyl donor, are nonspecific. Other dietary lipotropes and methyl donors can at least partially substitute for choline in these roles and, therefore, influence the body's need for choline.

Finally, no specific, easily measurable symptom of choline deficiency exists. It is well recognized that deposition of fat is a common response by the liver to insult or injury inflicted by a number of hepatotoxic agents as well as to lack of dietary lipotropic factors. Some of the conditions leading to fatty infiltration of the liver are listed in Table 5. Only from repeated liver biopsies can the magnitude of fat deposition be estimated by either histologic or chemical means. Ethical considerations, as well as expense, limit the use of this procedure. An excellent review by Griffith et al. can be referred to for detailed information about all aspects of choline and choline deficiency in different animal species.[2]

The complex interrelationships between choline and other dietary factors make production of uncomplicated choline deficiency extremely difficult and have discouraged any attempts to experimentally produce such a condition in man. Instead, major research efforts have focused on certain disease states which exhibit lesions of the liver resembling those produced by choline deficiency in animals — the fatty liver and cirrhosis. The experimental production in animals of hepatic cirrhosis by diets deficient in lipotropic factors and the prevention of this condition by choline or other lipotropes lead to speculation that choline deficiency may be a contributing factor in the development of fatty liver and cirrhosis in man.[4-7]

Most of the clinical studies concerned with choline or other lipotropic factors have been conducted using patients who have alcoholic cirrhosis of the liver. The mechanism of the development of cirrhosis remains unknown, despite the enormous amount of research on the topic. The available evidence seems to point to a complex etiology, with multiple nutritional deficiencies combined with the possible toxic effect of alcohol itself as the causative factors. However, differences still exist in interpretation of the experimental results.[8,9]

Most fatty livers are known to be easily reversible by elimination of the causative agent or condition and usually are not considered as a primary object of therapy. The same is true with acute alcoholic fatty liver, although when the consumption of alcohol becomes chronic, the condition of the liver becomes of primary concern. Even though the relationship between deposition of fat in the liver and the cirrhotic process is not completely understood, fatty infiltration seems to precede cirrhosis. For this reason, it has been tempting to assume that a fatty liver is a precursor of cirrhosis, even though not all fatty livers become cirrhotic.[9]

Cirrhosis of the liver had an extremely poor prognosis until the 1940s, when Patek and co-workers demonstrated the therapeutic value of a balanced diet which provides adequate protein and other nutrients for the regeneration of liver tissue.[10-12]

The effectiveness of choline and methionine in the prevention of cirrhosis in animals encouraged clinicians to test their therapeutic value in the treatment of cirrhosis in man. A number of reports on such trials were published between 1940 and 1960. Although no conclusive evidence was obtained to establish choline as either a causative or therapeutic agent in this condition, a brief reference to some of the findings of these studies is included. Because of the wide variation in experimental conditions employed, it is impossible to summarize the results in tabular form.

In the early 1940s, several groups reported improvement in patients with cirrhosis of the liver when choline supplements were used in addition to a nutritious diet.[13-15] The conclusions were based on clinical improvement and favorable changes observed in the blood picture and liver function tests shortly after therapy began. As no matched control groups treated with the diet alone were used, the beneficial effect of the supplemental choline can be questioned, even though some patients were reportedly unresponsive to the diet alone and responded rapidly after addition of choline.[13]

Several other studies have compared the rates of morbidity and mortality of groups treated with diet alone or with diet and lipotropic agents* and observed favorable results from the lipotropic therapy.[16-18] Patients with enlarged fatty liver were found to benefit most from the combined therapy, as assessed by histologic changes in repeated liver biopsies as well as by clinical and functional improvement.[17,18] The biopsy samples revealed loss of excess fat from the liver as a response to the therapy but showed no evidence that the cirrhotic process was either arrested or reversed.[18,19]

In some studies, histologic improvement (removal of fat) was found to be correlated with clinical and functional improvements.[19-21] However, other studies have reported discrepancies between these methods of assessment.[18]

The results from a series of studies utilizing partially purified, incomplete diets, as well as balanced diets, revealed some unexpected findings. Clinical and functional improvement was observed with a practically protein-free diet (if it was otherwise adequate), but protein or amino acids were necessary for the removal of fat deposits, as determined histologically.[22-24] A purified diet**resulted in no improvement when given to three patients, but all responded rapidly when changed to an adequate normal diet.[24] When the same purified diet was supplemented with 6.5 g of choline dihydrogen citrate, only two of the five patients treated showed some improvement, whereas all five improved rapidly when they were placed on an isocaloric diet containing 50 g of protein.[25] It was unfortunate that the basic diet was changed instead of replacing some of the glucose by a mixture of amino acids. Nevertheless, this series of studies gave added evidence to support the therapeutic efficacy of a balanced diet in the treatment of cirrhosis of the liver and established the necessity of protein in the lipotropic action of diet. Even though two patients showed some response to choline, the results lend equal support to the views of many investigators who have concluded that choline has little, if any, therapeutic value in

---

*   Choline, cystine, and methionine, individually or in combination.
** Glucose solution with vitamins and minerals, 1600 kcal.

the treatment of alcoholic cirrhosis beyond that provided by a balanced diet, which itself contains considerable amounts of choline.[19,26-30] On the other hand, one cannot expect choline to be effective in a diet which is inadequate in methionine or other amino acids and calories, as has been pointed out by Best and Lucas.[30] However, it has been found useful in initial treatment of seriously ill patients until their appetites have improved enough to allow ingestion of an adequate diet.[17,21,29]

Attempts to use choline in the treatment of fatty liver and cirrhosis in children with severe malnutrition have been equally inconclusive, even though some beneficial effects of lipotropic therapy have been reported.[31,32]

Indirect evidence of choline deficiency has been derived from studies using radioactive phosphorous in normal and cirrhotic patients. After a large dose of choline, a significantly higher turnover of plasma phospholipids was detected in cirrhotic patients with fatty livers than in normal subjects or cirrhotic patients without fatty infiltration of the liver, as judged from biopsy samples.[33,34] After 2 months of treatment which resulted in functional and clinical improvement, phospholipid turnover decreased and was comparable to that of the controls.

Even though the reviewed evidence does not allow one to conclude that choline deficiency is involved in the pathogenesis of liver disease in man, neither does it enable one to definitely deny such involvement. Lack of a definitive answer is due to the many difficulties that plague the design and interpretation of studies with human subjects, especially clinical investigations.

Table 1
TOTAL CHOLINE CONTENT
OF ANIMAL PRODUCTS

Product	Choline chloride (mg/g)	
	Fresh	Dry
Pig		
Adrenals	5.88	18.35
Liver[a]	5.52	18.35
Spinal cord	4.27	13.67
Brain	3.75	18.20
Pancreas	3.29	12.60
Kidney, no. 1	3.17	14.10
Kidney, no. 2	2.56	13.06
Ovary	2.78	17.38
Heart	2.31	11.16
Spleen	2.08	10.01
Small intestine	1.65	14.86
Tongue, no. 1	1.39	5.41
Tongue, no. 2	1.36	4.86
Shoulder, no. 1	1.05	2.29
Shoulder, no. 2	0.86	2.03
Ham	0.88	2.00
Chops	0.77	1.81
Lard	−	0.05

[a] An average value obtained from triplicate analyses on five different samples (pig liver, range 4.70 to 6.19, fresh basis; beef liver, range 4.86 to 7.08, fresh basis).

Table 1 (continued)
TOTAL CHOLINE CONTENT
OF ANIMAL PRODUCTS

Product	Choline chloride (mg/g)	
	Fresh	Dry
Chicken		
Egg yolk	17.13	32.81
Liver	3.42	12.50
Heart[b]	2.36	10.40
Kidney	2.23	11.32
Egg albumen	—	0.05
Lamb		
Kidney	3.60	17.82
Shoulder	1.19	3.07
Chops	1.07	3.27
Beef		
Veal liver	6.52	22.72
Beef liver	6.30	20.47
Veal kidney	3.48	15.00
Beef kidney	3.33	16.32
Veal rib roast	1.13	3.44
Beef roundsteak	0.95	3.53
Beef rib roast	0.82	2.44
Milk		
Skim milk powder	1.59	1.63
Whole milk powder	1.07	1.10
Cheddar cheese	0.48	0.70
Fresh milk[c]	0.147	1.14
Commercial casein	—	0.05
Butter	—	0.05
Fish		
Fish meal	3.29	3.47
Trout muscle	0.87	4.89
Red snapper muscle	0.84	4.12
Cod liver oil	—	0.05
Miscellaneous		
Liver extract	15.93	16.36
Extracted liver residue	4.39	4.50
Liver sausage	2.67	5.52
Tankage	2.31	2.65
Meat meal, no. 1	1.62	1.73
Meat meal, no. 2	1.30	1.42
Bologna sausage	0.71	2.38

[b] An average value obtained from triplicate analyses on five different samples (range 2.31 to 2.38, fresh basis).
[c] An average value obtained from triplicate analyses on six samples of fresh milk taken at semiweekly intervals (range 0.142 to 0.157, fresh basis).

From Engel, R. W., *J. Nutr.*, 25, 441–446, 1943. With permission.

Table 2
TOTAL CHOLINE CONTENT
OF PLANT PRODUCTS

	Choline chloride (mg/g)	
Product	Fresh	Dry
Cereal grains		
Defatted wheat germ	4.23	4.53
Raw wheatgerm, no. 1	4.10	4.40
Raw wheatgerm, no. 2	4.03	4.32
Raw corngerm stock	1.60	1.78
Rolled oats	1.51	1.63
Wheat shorts	1.48	1.63
Wheat bran	1.43	1.63
Barley	1.39	1.55
Rice polish	1.26	1.36
Oats	0.94	1.00
Wheat	0.92	1.01
Polished rice	0.88	1.02
Molasses (blackstrap)	0.86	—
Wheatena	0.62	0.68
White flour	0.52	0.57
Corn meal (unbolted)	0.42	0.47
Yellow corn	0.37	0.41
Corn meal (bolted)	0.10	0.11
Nonleafy vegetables (sun-dried)		
Snapbeans	3.40	3.81
Green soybeans	3.00	3.32
English peas	2.63	2.90
Cowpeas	2.57	2.84
Asparagus	1.28	1.47
Vegetable Oils		
Hydrogenated coconut oil	—	0.05
Oleomargarine	—	0.05
Refined corn oil	—	0.05
Refined soybean oil	—	0.05
Other seeds		
Cottonseed meal, no. 1 (7.0% fat)	3.50	3.76
Cottonseed meal, no. 2 (7.5% fat)	3.25	3.51
Soybean meal (2.5% fat)	3.45	3.75
Mature soybeans (19.5% fat)	3.40	3.58
Cottonseed kernels (36% fat)	2.98	3.19
Edible peanut meal	2.35	2.52
Peanut meal (6% fat)	2.26	2.44
Spanish peanuts (43% fat)	1.67	1.74
Runner peanuts (45.5% fat)	1.57	1.65
Peanut butter	1.45	1.48
Pecans	0.50	0.53
Root crops (sun-dried)		
Irish potatoes	1.06	1.31
Carrots	0.95	1.12
Turnips	0.94	1.11
Sweet potatoes	0.35	0.36
Leafy material (sun-dried)		
Mustard tops	2.52	2.77
Young cabbage	2.51	2.90
Turnip tops	2.45	2.69
Spinach	2.38	2.75
Rape	2.30	2.86
Pokeweed	2.28	2.64
Alfalfa leaf meal, no. 1	1.43	1.55
Alfalfa leaf meal, no. 2	1.22	1.31

From Engel, R. W., *J. Nutr.*, 25, 441—446, 1943. With permission.

Table 3
TOTAL CHOLINE CONTENT OF MEATS

| | Choline (mg/100 g) | | | |
| | Fresh | | Dry | |
Sample	Range	Average	Range	Average
Veal				
Leg	95–108	102	366–432	389
Roast leg	125–141	132	338–392	360
Shoulder	83–100	93	268–373	337
Roast shoulder	133–143	139	310–376	343
Sirloin chop	87–105	96	242–404	317
Braised chop	128–157	140	242–342	285
Shoulder chop	92–101	97	307–422	376
Braised chop	149–156	154	317–400	366
Stew meat	94–100	96	336–400	367
Cooked stew	137–149	142	360–378	370
Lamb				
Leg	75–92	84	262–317	290
Roast	122–124	123	284–295	290
Sirloin chop	75–77	76	179–198	189
Broiled chop	100–126	113	204–252	228
Stew meat	76–82	79	222–230	226
Cooked stew	116–128	122	247–291	269
Pork				
Ham	101–129	120		
Cured ham	98–129	122		
Beef				
Liver	470–570	510		
Round	65–70	68		
Tongue	108	108		
Heart	170	170		
Braised heart	200–275	238		
Kidney	240–284	262		
Brain	399–420	410		
Miscellaneous				
Bologna	60	60		
Frankfurters	57	57		
Pork links	48	48		
Canadian bacon	80	80		

From McIntire, J. M., Schweigert, B. S., and Elvehjem, C. A., *J. Nutr.*, 28, 219–223, 1944. With permission.

Table 4
TOTAL CHOLINE CONTENT OF GRAINS AND CEREAL PRODUCTS

Variety	Milligrams choline chloride/100 g[a]	Milled fractions	Milligrams choline chloride/100 g[a]
Hard spring wheat	91	Whole wheat	102
Hard winter wheat	79	Germ	354
Soft winter wheat	88	Bran	153
Soybeans	237	Low-grade flour	69
Oats	114	Bleached low-grade flour	69
Barley	110	Clear flour	61
Flax	107	Bleached clear flour	62
		Patent flour	70
		Bleached patent flour	69

[a] On a moisture-free basis.

From Glick, D., *Cereal Chem.*, 22, 95–101, 1945. With permission.

Table 5
MAJOR RECOGNIZED CAUSATIVE FACTORS IN FATTY LIVERS

etary deficiency or imbalance
    Deficiency of choline or its precursors (betaine, methionine, thetins)
    Deficiency of essential fatty acids
    Deficiency of protein
    Amino acid imbalance
    Excessive dietary fat and/or cholesterol
    Excessive food intake (calories)
isons
    Chemicals (carbon tetrachloride, chloroform, ethionine, ethanol (?), orotic acid, phosphorus, arsenic, lead, etc.)
    Bacterial toxins
epatic anoxia
sufficiency of insulin
polytic hormones (growth hormone, adrenocorticotrophic hormone, thyroid-stimulating hormone, glucagon, epi-
    nephrine, norepinephrine)
inical conditions in which one or more of the above situations occur
    Chronic debilitating diseases
    Alcoholism
    Diabetes mellitus
    Protein-calorie malnutrition
    Pregnancy
    Pernicious anemia
    Infections (tuberculosis, yellow fever, etc.)

dapted from Lucas, C. C. and Ridout, J. H., in *Progress in the Chemistry of Fats and other Lipids,* Vol. X, Holman, R. T.,
l., Pergamon Press, New York, 1967, 15–16.

REFERENCES

1. Lucas, C. C. and Ridout, J. H., Fatty livers and lipotropic phenomena, Part 1, in *Progress in the Chemistry of Fats and Other Lipids,* Vol. X, Holman, R. T., Ed., Pergamon Press, New York, 1967, 7–8.
2. Griffith, W. H., Nyc, J. F., Hartroft, W. S., and Porta, E. A., Choline, in *The Vitamins,* Vol. 3, 2nd ed., Sebrell, W. H., Jr. and Harris, R. S., Eds., Academic Press, New York, 1971, 3–154.
3. Food and Nutrition Board, National Research Council, *Recommended Dietary Allowances,* National Academy of Sciences, Washington, D. C., 1974, 65.
4. György, P. and Goldblatt, H., Experimental production of dietary liver injury (necrosis, cirrhosis) in rats, *Proc. Soc. Exp. Biol. Med.,* 46, 492–494, 1941.
5. Blumberg, H. and McCollum, E. V., The prevention by choline of liver cirrhosis in rats on high fat, low protein diets, *Science,* 93, 598–599, 1941.
6. Lillie, R. D., Daft, F. S., and Sebrell, W. H., Cirrhosis of liver in rats on deficient diet and effect of alcohol, *Public Health Rep.,* 56, 1255–1258, 1941.
7. György, P. and Goldblatt, H., Observations on conditions of dietary hepatic injury (necrosis, cirrhosis) in rats, *J. Exp. Med.,* 75, 355–368, 1942.
8. Hartroft, W. S. and Porta, E. A., Alcohol, food factors, and liver disease, *Gastroenterology,* 64, 350–352, 1973.
9. Lieber, C. S., Alcohol and nutrition, *Nutr. News,* 39, 9–12, 1976.
10. Patek, A. J., Jr., Treatment of alcoholic cirrhosis of the liver with high vitamin therapy, *Proc. Soc. Exp. Biol. Med.,* 37, 329–330, 1937.
11. Patek, A. J., Jr. and Post, J., Treatment of cirrhosis of the liver by a nutritious diet and supplements rich in vitamin B complex, *J. Clin. Invest.,* 20, 481–505, 1941.
12. Patek, A. J., Jr., Dietary treatment of Laennec's cirrhosis, with special reference to early stages of the disease, *Bull. N. Y. Acad. Med.,* 19, 498–506, 1943.
13. Broun, G. O. and Muether, R. O., Treatment of hepatic cirrhosis with choline chloride and diet low in fat and cholesterol, *JAMA* (Abstr.), 118, 1403, 1942.
14. Russakoff, A. H. and Blumberg, H., Choline as an adjuvant to the dietary therapy of the cirrhosis of the liver, *Ann. Intern. Med.,* 21, 848–862, 1944.
15. Barker, W. H., Modern treatment of the cirrhosis of the liver, *Med. Clin. North Am.,* 29, 273–293, 1945.
16. Morrison, L. M., Response of cirrhosis of the liver to an intensive combined therapy, *Ann. Intern. Med.,* 24, 465–476, 1946.
17. Beams, A. J., The treatment of cirrhosis of the liver with choline and cystine, *JAMA,* 130, 190–194, 1946.
18. Steigman, F., Efficacy of lipotropic substances in treatment of cirrhosis of liver, *JAMA,* 137, 239–242, 1948.
19. Franklin, M., Salk, M. R., Steigman, F., and Popper, S., Clinical, functional and histologic responses of fatty metamorphosis of human liver to lipotropic therapy, *Am. J. Clin. Pathol.,* 18, 273–282, 1948.
20. Beams, A. J. and Endicott, E. T., Histologic changes in the livers of patients with cirrhosis treated with methionine, *Gastroenterology,* 9, 718–735, 1947.
21. Leevy, C. M., Zinke, M. R., White, T. J., and Gnassi, A. M., Clinical observations on the fatty liver, *Arch. Intern. Med.,* 92, 527–541, 1953.
22. Eckhardt, R., Faloon, W. M., and Davidson, C. S., Improvement of active liver cirrhosis in patients maintained with amino acids intravenously as the source of protein and lipotropic substances, *J. Clin. Invest.,* 28, 603–614, 1949.
23. Eckhardt, R. D., Zamcheck, N., Sidman, R. L., Gabuzda, G. J., and Davidson, C. S., Effect of protein starvation and of protein feeding on the clinical course, liver function and liver histochemistry of three patients with active alcoholic cirrhosis, *J. Clin. Invest.,* 29, 227–237, 1950.
24. Phillips, G. B., Gabuzda, G. J., and Davidson, C. S., Comparative effects of a purified and an adequate diet on the course of fatty cirrhosis in the alcoholic, *J. Clin. Invest.,* 31, 351–356, 1952.
25. Phillips, G. B. and Davidson, C. S., Nutritional aspects of cirrhosis in alcoholism — Effect of purified diet supplemented with choline, *Ann. N. Y. Acad. Sci.,* 57, 812–830, 1954.
26. Volwiler, W., Jones, C. M., and Mallory, T. B., Criteria for the measurement of results of treatment in fatty cirrhosis, *Gastroenterology,* 11, 164–182, 1948.
27. Chalmers, T. C. and Davidson, C. S., Survey of recent therapeutic measures in cirrhosis of the liver, *N. Engl. J. Med.,* 240, 449–455, 1949.
28. Kessler, B. J., Seife, M., and Lisa, J. R., Use of choline supplements in fatty metamorphosis of the liver, *Arch. Intern. Med.,* 86, 671–681, 1950.

29. Gabuzda, G. J., Fatty liver in man and the role of lipotropic factors, *Am. J. Clin. Nutr.*, 6, 280–293, 1958.

30. Best, C. H. and Lucas, C. C., Choline malnutrition, in *Clinical Nutrition*, 2nd ed., Jolliffe, N., Ed., Harper, New York, 1962, 227–260.

31. Gillman, T., Gillman, J., Inglis, J., Friedlander, L., and Hammar, E., The substitution of whole stomach extract for vitamins in the treatment of malignant infantile pellagra, *Nature* (Abstr.), 154, 210, 1944.

32. Meneghello, J. and Neimeyer, H., Liver steatosis in undernourished Chilean children. III. Evaluation of choline treatment with repeated liver biopsies, *Am. J. Dis. Child.*, 80, 905–910, 1950.

33. Cayer, D. and Cornatzer, W. E., The use of radioactive phosphorus in measuring plasma phospholipid in patients with cirrhosis of the liver. The effects of methionine treatment, *J. Clin. Invest.* (Abstr.), 27, 528, 1948.

34. Cayer, D. and Cornatzer, W. E., The use of lipotropic factors in the treatment of liver disease *Gastroenterology*, 20, 385–402, 1952.

EFFECTS OF FOLATE DEFICIENCY IN MAN

A. V. Hoffbrand

Folate deficiency is probably the most common vitamin deficiency in the world. A number of recent major reviews of the effects of folate deficiency in man have been published. These include the books by Chanarin,[1] Blakley,[2] and the journal *Clinics in Haematology* (edited by A. V. Hoffbrand)[3] and the "Symposium on Vitamin B_{12} and Folate" (edited by V. Herbert).[4] Other relevant reviews include those of Herbert,[5] Mollin and Waters,[6] Hoffbrand,[7] and Herbert and Tisman.[8]

Many subjects who show biochemical evidence of folate deficiency do not have clinical or hematological features, however. The degree of deficiency in them is too slight to impair metabolic reactions in which folates are involved. Folates are required as coenzymes for many biochemical reactions in man involving transfer of single-carbon units, but the clinical effects of this vitamin deficiency largely occur only when one of these reactions, thymidylate synthesis, is impaired beyond a critical level. Although folates are also required in amino acid metabolism and purine synthesis, protein and RNA synthesis are not seriously disturbed by folate deficiency in humans. The most rapidly proliferating tissues of the body have the greatest requirement for DNA synthesis and, thus, are principally affected when deficiency of some severity occurs. These include bone marrow, the epithelial cell surfaces, and the gonads; the clinical manifestations are largely due to dysfunction of these tissues. However, a variety of other clinical effects of the deficiency have been reported with variable frequency. These are included in the lists of established and possible clinical manifestations of the deficiency in Tables 1, 2, and 3, described below.

CLINICAL FEATURES

Apart from features due to megaloblastic anemia or to abnormalities of the epithelial cell surfaces, the only well-established effects of folate deficiency are sterility in both men and women and, in a few percent of cases, widespread reversible melanin pigmentation of the skin, mainly affecting the skin creases and nail-beds[9] (see Table 1).

The features vary from case to case, some showing symptoms due to severe anemia, thrombocytopenia, or gastrointestinal disturbance and others having no symptoms or signs at all. Symptoms of anemia may have been present for weeks or months before diagnosis, depending on the severity of anemia, its speed of onset and age of the patient, older subjects tending to have more severe symptoms, particularly due to cardiovascular system abnormalities (Table 2). In some patients, symptoms of anemia occur when an intercurrent infection precipitates a fall in hemoglobin. Hemorrhage, easy bruising, or skin purpura is a prominent feature in a small proportion of patients, particularly alcoholics, and usually only occurs when the platelet count falls to less than $40 \times 10^9/1$. Mild jaundice occurs in patients with moderate or severe anemia, giving the patient a lemon yellow tint (see Figure 1)* and mild fever; apparently due to the excessive breakdown of hemopoietic cells again only occurs in patients with a substantial degree of anemia.

Symptoms and signs related to the gastrointestinal tract are next most common, particularly sore tongue (see Figure 2), angular cheilosis[10] (see Figure 3), loss of appetite, or diarrhea. The tongue may appear beefy red and shiny or smooth, pale, and atrophic. Recurrent aphthous ulceration has been ascribed to folate deficiency, but this is not established[11] although it is clear that methotrexate and aminopterin may cause severe

*Figures 1 to 6 appear following page 80.

Table 1
EFFECTS OF FOLATE DEFICIENCY IN MAN

Definite effects
    Megaloblastic anaemia
    Macrocytosis of epithelial cell surfaces, e.g., buccal, respiratory, cervix
      uteri, bronchial, bladder
    Sterility in both males and females
    Skin pigmentation
Possible effects
    Abnormalities of pregnancy
      Postpartum hemorrhage
      Antepartum hemorrhage
      Congenital malformations
      Prematurity
    Reduced liver regeneration
    Decreased growth in children
    Reduced bone regeneration
    Neurological effects (see Table 3)

ulceration of the G.I. tract; and nonrecurrent buccal and lingual ulceration may occur acutely together with glossitis in clinical deficiency.

On the other hand, it is still unclear how far nutritional folate deficiency affects small intestinal function. Ileal dysfunction assessed by vitamin B_{12} absorption with intrinsic factor seems to be impaired in some cases.[12] However, whereas some workers have shown structural and functional jejunal changes in nutritional folate deficiency which can be reversed by folate therapy,[13-15] others have not confirmed this.[16] Even in folate-replete subjects, folate therapy has been reported to increase the levels of some glycolytic enzymes in the jejunum.[17] There is a complicated relation between folate deficiency and tropical sprue. There is some evidence that nutritional folate deficiency may predispose to the disease and that deficiency due to reduced diet and to malabsorption may aggravate the small intestinal lesion. In early cases, folate therapy appears to improve intestinal function as well as correcting megaloblastic anemia. In later stages of the disease, however, the intestinal lesion responds less well to folic acid even though the patient is folate deficient, and antibiotic therapy and often vitamin B_{12} therapy are needed as well as folate (see References 18 to 21).

The other epithelial changes produced by folate deficiency may be seen if exfoliative cytology or biopsies are examined. Of these, the changes in the cervix uteri are most important, as they can be mistaken for carcinoma *in situ*.[22,23] Whether localized changes of folate deficiency can occur in the epithelia without hematological changes seems unlikely, but this has been suggested.

The biochemical basis for the widespread melanin pigmentation that occurs in a small proportion of patients (see Figure 4) is unclear. This reverses within a few weeks of folic acid therapy, and has also been described in vitamin B_{12} deficiency. The mucous membranes are not usually affected.

The other clinical effects listed in Table 1 can less certainly be ascribed to folate deficiency. A large number of studies have suggested that folate deficiency may cause a variety of complications of pregnancy.[24] Methotrexate is undoubtedly a powerful abortifacient. However, most controlled studies have not confirmed that folate deficiency as normally clinically encountered causes these complications, with the possible exception of prematurity.[25,26]

In alcoholics with liver damage and folate deficiency, studies using labeled thymidine uptake into liver biopsies have suggested that folate deficiency inhibits liver regeneration.[3,27] Long-continued methotrexate therapy in psoriasis or childhood leukemia appears to have caused liver fibrosis and even cirrhosis.[28] Only one study has suggested

Table 2
CLINICAL FEATURES OF
FOLATE DEFICIENCY

Symptoms
    Anemia
        Shortness of breath
        Fatigue
        Palpitations
        Headaches
        Vertigo
        Angina
        Intermittent claudication
        Edema
    Gastrointestinal disturbance
        Sore mouth
        Sore tongue
        Loss of appetite
        Loss of weight
        Diarrhea
        Vomiting
        Dysphagia
        Indigestion
    Other
        Weakness
        Tiredness
        Loss of energy and drive
        Irritability
        Forgetfulness
        Infertility
        Easy bleeding
    Symptoms of underlying or associated condition
        e.g, hemolytic anemia, malabsorption syndrome
Signs
        Pallor of mucous membranes
        Jaundice
        Melanin pigmentation
        Purpura or petechiae
        Fever (to 38.5°C)
        Glossitis (clean, shiny, smooth or beefy red)
        Angular cheilosis
        ? Aphthous ulceration[a]
        Tachycardia
        Hypotension
        Systolic murmur
        Features of congestive heart failure
        Hepatomegaly
        Splenomegaly
        Optic fundal hemorrhage
        ? Neurological features (see Table 3)[a]

[a] ? denotes not definitely established.

that folate deficiency in children (with sickle cell anemia) inhibits growth.[29] No good supporting evidence has been found for this, however. Low serum alkaline phosphatase ascribed to reduced bone regeneration has been described in untreated vitamin B_{12} deficiency.[30] A similar effect of folate deficiency on serum alkaline phosphatase has not been established but is likely in view of the similar effects of the two deficiencies on cell turnover.

It is still not clear whether or not folate deficiency causes histologically demonstrable

Table 3
FOLATE AND NEUROLOGICAL DISEASE: REPORTED ASSOCIATIONS[8,35,36]

Folate deficiency
    Mental effects
        Dementia
        Decreased drive, alertness
        Irritability
        Schizophrenia-like illness
        Retardation
    Neurological effects
        Protection from fits
        Peripheral neuropathy
        Spinal cord damage
        Brain damage (due to intrathecal methotrexate)
Folate excess
        Precipitation or aggravation of neuropathy due to vitamin B_{12} deficiency
        Precipitation of fits in epileptics
        Psychological disturbances

neurological damage, although mild psychiatric changes such as irritability and forgetfulness are frequent. A definite association between intrathecal methotrexate administration and brain damage has been found.[31] Also there is a definite association between inborn errors of metabolism and mental retardation dilatation, of the cerebral ventricles, and intracerebral calcification[32,33] but whether folate lack causes the CNS defects is not certain. It is also clear that loss of appetite and hence reduced folate intake may occur in a variety of chronic disabling neurological diseases. Whether folate deficiency can cause any of the other neurological effects listed in Table 3 is debatable. Many thousands of severely folate-deficient patients have been seen without any neurological defect and odd case reports have not yet established that folate deficiency actually causes the neurological lesions present, despite recent suggestive studies.[34]

Reynolds and co-workers have established much circumstantial evidence that folate deficiency induced by anticonvulsant therapy may help to protect patients with epilepsy from fits and that folate therapy may increase fit-frequency and may even precipitate status epilepticus.[35] However, double-blind trials of folate therapy have not shown a significant effect of folic acid on fit-frequency in epileptics receiving anticonvulsant drugs.[36] The subject has been reviewed in detail.[8,35]

HEMATOLOGICAL EFFECTS

The earliest, most consistent and prominent manifestation of the deficiency are the red cell and white cell changes. Macrocytosis occurs before anemia develops, the mean cell volume is raised usually to above 100 fl, and hypersegmentation of polymorph nuclei appears (i.e., polymorphs with six or more nuclear lobes are seen) (see Table 4 and Figure 5). When anemia is moderately severe, leucopenia and thrombocytopenia may also occur, but the leucocyte level rarely falls below $1.5 \times 10^9/l$ and platelets rarely below $40 \times 10^9/l$, except in alcoholics. The differential leucocyte count remains normal, an equal fall occurring in polymorphs and lymphocytes. The lymphocytes appear normal, but if made to proliferate by stimulation with phytohemagglutinin, they show similar morphological and biochemical defects to the bone marrow hemopoietic cells.[37]

The severity of the bone marrow changes is related to the degree of anemia and in nonanemic cases may be difficult to recognize. Asynchronous development of nucleus and cytoplasm and a stippled appearance of the nuclear chromatin are the earliest changes

Table 4
FOLATE DEFICIENCY IN MAN:
PERIPHERAL BLOOD CHANGES

Usual cases
   Anemia
   Oval macrocytosis
   Anisocytosis, poikilocytosis
   Hypersegmented neutrophils
Severe cases only
   Leucopenia
   Thrombocytopenia
   Presence of nucleated red cells and granulocyte pre-
     cursors
   Presence of Howell-Holly bodies, basophilic stippling,
     Cabot rings

Table 5
SEVERE FOLATE DEFICIENCY:
BONE MARROW ABNORMALITIES[1,38]

Morphological changes
   Expansion down long bones and to extramedullary
     sites
   Increased cellularity[a]
   Increased erythroid-myeloid ratio[a]
   Increased size of cells
   Asynchronous nuclear and cytoplasmic development
     (delay in nuclear maturation) of erythroblasts
   Fine, lacey, stippled nuclear appearance of erythro-
     blasts
   Increased nuclear remnants, dyserythropoiesis, mitosis,
     multinucleate cells, dying cells, erythrophagocytosis
   Increased proportion of primitive cells (increase in
     colony-forming units in agar)
   Increased and coarse intracytoplasmic siderotic gran-
     ules; rarely, ringed sideroblasts
   Giant and abnormally shaped metamyelocytes
   Hypersegmented neutrophils
   Hypersegmented megakaryocytes

[a] Rare cases show general hypoplasia or pure erythroid
hypoplasia.

in the developing red cells (see Figure 6). Giant, abnormally shaped metamyelocytes and hypersegmented polymorphs are the earliest abnormalities in the white cell series. In more anemic cases, the marrow becomes hypercellular with expansion of hemopoietic tissue down the long bones and in some cases, with extra-medullary hemopoiesis in the liver, spleen, and other organs. The abnormalities that may be seen in these severe cases are listed in Table 5. Iron deficiency may mask the changes in the developing erythroblasts but does not affect the white cell abnormalities.

CHROMOSOME CHANGES

The chromosome changes listed in Table 6 have been described in bone marrow, dividing lymphocytes, and proliferating epithelial cells. Although abnormalities of histones have been described, the primary fault appears to be within DNA itself.

Table 6
SEVERE FOLATE DEFICIENCY:
EFFECTS ON CHROMOSOMES AND CELL CYCLE

Discernible changes	Ref.
Chromosome changes	39, 40
Despirallization (incomplete contraction)	
Exaggeration of centromeric constrictions and normal sites of secondary constrictions	
Random breaks	
Elongation	
DNA changes	41
Smaller fragments than normal on alkaline sucrose gradient	
Normal repair from X-ray damage	
Normal major base composition	
Normal proportion of methylated bases	
Decrease of biosynthesis, acetylation, and methylation of arginine-rich histone	42
Disturbances of the cell cycle	43
Increase in cells in G_2	
Presence of cells with intermediate DNA content (between 2c and 4c)[a] not in DNA synthesis ("U" cells)	
Increased proportion of cells in prophase of the mitotic cycle; changes especially marked in the early polychromatophilic erythroblasts	
? Prolongation of "S" phase	
Electron-microscope changes	44
Nuclear	
Reduced quantity and spongy appearance of heterochromatin and membrane-bound nuclear clefts	
Irregularly shaped, fragmenting nuclei	
Intranuclear inclusion	
Partial absence and myelinization of nuclear membrane and separation from nucleus	
Separation of two layers of nuclear membrane ribosome depends on outer layer	
Cytoplasmic	
Long strands of endoplasmic reticulum	
Autophagic vacuoles	
Annulate lamellae	
Free ferritin molecules increased	
Iron-laden and degenerating mitochondria	
Inclusions of nuclear material	
Clustering of cytoplasmic organelles near nucleus	
Reduced ribosome content	

[a] c: an arbitrary figure for cell DNA content.
[b] ?: not definitely established as caused by folate deficiency.

However, the exact fault is uncertain. The chromosomes are randomly affected and the base composition of the DNA that is formed is normal.[45] Presumably, the fault arises because of precursor starvation (shortage of thymidine triphosphate) which is caused by folate deficiency and impairs replication of DNA during the "S" phase of the cell cycle. Hoffbrand et al.[41] have suggested that there is excess initiation of new replicons over the cell's ability to elongate these and thus complete replicating segments. Further studies are needed in this area.

Biochemical Defects

Many of the biochemical changes and the disturbances of iron metabolism in severe

Table 7
SEVERE FOLATE DEFICIENCY: BIOCHEMICAL CHANGES[1,3,7]

Plasma
  Increase
    Bilirubin (unconjugated), iron, ferritin, lactic acid dehydrogenase (especially
    LDH_1 and LDH_2 isoenzymes), phosphohexoisomerase, β-hydroxybuty-
    rate dehydrogenase, isocitric dehydrogenase, glucose-6-phosphate dehydro-
    genase, 6-phosphogluconic dehydrogenase malic dehydrogenase, lysozyme,
    glutamate-oxaloacetic transaminase, phenol derivatives
  Decrease
    Haptoglobin, alkaline phosphatase, cholesterol, immunoglobulins, cho-
    linesterase
  May be positive
    Methemalbumin, fibrin degradation products
Urine
  Increase
    Urobilinogen, aminoimidazole carboxamide (AIC) excretion, uric acid,
    Figlu excretion, hemosiderin, fibrin degradation products, hydroxyphenyl
    compounds, taurine
Stool
  Increase
    Stercobilinogen
Iron studies
  Rapid iron clearance
  Increased iron turnover
  Decreased red cell iron utilization

Table 8
RED AND OTHER CELL ENZYMES[1,7,49]

Raised
  Aldolase, glyceraldehyde-3-phosphate isomerase, glyceraldehyde-3-phosphate
  dehydrogenase, lactate dehydrogenase, pyruvate kinase, 3-phosphoglycerate-
  1-kinase, glucose-6-phosphate dehydrogenase, methemoglobin reductase,
  phosphohexoisomerase, malic dehydrogenase, isocitric dehydrogenase, gluta-
  mic oxaliocetic transaminase, transketolase, dehydroascorbic acid reductase
  Thymidine kinase, aspartate carbamyltransferase, dihydro-orotase
Other red cell changes
  Direct Coombs' positive (complement only)
  Increase in fetal hemoglobin
  Increased agglutinability by anti-I and anti-i[a]

[a] I, i: antigens on the red cell surface.

folate deficiency are due to ineffective hemopoiesis, i.e., death of red-cell, white-cell, or platelet precursors within the marrow or extra-medullary organs. This is the main cause of the raised bilirubin, raised LDH,[46,47] and raised lysozyme[48] in serum (see Tables 7 and 8).

Laboratory Diagnosis

This usually depends on recognition of the blood changes due to the deficiency and then distinguishing vitamin B_{12} or folate deficiency as the cause (see Table 9). (Other causes of megaloblastic anemia are given in Table 13.) Vitamin B_{12} deficiency is excluded by measuring the serum vitamin B_{12} level or carrying out a therapeutic trial. A positive diagnosis of folate deficiency is usually made by the finding of subnormal serum and red-cell folate levels. Provided the result is not falsely low because of inhibition of the microbiological assay by drugs in the serum, the serum folate is an accurate guide to the

Table 9

LABORATORY TESTS FOR FOLATE DEFICIENCY

	Ref.
Commonly in use	
Serum folate	
Microbiological assay	50–54
Isotope assay (normal range 3.0–15.0 ng/ml)	
Red cell folate	
Microbiological assay	54–56
Isotope assay (normal range 160–640 ng/ml)	
Deoxyuridine suppression of ^{3}H-thymidine or ^{125}I-iodouridine incorporation into DNA (marrow or phytohemagglutinin-transformed lymphocytes)	37, 57–59
Correction by methyltetrahydrofolate, but not by vitamin B_{12}	
Therapeutic trial; hematological response to 50–200 μg folic acid daily	60
Not normally carried out	
Excretion of formiminoglutamic acid (Figlu) after histidine loading	61
Clearance of i.v. injected folic acid	62
Excretion of aminoimidazole carboxamide (AIC)	63
Incorporation of labeled formate into serine	64, 65
Oxidation of β-carbon of serine into respiratory CO_2	66
Folate absorption tests	67

presence of folate deficiency. However, equally low results may be obtained in patients with extremely severe deficiency and in nondeficient subjects whose intake of folate has fallen for only a few days. The serum folate is normal or raised in vitamin B_{12} deficiency uncomplicated by folate deficiency.

The red-cell folate is a better guide than serum folate to tissue folate stores. A subnormal result normally means folate deficiency of some standing and severity. However, the red-cell folate concentration also falls in vitamin B_{12} deficiency, so that a low level alone does not distinguish the two. Thus, the red-cell folate assay is better used in conjunction with the serum vitamin B_{12} and folate assays. Both serum and red-cell folate may be measured by isotope dilution techniques using a binding protein derived from milk and radioactive labeled folic acid or methyltetrahydrofolate. These assays have the advantage of speed and are not affected by drugs, e.g., antibiotics, tranquilizers, antimitotic agents which have been found to inhibit the growth of *Lactobacillus casei.* The accuracy of the radioassays has yet to be proved, however.

Full therapeutic trials with a preliminary observation period, diet of low folate and vitamin B_{12} content, and therapy with a physiological dose of vitamin B_{12} (1 to 2 μg), a folic acid (50 to 200 μg) and frequent measurement of reticulocyte response and hemoglobin, is most valuable in analyzing the cause of megaloblastic anemia in patients with low serum levels of both vitamin B_{12} and folate but is too tedious for routine use.

The deoxyuridine suppression test is thought to be an indirect measure of thymidylate synthetase activity. The test is abnormal in megaloblastic anemia due to vitamin B_{12} or folate deficiency and correlates with the severity of the morphological changes. The test is normal in patients with megaloblastic anemia not associated with a defect of thymidylate synthesis. Differential correction of the suppression test can be used to distinguish vitamin B_{12} and folate deficiency.

CAUSES OF FOLATE DEFICIENCY

In most patients with folate deficiency, a combination of factors leads to negative folate balance. Poor diet is usually the major cause, since few diseases cause such malabsorption or excess of folate utilization of such severity that a good intake of the vitamin cannot overcome losses. It is likely that severe folate deficiency may occur in tropical sprue, gluten-induced enteropathy, and congenital specific malabsorption of folate, despite a normal dietary folate content. In all the other conditions listed in Table 10, however, when folate deficiency severe enough to cause megaloblastic anemia occurs, poor diet is a factor. Normal adult dietary intake of folate ranges from about 400 to 1000 μg of which from 50 to 80% is absorbed. The more folate eaten, the more absorbed with

Table 10
CAUSES OF FOLATE DEFICIENCY[1,7,68]

Diet
> Poverty, psychiatric, special institutions, alcoholics, special diets, slimming diets, gastric operations, faulty food preparation, chronic debilitating disease, intensive care units, goat's milk
> Associated with scurvy, kwashiorkor, protein-calorie malnutrition

Malabsorption
> Major factor
>> Celiac disease, adult celiac disease, celiac disease with dermatitis herpetiformis, tropical sprue, jejunal resection, congenital malabsorption of folate
> Minor factor or single reports
>> Crohn's disease, Whipple's disease, scleroderma, amyloid, diabetic enteropathy, cardiac failure, lymphoma

Increased utilization
> Physiological
>> Pregnancy (especially multiple)
>> Lactation
> Pathological
>> Hematological
>>> Hemolytic anemias (especially sickle cell anemia, thalassemia major, auto-immune, hereditary spherocytosis), myelosclerosis, sideroblastic anemia
>> Malignant
>>> Carcinoma, lymphoma, myeloma, leukemia
>> Inflammatory
>>> Rheumatoid arthritis, tuberculosis, Crohn's disease, chronic malaria, exfoliative dermatitis, widespread psoriasis
> Metabolic
>> Homocystinuria
>> ? Lesch-Nyhan syndrome

Increased folate loss
> Liver damage, congestive heart failure, chronic hemodialysis or peritoneal dialysis

Drugs
> Mechanism unknown
>> Anticonvulsants (diphenylhydantoin, primidone), barbiturates, cycloserine, ? nitrofurantoin, ? tetracycline, ? glutethimide
> Malabsorption
>> Cholestyramine, sulphasalazine, ? paraminosalicylate
> Uncertain
>> Alcohol, ? oral contraceptives

Note: In all cases, ? denotes not definitely established as a cause of folate deficiency.

probably little fall in the proportion absorbed over the normal dietary range. Normal adult daily requirements are 100 to 200 μg and stores about 10 to 15 mg. Requirements in normal pregnancy are increased to 300 to 400 μg daily and probably to around this figure in diseases with increased cell turnover, e.g., severe chronic hemolysis and myelosclerosis, when degradation of folate is increased. In some instances, only single case reports have described the clinical association for which there is no experimental proof and in many of the conditions, particularly the drug-associated folate deficiency, it is not definitely established that the drug has an antifolate action.[69,70]

DISTURBANCES OF FOLATE METABOLISM

A number of congenital disturbances of folate metabolism have been described, most associated with brain defects and mental retardation[32,33] (see Table 11). Two are reasonably well established, formiminoglutamic acid (Figlu) transferase deficiency and 5,10-methylenetetrahydrofolate reductase.[71]

A number of others have been described in Japanese babies by Arakawa and colleagues,[32] but have not been reported elsewhere. Fuller biochemical details are needed of these or future cases before they can be accepted as proven. A recent paper[72] described two children with dihydrofolate reductase deficiency; however, the second case has been reexamined by the author and rediagnosed as transcobalamin II deficiency. On the other hand, sufficient numbers of patients with congenital specific malabsorption of folate have been described for this condition to be accepted.[68]

Inhibition of dihydrofolate reductase by methotrexate causes similar effects to nutritional folate deficiency. If toxic doses of methotrexate are given, clinical effects due to inhibition of cell proliferation in the gastrointestinal tract and of platelets or polymorphs are felt first; megaloblastic anemia is a later event. Pyrimethamine has a tenfold, triemterene a 100-fold and trimethoprim 100,000-fold weaker affinity for human dihydrofolate reductase than methotrexate.[70]

The other major disturbance of folate metabolism is that caused by vitamin B_{12} deficiency. Patients with vitamin B_{12} deficiency show all the clinical, hematological, and many of the biochemical features of folate deficiency. This is thought to be due to disturbance in folate metabolism whereby folate is trapped as methyl-tetrahydrofolate (the form in plasma and synthesized by the small intestine from dietary folate) and cells become depleted of folate coenzymes. There is some debate of the exact mechanism.[73,74] Perhaps the most likely explanation is failure of formation of intracellular

Table 11
DISTURBANCES OF FOLATE METABOLISM

Changes	Ref.
Inborn errors of metabolism	32, 33
Defects of enzyme	
Figlu transferase, homocysteine-methionine, methyltransferase, cyclohydrase, dihydrofolate reductase, methenyltetrahydrofolate reductase	
Acquired	
Drugs	69, 70
Dihydrofolate reductase inhibitors	
Methotrexate, aminopterin, pyrimethamine, DDMP, DDEP, trimethoprim, triamterene, ? pentamidine[a]	
Vitamin B_{12} deficiency	73—75

[a] ?, not definitely established.

polyglutamate forms of folate due to depletion of the necessary intracellular tetrahydro-folate substrate for folate polyglutamate synthesis.[75]

TREATMENT OF FOLATE DEFICIENCY

It is usual to treat with 5-mg tablets of pteroylglutamic acid (folic acid) one to three times daily (see Table 12). This is vastly more than daily requirements, the excess being excreted in urine and probably bile. Even in patients with malabsorption of folate, sufficient folate is absorbed from these large doses to obtain a response and replenish body stores. It is likely that liver stores will be replenished within days, but it is usual to continue treatment for about 4 months and then make the decision to stop therapy or continue, depending on the underlying cause of the deficiency. Other therapy for patients with severe megaloblastic anemia that may be needed includes treatment of an intercurrent infection which often precipitates the anemia in patients with subclinical deficiency, treatment of congestive heart failure which is more likely to be present in the elderly and platelet transfusions for patients with hemorrhage or purpura due to thrombocytopenia. Blood transfusion should be avoided. If absolutely essential, 1 to 2 units of packed cells should be given very slowly, with careful monitoring of the jugular venous pressure and patient for cough or moist chest sounds. Oral potassium supplements are not needed routinely but should be given to severely anemic patients or those with heart failure during therapy with folic acid.

For patients who need therapy after the 4-month period, e.g., patients with chronic hemolytic anemias or myelosclerosis, 5 mg folic acid by mouth once a week is usually sufficient. Patients with continuing severe malabsorption of folate, e.g., celiac disease not responding to a gluten-free diet, are best maintained on 5 mg folic acid daily.

In all patients receiving folate therapy, it is important to exclude vitamin B_{12} deficiency, and for those on regular folic acid, it is advisable to measure the serum vitamin B_{12} level once a year, since folate therapy masks megaloblastic anemia due to vitamin B_{12} deficiency but does not correct the neurological defect and indeed may aggravate or precipitate this.[76,77]

OTHER CAUSES OF MEGALOBLASTIC ANEMIA

Megaloblastic anemia may arise from defects in DNA synthesis, other than those

Table 12
TREATMENT OF FOLATE DEFICIENCY

Folic (pteroylglutamic) acid
  Initial therapy
    Replenishment of body stores
      Folic acid 5–15 mg daily by mouth
      Continue after 4 months if underlying cause not corrected
    Indications for prophylactic therapy
      Pregnancy (extra requirement ca. 300 μg daily)
      Prematurity
      Chronic dialysis
    Intensive care units, if necessary
      Severe chronic hemolysis, e.g., sickle cell anemia, thalassemia major,
        chronic myelosclerosis
    Contraindications
      Vitamin B_{12} deficiency
      Malignancy
Folinic acid (5-formyltetrahydrofolate)
  Prevention or reversal of toxicity of methotrexate or other dihydrofolate reductase inhibitors

Table 13
CAUSES OF MEGALOBLASTIC ANEMIA OTHER THAN FOLATE DEFICIENCY OR DEFECTS OF FOLATE METABOLISM[1,7,41,78]

Congenital defects of DNA synthesis
    Orotic aciduria
    Lesch-Nyhan syndrome[a]
    Responding to large doses of vitamin B_{12} and folate
    Responding to thiamine (single case report)
    Congenital dyserythropoietic anemia[a]
Acquired defects of DNA synthesis
    Drugs, e.g., cytosine arabinoside, hydroxyurea, 5-fluorouracil, 6-mercaptopur-
    ine, azathioprine
    Erythroleukemia[a], other myeloid leukemias[a]
    Primary acquired sideroblastic anemia[a]
    ? Vitamin E deficiency[b]

[a] Some cases only.
[b] ?, not definitely established.

caused by folate deficiency or by disturbances of folate metabolism (see Table 13). In some cases, the site of the defect is known — e.g., in congenital orotic aciduria, the activity of two enzymes (orotidylic pyrophosphorylase and orotidylic decarboxylase) concerned in the early stages of pyramidine synthesis is reduced or absent. It is also known at which points many of the antimetabolite drugs inhibit DNA synthesis. The site of the DNA defect in some of the other conditions listed in Table 13 such as erythroleukemia or acquired sideroblastic anemia is, however, not known.

REFERENCES

1. **Chanarin, I.,** *The Megaloblastic Anaemias,* Blackwell, Oxford, 1969.
2. **Blakley, R. L.,** *The Biochemistry of Folic Acid and Related Pteridines,* North-Holland, Amsterdam and London, 1969.
3. **Hoffbrand, A. V.,** Megaloblastic anaemia, *Clinics Haematol.,* 5(3), 1976.
4. **Herbert, V.,** Symposium on vitamin B_{12} and folate, *Am. J. Med.,* 48, 539—609, 1970.
5. **Herbert, V.,** Folic acid deficiency in man, *Vitam. Horm.* (New York), 26, 525—533, 1968.
6. **Mollin, D. L. and Waters, A. H.,** Nutritional megaloblastic anaemia, *Symp. Swed. Nutr. Found.,* 6, 121, 1968.
7. **Hoffbrand, A. V.,** The megaloblastic anaemias, in *Recent Advances in Haematology,* Goldberg, A. and Brain, M. C., Eds. Churchill Livingstone, London, 1971, 1—76.
8. **Herbert, V. and Tisman, G.,** Effects of deficiencies of folic acid and vitamin B_{12} on central nervous system function and development, in *Biology of Brain Dysfunction,* Gaull, G., Ed., Plenum Press, New York, 1973, 373—392.
9. **Fleming, A. F. and Dawson, I.,** Pigmentation in megaloblastic anaemia, *Br. Med. J.,* 2, 236, 1972.
10. **Rose, J. A.,** Folic-acid deficiency as a cause of angular cheilosis, *Lancet,* 2, 453, 1971.
11. Recurrent oral ulceration (leading article), *Br. Med. J.,* 3, 757, 1974.
12. **Scott, R. B., Kammer, R. B., Burgher, W. F., and Middleton, F. G.,** Reduced absorption of vitamin B_{12} in two patients with folic acid deficiency, *Ann. Intern. Med.,* 69, 111, 1968.
13. **Bianchi, A., Chipman, D. W., Dreskin, A., and Rosensweig, N. S.,** Nutritional folic acid deficiency with megaloblastic changes in the small bowel epithelium, *N. Engl. J. Med.,* 282, 859—861, 1970.
14. **Dawson, D. W.,** Partial villous atrophy in nutritional megaloblastic anaemia corrected by folic acid therapy, *J. Clin. Pathol.,* 24, 131—135, 1971.
15. **Berg, N. O., Dahlquist, A., Lindberg, T., Lindstrand, K., and Norden, A.,** Morphology, dipeptidases and disaccharidases of small intestinal mucosa in vitamin B_{12} and folic acid deficiency, *Scand. J. Haematol.,* 9, 167, 1972.

16. Winawer, S. J., Sullivan, L. W., Herbert, V., and Zamcheck, N., The jejunal mucosa in patients with nutritional folate deficiency and megaloblastic anemia, *N. Engl. J. Med.*, 272, 892–895, 1965.

17. Rosensweig, N. S., Herman, R. H., Stifel, F. B., and Herman, Y. J., Regulation of human jejunal glycolytic enzymes by oral folic acid, *J. Clin. Invest.*, 48, 2038, 1969.

18. Klipstein, F. A., Tropical sprue, *Gastroenterology*, 54, 275, 1968.

19. Klipstein, F. A., Folate in tropical sprue, *Br. J. Haematol.*, Suppl. 23, 119–133, 1972.

20. Baker, S. J., Tropical sprue, *Br. Med. Bull.*, 28, 87, 1972.

21. Wellcome Trust Collaborative Study, *Tropical sprue and megaloblastic anaemia*, Churchill Livingstone, Edinburgh, 1971.

22. Boddington, M. M. and Spriggs, A. I., The epithelial cells in megaloblastic anaemias, *J. Clin. Pathol.*, 12, 228, 1959.

23. Blackledge, D. and Goodall, H. B., Changes in the cervico-vaginal epithelium in gestational megaloblastic anaemia, *Aust. N.Z. J. Obstet. Gynaecol.*, 13, 241, 1973.

24. Hibbard, B. M., Hibbard, E. D., and Jeffcoate, T. N. A., Folic acid and reproduction, *Acta Obstet. Gynecol. Scand.*, 44, 375–400, 1965.

25. Giles, C., An account of 335 cases of megaloblastic anaemia of pregnancy and the puerperium, *J. Clin. Pathol.*, 19, 1–4, 1966.

26. Baumslag, N., Edelstein, T., and Metz, J., Reduction of incidence of prematurity by folic acid supplementation in pregnancy, *Br. Med. J.*, 1, 16, 1970.

27. Leevy, C. M., Hepatic DNA synthesis in man, *Medicine* (Baltimore), 45, 423, 1966.

28. Schein, P. S. and Winokur, S. H., Immunosuppressive and cytotoxic chemotherapy: long term complications, *Ann. Intern. Med.*, 82, 84–95, 1975.

29. Watson-Williams, E. J., *East Afr. Med. J.*, 39, 213–219, 1962.

30. Van Dommelen, C. K. V. and Klaassen, C. H. L., Cyanocobalamin-dependent depression of the serum alkaline phosphatase level in patients with pernicious anemia, *N. Engl. J. Med.*, 271, 541, 1964.

31. Kay, H. E. M., Knapton, P. J., O'Sullivan, J. P., Wells, D. G., Harris, R. F., Innes, E. M., Stuart, J., Schwartz, F. C. M., and Thompson, E. N., Severe neurological damage associated with methotrexate therapy, *Lancet*, 2, 542, 1971.

32. Arakawa, T., Congenital defects in folate utilization, *Am. J. Med.*, 48, 594–598, 1970.

33. Erbe, R. W., Inborn errors of folate metabolism, *N. Engl. J. Med.*, 293, 753–757 and 807–812, 1975.

34. Manzoor, M. and Runcie, J., Folate-responsive neuropathy: report of ten cases, *Br. Med. J.*, 2, 1176–1178, 1976.

35. Reynolds, E. H., Neurological and psychiatric aspects of vitamin B_{12} and folate deficiencies, *Clinics Haematol.*, 5, 661–696, 1976.

36. Mattson, R. H., Gallagher, B. B., Reynolds, E. H., and Glass, D., Folate therapy in epilepsy. A controlled study, *Arch. Neurol.* (Chicago), 29, 78–81, 1973.

37. Das, K. C. and Hoffbrand, A. V., Lymphocyte transformation in megaloblastic anaemia: morphology DNA synthesis, *Br. J. Haematol.*, 19, 459, 1970.

38. Dacie, J. V. and White, J. C., Erythropoiesis with particular reference to its study by biopsy of human bone marrow: a review, *J. Clin. Pathol.*, 2, 1, 1949.

39. Heath, C. W., Cytogenetic observations in vitamin B_{12} and folate deficiency, *Blood*, 27, 800, 1966.

40. Menzies, R. C., Crossen, P. E., Fitzgerald, P. H., and Gunz, F. W., Cytogenetic and cytochemical studies on marrow cells in B_{12} and folate deficiency, *Blood*, 28, 581, 1966.

41. Hoffbrand, A. V., Ganeshaguru, K., Hooton, J. W. L., and Tripp, E., Megaloblastic anaemia: initiation of DNA chain elongation as the underlying mechanism, *Clinics Haematol.*, 5, 727–745, 1976.

42. Kass, L., Acetylation and methylation of histones in pernicious anemia, *Blood*, 44, 125–129, 1974.

43. Wickramasinghe, S. N., Some aspects of abnormal erythropoiesis and granulocytopoiesis, in *Human Bone Marrow*, Blackwell, Oxford, 1975, chap. 9.

44. Wickramasinghe, S. N. and Bush, V., Election microscope and high resolution autoradiographic studies of megaloblastic erythropoiesis, *Acta Haematol.*, 57, 1–14, 1977.

45. Hoffbrand, A. V. and Pegg, A. E., DNA base composition in normoblastic and megaloblastic marrow, *Nature (London) New Biol.*, 235, 187, 1972.

46. Emerson, P. M. and Wilkinson, J. H., Lactate dehydrogenase in diagnosis and assessment of response to treatment of megaloblastic anemia, *Br. J. Haematol.*, 12, 678, 1966.

47. Hoffbrand, A. V., Kremenchuzky, S., Butterworth, P. J., and Mollin, D. L., Serum lactic dehydrogenase activity and folate deficiency in myelosclerosis and other haematological diseases, *Br. Med. J.*, 1, 577, 1966.

48. Perillie, P. E., Kaplan, S. S., and Finch, S. C., Significance of changes in serum muramidase activity in megaloblastic anemia, *N. Engl. J. Med.*, 277, 10, 1967.

49. **Stuart, J. and Skowron, P. N.,** A cytochemical study of marrow enzymes in megaloblastic anaemia, *Br. J. Haematol.,* 15, 443, 1968.

50. **Baker, H., Herbert, V., Frank, O., Pasher, I., Hutner, S. H., Wasserman, L. R., and Sobotka, H.,** A microbiological method for detecting folic acid deficiency in man, *Clin. Chem.* (New York), 5, 275, 1959.

51. **Waters, A. H. and Mollin, D. L.,** Studies on the folic acid activity of human serum, *J. Clin. Pathol.,* 14, 335–344, 1971.

52. **Herbert, V.,** The aseptic addition method *L. casei* assay of folate activity in human serum, *J. Clin. Pathol.,* 19, 12, 1966.

53. **Rothenberg, S. P. and daCosta, M.,** Folate binding proteins and radioassay of folate, *Clinics Haematol.,* 5, 569–587, 1976.

54. **Longo, D. L. and Herbert, V.,** Radioassay for serum and red cell folate, *J. Lab. Clin. Med.,* 88, 138–151, 1976.

55. **Hoffbrand, A. V., Newcombe, B. F. A., and Mollin, D. L.,** Method of assay of red cell folate and the value of the assay as a test for folate deficiency, *J. Clin. Pathol.,* 19, 17–28, 1966.

56. **Schreiber, C. and Waxman, S.,** Measurement of red cell folate levels by ^{3}H-pteroylglutamic acid (^{3}H-PteGlu) radioassay, *Br. J. Haematol.,* 27, 551, 1974.

57. **Killmann, S. A.,** Effect of deoxyuridine on incorporation of triated thymidine: difference between normoblasts and megaloblasts, *Acta Med. Scand.,* 175, 483, 1964.

58. **Metz, J., Kelly, A., Swett, V. C., Waxman, S., and Herbert, V.,** Deranged DNA synthesis by bone marrow from vitamin B_{12}-deficient humans, *Br. J. Haematol.,* 14, 575, 1968.

59. **Herbert, V., Tisman, G., Le Teng Go, and Brenner, L.,** The dU suppression test using ^{125}I-UdR to define biochemical megaloblastosis, *Br. J. Haematol.,* 24, 713–723, 1973.

60. **Marshall, R. A. and Jandl, J. H.,** Responses to "physiologic" doses of folic acid in the megaloblastic anaemias, *Arch. Intern. Med.,* 105, 353, 1960.

61. **Luhby, A. L. and Coopermann, J. M.,** Folic acid deficiency in man and its interrelationship with vitamin B_{12} metabolism, *Adv. Metab. Disord.,* 1, 263, 1964.

62. **Chanarin, I., Mollin, D. L., and Anderson, B. B.,** The clearance from the plasma of folic acid injected intravenously in normal subjects and patients with megaloblastic anaemia, *Br. J. Haematol.,* 4, 435, 1958.

63. **Herbert, V., Streiff, R. R., Sullivan, L. W., and McGeer, P. L.,** Deranged purine metabolism manifested by aminoimidazole carboxamide (A.I.C.) excretion in megaloblastic anaemias, haemolytic anaemia and liver disease, *Lancet,* 2, 445–446, 1964.

64. **Ellegaard, J. and Esmann, V.,** Folate deficiency in pernicious anaemia measured by determination of decreased serine synthesis in lymphocytes, *Br. J. Haematol.,* 24, 571–577, 1973.

65. **Ellegaard, J. and Esmann, V.,** Folate activity of human lymphocytes determined by measurements of serine synthesis, *Scand. J. Clin. Lab. Invest.,* 31, 1–9, 1973.

66. **DeGrazia, J. A., Fish, M. B., Pollycove, M., Wallenstein, R. O., and Hollanden, L.,** Oxidation of the beta carbon of serine in human folate and vitamin B_{12} deficiency, *J. Nucl. Med.,* 10, 329, 1969.

67. **Freedman, D. S., Brown, J. P., Weir, D. G., and Scott, J. M.,** The reproducibility and use of the initiated folic acid urinary excretion test as a measure of folate absorption in clinical practice: effect of methotrexate as absorption of folic acid, *J. Clin. Pathol.,* 26, 261–267, 1973.

68. **Rosenberg, I. H.,** Absorption and malabsorption of folates, *Clinics Haematol.,* 5, 589–618, 1976.

69. **Stebbins, R., Scott, J., and Herbert, V.,** Drug-induced megaloblastic anaemias, *Semin. Hematol.,* 10, 235–251, 1973.

70. **Stebbins, R. and Bertino, J. R.,** Megaloblastic anaemia produced by drugs, *Clinics Haematol.,* 5, 619–630, 1976.

71. **Freeman, J. M. and Finkelstein, J. D.,** Folate-responsive homocystinuria and "schizophrenia," *N. Engl. J. Med.,* 292, 491, 1975.

72. **Tauro, G. P., Danks, D. M., Rowe, P. B., Van der Weyden, M. B., Schwarz, M. A., Collins, V. L., and Neal, B. W.,** Dihydrofolate reductase deficiency causing megaloblastic anaemia in two families, *N. Engl. J. Med.,* 294, 466–470, 1976.

73. **Das, K. C. and Herbert, V.,** Vitamin B_{12}-folate interrelations, *Clinics Haematol.,* 5, 697–725, 1976.

74. **Perry, J., Lumb, M., Laundy, M., Reynolds, E. H., and Chanarin, I.,** Role of vitamin B_{12} in folate coenzyme synthesis, *Br. J. Haematol.,* 32, 243–248, 1976.

75. **Hoffbrand, A. V.,** Synthesis and breakdown of natural folates (folate polyglutamates), *Prog. Hematol.,* 9, 85–105, 1975.

76. **Heinle, R. W. and Welch, A. D.,** Folic acid in pernicious anaemia: failure to prevent neurologic relapse, *JAMA,* 133, 739, 1947.

77. **Meyer, L. M.,** Folic acid in the treatment of pernicious anemia, *Blood,* 2, 50, 1947.

78. **Cooper, B. A.,** Megaloblastic anaemia in childhood, *Clinics Haematol.,* 5, 631–659, 1976.

NUTRIENT DEFICIENCIES IN MAN: VITAMIN B_{12}*

S. P. Rothenberg and R. Cotter

INTRODUCTION

The objective of this handbook on nutritional aspects of vitamin B_{12} in man is to present a concise and systematically organized compendium of up-to-date and generally acceptable information. Toward this end, the fundamental facts about the chemistry, physiology of absorption and transport, and metabolic functions of this vitamin have been summarized so that the reader will be better able to appreciate the pathogenesis of vitamin B_{12} deficiency and the clinical sequelae that accompany it. Controversial aspects of this subject have been avoided wherever possible. Where such discussion could not be avoided (i.e., methyl-folate trap hypothesis of vitamin B_{12} deficiency), the viewpoint which appears to be accepted by most students of this subject has been presented.

This handbook is not intended to be an exhaustive review of the subject of vitamin B_{12} since such a goal would have required several volumes of text and references. We have used selected references as support rather than specific facts or statistical information. Review articles covering major areas of vitamin B_{12} metabolism are indicated in footnotes.

HISTORY

As early as 1855, Addison described a fatal idiopathic anemia with neurological involvement of unknown etiology for which there was no successful treatment.[1] In 1860, Flint was astute enough to recognize that "Addisonian anemia" was associated with defective gastric secretion.[2] In the early 1900s, Whipple showed that a diet rich in liver and kidney was effective therapy for anemia induced by chronic bleeding in dogs.[3] Minot and Murphy, in the 1920s, also showed that a diet rich in liver and kidney was effective in patients suffering from Addison's pernicious anemia.[4] Castle then provided definitive evidence that deficiency of a nutritional factor alone was insufficient to explain the cause of Addisonian anemia. He proved this by demonstrating that a heat labile component of normal gastric juice could facilitate the nutritional response and concluded that an "intrinsic factor" was a normal component of gastric juice.[5]

Further progress in understanding vitamin B_{12} deficiency was hindered by the inability to induce the disease in experimental animals. This necessitated the use of patients with pernicious anemia to test the efficacy of various liver factors. Finally in 1948, Smith, in England,[6] and Rickes and associates, in the U.S.,[7] isolated, in crystalline form, a factor from the liver which produced remission of pernicious anemia. They named this factor vitamin B_{12}. The chemical structure of vitamin B_{12} was established using X-ray diffraction by Hodgkin's group.[8]

CHEMICAL STRUCTURE AND BIOCHEMICAL FUNCTION

Vitamin B_{12} (Figure 1) resembles a prophyrin structure consisting of four pyrrole-type units coupled directly to each other, with the inner nitrogen atom of each pyrrole coordinated with a single atom of cobalt. The basic tetrapyrrole structure is the corrin nucleus, which positionally is a planar structure coupled below to the nucleotide 5,6-dimethylbenzimidazole and above to CN or some other derivative. The base is

* This work was supported by grant no. AM 06045 from the National Institutes of Health.

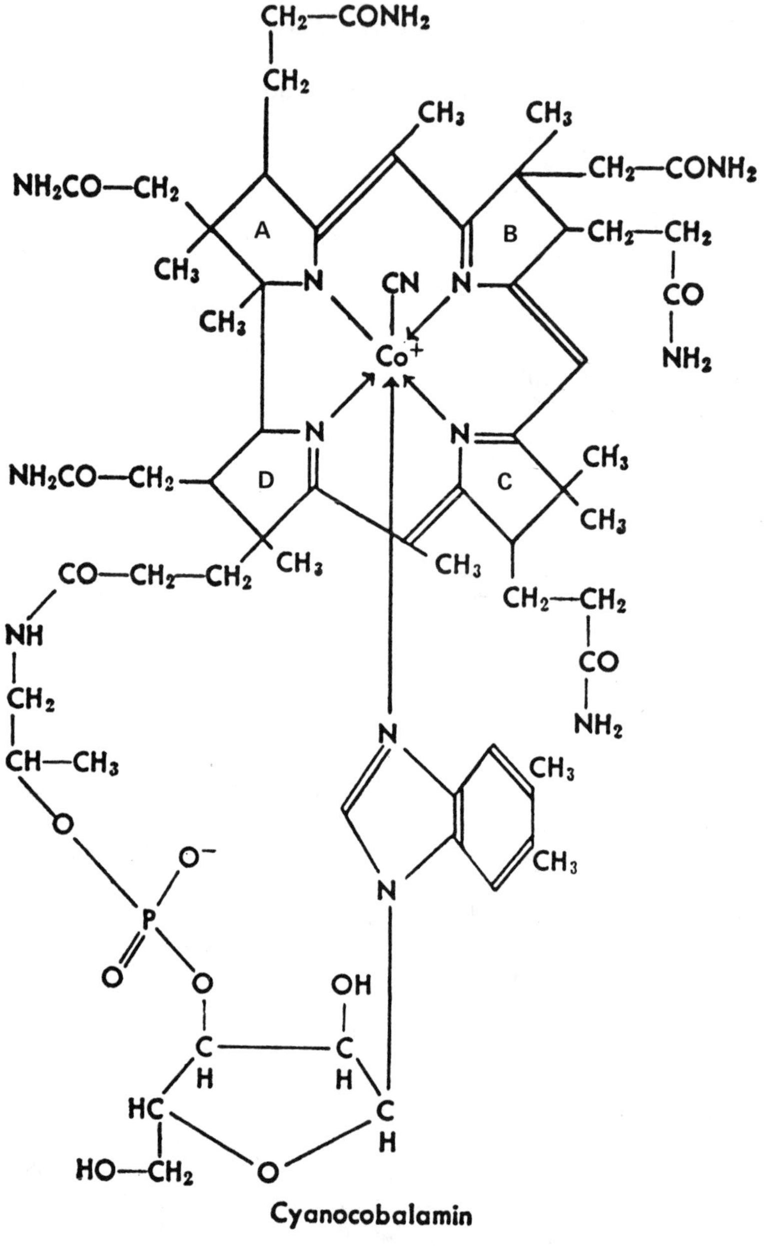

Cyanocobalamin

FIGURE 1.   Structure of vitamin B_{12}. The corrin ring, comprised of the four
pyrrolic groups, lies between the 5,6-dimethylbenzimidazolyl nucleotide below
and CN group above, both of which are attached to the central cobalt atom. In
the biologically active forms of vitamin B_{12}, 5-deoxyadenosine (deoxyadenosyl-
cobalamin) or a methyl group (methylcobalamin) substitutes for the CN grouping
above the corrin ring. (From West, E. S., Todd, W. R., Mason, H. S., and Von
Bruggen, J. T., *Textbook of Biochemistry*, Macmillan, New York, 1966, 740.
With permission.)

coupled directly to the cobalt atom, and an ester linkage from the phosphate group of the
nucleotide to the proprionic acid group of the D ring of the corrin nucleus adds further
stability to the molecule. Cyanide, which lies above the planar ring and is attached to the
cobalt atom, is actually an artifact of isolation and is replaced by deoxyadenosine and a
methyl group in the active coenzyme forms of the vitamin.

Vitamin B_{12} is an essential vitamin in man, with 1 to 2 μg thought to be the normal daily requirement. This vitamin, which can only be synthesized by certain bacteria, is an important cofactor for a number of enzymatic reactions in mammalian cells. In man, the following coenzyme functions for vitamin B_{12} have been defined:

1. Methyl group carrier for the enzyme methyltetrahydrofolate-homocysteine methyltransferase, which catalyzes the conversion of homocysteine to methionine. N^5-methyltetrahydrofolate is the methyl donor, and tetrahydrofolate is also a product of this reaction. Since this is a primary pathway for the metabolism of N^5-methyltetrahydrofolate, cobalamin deficiency actually impairs the generation of tetrahydrofolate. This is the fundamental biochemical link which interrelates vitamin B_{12} and folic acid metabolism. There is also evidence, albeit less definitive, that vitamin B_{12} and folate may be related in two other ways. First, the folylpolyglutamate concentration of liver (in rat) and erythrocytes of human blood are low in B_{12} deficiency, suggesting that cobalamin participates in some way in polyglutamate synthesis. Second, the low tissue folate concentration in B_{12} deficiency suggests that B_{12} may be necessary for the uptake of folate by tissues.

2. Metabolism of three carbon units (e.g., proprionate) which results from catabolism of the amino acids valine and isoleucine via proprionate-CoA to methylmalonyl-CoA and finally to succinyl-CoA. Deoxyadenosyl-B_{12} (coenzyme B_{12}) is the cofactor for the enzyme methylmalonyl-CoA mutase in the following reaction:

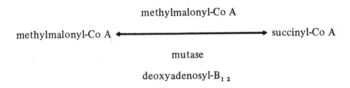

Deficiency of vitamin B_{12} results in the accumulation of methylmalonic acid, which is excreted in the urine. This metabolic derangement can be accentuated by the administration of valine. For a more in depth study of the biochemistry of vitamin B_{12} see References 32 to 34.

PHYSIOLOGY OF VITAMIN B_{12}
ABSORPTION AND TRANSPORT

Absorption of vitamin B_{12} in man and most animals is a complex process requiring the interaction of a number of specific macromolecules. Vitamin B_{12} in the diet is bound to food proteins. In the stomach, the combined effect of low pH and peptic digestion releases the vitamin, which is then bound to the first specific macromolecule, intrinsic factor (IF). This factor is a glycoprotein synthesized and secreted by the parietal cells of the gastric mucosa. The formation of this IF-B_{12} complex serves to protect the vitamin from bacterial utilization and/or degradation as it traverses the lumen of the small intestine to the terminal ileum, where absorption occurs. In this area of the small intestine, the IF moiety of the IF-B_{12} complex attaches to a second and specific macromolecule — a receptor protein on the microvillus membranes of the brush borders of the intestinal epithelial cells. The binding of the IF-B_{12} complex to the ileal receptors is not an energy-dependent process but does require Ca^{++}, which may either form a salt bridge between the receptor and IF or serve to stabilize the conformational binding site of the receptor. The binding of the IF-B_{12} complex to this receptor is optimum between pH 6.5 and 9.0.

The transport of B_{12} from the receptor-IF-B_{12} complex through the epithelial cell to the portal blood is a process requiring 3 to 4 hr; however, the precise mechanism has not yet been elucidated. During the absorptive process, B_{12} accumulates first on the mitochondria in a bound form. Studies in our laboratory indicate that a major fraction of the mitochondrial B_{12} is bound to a protein which reacts immunologically with antibodies to IF. Some, but not all, of this mitochondrial B_{12} is converted to deoxyadenosyl-B_{12}, but this is not a necessary step in the absorption of vitamin B_{12}.

The mechanism for release of B_{12} from the mitochondria is unknown, but ultimately the vitamin appears in the blood initially bound to a B_{12} binding protein called transcobalamin-II (TC-II).

Vitamin B_{12} Transport

Following the administration of a single oral dose of $[^{57}Co]B_{12}$, plasma radioactivity reaches a peak in about 8 hr. TC-II is the physiological plasma transport protein for B_{12} and makes up 80 to 100% of the unsaturated B_{12} binding capacity of normal plasma. TC-II, unlike IF, is not a glycoprotein, has a mol wt of 38,000, and is probably synthesized in part by the liver and in part by other as of yet unidentified sites. TC-II has been shown to deliver B_{12} to various tissues, such as liver, kidney, spleen, heart, lung, and small intestine. The mode of delivery of B_{12} to cells probably follows this sequence: The TC-II-B_{12} complex attaches to a specific receptor on the cell surface by means of a Ca^{++}-dependent, nonenergy-requiring process. The TC-II-B_{12} complex is then taken into the cell intact by means of pinocytosis, a process which is temperature- and energy-dependent. The pinocytotic vesicles then fuse with the lysosomes, and the proteolytic enzymes of the lysosomes digest TC-II, in turn releasing B_{12} into the cell. This released B_{12} is found in the cytosol and, at a later time period, in the mitochondria. The absolute requirement of TC-II for the uptake of B_{12} by the cells has been established by the B_{12}-deficiency state which follows congenital absence of this transport protein.

TC-II is not the only B_{12} binding protein in plasma. There exist two others; transcobalamin I (TC-I) and the granulocyte R binder, which is part of a much larger group of B_{12} proteins found in many tissues and secretions of the body and called R proteins because they have more rapid electrophoretic mobility than IF. TC-I in plasma carries 80 to 90% of the endogenous B_{12} in normal plasma and is usually 80 to 90% saturated. The site of synthesis of TC-I was originally thought to be the granulocytes. However, recent evidence and elucidation of TC-I structure now cast some doubt on this assumption. The physiological role of TC-I in B_{12} metabolism has not as yet been defined. Although it binds B_{12}, it is probably not a physiologic transport protein because it does not easily release B_{12}, and the half-life of the TC-I-B_{12} complex is about 10 days. This is in contrast to the half-life of TC-II-B_{12}, which is approximately 1 hr.

The granulocyte R binder is synthesized, as the name implies, in the circulating granulocyte. It has been shown to play a role in B_{12} transport, rapidly and specifically delivering B_{12} to the liver and with a plasma half-life of less than 5 min. This function may be important in recovering B_{12} from necrotic tissue during periods of infection and tissue trauma. Another B_{12}-binding R protein that has been found in plasma, transcobalamin-III (TC-III), may be an artifact generated by granulocytes during blood collection and separation.

Vitamin B_{12}-binding R proteins have been found in numerous tissues and secretions, including saliva, gastric mucosa, gastric juice, bile, cerebrospinal fluid, seminal plasma, urine, lymph, liver, and milk, to name a few from a continually growing list. The R proteins have been shown to be similar in structure and composition. They are immunologically cross-reactive when exposed to anti-R antisera, and amino acid analyses of these binders isolated from different sources show very little variation in amino acid composition. Studies employing saliva and milk R proteins have shown both to have identical N-terminal sequences, and carbohydrate analyses indicate they are composed of similar monosaccharide components. Variations of each binder appear to depend on the content of various monosaccharides within the protein itself. The physiological role of B_{12}-binding R proteins is unclear. They may play a role in B_{12} transport in a local or limited way within a cell, tissue, organ, or system. It has also been postulated that R proteins may serve an antibacterial role in two ways; first, by sequestering B_{12} and thus preventing its utilization by bacteria. Support of the latter hypothesis comes from recent studies which show granulocyte R protein can leach B_{12} from bacteria. Second, the binders may function as a screening mechanism to sequester nonphysiologic B_{12} analogues synthesized by bacteria. Such B_{12} "antimetabolites" could otherwise compete with metabolically reactive B_{12} coenzymes. There is extensive literature concerning vitamin B_{12} absorption and transport; only the highlights are herein presented. For greater in depth information, see References 35 to 37.

NUTRITIONAL ASPECTS OF VITAMIN B_{12} METABOLISM

The mammalian cell cannot synthesize the basic corrin-nucleotide structure of the metabolically active cobalamin coenzymes, so an adequate intake of the vitamin must be assured from dietary sources. In nature, vitamin B_{12} is synthesized only by micro-organisms, and it is absent from uncontaminated plant life. In some instances, bacteria growing in the roots of leguminous plants may produce vitamin B_{12}, which is retained in the plant food when ingested. In general, however, cereals and most fruits and vegetables contain no vitamin B_{12}.[9]

The dietary supply of vitamin B_{12} comes from products of animal life. Table 1 lists the sources and range of vitamin content of certain foods. With normal intestinal absorption, the recommended daily nutritional allowance of vitamin B_{12} is 2 μg.[10] This

Table 1
VITAMIN B_{12} CONTENT OF SOME FOODS

Food	Concentration (μg/100 g wet weight)
High	>10
Liver	
Lamb	
Kidney	
Heart	
Clams and oysters	
Medium	3—10
Dried fat-free milk	
Crabs	
Salmon	
Sardines	
Egg yolk	
Moderate	1—3
Mussels	
Lobster	
Scallops	
Flounder	
Tuna	
Fermenting cheese	
Low	<1
Whole milk	
Cream	
Cottage cheese	

insures adequate assimilation to replace the daily loss of vitamin B_{12}, which is less than 0.1% of body stores.

Cyanocobalamin, although biologically active, is not the natural form of vitamin B_{12} in food. Rather, the unstable deoxyadenosyl-B_{12} and methyl-B_{12} are the two enzymatically functional forms of vitamin B_{12} predominant in foods, and they are coupled to polypeptides. Cyanocobalamin is a more stable structure of the vitamin, which forms in the presence of cyanide during the isolation and purification procedure and generally is resistant to moderate heating unless at high pH or in the presence of excess ascorbate.

MEASUREMENT OF VITAMIN B_{12}

Microbiological and radiometric assay procedures are the only methods that have sufficient sensitivity to measure the low concentration of vitamin B_{12} in blood and other biological fluids. Microbiologic assay employs organisms which are dependent on vitamin B_{12} for growth, such as *Euglena gracilis, Lactobacillus leichmannii,* and *Ochromonas malhamvisis.*[11]

In the past several years, radioassay procedures have replaced microbiological methods in most clinical laboratories because they are less tedious and are not affected by factors which may inhibit the growth of microorganisms, such as antibiotics or cancer chemotherapeutic agents. Many radioassay procedures have been described, and these have also been recently reviewed.[12]

The normal range for cyanocobalamin in human serum (or plasma) is 200 to 800 pg/ml (0.2 to 0.8 μg/l or 1.4 to 5.9 $\times$ 10^{-10} M). Radioassay values are generally higher than those obtained by microbiological methods,[12] and interpretation of deficiency states must provide for this. In our laboratory, normal serum B_{12} concentration ranges from 200 to 800 pg/ml. Values between 150 and 200 pg strongly suggest B_{12} deficiency, although this concentration range is frequently seen in pregnant women whose B_{12} stores are not necessarily below normal. We have also observed patients with unequivocal pernicious anemia who have serum B_{12} concentrations between 150 and 175 pg/ml. B_{12} concentrations below 100 pg/ml, whether measured by radioassay or microbiological procedures, almost always indicate a deficiency state (except in the rare congenital disorder of absence of circulating TC-I).

VITAMIN B_{12} DEFICIENCY

A deficiency of any nutrient substrate can be evaluated from a quantitative as well as functional consideration. A quantitative deficiency exists when the measured parameter (serum or tissue concentration) is below the confidence limits of the established norm. With nutrition, a quantitative deficiency of a nutrient precedes the functional deficiency, which may not be clinically evident for a considerable time after the parameter is measured below normal. Functional deficiency may occur without quantitative deficiency, which occurs, for example, with the interference of a vitamin by an antimetabolite.

A functional deficiency is the development of an impaired biochemical pathway as a result of the quantitative deficiency of the nutrient coenzyme (or a metabolic block). Frequently, one can detect a functional deficiency in the metabolic pathway of a particular substrate, using biochemical tests, before clinical evidence of the deficiency state becomes apparent.

Normal serum vitamin-B_{12} levels are maintained by absorption and by mobilization of the vitamin from hepatic stores. In malabsorption, the serum B_{12} concentration may be slightly below normal even though the liver contains a considerable amount of B_{12}

Table 2
MAJOR CAUSES OF
VITAMIN B_{12} DEFICIENCY

Nutritional deficiency
  Vegan
  Extreme deficiency of animal protein
Failure of absorption
  Disorders of the stomach
    Atrophic gastritis and gastric atrophy (Addisonian
      pernicious anemia)
    Gastrectomy and toxic injury
    Congenital failure to secrete intrinsic factor
    Secretion of abnormal intrinsic factor molecule
  Disorders of small intestine
    Intestinal diverticuli
    Entero-entero fistulae
    Scleroderma
    Intestinal bypass surgery
    Regional ileitis
    Tropical sprue
    Nontropical sprue
  Drugs and toxins
    Ethanol
    p-Aminosalicylic acid
    Colchicine
    Neomycin
  Fish tapeworm (Diphyllobothriasis)
  Chronic pancreatitis
  Specific disorder of B_{12} absorption
    Imerslund-Gräsbeck syndrome
Defect in transport
  Deficiency of TC-II

(normal is 0.6 to 1.5 $\mu g/g$ wet weight). In general, however, when the serum B_{12} concentration falls below 120 pg/ml, the B_{12} concentration in the liver is below 0.6 $\mu g/g$ wet weight.[13,14] Thus, it is evident that the simplest test to estimate B_{12} deficiency in hepatic stores is measuring serum B_{12} concentration.

In the presence of normal B_{12} absorption, the serum B_{12} will be normal or even high, with low hepatic B_{12} when there is liver disease.[13] In most of these instances, there is no functional evidence of B_{12} deficiency (hematopoietic or neural), although there is a possibility that liver injury results in the release of a B_{12}-binding protein (R-type), which interferes with the transport of B_{12} by TC-II. Such an R-type binder has been identified in the serum of patients with hepatocellular carcinoma.[15]

Causes of vitamin B_{12} deficiency can be easily appreciated with an understanding of the mechanism for assimilation of the vitamin from food into active coenzyme forms. This process requires

1. Exogenous source of the vitamin
2. Gastric secretory function
3. Intact intestinal transit to the distal ileum
4. Normal structure and function of the intestinal epithelial cells
5. TC-II transport protein.

Table 2 lists the etiology of vitamin B_{12} deficiency in man, and each cause can be referred to a perturbation of one of the above physiologic mechanisms required to convert food B_{12} to a functioning intracellular coenzyme. This list omits speculative causes of B_{12} deficiency in man. For example, increased metabolic requirements or utilization of B_{12} such as might be conjectured in hyperthyroidism, hemolytic anemia, or malignant diseases have not been definitely established as causes of vitamin B_{12} deficiency in the absence of a defect in one of the above mechanisms. Even with pregnancy, where parasitism by the fetus is an a priori fact and maternal serum B_{12} concentration is frequently below normal, maternal tissue stores of vitamin B_{12} are generally normal.

In our experience with the patient population of Metropolitan Hospital, New York City, pernicious anemia probably accounts for 80 to 90% of the cases of clinical vitamin B_{12} deficiency, with intestinal disorders of absorption accounting for most remaining cases. Clearly, statistics from other areas of the world where food fadism, nutritional deficiency, or tropical sprue are more common would be different.

With an understanding of the physiology of vitamin B_{12} absorption, transport, and metabolism, the specific disorders listed in Table 2 are self-explanatory. Comment, however, about some of the categories may be helpful.

Nutritional Deficiency

Only a strict vegetarian diet for a prolonged period of time will lead to vitamin B_{12} deficiency. In some areas of India, religion and poverty result in a basic diet of polished rice, cereals, some vegetables, and fruit, with an inadequate intake of milk and eggs. In Western societies, such diets are seen in Vegans, — a cult of food fadists who shun all forms of animal products. It is interesting that although serum B_{12} concentration decreases in the first few years following the initiation of such a diet, clinical evidence of vitamin B_{12} deficiency is not common. Proof that a low serum vitamin B_{12} concentration is due to poor nutritional intake requires, fundamentally, a response to small dietary amounts of vitamin B_{12} (2 to 5 μg/day).

It is of interest that the severe nutritional deficiency states, kwashiorkor (protein malnutrition), and marasmus (protein-calorie malnutrition) are not usually associated with vitamin B_{12} deficiency. In fact, serum vitamin B_{12} levels may be elevated even though the concentration of B_{12} in the liver is probably lower than normal if fatty changes have occurred. It is likely that infants and young children with these severe deficiencies do not live long enough to reach the level of B_{12} deficiency required for hematologic and neurologic disturbances.

Failure of Absorption

Causes

The most common causes of absorption failure are atrophic gastritis and gastric atrophy with secondary achylia gastrica, as occurs in true Addisonian pernicious anemia. The etiology of this disease is unknown, although there is substantial evidence that an autoimmune process may be contributory, if not fundamental, in a large percentage of these patients.[16] As in most patients with atrophic gastritis who show no evidence of intrinsic factor (IF) deficiency, the serum of 60 to 70% of these patients contains antibodies to the parietal cells of the gastric mucosa, which can be demonstrated by immunofluorescence.[17] However, patients with pernicious anemia usually have one or two types of serum antibody which react with intrinsic factor.[18] One type of antibody blocks the binding of vitamin B_{12} and is called Type I or blocking antibody, and the second antibody binds the IF-B_{12} complex and is called Type II or binding antibody. In our experience, approximately 70 to 75% of patients with pernicious anemia have Type I, and 50% have Type II.[19] It is unusual to find Type II antibody in the serum without Type I. Of interest is the observation that the younger the patient, the more likely that anti-IF antibodies will be found. Although pernicious anemia is a disease of the elderly, it is not uncommon in persons below the age of 50 years (approximately 20% of cases). We have only observed one patient below the age of 50 years with established pernicious anemia who did not have serum antibodies to IF and yet did not have hypogammaglobulinemia.

Pernicious anemia is most common in people of northern Europe, particularly the Scandinavian countries. At Metropolitan Hospital, New York City, we have seen a substantial number of black patients with this disease. In the U.S., the frequency of pernicious anemia in persons admitted to hospitals is 77.8 cases per 100,000 white persons and 26 cases per 100,000 black persons.[20]

Pernicious anemia may have a familial tendency, and there are many reports of the disease in more than one member of a family and in successive generations.[21] Studies to investigate the autoimmune etiology of pernicious anemia have also revealed that antibodies to gastric parietal cells may be found in 36 to 50% of blood relatives of patients with this disease.[22,23] Such antibodies are more commonly found in women than men.

Pernicious anemia has a tendency to coexist with other diseases. The most important is gastric carcinoma, which is probably the sequela of longstanding atrophic gastritis. The incidence of gastric carcinoma in pernicious anemia patients is about three times higher than in the general population, and it is more common in males than females.

Pernicious anemia has also been associated with the myeloproliferative diseases, such as polycythemia vera and chronic myelogenous leukemia, thyroid disease (both Graves disease and hypothyroidism), hypoparathyroidism, and Addison's disease of the adrenal gland. A common thread between the diseases of these endocrine organs and pernicious anemia is the tendency to form autoantibodies against structures of the glands and the parietal cells of the stomach.

A number of patients have been described with pernicious anemia and primary-acquired hypogammaglobulinemia. These patients are usually in the younger age group for pernicious anemia and do not develop anti-IF antibodies.

Impaired absorption of vitamin B_{12} also occurs following gastrectomy,[24] usually increasing with time until approximately 25% of patients have impaired absorption 8 to 25 years after the operation and 40% develop subnormal serum B_{12} levels.[25] This impaired absorption of radioactive B_{12}, when tested in the fasting state, is often corrected if the test is done with the administration of food or histamine, suggesting that the actual number of patients with impaired vitamin B_{12} absorption following gastrectomy may be overestimated. Furthermore, only about 50% of patients with impaired B_{12} absorption following gastrectomy develop quantitative B_{12} deficiency.

The cause of impaired B_{12} absorption following gastrectomy is unknown. It is not simply due to decreased secretion of intrinsic factor, since only a small percentage of the normal IF secretion is required to normalize the urinary excretion test for B_{12} absorption. In general, the greater the extent of surgical removal of the stomach, particularly the fundus, the more impaired the absorption. Since impaired absorption increases with post-operative time, delayed gastric atrophy may play a role. Addition of exogenous IF will improve the urinary excretion test, but it usually does not fall into the normal range, and, in some instances, there may not even be an improvement. Accordingly, other factors such as postgastrectomy malabsorption, bacterial overgrowth, and impaired enterohepatic circulation of B_{12} may contribute to impaired B_{12} absorption.

Congenital absence of intrinsic factor is an autosomal recessive disorder, and evidence of quantitative vitamin B_{12} deficiency occurs in the first 2 years of life. There is no other abnormality of gastric secretory function, and gastric biopsy reveals normal mucosa.

Recently, a functionally deficient IF molecule was identified as a cause of congenitally impaired vitamin B_{12} absorption.[26] The gastric juice reacted with natural anti-IF antibody but did not promote the absorption of vitamin B_{12}.

Disorders of the Small Intestine

Any alteration of the IF-B_{12} complex during transit from the stomach and proximal small intestine to the distal ileum may interfere with vitamin B_{12} absorption. Intestinal diverticuli, blind loops caused by pathologic or surgical entero-enteric anastomosis, and impaired motility (as occurs with scleroderma) will permit bacterial overgrowth to occur, and these bacteria compete with IF for the B_{12} molecule. Steatorrhea need not be present, and the B_{12} malabsorption can be corrected by the administration of antibiotics.

Intestinal bypass surgery has recently been used for treatment of massive obesity. In addition to other absorptive alterations (particularly for fat), impaired B_{12} absorption may occur if the surgery excludes the distal 24 cm of ileum from intestinal continuity.

An injury to the ileal mucosa which alters the brush border structure containing the receptor for intrinsic factor will depress vitamin B_{12} absorption. This is the primary defect in regional ileitis and tropical and nontropical (gluten-sensitive) sprue. A significant number of patients with these disorders develop quantitative deficiency of vitamin B_{12}.

In some instances of severe vitamin B_{12} (as in pernicious anemia) or folate deficiency (as in malnutrition), a secondary change occurs in the ileal mucosa because of impaired turnover of the intestinal epithelial cells. This can lead to impaired vitamin B_{12} absorption, which is corrected only after the administration of a sufficient amount of vitamin (B_{12} or folic acid) to permit generation of a new epithelial mucosa.

Drugs and Toxins

Excessive alcohol, *p*-aminosalicylic acid (PAS), and colchicine can impair the absorption of vitamin B_{12}. A specific pathophysiologic mechanism for each agent has not been defined, but they probably differ for PAS and colchicine. Whereas colchicine can depress absorption of B_{12} after 7 days, several weeks of PAS are required to observe a similar effect. Extensive studies with colchicine have not demonstrated its effect on IF secretion or on attachment of the IF-B_{12} complex to the ileal receptor of the guinea pig in vitro.[27] Therefore, colchicine probably effects the intracellular B_{12} transport from the intestinal lumen to the portal blood.

Alcohol has also been implicated as a direct cause of impaired vitamin B_{12} absorption, but it is probably not a significant cause of quantitative deficiency of this vitamin. In our municipal hospital, where serum folate and B_{12} concentration are measured in most anemic alcoholic patients, low values for vitamin B_{12} are unusual.

Diphyllobothrium latum

Fish tapeworm infestation (Diphyllobothrium latum) is a well-recognized cause of impaired vitamin B_{12} absorption. The worm sequesters vitamin B_{12} as the vitamin progresses through the small intestine, and deficiency is more likely to occur when the worms reside in the jejunum. Vitamin B_{12} absorption returns to normal when the worms are expelled. The disease is most common in Finland, but since the tapeworm is a parasite of fresh water fish, human infestation can be found in scattered areas of Europe and the Great Lakes region of North America.

Chronic Pancreatitis

Chronic pancreatitis is another cause of impaired vitamin B_{12} absorption, but the mechanism is unknown. It is not due to abnormal pH in the terminal ileum region, nor is it due to a deficiency of ionic calcium. Intrinsic factor secretion is also normal. The defective absorption can be corrected by pancreatic extract or by the administration of food at the time of the test. In spite of this abnormality, which can be demonstrated in approximately 30% of patients with this disease, evidence for quantitative deficiency of this vitamin in these patients is virtually unknown.

Imerslund-Graisbeck Syndrome

A congenital disorder of defective vitamin B_{12} absorption has been described and is known as the Familial Selective Malabsorption for Vitamin B_{12} (Imerslund-Gräsbeck syndrome). Quantitative evidence for vitamin B_{12} deficiency usually occurs in the first few years of life, and the affected children also have proteinuria. Although it was generally presumed that the disorder was probably caused by deficiency of the receptor protein on the brush border of the ileal epithelial cell, a recent study in one patient demonstrated that the receptor functioned normally in homogenates of biopsy material.[28]

Abnormality of B_{12} Transport

Defective transport is another cause of vitamin B_{12} deficiency. Two female siblings born without TC-II developed severe megaloblastic anemia which responded only to massive doses of parenteral B_{12}.[29] The serum concentration of B_{12} in these patients was not commensurately low because TC-I, which carries most of the circulating endogenous B_{12}, was normal. However, this B_{12} binding protein does not release its B_{12}. As discussed previously, TC-II is the physiologic B_{12} transport protein and facilitates uptake of B_{12} into cells by attaching to a specific receptor on the cell membrane. In the absence of this protein, B_{12} must enter the cell by gradient passive diffusion and consequently, large doses must be administered at frequent intervals.

PATHOLOGY

Histologic changes observed in tissues as a consequence of vitamin B_{12} deficiency, except for those of the nervous system, can be ascribed to the known biochemical function of this vitamin, which is interrelated with folate metabolism and nucleic synthesis. As discussed previously, vitamin B_{12} is a cofactor for the methyltetrahydro-folate-homocysteine methyltransferase reaction by which the methyl group of N^5-methyltetrahydrofolate transfers to homocysteine via the cobalamin intermediary in the biosynthesis of methionine. Since this is the main metabolic pathway for the generation of tetrahydrofolate, deficiency of vitamin B_{12} leads to insufficient formation of the reduced folate coenzymes necessary for carbon unit transfer. The direct effect of impairment of this pathway, with respect to cell replication, is the deficiency of 5, 10-methylenetetrahydrofolate which is necessary for *de novo* thymidylate synthesis from deoxyuridine. This biochemical perturbation of vitamin B_{12} deficiency can be overcome by administering greater than normal amounts of folic acid (more than 400 μg/day).

Tissues most affected by vitamin B_{12} deficiency are those with high turnover rates because of inadequate DNA synthesis. RNA synthesis usually continues, and the cell enlarges, with a stretching out of the chromatin structure of the nucleus into the "megaloblast." This classical appearance is seen in bone marrow (see Figure 2).

The same fundamental changes may be seen in other replicating cells, particularly of the epithelial lining. Morphologically, similar changes have been found in the scraping of buccal mucosa, gastric mucosa, small bowel biopsy, and vaginal cells.

Biochemically, the effect of vitamin B_{12} deficiency on the nervous system cannot be as easily explained. The morphologic alteration primarily affects the white matter of the central nervous system. Since this lesion is not reversed and may even be worsened by the administration of folic acid, the biochemical pathogenesis must be by some pathway other than that interrelated with folate. The disturbance in proprionate metabolism and its pathway through methylmonyl-CoA to succinate, an abnormality which is specific for vitamin B_{12} deficiency, probably plays some role. In vitro studies with peripheral nerve tissue have demonstrated decreased total fatty acid synthesis and abnormal synthesis of odd chain fatty acids (C-15, C-17) in B_{12} deficiency, and this may be related to excessive accumulation of proprionate.

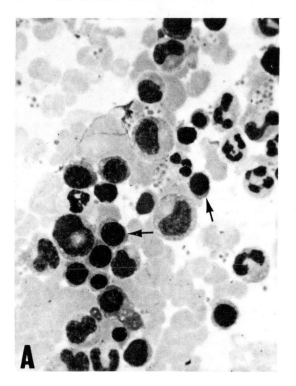

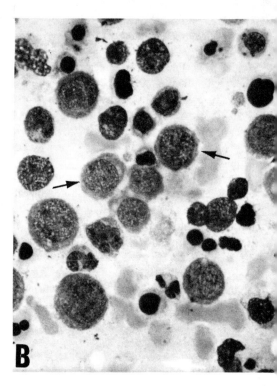

FIGURE 2A. Normal bone marrow smears. Arrow points to condensed nuclear material of a normal orthochromatic erythroblast.

FIGURE 2B.   Megaloblastic bone marrow from a patient with pernicious anemia. Arrow points to an orthochromatic megaloblastic erythroblast which contains lacy nuclear chromatin material.

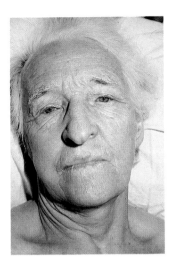

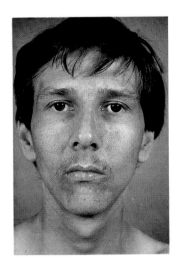

FIGURE 1. Mild jaundice, pallor of mucous membranes due to folate deficiency (female of 72 with nutritional megaloblastic anemia).

FIGURE 4. Melanin pigmentation of the skin which disappeared with folic acid therapy. Male patient, age 24, with megaloblastic anemia due to folate deficiency.

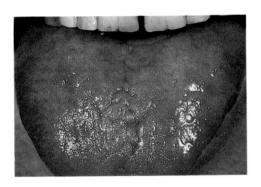

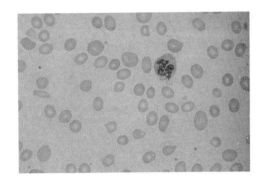

FIGURE 2. Glossitis in a 33-year-old female patient with megaloblastic anemia due to folate deficiency (adult celiac disease).

FIGURE 5. Peripheral blood in patient illustrated in Figures 2 and 3 of 5.3 g/dl, mean corpuscular volume 123 fl, white cell count $3.2 \times 10^9/1$, platelets $90 \times 10^9/1$, reticulocytes 1.2%.

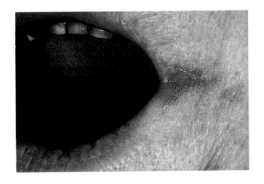

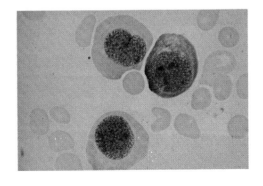

FIGURE 3. Angular cheilosis in same patient as Figure 2.

FIGURE 6. Bone marrow megaloblasts from the same patient.

CLINICAL MANIFESTATIONS OF VITAMIN B$_{12}$ DEFICIENCY

Symptoms which are directly caused by deficiency of this vitamin are related to its function in cell replication and synthesis of myelin of the nervous system. Table 3 lists the most frequent clinical findings. Anemia is the most common, and since it is slow in progression, it is not unusual for patients to have a hemoglobin concentration as low as 3 to 4 g/dl before seeking medical assistance and then with only a complaint of moderate weakness. In more severe cases, granulocytopenia and thrombocytopenia also occur, and we have observed total white and platelet concentrations in peripheral blood as low as 1000/mm^3 and 10,000/mm^3, respectively. Such patients may also develop infections and bleeding tendency.

Disturbance of the alimentary tract leads to glossitis, which causes a burning tongue. Diarrhea is also a frequent complaint. Weight loss occurs primarily in elderly patients who become so severely anemic that anorexia and profound weakness prevent them from ingesting a calorically sufficient diet. Weight loss may also occur in severe cases, when intestinal malabsorption is a complication.

Clinical disturbance of the central nervous system is now very rare, since the disease is discovered much earlier than in previous times. Paresthesias, which indicates peripheral nerve pathology, is more commonly seen. Signs of central nervous system involvement are organic psychosis, loss of position sense, ataxic gait, and loss of deep tendon reflexes (posterior spinal cord tracts) followed by muscle weakness, spasticity, and pathologic reflexes (lateral spinal cord tracts).

Physical examination serves to confirm the system disturbance obtained by history. Thus, there is often marked pallor, slight scleral icterus because of the indirect hyperbilirubinemia (see laboratory findings), and a red or smooth tongue, which is not coated because of atrophy of the papillae. Lymphadenopathy and a marked hepato-splenomegaly are usually lacking and, if present, should suggest another, or complicating, hematologic disease. In fact, fewer than 5% of our patients with pernicious anemia develop palpable splenomegaly.

Signs of posterior column disease include loss of vibratory and position sense and loss

Table 3
MAJOR CLINICAL MANIFESTATIONS
OF VITAMIN B$_{12}$ DEFICIENCY

System		Clinical disturbance
Hematopoiesis		
Red cells $\longrightarrow$ anemia		Weakness
Granulocytes $\longrightarrow$ leukopenia		Infection
Platelets $\longrightarrow$ thrombopenia		Purpura
Alimentary tract		Burning tongue
		Malabsorption
		Diarrhea
Nervous system		
Central		
Posterior columns		Psychosis; loss of position sense
Lateral		Paralysis and spasticity
Peripheral		Paresthesias
		Hypesthesias

of deep tendon reflexes. Paralysis with spasticity and positive plantar reflex indicates lateral (pyramidal) tract involvement.

LABORATORY FINDINGS IN VITAMIN B_{12} DEFICIENCY

Hematologic Abnormalities
Peripheral Blood

Macrocytic anemia — The mean corpuscular volume (MCV) of erythrocytes is usually greater than 96 μ^3. The more severe the deficiency and anemia, the greater the MCV; values of 160 μ^3 have been observed. The mean corpuscular hemoglobin (MCH) concentration is also elevated, with values usually greater than 33 pg (values greater than 40 pg are not infrequent). The cells are generally normochromic, so the mean corpuscular hemoglobin concentration (MCHC) is usually normal. Comparison of normal and macrocytic peripheral smears is shown in Figure 3.

Dimorphic anemias, particularly iron deficiency and vitamin B_{12} deficiency, may result in normalization of the red cell indices, as enumerated by automated counters used in most clinical laboratories. It is, therefore, important to carefully examine the peripheral blood smear of all anemic patients for evidence of the macrocytosis and macroovalocytosis so characteristic of vitamin B_{12} deficiency.

Granulocytopenia — This occurs after the anemia is apparent and may be quite severe. The granulocytes could appear multilobed (5 to 10 lobes/cell) even before they are reduced in number, and this may be one of the early findings. Usually, in normal blood, there are no granulocytes with six lobes and no more than two or three granulocytes with five lobes. Increase in the number of lobes must be determined by direct examination of the smear.

Thrombocytopenia — A decrease in the number of platelets occurs in the more severe cases. Morphologically, the platelets frequently appear bizarre, with many giant forms.

Bone Marrow

The marrow cavity of either the sternal bone or posterior iliac crest is easily penetrated, and abundant spicules are usually obtained. Microscopic examination of Wright-Giemsa-stained smears shows the characteristic megaloblastic hematopoiesis (Figure 2). The severity of this morphologic abnormality is usually inversely related to the severity of the anemia. The erythroid precursors are abundant, and the "megaloblastic" appearance is characterized by large cells with very lacy chromatin strands that have distinct parachromatin spaces between them. Similar chromatin strands are seen in the very early basophilic precursors as well as the late orthochromatic erythroblasts. The granulocyte nuclei also may have a similar appearance, and very large myelocytes and metamyelocytes are characteristically seen. Megakaryocytes may be diminished in number, but the nuclei may also have a similar chromatin structure.

Characteristic morphologic abnormalities may be very subtle in cases of mild B_{12} deficiency. In such cases, an experienced hematologist will be more likely than a laboratory technologist to recognize early changes. In addition, the presence of concomitant iron deficiency may retard the full expression of the megaloblastic appearance of the erythroblasts. It should be noted that neither the severity of the hematologic disturbance nor the appearance of the megaloblasts will serve to distinguish vitamin B_{12} from folic acid deficiency.

Biochemical Abnormalities

Biochemical disturbances are (1) directly secondary to the vitamin B_{12} deficiency and (2) secondary to the impaired synthesis of DNA, which results in the megaloblastic maturation of proliferating cells.

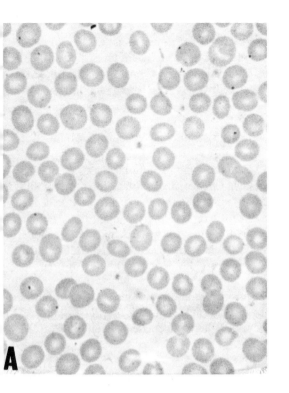

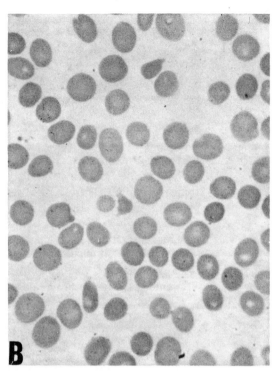

FIGURE 3A.   Normal peripheral blood smear.

FIGURE 3B.   Peripheral blood smear from a patient with pernicious anemia, showing macrocytosis (large red cells) and poikilocytosis (irregularly shaped red cells).

Direct Effects of Vitamin B_{12} Deficiency

Methylmalonic acid excretion — This is secondary to the impaired metabolism of three carbon units (e.g., proprionate) because of the lack of the deoxyadenosyl-B_{12} coenzyme, which is a necessary cofactor for conversion of methylmalonyl-Co A to succinyl-Co A via the enzyme methylmalonyl mutase. When this pathway is stressed by administration of the amino acid valine, there is excessive excretion of methylmalonic acid. Impairment of this pathway also leads to increased basal excretion of proprionate and decreased urinary excretion of succinate. This defect is specific for vitamin B_{12} deficiency and does not occur in folate deficiency.

Formiminoglutamate excretion (FIGLU) — Excretion of this intermediary of histidine metabolism occurs because of the lack of sufficient tetrahydrofolate receptor for the formimino group. There is also, generally, an increased excretion of urocanic acid, the precursor of FIGLU. The deficiency of tetrahydrofolate is secondary to impaired transfer of the methyl group from N^5-methyltetrahydrofolate to homocysteine to generate methionine and, at the same time, tetrahydrofolate (methyl trap). Obviously, the consequences of this impaired pathway (tetrahydrofolate deficiency) will also occur in folate deficiency.

Amino-imidazole carboxamide (AIC) — This metabolite is an intermediary of purine biosynthesis, and it undergoes formylation to form inosinic acid. This reaction is impaired in both vitamin B_{12} and folic acid deficiency, and there is increased excretion of AIC in both conditions.

Biochemical Disturbance Secondary to Impaired DNA Synthesis

Indirect hyperbilirubinemia — This is secondary to the ineffective erythropoiesis, which results in greater synthesis and greater intramedullary breakdown of hemoglobin.

Increased excretion of urobilinogen — This is also secondary to an increased synthesis and degradation of heme.

Increased serum iron — This is secondary to an increase in the plasma iron transport rate, which is required to meet the increased rate of erythropoiesis.

Increased lactate dehydrogenase (LDH)— This enzyme is released from megaloblastic erythroblasts which undergo intramedullary destruction. The higher the plasma LDH, the more severe the megaloblastic disturbance and anemia and, generally, the more severe the vitamin B_{12} deficiency. Of the LDH isoenzymes, LDH-1 (the electrophoretically faster moving form) is increased the most.

Other erythrocyte enzymes, such as malate dehydrogenase, 6-phosphogluconate dehydrogenase, and α-hydroxobutyrate dehydrogenase, are also usually increased to high levels inversely proportional to the degree of anemia. Lesser elevations have been observed for aldolase, phosphohexose, isomerase, and isocitrate dehydrogenase. During the deficiency state, there is usually a negative nitrogen balance and increased excretion of amino acids.

DIAGNOSTIC TESTS FOR B_{12} DEFICIENCY

The finding of a serum vitamin B_{12} concentration between 150 and 200 pg/ml by radioassay or between 100 and 150 pg/ml by microbiological assay is suggestive of vitamin B_{12} deficiency. Exceptions to this are (1) pregnancy, where there is a more rapid transfer of B_{12} to the fetal circulation, thus lowering the concentration of B_{12} in maternal blood and (2) early malabsorption, where deficient absorption of B_{12} may lower serum B_{12} 20 to 30% before tissue stores are actually depleted. When serum B_{12} concentration is below 150 pg/ml by radioassay and below 100 pg/ml by microbiological assay, vitamin B_{12} deficiency is most likely.

Depending on the basic etiology of the vitamin B_{12} deficiency, serum folate may be normal or high because of failure to properly utilize the circulating N^5-methyl-etrahydrofolate or it may be low because of concomitant nutritional deficiency or malabsorption for folic acid.

It is important to note that erythrocyte folate concentration, which is usually considered an excellent measure of tissue stores of folate, may be low in vitamin B_{12} deficiency because uptake of folate by erythrocyte precursors requires adequate cobalamin function. Thus, finding low serum vitamin B_{12}, low erythrocyte folate, and normal or high serum folate concentrations is not unusual in vitamin B_{12} deficiency without folate depletion.

The only other test which is specifically for diagnosis of vitamin B_{12} deficiency is increased methylmalonic acid excretion following the administration of valine. The measurement of methylamalonic acid, however, has not yet been sufficiently simplified for routine clinical application. A therapeutic response (e.g., reticulocytosis) following the parenteral administration of 1 to 2 μg of vitamin B_{12} daily for 5 days is diagnostic of vitamin B_{12} deficiency.

DIAGNOSTIC TESTS FOR THE ETIOLOGY OF VITAMIN B_{12} DEFICIENCY

Once the vitamin B_{12} deficiency has been established, it is necessary to determine the cause, i.e., pernicious anemia or intestinal disease (causes listed in Table 2). Systematic evaluation will usually indicate the cause which, in most cases, is secondary to lack of intrinsic factor.

Anti-intrinsic factor antibody — Testing is now routine in most laboratories which do radioassays for vitamin B_{12}. A positive test for this antibody in a patient with low serum vitamin B_{12} concentration is virtually diagnostic of pernicious anemia.

Gastric analysis — This is necessary when the test for anti-intrinsic factor antibody is negative. There is achlorhydria, which is not corrected by histamine or pentagastrin administration. The intrinsic factor can be measured by radioimmunoassays, and the concentration is less than 5 ng (usually zero) of B_{12} binding per milliliter of gastric juice.

Contrast Barium Studies

If the etiology of the vitamin B_{12} deficiency has not been defined by these appropriate tests, evaluation of intestinal structure and transit by contrast barium studies is indicated.

Specific radio-B_{12} absorption studies — It is best to delay these until after the patient has been treated because, in the presence of severe vitamin B_{12} deficiency, there may be a structural alteration of the intestinal mucosa, which can impair absorption of vitamin B_{12}. As a consequence, a patient with pernicious anemia may appear to have malabsorption (i.e., absorption not corrected by the administration of exogenous IF). The test is performed by administering 0.5 to 1.0 μg of radioactive B_{12} (usually [^{57}Co] B_{12}) to a fasting patient, and the percentage of the administered dose which is absorbed is determined by measuring

1. Fecal excretion over the next 2 to 3 days
2. Urinary excretion over the next 48 hr
3. Plasma radioactivity 8 hr after administration
4. Hepatic radioactivity by external scanning over the liver 5 to 7 days later
5. Total body counting 1 week later

Of the methods listed, the most commonly used is the urinary excretion procedure (method of Schilling).[31] In this test, about 1 hr following the oral administration of radioactive B_{12}, a large parenteral dose (1000 μg) of nonradioactive B_{12} is given. This serves to saturate tissue binding stores and circulating B_{12} binding protein, and this promotes the urinary excretion of the absorbed radioactive B_{12}. With normal renal function, most of the radioactivity appears in the urine during the first 24 hr. With renal insufficiency, there may be a delay of excretion, and it is best to collect the urine for two 24-hr periods.

Normal subjects excrete more than 8.0% of the administered dose in the urine. In patients with pernicious anemia, less than 5.0% (usually less than 3.0%) is excreted, and if the test is repeated with a source of intrinsic factor (commercial preparations are available), the urinary excretion is normalized. If the vitamin B_{12} deficiency is due to intestinal disease or infestation with *D. latum,* the addition of intrinsic factor will not normalize the test. In such cases, it is necessary to thoroughly investigate the gastrointestinal tract by absorption studies for other nutrients (fat, sugar, carotene, etc.), structural abnormalities (diverticulae, abnormal mucosa, enteritis, etc.), or fish tapeworm.

TREATMENT

The objectives of treatment of vitamin B_{12} deficiency are 1. the administration of sufficient vitamin B_{12} to replenish all stores and, 2. correction of the underlying disorder which led to the deficiency state.

The first requirement is easily satisfied, since vitamin B_{12} is readily available in injectable form. Sufficient vitamin B_{12} should be given to replenish total stores, estimated to be 5000 to 10,000 μg. A single dose of 100 μg of vitamin B_{12} will correct hematologic abnormalities completely. When the dose is 100 to 1000 μg, a large percentage is excreted in the urine, but a larger absolute amount is retained in the body as compared to that retained with a smaller dose. For example, 80% of 10 μg and 10% of 1000 μg may be retained after injection, yet the absolute amount retained after the larger dose is 100 μg as compared to 8 μg for the smaller dose. Accordingly, we initiate therapy with 1000 μg daily for the first 7 days if the patient has been hospitalized for a severe B_{12} deficiency state. Therapy is then continued with 1000 μg/week for a few months. This usually amounts to about 20,000 μg of total B_{12}, and probably all the stores become fairly well replenished. A therapeutic program that replenishes stores more slowly, 1000 μg/month, is also acceptable after the initial few days of therapy. However, this schedule must be continued for at least 20 months.

If the underlying disorder cannot be corrected (e.g., pernicious anemia, vegan, etc.), a monthly maintenance of 100 to 1000 μg will be adequate. Many of our patients with pernicious anemia receive 1000 μg every 2 to 3 months after stores are replenished, and their serum B_{12} remains above 200 pg/ml.

THERAPEUTIC RESPONSE

Hematologic abnormalities should all be corrected. If the hemoglobin does not completely return to normal, the patient should be evaluated for a secondary problem, such as iron deficiency or another disease process. Then, other symptoms, such as those secondary to glossitis, gastrointestinal malfunction, or peripheral neuropathy, should completely disappear.

Disorders of the central nervous system frequently improve, particularly organic brain dysfunction. The symptoms secondary to structural demyelinization are slow to reverse; however, we have seen remarkable improvement in a patient with almost total loss of posterior spinal tract function. Lateral tract impairment usually is not reversed, and an element of spastic paralysis remains.

REFERENCES

1. Addison, T., *On the Constitutional and Local Effects of Disease on the Suprarenal Capsules*, S. Higley, London, 1855, 43.

2. Flint, A., A clinical lecture on anemia, *Am. Med. Times,* 1, 181–186, 1860.

3. Whipple, G. H., Pigment metabolism and regeneration of hemoglobin in the body, *Arch. Intern. Med.,* 29, 711–731, 1922.

4. Minot, G. R. and Murphy, W. P., Treatment of pernicious anemia by a special diet, *JAMA,* 87, 470–476, 1926.

5. Castle, W. B., Observation on the etiologic relationship of achylia gastrica to pernicious anemia, *Am. J. Med. Sci.,* 78, 748–764, 1929.

6. Smith, E. L., Purification of anti-pernicious anemia factors from the liver, *Nature* (London), 161, 638–639, 1948.

7. Rickes, E. L., Brink, N. G., Koniuszy, R. F., Wood, T. R., and Folkers, K., Crystalline vitamin B_{12}, *Science,* 107, 396–397, 1948.

8. Hodgkin, D. C., Kamper, J., MacKay, M., Pickworth, J., Trueblood, K. N., and White, J. G., Structure of vitamin B_{12}, *Nature* (London), 178, 64–66, 1956.

9. McCance, R. A. and Widdowson, E. M., The Composition of Foods, Special Report Series Medical Research Council, 297, 1960.

10. FAO-WHO Expert Group, Requirements of Ascorbic Acid, Vitamin D, Vitamin B_{12}, Folate, and Iron, WHO Technical Report Series, 452, 1970.

11. Chanarin, I., *The Megaloblastic Anemias,* F. A. Davis, Philadelphia, 1969, 192–229.

12. Rothenberg, S. P., Application of competitive ligand binding for radioassay of vitamin B_{12} and folic acid, *Metabolism,* 22, 1075–1082, 1973.

13. Joske, R. A., The vitamin B_{12} content of human liver tissue obtained by aspiration biopsy, *Gut,* 4, 231–235, 1963.

14. Stahlberg, K. G., Radner, S., and Norden, A., Liver B_{12} in subjects with and without vitamin B_{12} deficiency. A quantitative and qualitative study, *Scand. J. Haematol.,* 4, 312–320, 1967.

15. Burger, R. L., Waxman, S., Gilbert, H. S., Mehlman, C. S., and Allen, R. H., Isolation and characterization of a novel vitamin B_{12}-binding protein associated with hepatocellular carcinoma, *J. Clin. Invest.,* 56, 1262–1270, 1975.

16. Doniach, D., Roitt, I. M., and Taylor, K. B., Autoimmune phenomena in pernicious anemia, *Br. Med. J.,* 1, 1374–1379, 1963.

17. Fisher, J. M. and Taylor, K. B., A comparison of autoimmune phenomena in pernicious anemia and chronic atrophic gastritis, *N. Engl. J. Med.,* 272, 499–503, 1965.

18. Schade, S. G., Abels, J., and Shilling, R. F., Studies on antibody to intrinsic factor, *J. Clin. Invest.,* 46, 615–620, 1967.

19. Rothenberg, S. P., Kantha, K. R., and Ficarra, A., Autoantibodies to intrinsic factor: Their determination and clinical usefulness, *J. Lab. Clin. Med.,* 77, 476–484, 1971.

20. Jones, D. E. and Dowling, W. F., A comparative study of pernicious anemia in the negro and white races, *Tenn. State Med. Assoc. J.,* 50, 362–363, 1957.

21. McIntyre, P. A., Hahn, R., Conley, C. L., and Glass, B., Genetic factors in predisposition to pernicious anemia, *Bull. Johns Hopkins Hosp.,* 104, 309–342, 1959.

22. Ardeman, S., Chanarin, I., Jacobs, A., and Griffiths, L., Family study in addisonian pernicious anemia, *Blood,* 27, 599–610, 1966.

23. Doniach, D., Roitt, I. M., and Taylor, K. B., Autoimmunity in pernicious anemia and thyroidites. A family study, *Ann. N.Y. Acad. Sci.,* 124, 605–625, 1965.

24. Lous, P. and Schwartz, M., The absorption of vitamin B_{12} following partial gastrectomy, *Acta Med. Scand.,* 164, 407–417, 1959.

25. Shafer, R. B., Ripley, D., Gwain, W. R., Mahmud, K., and Doscherholmen, A., Hematologic alterations following partial gastrectomy, *Am. J. Med. Sci.,* 266, 240–248, 1973.

26. Katz, M., Lee, S. K., and Cooper, B., Vitamin B_{12} malabsorption due to a biologically inert intrinsic factor, *N. Engl. J. Med.,* 287, 425–429, 1972.

27. Webb, D. I., Chodas, R. B., Mahar, C. Q., and Falcon, W. W., Mechanism of vitamin B_{12} malabsorption in patients receiving colchicine, *N. Engl. J. Med.,* 279, 845–850, 1968.

28. Mackenzie, I. L., Donaldson, R. M., Jr., Trier, J. S., and Mathan, V. I., Ileal mucosa in familial selective vitamin B_{12} malabsorption, *N. Engl. J. Med.,* 286, 1021–1025, 1972.

29. Hakami, N., Neiman, P. E., Canellos, G. P., and Lozerson, J., Neonatal megaloblastic anemia due to inherited transcobalamin II deficiency in 2 siblings, *N. Engl. J. Med.,* 285, 1163–1170, 1971.

30. Frenkel, E. P., Abnormal fatty acid metabolism in peripheral nerves of patients with pernicious anemia, *J. Clin. Invest.,* 52, 1237–1245, 1973.

31.   **Schilling, R. F.,** Intrinsic factor studies. II. The effect of gastric juice on the urinary excretion of radioactivity after the oral administration of radioactive vitamin B_{12}. *J. Lab. Clin. Med.,* 42, 860–866, 1953.

32.   **Stadtman, T. C.,** Vitamin B_{12}, *Science,* 171, 859–867, 1971.

33.   **Baker, H. A.,** Corrinoid-dependent enzymic reactions, *Annu. Rev. Biochem.,* 41, 55–90, 1972.

34.   **Lehninger, A.,** *Biochemistry,* Worth, New York, 1975.

35.   **Toskes, P. P. and Deren, J. J.,** Vitamin B_{12} absorption and malabsorption, *Gastroenterology,* 65, 662–683, 1973.

36.   **Glass, G. B. J.,** *Gastric Intrinsic Factor and Other Vitamin B_{12} Binders,* Thieme, Stuttgart, 1974.

37.   **Allen, R. H.,** Human vitamin B_{12} transport proteins, in *Progress in Hematology,* Brown, E. B., Ed., 9, 1975, 57–84.

EFFECT OF NUTRIENT DEFICIENCIES IN MAN: INOSITOL

J. Kroes

There is no known requirement for dietary inositol in man, nor have any signs or symptoms of deficiency been described. However, there is also no evidence which rules out a need for dietary inositol. Several in vitro studies have in fact demonstrated both a requirement[1,2] for and synthesis[2-4] of inositol by human cells. There have also been a few studies which suggest dietary inositol can function as a lipotrobe. It is conceivable, therefore, that the endogenous supply of inositol may not meet needs under special conditions.

Eagle et al. was the first to demonstrate a requirement of inositol by human cell lines.[1] Growth retardation, cytopathogenic changes, and eventual death occurred in all 18 normal and malignant cell lines investigated. From 5 to $10 \times 10^{-7} M$ inositol permitted 50% maximum growth. In at least two of the inositol-dependent cell lines, there was a minor but significant synthesis of inositol.[2] Thus, at least in tissue culture, cells required exogenous inositol in the presence of endogenous synthesis.

The possibility of intestinal synthesis of inositol[5] along with the wide distribution of inositol in animal and vegetable foods makes a deficiency extremely difficult to demonstrate. This is compounded by the increasing use of antibiotics which could affect the microbiological synthesis of inositol in the lumen of the intestine.

Inositol exists in nature in both free and combined forms, with plants being our main dietary source. Phytic acid, the hexaphosphate ester, is the main compound in plants. In tissue culture, phytic acid and inositol monophosphate had 4% and 18%, respectively, of the growth activity of free inositol.[1] The availability of phytic acid for absorption is dependent on the level of divalent cations in the diet and the presence of phytase. Barley, wheat, and rye contain exogenous phytase, while maize and oats do not. Even without exogenous phytase, appreciable amounts of phytate can be cleaved to inositol.[6] Abnormally high levels of calcium can decrease the digestibility of phytate, as high levels of phytate may decrease the absorption of calcium.[7] Williams has calculated that a mixed diet furnishing 2500 cal supplies approximately 1 g of inositol per day.[8] Thus, he estimated the maximum possible requirement of dietary inositol as 1 g/day.

Several early studies have suggested a possible role for exogenous inositol as a lipotrobe. Abels et al. studied patients with gastrointestinal cancer whose livers were heavily infiltrated with fats.[9,10] Only 280 mg of inositol given 10 hr preoperatively reduced heptic lipids from 16.4 to 8.2 g/100 g wet liver. No subsequent studies have controlled inositol or other dietary constituents to confirm these observations in man.

There have also been several poorly controlled studies which observed a lowering of serum lipids during the administration of inositol.[11] These studies made no attempt to control diets which are a major affector of serum lipids. In several of the studies, inositol was administered along with other lipotrobes.[12,13] However, no significant effect of exogenous inositol on serum lipids has been substantiated.[14-16]

One can only conclude that well controlled studies on dietary inositol in man are lacking. However, its importance in metabolism is evidence by its ubiquitous occurrence as phosphatidyl inositol in the membranes of human tissues, its requirement by human cell lines in tissue culture, and its high content in human milk.[17] Thus, it would seem prudent at this time to assume its essentiality, at least in infant formulas and perhaps in other special conditions, until proven otherwise.

REFERENCES

1. **Eagle, H., Oyama, V., Levy, M., and Freeman, A.,** Myo-inositol as an essential growth factor for normal and malignant human cells in tissue culture, *J. Biol. Chem.,* 229, 191–205, 1957.
2. **Eagle, H., Agronoff, B., and Snell, E.,** The biosynthesis of meso-inositol by cultured mammalian cells, and the parabiotic growth of inositol-dependent and inositol-dependent strains, *J. Biol. Chem.,* 235, 1891–1893, 1960.
3. **Needham, J.,** Studies on inositol. II. The synthesis of inositol in the animal body, *Biochem. J.,* 18, 891–893, 1924.
4. **Hauser, B. and Finelli, V.,** The biosynthesis of free and phosphatide myo-inositol from glucose by mammalian tissue slices, *J. Biol. Chem.,* 238, 3224–3228, 1963.
5. **Woolley, D.,** Synthesis of inositol in mice, *J. Exp. Med.,* 75, 277–281, 1942.
6. **Cruckshank, E., Duckworth, J., Kosterlitz, H., and Warnock, G.,** *J. Physiol.* (London), 104, 41–45, 1945.
7. **Collumbine, H., Basnayak, V., Lemottee, J., and Wickramanayake, T.,** *Br. J. Nutr.,* 4, 101–105, 1950.
8. **Williams, R.,** The appropriate vitamin requirements of human beings, *JAMA,* 119, 1, 1942.
9. **Abels, J., Kupel, C., Pack, G., and Rhoads, C.,** Metabolic studies in patients with cancer of gastro-intestinal tract. XV. Lipotropic properties of inositol, *Proc. Soc. Exp. Biol. Med.,* 54, 157–158, 1943.
10. **Abels, J., Ariel, I., Murphy, H., Pack, G., and Rhoads, C.,** Metabolic studies in patients with cancer of gastro-intestinal tract. IX. Effects of dietary constituents upon the chemical composition of the liver, *Ann. Intern. Med.,* 20, 580–588, 1941.
11. **Leinwand, I. and Moore, D.,** Simultaneous studies on the serum lipid and the electrophoretic pattern of the serum proteins in man: action of inositol and other substances, *Am. Heart J.,* 38, 467, 1949.
12. **Sherber, D. and Levites, M.,** Effect on cholesterol metabolism of a polysorbate 80-choline-inositol complex (Monichol), *JAMA,* 152, 682–687, 1951.
13. **Abrahamson, E.,** Hypercholesterolemia and senescence, *Am. J. Dig. Dis.,* 19, 186–191, 1952.
14. **Wilkinson, C.,** Effects of lipotropes and sistosterol on the level of blood lipids and the clinical course of angina pectoris, *J. Am. Geriatr. Soc.,* 3, 381–388, 1951.
15. **DeWind, L., Michaels, D., and Kinsell, L.,** Lipid studies in patients with advanced diabetic atherosclerosis, *Ann. Intern. Med.,* 37, 344–351, 1952.
16. **McKibben, J. and Brewer, D.,** Effect of inositol feeding on inositol phosphatides and other lipids of human blood plasma, *Proc. Soc. Exp. Biol. Med.,* 84, 386–388, 1953.
17. **Ogasa, K., Kuboyama, M., Kiyosawa, I., Suzuki, T., and Itoh, M.,** The content of free and bound inositol in human and cow's milk, *J. Nutr. Sci. Vitaminol.,* 21, 129–135, 1975.

NUTRIENT DEFICIENCIES IN MAN: VITAMIN C

R. W. Vilter

Scurvy is a clinical syndrome due to vitamin C (ascorbic acid) deficiency. Most animals are able to synthesize vitamin C from glucose by the following reactions.[1,2]

$$\text{D-Glucuronate} \xrightleftharpoons{\text{NADPH}} \text{L-gulonate} \xrightleftharpoons{\text{NAD}} \text{(3 keto-L-gulonate)} \longrightarrow \text{L-ascorbate}^{[1,2]}$$

Primates (including man), guinea pigs, the red-vented bulbul, the fish-eating bat, the rainbow trout, and the coho salmon[3] are dependent on exogenous sources for the vitamin, because they lack the enzyme system necessary for the conversion of 3-keto-L-gulonate to L-ascorbate.

The disease may be fatal, particularly in infants and otherwise debilitated adults. Though it is rare today, it does occur in artificially fed infants, alcoholic addicts, and elderly bachelors who eat alone or in restaurants, avoiding citrus fruits and tomatoes. It was much more common in days gone by and is one of the earliest recorded diseases. A historical resume is given below:

Scurvy — Historical Resume

1550 B.C.	Scurvy described in Ebers Papyrus (Thebes)[4]
600 B.C.	Hippocrates described soldiers afflicted with scurvy[5]
1200 A.D.	Crusades weakened by scurvy[6]
1492–1600	World exploration threatened by scurvy[6,7]
	Magellan lost four fifths of his crew
	Vasco da Gama lost 100 of 160 men
1536	Jacques Cartier's expedition immobilized by scurvy, but he learned from the Indians the curative value of pine needles and bark[8]
1593	Sir Richard Hawkins used oranges and lemons to treat scurvy in the British Navy[9]
	Ponsseus referred to the therapeutic use of scurvy grass, water cress, and oranges[10]
1600	British Naval Captain Lancaster demonstrated the preventive value of lemon juice[11]
1650	Infantile scurvy described by Glisson, but was confused with rickets[12]
17th century	Lime juice used experimentally on ships of the East India Company
1734	Backstrom related scurvy to a deficiency of fresh fruits and vegetables[11]
1740–1744	Lord Anson lost three fifths of his crew of 1950 men due to scurvy[13]
1747	James Lind performed controlled shipboard experiment on the preventive effect of oranges and lemons (published 1753)[10]
1768–1771	Captain James Cook demonstrated that prolonged sea voyages were possible without ravages of scurvy[14,15]
1789	William Stark induced scurvy in himself by a diet of bread and water for 60 days[16]
1795	Lemon juice made a regular ration in British Navy[6]
1854	Lemon juice made a regular ration in British Merchant Marines[6]
1863	Scurvy epidemic in the opposing armies during the American Civil War[17]
1883	Sir Thomas Barlow differentiated infantile scurvy from rickets[18]
1895	Antiscorbutic ration became official in U.S. Army[17]
1900	Boiling and pasteurization of infant formulae increased incidence of scurvy
1906	Hopkins suggested that infantile scurvy was a deficiency disease[19]
1907	Holst and Frolich produced scurvy in guinea pigs by feeding deficient diet[20]
1928	Szent-Gyorgyi synthesized hexuronic acid[21,22]
1932	Waugh and King isolated hexuronic acid from lemons and identified it as vitamin C[23,24]
1933	Haworth determined structure of vitamin C[25]
1933	Reichstein synthesized vitamin C[26,27]

The manifestations of scurvy (given below) appear insidiously, usually after 5 to 6 months of severe deprivation of vitamin C. In infants, scurvy occurs most commonly between the 5th and 24th month of age, with the peak between the 8th and 11th months. In adults, the disease may occur at any time, but is found most often in neglected or alcoholic elderly males, often bachelors (thus the designation, bachelor scurvy).

Symptoms and Signs of Scurvy in the Adult[28-30] in Approximate Order of Occurrence

Weakness, fatigue, listlessness
Neurotic behavior[31]
Aching in bones, joints, and muscles
Dry, rough skin
Perifollicular hyperkeratotic papules
Fragmented, coiled, ingrown hairs
Perifollicular hemorrhages
Petechiae and ecchymoses, arms, legs, and conjunctivae[32]
Hemorrhages in muscles causing brawny induration
Hemarthroses, especially involving large joints
Subungual splinter hemorrhages in palisades
Swollen, blue-red, friable gums where teeth are present
Sjögren's syndrome[33]
Teeth loosen and fall out
Neuritis due to nerve hemorrhages[34]
Wounds fail to heal and scars become hemorrhagic and may break down
Sclerae become icteric, the skin a sallow yellow-brown color
Temperature of 101 to 102°F
Peripheral stasis of blood and cyanosis of extremities
Cheyne-Stokes respirations, convulsions, shock, and death

Symptoms and Signs of Scurvy in the Infant[35,36] in Approximate Order of Occurrence

Listlessness, loss of appetite, irritability, and failure to thrive
Assumption of the fetal position
Tender extremities; pseudoparalyses
Apprehension when handled
Subperiosteal hemorrhages
Scorbutic rosary
Petechiae and purpura
Blue-red, swollen gums around erupting teeth
Red blood cells in urine, stool, and spinal fluid
Hemorrhages into brain and other viscera
Dyspnea, cyanosis, shock, death

Roentgenological Evidence of Scurvy in the Infant[37-39]

The Corner Sign
    Ankles and wrists most often affected
    Defect at anterior corner of lower end of tibia
    Defect at outer corner of lower end of radius
    Defect becomes a cleft or crevice underneath the epiphyseal line
The White Line
    Dense zone of provisional calcification at the epiphyseal end of
       diaphysis of tibia and radius
The zone of rarefaction
    The white line accentuated by zone of rarefaction shaftward
The halo epiphysis
    Similar involvement of the epiphysis
Ground glass appearance
    Cortex of the bones become osteoporotic
Scorbutic rosary
    Separation at the swollen bulbous costochondral junctions
      with sinking of sternum inward
Laminated calcification of subperiosteal hemorrhages
    Appears after treatment with vitamin C

In the adult patient with scurvy, osteoporosis with collapse of vertebrae and patchy or diffuse demineralization of the long bones is the most common sign. Destructive changes in large joints secondary to recurrent hemarthroses may be found also.[40]

Numerous laboratory tests have been proposed to confirm a diagnosis of scurvy or to determine the ascorbic acid status of patients. Many of these are "load tests," based upon the principle of "tissue saturation." This term implies a finite body pool of ascorbic acid that can be "filled" or "saturated" by giving enough ascorbic acid. Since the renal threshold for ascorbic acid is 1.4 mg%,[41] any excess over this amount will be excreted and can be measured in the urine. The greater the amount already in the pool, the higher the plasma level will rise and the more ascorbic acid will be excreted in the urine, following a test dose of ascorbic acid given either orally or parenterally. Table 1 lists tests that are compatible with a diagnosis of scurvy. It must be remembered, however, that the Rumpel-Leede test may be positive in many types of vascular purpura and that, months before any clinical manifestations of scurvy are evident, plasma and whole blood ascorbic acid levels reach 0. Estimation of the body pool and the catabolic rate of ascorbic acid (2.6%/day) using ^{14}C-L-ascorbic acid as a label is the most dependable technique, but it is available only in research laboratories.[42]

Unsupplemented scorbutogenic milk diets of infants lack iron also. As the infants' rapidly expanding red-cell mass demands more iron than is available, microcytic hypochromic anemia appears. The iron deficiency, shown in Table 2, is exaggerated by blood loss due to scurvy and reduction in the absorption of food iron due to lack of vitamin C.[53]

Milk diets also provide minimal amounts of folic acid. In addition, the reduction potential of ascorbic acid is probably important for the protection of folic acid reductase, an enzyme necessary to maintain tetrahydrofolic acid in reduced, and therefore active, form.[54,55] Marginal amounts of folic acid are insufficient in patients with scurvy, and megaloblastic anemia occurs.[56] Metabolic interrelationships of vitamin C-vitamin B_{12} metabolism have been postulated, since patients with pernicious anemia and vitamin C deficiency frequently have not responded to usual doses of therapeutic agents containing vitamin B_{12} until vitamin C is given.[57] Patients with pernicious anemia have been reported to have low plasma and buffy-coat ascorbic acid levels until vitamin B_{12} is given.[58,59]

In adults with scurvy, megaloblastic anemia occurs for similar reasons, and will

Table 1
TESTS COMPATIBLE WITH A DIAGNOSIS OF SCURVY[43]

Rumpel-Leede	+
Plasma Ascorbic Acid Level	0
Whole Blood Ascorbic Acid Level	0
Buffy Coat Ascorbic Acid Level	<4 mg %
Ascorbic Acid Saturation Tests[a]	

Amount of ascorbic acid	Route	Frequency of measurement post ascorbic acid administration	Values consistent with severe ascorbic acid depletion		Ref.
15 mg/kilo	Oral	Urine and plasma hourly for 5 hr	Maximum Plasma level <0.4 mg%	Urinary Excretion 0	44
1000 mg (adults)	i.v.	Urine collection 5 hr post injection	<100 mg >400 mg	(Severe depletion) (Adequate saturation)	45
100 mg	i.v.	Urine collection 3 hr post injection	Normal >50% test dose Depleted <15% test dose Scorbutic <5% test dose		41

[a] Ascorbic acid body pool < 300 mg (measured with ^{14}C-L-ascorbic acid; ascorbic acid catabolic rate < 9 mg/day.[42]

Note: Abnormalities of blood and bone marrow occur frequently in patients with scurvy and have been studied extensively.

Table 2
ANEMIA IN PATIENTS WITH SCURVY

Peripheral blood morphology	Bone marrow morphology	Etiology
Infancy and Childhood[46-48]		
Microcytic, hypochromic	Normoblastic	Chronic blood loss Iron deficiency
Macrocytic	Megaloblastic	Folic acid deficiency
Normocytic	Normoblastic	Chronic inflammatory diseases
Adults[49-52]		
Normocytic, normochromic	Normoblastic	Blood loss into hematomas Possible direct effect of Vitamin C deficiency
Macrocytic	Megaloblastic	Folic acid deficiency

respond to folic acid even though scurvy persists.[60] Ascorbic acid will cure the scurvy, but unless a minimal amount of folic acid is available, anemia and megaloblastosis are unaffected.

Though patients with mild scurvy frequently do not have anemia, patients with more severe scurvy have normocytic or slightly macrocytic and normochromic anemia.[29] Reticulocytosis of 10 to 15% is common and the bone marrow, which may be either moderately hyper- or hypoplasic, is normoblastic. Treatment with ascorbic acid increases the number of reticulocytes and normoblasts further as the anemia begins to respond. Though blood loss externally and into tissues undoubtedly plays a causative role, and increased hemolysis has been suggested,[61,62] there is a possibility, still unproved, that ascorbic acid deficiency adversely affects the development of red cell precursors. White blood cells usually are normal unless infection had induced leucocytosis. Platelets usually are present in normal numbers, but abnormalities have been recognized in platelet adhesion and aggregation.[63-66] Other laboratory abnormalities that have been noted in patients with scurvy are listed in the following table.

		Ref.
Urine		
Positive test for "tyrosyl derivatives" (homogentisic acid, p-hydroxyphenylpyruvic acid, p-hydroxy-phenyllactic acid) after a loading dose of Tyro-sine[67-71]		29
Urobilinogen	>3.5 mg/day	
RBC	+	
Protein	+	
Stool		
Guaiac test	+	
Urobilinogen	>250 mg/day	29
Blood		
Indirect reacting bilirubin	1.2 mg %	29
Alkaline phosphatase	low or low-normal	72
Electolytes and other chemical tests	compatible with malnutrition	

Scurvy has been induced in volunteers, and much has been learned about the body pool, depletion rate, blood, urine, and tissue levels at various stages of depletion and the length of time that must elapse before clinical manifestations of scurvy appear in subjects on a vitamin C-free diet. In the 18th century, William Stark, a physician, subsisted on bread and water, developing the swollen bleeding gums of scurvy in 60 days.[16] While a surgical house officer, Crandon induced scurvy in himself after living on a vitamin C-deficient diet for 134 days[73] and demonstrated poor wound healing. In 1944, Pijoan and Lozner reported on scurvy induced in human beings by a vitamin C-deficient diet.[74] Petechiae, perifollicular hemorrhages, and tender, swollen, bleeding gums developed after 150 days. In 1953, the Sheffield Study reported that follicular hyperkeratosis appeared in human subjects after 110 days on a vitamin C-deficient diet. Perifollicular hemorrhages appeared after 182 days.[75] In 1969, Hodges and associates induced scurvy in volunteers maintained on a diet completely free of vitamin C.[30] The chronology of two of these studies is summarized in Tables 3 and 4.

Baker and associates[42,76] labeled the ascorbic acid body pool of volunteers with ^{14}C-L-ascorbic acid when they were well nourished and determined the body pool size and ascorbic acid catabolic rate during depletion brought about by subsistence on a vitamin C-free diet. The data that they accumulated allowed them to construct an

Table 3
EXPERIMENTAL SCURVY IN ADULTS[73]

		Vitamin C (mg%)	
Days	Clinical signs	Plasma	Buffy coat
Vitamin C-deficient Diet[a]			
17		0.1	28
42		0.0	10
82	Fatigue	0.0	4
90	Normal wound healing	0.0	0.0
134	Perifollicular hyperkeratoses	0.0	0.0
	Ingrown, fragmented hairs	0.0	0.0
	Dryness and flaking of skin	0.0	0.0
161	Perifollicular hemorrhages	0.0	0.0
180	Minimal swelling of gums	0.0	0.0
182	Poor wound healing	0.0	0.0
Supplemented Vitamin C Diet (1000 mg/day)			
1			3
2			16
3			28
6	Petechiae faded; skin moist and smooth		

[a] Diet consisted of bread, meat or eggs, cornflakes, cake, butter.

Table 4
EXPERIMENTAL SCURVY IN ADULTS ON A VITAMIN C-FREE FORMULA DIET[a]

Days	Clinical signs	Urine	Plasma
26	Petechiae	0	—
43	Gingival hemorrhages	—	—
45	Perifollicular hemorrhages	—	—
56	—	—	(±)
60–80	Follicular hyperkeratoses and congestion	—	—
76–91	Congested swollen gums and interdental papillae	—	—
84	Bulbar conjunctival hemorrhages	—	—

[a] Diet consisted of vitamin-free casein, sucrose, peanut oil, cocoa butter, safflower oil, mineral and vitamin mixture lacking vitamin C.

From Hodges, R. E., Baker, E. M., Hood, J., Sauberlich, H. E., and March, S. C., *Am. J. Clin. Nutr.,* 22, 535–548, 1969.

ascorbic acid depletion curve which is reproduced in Figure 1. It indicates a 1500 mg or greater body pool, daily catabolism of 45 mg when saturated, a catabolic rate of 3% of the body pool per day. After 55 to 60 days on the diet, the body pool fell to 300 mg and the daily catabolism to 9 mg. These figures suggest that 40 to 60 mg/day will maintain the saturated state and that 10 mg/day will prevent scurvy.

PATHOPHYSIOLOGICAL AND CHEMICAL ASPECTS OF SCURVY

The lesions of scurvy occur in tissues of mesenchymal origin particularly in newly

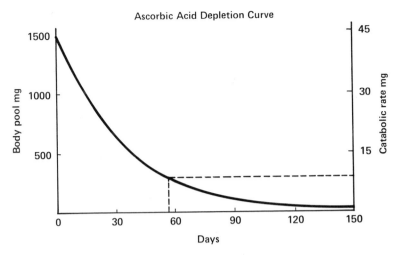

FIGURE 1.   Curve of ascorbate pool derived from data of nine men whose body pool of ascorbate was labeled with ^{14}C-L-ascorbic acid. They were then fed a diet devoid of vitamin C. Initially, the body pool averaged 1500 mg. The average daily rate of catabolism was 3% of the existing body pool. Thus, the maximal rate of catabolism approximated 45 mg/day. When the body pool fell below 300 mg total and the catabolic rate below 9 mg/day, signs of scurvy began to appear (about 55 days). From the curve, one can estimate the approximate body pool size from the dose. Thus, with a daily intake of 30 mg, the pool size should be about 1000 mg. (From Hodges, R. E. and Baker, E. M., in *Modern Nutrition in Health and Disease*, Goodhart, R. S. and Shils, M. E., Eds., Lea & Febiger, Philadelphia, 1973, chap. 5. With permission.)

formed connective tissue.[77,78] Similar lesions also occur in developing teeth, growing bones, and blood vessels. Ground substance secreted by fibroblasts is defective; therefore, the tensile strength of newly formed fibrous tissue in wounds is poor. Wounds may fail to heal or, if recently healed, may break down again.

In growing bones, bone matrix, or osteoid derived from osteoblasts, a specialized fibroblast is defective. At the diaphyseal end of long bones, the cartilage cells of the epiphyseal plate continue to proliferate and line up normally in rows. The cartilage between the rows is calcified and compressed; however, osteoblasts do not migrate into this zone of provisional calcification, and it becomes compressed and brittle (the white line seen in X-rays). Instead, proximal to this zone, loose connective tissue, probably containing cells that should be functioning as osteoblasts, forms a zone of rarefaction; the osteoid thus produced is defective, and normal ossification does not occur. Similar loose connective tissue appears under the periosteum. Cracks occur easily in the zone of rarefaction, most often laterally where the shaft of the bone, the cartilage, and the loose connective tissue under the periosteum are juxtaposed (the corner sign as seen in the X-ray). Where these cracks occur, hemorrhages may develop. These may spread in the abnormally loose subperiosteal connective tissue and strip the periosteum back from the bone (the subperiosteal hemorrhage). Fractures and subluxations at costochondral junctions occur, forming the scorbutic rosary. Growth of dentine in developing teeth ceases, and the pulp becomes separated from the dentine by liquid derived from odontoblasts (like the osteoblasts, a specialized fibroblast). The dentine becomes porous, alveolar bone becomes osteoporotic, and teeth loosen and fall out. In the adult, the main boney change is osteoporosis due to poor bone regeneration. The alkaline phosphatase level in the blood tends to be low in infants,[72] but usually is normal in adults.

As in scorbutic guinea pigs, the synthesis of collagen probably is impaired because of a defect in the hydroxylation of proline and lysine to hydroxyproline and hydroxylysine, after these amino acids have been incorporated into the peptide chains of collagen.[79-87] Whether this occurs before or after release from the ribosomes has not been determined with certainty. Ascorbic acid appears to act as a cofactor for peptidyl proline and lysine hydroxylases,[88] possibly by keeping copper and iron in a reduced state.[89,90] A defect in the formation of mucopolysaccharides and glycoproteins has been postulated.[91,92] Such a defect might adversely effect basement membranes, particularly of capillaries.[93-95] Collagen formed in scorbutic animals is not as effective a platelet-aggregating agent as is normal collagen.[84] Though the bleeding tendency in scurvy still is not thoroughly understood, it may be due to the combination of defective vascular wall, defective ground substance in basement membrane, and defective platelet aggregation to all agents, but particularly to abnormal collagen.

Defective collagen may explain the poor wound healing in scorbutic patients which has led to the recommendation of high doses of vitamin C in pre- and postoperative surgical patients.[96-98] Substantial proof of efficacy is lacking.

Several other defective hydroxylation reactions have been identified:[99-103]

$$\text{Tyrosine} \longrightarrow \text{Homogentisic acid}$$

$$\text{Tryptophane} \longrightarrow \text{5-Hydroxytryptophane}$$

Defects in both of these reactions might interfere with the production of chemical mediators of vascular tone such as norepinephrine and serotonin.[104-107] An abnormality in vascular tone and in vascular responsiveness has been recognized in scurvy.[108,109] These abnormalities also could be involved in the bleeding diathesis as well as in the postural hypotension and sudden death that have been reported in patients with scurvy. The adrenal gland is very rich in vitamin C. In scurvy, this supply is depleted severely. It has been postulated[110-112] that this depletion results in deficiency of adrenal cortico-steroid hormones; however, such deficiencies have not been demonstrated in scorbutic patients.[113,114]

It has been noted that animals[115] and humans[116] with scurvy are susceptible to infections, but no abnormalities have been identified in the various immune systems or in the functional capacity of neutrophiles,[30,117,118] and efficacy of large doses of ascorbic acid in the prevention of infections has not been established.

Patients with scurvy lose protein in the urine; the serum albumen level tends to fall and the globulin level to rise.[118,119] The cholesterol level tends to be low, and to rise when ascorbic acid is given in some investigations,[118,120,121] but other studies suggest that scurvy is associated with high cholesterol levels, due to slowed cholesterol catabolism,[122] and that feeding high doses of vitamin C lowers cholesterol and triglycerides in hyperlipidemic persons.[123] To add to the confusion, a population survey failed to show any correlation between the blood ascorbic-acid level and the plasma cholesterol level.[124] The subject of ascorbate-cholesterol-lecithin interactions has been recently reviewed.[125] In any case, it is imperative to treat scurvy promptly because of the danger of sudden death.

Prevention and Treatment of Scurvy with Ascorbic Acid[30,75]

Body level	Dosage (mg/day)	Ref.
Minimum intake that will prevent scurvy	10	30, 75
Intake that will maintain tissue saturation[a]	45–50	42
Intake used in treatment of clinical scurvy	100 (three times per day)	

[a]There is no solid evidence that tissue saturation is beneficial to health. See Reference 126.

There is some experimental evidence showing that supplementation of the diet with excessive amounts of ascorbic acid may lead to conditioning wherein the excretion and destruction of the vitamin are increased. When the excessive doses are discontinued, a deficiency state may occur[127] until normal balance is reestablished. When 4 g or more of ascorbic acid is ingested daily, it will induce uricosuria and oxaluria, which may cause kidney stones to form. The uricosuria can be inhibited by small doses of aspirin or pyrazinamide.[128]

REFERENCES

1. Burns, J. J., Peyser, P., and Moltz, A., Missing step in guinea pigs required for biosynthesis of l-ascorbic acid, *Science,* 124, 1148–1149, 1956.
2. Grollman, A. P. and Lehninger, A. L., Enzymatic synthesis of L-ascorbic acid in different animal species, *Arch. Biochem. Biophys.,* 69, 458–467, 1957.
3. Johnson, C. L., Hammer, D. C., Halver, J. E., and Baker, E. M., *Fed. Proc. Fed. Am. Soc. Exp. Biol.* (Abstr.), 30, 521, 1971.
4. Papyrus, E., *Medical Writings Ca. 1550 B.C.;* quoted in Encyclopedia Britannica, Chicago, 1964.
5. Hippocrates, *The Genuine Works of Hippocrates Ca. 600 B.C.,* Vol. 1, Adams, F., Ed., printed for the Sydenham Society, London, 1849, 196, 267.
6. Major, R. H., *Classic Descriptions of Disease with Biographical Sketches of the Authors,* 3rd ed., Charles C Thomas, Springfield, Illinois, 1945, 585–615.
7. Hess, A. F., *Scurvy, Past and Present,* Lea & Febiger, Philadelphia, 1920.
8. Biggar, H., The Voyages of Jacques Cartier, *Canadian Public Archives,* Publication 11, 204, 1924.
9. *Encyclopedia Britannica,* 1st ed., Vol. 3, 1771, 106–110.
10. Lind, J., *A Treatise of Scurvy, In Three Parts Containing an Inquiry into the Nature, Causes and Cure of that Disease Together with Actual and Chronological Views of What Has Been Published on the Subject,* H. Miller in the Strand, London, 1753; republished by the University Press of Edinburgh, 1953.
11. Friedman, G. J. and Jolliffe, N., Vitamin C, malnutrition, and scurvy, in *Clin. Nutr.,* 2nd ed., Hueber Medical Division, Harper & Row, New York, 656–690, 1962.
12. Glisson, F., De Rachitide sive Morbo Pueriti qui vulgo, "The Rickets", dicitur. London, 1650; quoted by Tisdall, F. F. and Jolliffe, N., Vitamin C. Malnutrition and scurvy, in *Clin. Nutr.,* Jolliffe, N., Tisdall, F., and Cannon, P., Eds., Paul B. Hoeber, New York, 1950, 586–601.
13. Anson, G., *A Voyage Around the World in the Years 1740, 1,2,3,4,* Compiled by Richard Walter, M. D., Chaplain of H.M.S. Centurion, 3rd ed., Printed by the author, London, 1748.
14. Villiers, A., That extraordinary sea genius, Captain James Cook, *Nutr. Today,* 4(3), 8–16, 1969.
15. Kodicek, E. H. and Young, F. G., *Notes and Records of the Royal Society of London,* 24, 43, 1969.
16. Drummond, J. C. and Wilbraham, A., An 18th century experiment in nutrition, *Lancet,* 2, 459–464, 1935.
17. Hunt, C., in *Contributions Relating to the Causation and Prevention of Disease,* Flint, A., Ed., Published for the U.S. Sanitary Commission by Hurd and Houghton, New York, 1867, chap. 6.
18. Barlow, T., On cases described as "acute rickets," which are probably a combination of scurvy and rickets, *Medico-Chirurgical Transactions,* Vol. 66, London, 1883, 159; reprinted in *Arch. Dis. Child.,* 10, 223–252, 1935.
19. Hopkins, F. G., The analyst and the medical man, *The Analyst,* 31, 385–404, 1906.
20. Holst, A. and Frölich, T., Experimental studies relating to "ship-beri-beri" and scurvy, *J. Hyg.,* 7, 634–671, 1907.
21. Szent-Gyorgyi, A., Observations on function of peroxidase systems and chemistry of adrenal cortex: description of new carbohydrate derivative, *Biochem. J.,* 22, 1387–1409, 1928.
22. Svirbely, J. L. and Szent-Gyorgyi, A., Chemical nature of vitamin C, *Biochem. J.,* 26, 865–870, 1932.
23. Waugh, W. A. and King, C. G., Isolation and identification of vitamin C, *J. Biol. Chem.,* 97, 325–331, 1932.
24. King, C. G. and Waugh, W. A., Chemical nature of vitamin C, *Science,* 75, 357–358, 1932.

25. **Haworth, W. N. and Hirst, E. L.,** Synthesis of ascorbic acid, *J. Soc. Chem. Ind. London,* 52, 645–646, 1933.
26. **von Reichstein, T., Grüssner, A., and Oppenhauer, R.,** Die Synthese der d-Ascorbinsäure (d-Forme des C-Vitamins), *Helv. Chim. Acta,* 16, 561–565, 1933.
27. **von Reichstein, T., Grüssner, A., and Oppenhauer, R.,** Synthese der d- und l-Ascorbinsäure (C-Vitamins), *Helv. Chim. Acta,* 16, 1019–1033, 1933.
28. **Sebrell, W. H., Jr. and Harris, R. S.,** *The Vitamins: Chemistry, Physiology, Pathology, Methods,* Academic Press, New York, 1967, chap. 2, 305.
29. **Vilter, R. W., Woolford, R. M., and Spies, T. D.,** Severe scurvy, a clinical and hematologic study, *J. Lab. Clin. Med.,* 31, 609–630, 1946.
30. **Hodges, R. E., Baker, E. M., Hood, J., Sauberlich, H. E., and March, S. C.,** Experimental scurvy in man, *Am. J. Clin. Nutr.,* 22, 535–548, May 1969.
31. **Kinsman, R. A. and Hood, J.,** Neurotic triad in scurvy, *Am. J. Clin. Nutr.,* 24, 455–464, 1971.
32. **Hood, J. and Hodges, R. E.,** Ocular lesions in scurvy, *Am. J. Clin. Nutr.,* 22, 559–567, May 1969.
33. **Hood, J., Burns, C. A., and Hodges, R. E.,** Sjögren's syndrome in scurvy, *N. Engl. J. Med.,* 282, 1120–1124, 1970.
34. **Hood, J.,** Neuritis in scurvy due to nerve hemorrhages, *N. Engl. J. Med.,* 281, 1292–1293, 1969.
35. **Woodruff, C. W.,** in *Nutrition, a Comprehensive Treatise,* Vol. 2, Beaton, G. H. and McHenry, E. W., Eds., Academic Press, New York, 1964, 265–298.
36. **Woodruff, C.,** Infantile scurvy: the increase in the incidence of scurvy in the Nashville area, *JAMA,* 161, 448–456, 1956.
37. **Park, E. A., Guild, H. G., Jackson, D., and Bond, M.,** Recognition of scurvy with especial reference to early X-ray changes, *Arch. Dis. Child.,* 10, 265–294, 1935.
38. **Ellenbogen, L. S., Aldrich, C. C., and Green, M.,** Roentgen findings in the diagnoses and management of infantile scurvy, *J. Med. Soc. N. J.,* 48, 73–76, 1951.
39. **McCann, P.,** The incidence and value of radiologic signs in scurvy, *Br. J. Radiol.,* 35, 683–686, October 1962.
40. **Joffe, N.,** Some radiological aspects of scurvy in the adult, *Br. J. Radiol.,* 34, 429–437, 1961.
41. **Ralli, E. and Sherry, S.,** Adult scurvy and metabolism of vitamin C, *Medicine* (Baltimore), 20, 251–340, 1941.
42. **Baker, E. M., Hodges, R. E., Hood, J., Sauberlich, R. E., and March, S. C.,** Metabolism of ascorbic-l-^{14}C acid in experimental scurvy, *Am. J. Clin. Nutr.,* 22, 549–558, 1969.
43. **Vilter, R. W.,** Effects of ascorbic acid deficiency in man, in *The Vitamins: Chemistry, Physiology, Pathology, Methods,* Vol. 1, Sebrell, W. H. and Harris, R. S., Eds., Academic Press, New York, 1967, 471–473.
44. **Wolfer, J. A., Farmer, C. J., Carroll, W. W., and Manshardt, D. O.,** Experimental study in wounds healing in vitamin C-depleted human subjects, *Surg. Gynecol. Obstet.,* 84, 1–15, 1947.
45. **Wright, I. S., Lilienfeld, A., and MacLenathen, E.,** Determination of vitamin C saturation: five hour test after intravenous dose, *Arch. Intern. Med.,* 60, 264–271, 1937.
46. **Parsons, L. G. and Smallwood, W. C.,** Studies in anaemia of infancy and early childhood; anaemia of infantile scurvy, *Arch. Dis. Child.,* 10, 327–336, 1935.
47. **Zuelzer, W. W., Hutaff, L., and Apt, L.,** Relationship of anemia and scurvy, *Am. J. Dis. Child.,* 77, 128, 1949.
48. **Zuelzer, W. W. and Ogden, F. N.,** Folic acid therapy in macrocytic anemias of infancy, *Proc. Soc. Exp. Biol. Med.,* 61, 176–177, 1946.
49. **Bronte-Stewart, R.,** The anemia of adult scurvy, *Q. J. Med.,* 22, 308–329, 1953.
50. **Cox, E. V., Meynell, M. J., Cooke, W. T., and Gaddie, R.,** Scurvy and anemia, *Am. J. Med.,* 32, 240–250, 1962.
51. **Cox, E. V., Meynell, M. J., Northam, B. E., and Cooke, W. T.,** The anemia of scurvy, *Am. J. Med.,* 42, 220–227, 1967.
52. **Goldberg, A.,** The anemia of scurvy, *Q. J. Med.,* 32, 51–64, 1963.
53. **Moore, C. V. and Dubach, R.,** Observations on absorption of iron from foods tagged with radioiron, *Trans. Assoc. Am. Physicians,* 64, 245–256, 1951.
54. **Welch, A. D., Nichol, C. A., Anker, R. M., and Boehne, J. W.,** The effect of ascorbic acid on the urinary excretion of citrovorum factor derived from folic acid, *J. Pharmacol. Exp. Ther.,* 103, 403–411, 1951.
55. **Stokes, P. L., Melikian, V., Leeming, R. L., Portman-Graham, H., Blair, J. A., and Cooke, W. T.,** Folate metabolism in scurvy, *Am. J. Clin. Nutr.,* 28(2), 126–129, 1975.
56. **Jandl, J. H. and Gabuzda, G. J., Jr.,** Potentation of pteroyl glutamic acid by ascorbic acid in anemia of scurvy, *Proc. Soc. Exp. Biol. Med.,* 84, 452–455, 1953.

57. Dyke, S. C., Della Vida, B. L., and Delikat, E., Vitamin C deficiency in "irresponsive" pernicious anemia, *Lancet*, 2, 278, 1942.

58. Wallerstein, R. O., Harris, J. W., and Gabuzda, G. J., Jr., Ascorbic acid deficiency in pernicious anemia, *Am. J. Med.*, 14, 532, 1953.

59. Cox, E. V., Gaddie, R., Mathews, D., Cooke, W. T., and Meynell, M. J., An interrelationship between ascorbic acid and cyanocobalamin, *Clin. Sci.*, 17, 681–692, 1958.

60. Zalusky, R. and Herbert, V., Megaloblastic anemia in scurvy with response to 50 γ folic acid daily, *N. Engl. J. Med.*, 265, 1033–1038, 1961.

61. Merskey, C., Survival of transfused red cells in scurvy, *Br. Med. J.*, 2, 1353–1356, 1953.

62. Goldberg, A., The anemia of scurvy, *Q. J. Med.*, 32, 51–64, 1963.

63. Cetingil, A. J., Ulutin, O. N., and Karaca, M., A platelet defect in a case of scurvy, *Br. J. Haematol.*, 4, 350–354, 1958.

64. McNicol, G. P., Douglas, A. S., and Wilson, P. A., Platelet abnormality in human scurvy, *Lancet*, 1, 975–978, 1967.

65. Born, G. V. R. and Wright, H. P., Diminished platelet aggregation in experimental scurvy, *J. Physiol.* (London), 197, 27p–28p, 1968.

66. Sahud, M. A. and Aggeler, P. M., Utilization of ascorbic acid during platelet aggregation, *Proc. Soc. Exp. Biol. Med.*, 134, 13–16, 1970.

67. Sealock, R. R. and Silberstein, H. E., Excretion of homogentisic acid and other tyrosine metabolites by vitamin C-deficient guinea pigs, *J. Biol. Chem.*, 135, 251–258, 1940.

68. Sealock, R. R., Goodland, R. L., Summerwell, W. N., and Brierly, J. M., Role of ascorbic acid in oxidation of L-tyrosine by guinea pig liver extracts, *J. Biol. Chem.*, 196, 761–767, 1952.

69. Turner, E. J. and Campbell, D. J., Scurvy, amino acids and chromatography, *Can. Med. Assoc. J.*, 84, 113–115, 1961.

70. Levine, S. Z., Marples, E., and Gordon, H. H., Defect in metabolism of tyrosine and phenylalanine in premature infants: identification and assay of intermediary products, *J. Clin. Invest.*, 20, 199–207, 1941.

71. Rogers, W. F. and Gardner, F. H., Tyrosine metabolism in human scurvy, *J. Lab. Clin. Med.*, 34, 1491–1501, 1941.

72. Dogramaci, I., Scurvy: survey of 241 cases, *N. Engl. J. Med.*, 235, 185–189, 1946.

73. Crandon, J. H., Lund, C. C., and Dill, D. B., Experimental human scurvy, *N. Engl. J. Med.*, 223, 353–369, 1940.

74. Pijoan, M. and Lozner, E. L., Vitamin C economy in human subject, *Bull. Johns Hopkins Hosp.*, 75, 303–314, 1944.

75. Bartley, W. H., Krebs, A., and O'Brien, J. R. P., A report on the Sheffield study, *Med. Res. Counc. G.B. Spec. Rep. Ser.*, No. 280., H.M. Stationery Office, London, 1953.

76. Hodges, R. E. and Baker, E. M., in *Modern Nutrition in Health and Disease*, Goodhart, R. S. and Shils, M. E., Eds., Lea & Febiger, Philadelphia, 1973, chap. 5, 245–255.

77. Wolbach, S. B. and Bessey, O. A., Tissue changes in vitamin deficiencies, *Physiol. Rev.*, 22, 233–289, 1942.

78. Follis, R. H., Jr., *Deficiency Diseases, Ascorbic Acid*, Charles C Thomas, Springfield, Illinois, 1958, 175–195.

79. Barnes, M. J., Studies in vivo on the biosynthesis of collagen and elastin in ascorbic acid deficient guinea pigs, *Biochem. J.*, 113, 387–397, 1969.

80. Kirchheimer, B., The influence of ascorbic acid deficiency in connective tissue, *Dan. Med. Bull.*, 16, 73–76, March 1969.

81. Windsor, A. C. and Williams, C. B., Urinary hydroxyproline in the elderly with low leucocyte ascorbic acid levels, *Br. Med. J.*, 1, 732–733, 1970.

82. Grant, M. E. and Prockop, D. J., The biosynthesis of collagen, *N. Engl. J. Med.*, 286, 194–199, 1972.

83. Barnes, M. J. and Kodicek, E., Biological hydroxylations and ascorbic acid with special regard to collagen metabolism, *Vitam. Horm.* (New York), 30, 1–43, 1973.

84. Caen, J. P. and Legrand, Y., Abnormalities in the platelet-collagen reaction, *Ann. N.Y. Acad. Sci.*, 201, 194–204, 1972.

85. Formation of connective tissue proteins, *Nutr. Rev.*, 3(1), 28–29, 1973.

86. Harwood, R., Grant, M. E., and Jackson, D. S., Influence of ascorbic acid on ribosomal patterns and collagen biosynthesis in healing wounds of scorbutic guinea pigs, *Biochem. J.*, 142(3), 641–651, 1974.

87. Robosova, B. and Chvapil, M., Relationship between the dose of ascorbic acid and its structural analogues and proline hydroxylation in various biological systems, *Connect. Tissue Res.*, 2(3), 215–221, 1974.

88. **Nigra, T. P., Friedland, M., and Martin, G. R.,** Controls of connective tissue synthesis: collagen metabolism, *J. Invest. Dermatol.,* 59, 44–49, 1972.

89. **Staudinger, H., Krisch, K., and Leonhäuser, S.,** Role of ascorbic acid in microsomal electron transport and the possible relationship to hydroxylation reactions, *Ann. N.Y. Acad. Sci.,* 92, 195–207, 1961.

90. **Goldberg, A.,** The enzymic formation of haem by the incorporation of iron into protoporphyrin: importance of ascorbic acid, ergothionine and glutathione, *Br. J. Haematol.,* 5, 150–157, 1959.

91. **Penny, J. R. and Balfour, B. M.,** Effect of vitamin C on mucopolysaccharide production in wound healing, *J. Pathol. Bacteriol.,* 61, 171–178, 1949.

92. **Krumdieck, C. and Butterworth, C. E., Jr.,** Ascorbate-cholesterol-lecithin interactions. Factors of potential importance in the pathogenesis of arteriosclerosis, *Am. J. Clin. Nutr.,* 27, 866–876, 1974.

93. **Friederici, H. H. R., Taylor, H., Rose, R., and Pirani, C. L.,** The fine structure of the capillaries in experimental scurvy, *Lab. Invest.,* 15, 1442–1458, 1966.

94. **Gore, I., Wada, M., and Goodman, M. L.,** Capillary hemorrhage in scorbutic guinea pigs – an electron microscopic study, *Arch. Pathol.,* 85, 493–502, 1968.

95. **Thaete, L. G. and Grim, J. N.,** Fine structural effects of L-ascorbic acid buccal epithelium, *Am. J. Clin. Nutr.,* 27(7), 719–727, 1974.

96. **Crandon, J. H., Lennihan, R., Jr., Mikal, S., and Reif, A. E.,** Ascorbic acid economy in surgical patients, *Ann. N.Y. Acad. Sci.,* 92, 246–267, 1961.

97. **Crandon, J. H., Landau, B., Mikal, S., Balmanno, J., Jefferson, M., and Mahoney, N.,** Ascorbic acid economy in surgical patients as indicated by blood ascorbic acid levels, *N. Engl. J. Med.,* 258, 105–113, 1958.

98. **Coon, W. W.,** Ascorbic acid metabolism in postoperative patients, *Surg. Gynecol. Obstet.,* 114, 522–534, 1962.

99. **Rogers, W. F. and Gardner, F. H.,** Tyrosine metabolism in human scurvy, *J. Lab. Clin. Med.,* 34, 1491–1501, 1949.

100. **Nakashima, Y., Suzue, R., Sanada, H., and Kawada, S.,** Effect of ascorbic acid on tyrosine hydroxylase activity in vivo, *Arch. Biochem. Biophys.,* 152(2), 515–520, 1972.

101. Ascorbate stimulation of tyrosine hydroxylase formation, *Nutr. Rev.,* 31(3), 93–94, 1973.

102. **Levine, S. C., Marples, E., and Gordon, H. H.,** A defect in the metabolism of tyrosine and phenylalanine in premature infants. Identification and assay of intermediate products, *J. Clin. Invest.,* 20, 199–207, 1941.

103. **Cooper, J. R.,** The role of ascorbic acid in the oxidation of tryptophan to 5-hydroxy-tryptophan, *Ann. N.Y. Acad. Sci.,* 92, 208–211, 1961.

104. **Hankes, L. V.,** Letter: Interrelationships of ascorbic acid and tryptophan metabolism, *Am. J. Clin. Nutr.,* 27(8), 770–771, 1974.

105. **Friedman, S. and Kaufman, S.,** 3,4-Dihydroxyphenylethylamine beta-hydroxylase: a copper protein, *J. Biol. Chem.,* 240, 552 pc-554pc, 1965.

106. **Levin, E. Y., Levenberg, B., and Kaufman, S.,** The enzymatic conversion of 3,4-dihydroxyl-phenylethylamine to norepinephrine, *J. Biol. Chem.,* 235, 2080–2086, 1960.

107. **Abboud, F. M., Hood, J., Hodges, R. E., and Mayer, H. E.,** Impairment of vascular responsiveness in patients with scurvy, *J. Clin. Invest.,* 49, 298–307, 1970.

108. **Lee, R. E.,** Ascorbic acid and the peripheral vascular system, *Ann. N.Y. Acad. Sci.,* 92, 295–301, 1961.

109. **Akers, R. P. and Lee, R. E.,** Nutritional factors in hemodynamics: importance of vitamin C in maintaining renal VEM systems, *Proc. Soc. Exp. Biol. Med.,* 82, 195–197, 1953.

110. **Pirani, C. L.,** Review: Relation of vitamin C to adrenocortical function and stress phenomena, *Metabolism,* 1, 197–222, 1952.

111. **Odumosu, A. and Wilson, C. W. M.,** The relationship between ascorbic acid concentration and cortisol production during the development of scurvy in the guinea pig, *Br. J. Pharmacol.,* 40, 548p–549p, 1970.

112. **Hodges, J. R. and Hotston, R. T.,** Suppression of adrenocorticotropic activity in the ascorbic acid deficient guinea pig, *Br. J. Pharmacol.,* 42, 595–602, 1971.

113. **Treager, H. S., Gabuzda, G. J., Zamchek, N., and Davidson, C. S.,** Response to adrenocortico-tropic hormone in clinical scurvy, *Proc. Soc. Exp. Biol. Med.,* 75, 517–520, 1950.

114. **Kitabchi, A. E. and Duckworth, W. C.,** Pituitary-adrenal axis evaluation in human scurvy, *Am. J. Clin. Nutr.,* 23, 1012–1014, 1970.

115. **Honjo, S., Takasaka, M., Fujiwara, T., Imaizumi, K., and Ogawa, H.,** Shigellosis in Cynomolgus monkeys (*Macaca irus*), *Jpn. J. Med. Sci. Biol.,* 22, 149–162, 1969.

116. **Scrimshaw, N., Taylor, C. E., and Gordon, J. E.,** Interactions of Nutrition and Infection, *W.H.O. Monogr. Ser.,* 57, Geneva 1968, 97–100.

117. **Kumar, M. and Axelrod, A. E.,** Circulating antibody formation in scorbutic guinea pigs, *J. Nutr.,* 98, 41–44, 1969.

118. **Hodges, R. E., Hood, J., Canham, J. E., Sauberlich, H. E., and Baker, E. M.,** Clinical manifestations of ascorbic acid deficiency in man, *Am. J. Clin. Nutr.,* 24, 432–443, 1971.

119. **Baker, E. M., Hodges, R. E., Hood, J., Sauberlich, H. E., March, S. C., and Canham, J. E.,** Metabolism of ^{14}C- and ^{3}H-labeled L-ascorbic acid in human scurvy, *Am. J. Clin. Nutr.,* 24, 444–454, 1971.

120. **Bronte-Stewart, B., Roberts, B., and Wells, V. M.,** Serum cholesterol in vitamin C deficiency in man, *Br. J. Nutr.,* 17, 61–68, 1963.

121. **Mumma, R. O.,** Ascorbic acid sulfate as a sulfating agent, *Biochim. Biophys. Acta,* 165, 571–573, 1968.

122. Ascorbic acid and the catabolism of cholesterol, *Nutr. Rev.,* 31, 154–156, 1973.

123. **Ginter, E.,** Vitamin C and plasma lipids, *N. Engl. J. Med.,* 294, 559–560, 1976.

124. **Elwood, P. C., Hughes, R. E., and Hurley, R. J.,** Ascorbic acid and serum cholesterol, *Lancet,* 2, 1197, 1970.

125. **Krumdieck, C. and Butterworth, C. E.,** Ascorbate-cholesterol-lecithin interactions: factors of potential importance in the pathogenesis of artherosclerosis, *Am. J. Clin. Nutr.,* 27, 866–876, 1974.

126. **Goldsmith, G. A.,** Human requirements for vitamin C and its use in clinical medicine, *Ann. N.Y. Acad. Sci.,* 92, 230, 1961.

127. **Schrauzer, G. N. and Rhead, W. J.,** Ascorbic acid abuse: effects on long term ingestion of excessive amounts on blood levels and urinary excretion, *Int. J. Vitam. Nutr. Res.,* 43(2), 201–211, 1973.

128. **Stein, H. B., Hasan, A., and Fox, I. H.,** Ascorbic acid induced uricosuria, *Ann. Intern. Med.,* 84, 385–388, April 1976.

Fat-soluble Vitamins

NUTRIENT DEFICIENCIES IN ANIMALS AND MAN: VITAMIN A

Compiled by Donald S. McLaren -

Table 1
LESIONS PRODUCED IN ANIMALS BY VITAMIN A DEFICIENCY

Abnormality	Animal
General	
Failure of appetite	Rat, fowl
Cessation of growth	Rat, fowl
Xerosis of membranes	Rat, germfree rat,[2] fowl
Decline in body weight	Rat, fowl
Infections	Rat, fowl
Death	Rat, fowl
Eyes	
Impairment of dark adaptation	Farm animals, rat
Electroretinogram abnormalities	Rat[3]
Photoreceptor damage	Rat,[3] house fly,[4] mosquito,[5] moth[6]
Xerophthalmia	Rat
Keratomalacia	Rat
Lens epithelium keratinization	Rat[7]
Optic nerve constriction, papilledema	Bovine (calves, negative older heifers)[8]
Respiratory system	
Metaplasia of nasal passages	Fowl
Pneumonia	Rat
Lung abscess	Rat
Intestinal tract	
Metaplasia of forestomach	Rat
Enteritis	Rat
Volvulus	Germfree rat[2]
Urinary tract	
Thickening of bladder wall	Rat, germfree rat[2]
Cystitis	Rat
Urolithiasis	Rat
Pyelitis	Rat, fowl
Nephrosis	Rat
Pus in ureters	Rat
Liver	
Metaplasia of bile ducts	Rat
Degeneration of Kupffer cells	Rat
Nervous system	
Incoordination	Rat, germfree rat,[2] bovine, pig
Paresis	Rat, pig
Nerve degeneration	Rat, dog, rabbit, bovine, pig, fowl[9]
Constriction at foramina	Bovine
Twisting of nerve	Bovine, rat
Hydrocephalus	Rabbit
Raised cerebrospinal fluid pressure	Bovine, pig[10]
Bone and teeth	
Defective modeling	Dog, bovine
Cancelous bone	Dog
Restriction of brain cavity	Dog
Narrowing of foramina	Bovine, dog
Enamel and dentinal hypoplasia	Rat[11]

Table 1 (continued)
LESIONS PRODUCED IN ANIMALS BY VITAMIN A DEFICIENCY

Abnormality	Animal
Reproductive system	
Degeneration of testis	Rat, germfree rat,[2] bovine[1 2]
Abnormal estrous cycle	Rat
Resorption of fetus	Rat
Congenital abnormalities	
Anophthalmos	Pig, rat
Microphthalmos	Pig, rat
Cleft palate	Pig, rat
Aortic arch deformities	Rat
Diaphragmatic defects	Rat
Kidney deformities	Rat
Hydrocephalus	Rabbit
Umbilical artery defects	Rat[1 3]
Cranial development delayed	Rat[1 4]
Odontogenesis delayed	Rat[1 4]
Choroid plexus defects	Rabbit[1 5]
Miscellaneous	
Cystic pituitary	Bovine
Impaired resistance to parasites	Chicken, rat
Untidy hair or feathers	Rat, germfree rat,[2] farm animals, birds
Decreased elasticity of lungs and aorta	Rabbit[1 6]
Hemoconcentration	Rat,[1 7] chicken[1 8]
Atrophy of thymus and spleen	Rat[1 9]

Note: Arranged mainly according to the organ or function affected. The animals mentioned are those in which the abnormality has been most extensively studied. The occurrence of the abnormality in other animals is not precluded.

Adapted from Moore, T., in *The Vitamins: Effect of Vitamin A Deficiency in Animals* Academic Press, New York, 1967, 248, with additional evidence referenced.

REFERENCES

1. **Moore, T.,** Effects of vitamin A deficiency in animals, in *The Vitamins: Chemistry, Physiology, Pathology, Methods,* Vol. 1, 2nd ed., Sebrell, W. H., Jr. and Harris, R. S., Eds., Academic Press, New York, 1967, 245–266.
2. **Rogers, W. E., Jr., Bieri, J. G., and McDaniel, E. G.,** Vitamin A deficiency in the germfree state, *Fed. Proc. Fed. Am. Soc. Exp. Biol.,* 30, 1773–1778, 1971.
3. **Dowling, J. E. and Wald, G.,** The role of vitamin A acid, *Vitam. Horm.,* 18, 515–541, 1960.
4. **Goldsmith, T. H., Barker, R. J., and Cohen, C. F.,** Sensitivity of visual receptors of carotenoid-depleted flies: a vitamin A deficiency in an invertebrate, *Science,* 146, 65–67, 1964.
5. **Brammer, J. D. and White, R. H.,** Vitamin A deficiency: effect on mosquito eye ultrastructure, *Science,* 163, 821–823, 1969.
6. **Carlson, S. D., Gemne, G., and Robbins, W. E.,** Ultrastructure of photoreceptor cells in a vitamin A-deficient moth (*Manduca sexta*), *Experientia* (Basel), 25, 175–177, 1969.
7. **Pirie, A. and Overall, M.,** Effect of vitamin A deficiency on the lens epithelium of the rat, *Exp. Eye Res.,* 13, 105–109, 1972.
8. **Hayes, K. C., Nielsen, S. W., and Eaton, H. D.,** Pathogenesis of the optic nerve lesion in vitamin A-deficient calves, *Arch. Ophthalmol.,* 80, 777–787, 1968.

9. Howell, J. M., Pitt, G. A. J., and Thompson, J. N., The development of lesions in vitamin A-deficient adult fowl, *Br. J. Exp. Pathol.*, 50, 181–186, 1969.

10. Sanger, V. L., Dehority, B. A., Grifo, A. P., and Nelson, E. C., A comparison of atlanto-occipital and lumber spinal fluid pressure measured simultaneously in adequate and vitamin A-deficient pigs, *Am. J. Vet. Res.*, 25, 1376–1379, 1964.

11. Conne, P. and Baume, L. J., Calcification de l'email et de la dentine de l'incisive du rat blanc carence en vitamine A: etude comparative histologique et microradiographique, *J. Can. Dent. Assoc.*, 35, 588–598, 1969.

12. Ghannam, S., Shehata, O., Deeb, S., and Al-Alily, H., The effect of vitamin A depletion on the vasa deferentia of young bulls, *Res. Vet. Sci.*, 10, 79–82, 1969.

13. Monie, I. W. and Khemmani, M., Absent and abnormal umbilical arteries, *Teratology*, 7, 135–141, 1973.

14. Baume, L. J., Franquin, J.-C., and Körner, W. W., The prenatal effects of maternal vitamin A deficiency on the cranial and dental development of the progeny, *Am. J. Orthod.*, 62, 447–460, 1972.

15. Witzel, E. W. and Hunt, G. M., The ultrastructure of the choroid plexus in hydrocephalic offspring from vitamin A deficient rabbits, *J. Neuropathol. Exp. Neurol.*, 21, 250–262, 1962.

16. Saksena, J. S., Mehta, J. A., and Naimark, A., The effect of vitamin A deficiency in rabbits on the elastic properties of the lung and thoracic aorta, *Can. J. Physiol. Pharmacol.*, 49, 127–133, 1971.

17. McLaren, D. S., Tchalian, M., and Ajans, Z. A., Biochemical and hematologic changes in the vitamin A deficient rat, *Am. J. Clin. Nutr.*, 17, 131–138, 1965.

18. Nockels, C. F. and Kienholz, E. W., Influence of vitamin A deficiency on testes, bursa fabricus, adrenal and hematocrit on cockerels, *J. Nutr.*, 92, 384–388, 1967.

19. Krishnan, S., Bhuyan, U. N., Talwar, G. P., and Ramalingaswami, V., Effect of vitamin A and protein-calorie undernutrition on immune responses, *Immunology*, 27, 383–392, 1974.

Table 2
LESIONS PRODUCED IN MAN
BY VITAMIN A DEFICIENCY

Eye and related structures[1,2]
 Night blindness (hemeralopia, nyctalopia)
 Impaired dark adaptation
 Defects in rod scotometry[3]
 Altered electroretinogram[4]
 Dyschromatopsia[2]
 Xerophthalmia[5]
  Xerosis of the bulbar conjunctiva (dryness, thickening, wrinkling, unwetability, pigmentation, diminished sensitivity)
  Bitot's spot (superficial patch of frequently foamy material in interpalpebral fissure nearly always temporal, usually bilateral, consisting of inspissated mucin and keratinized conjunctival epithelial cells; see also Table 3)
  Xerosis of the cornea (dryness, thickening, infiltration of stroma, unwetability, diminished sensitivity)
  Xerosis of the cornea with ulcer (any breach in the continuity of the epithelium and stroma)
  Keratomalacia (liquefaction, partial or total; single or multiple of cornea. Destruction and deformity result in impaired vision)

TABLE 2 (Continued)
LESIONS PRODUCED IN MAN
BY VITAMIN A DEFICIENCY

Sequelae (corneal scars; fine (nebula), medium (macula), dense (leucoma); total or partial, usually inferior and central; phthisis bulbi, ectasia of cornea, staphyloma, extrusion of lens)

Xerophthalmic fundus[6]

Enlarged pores of Meibomian glands

Dry skin of eyelids

Diminished lacrimation

Other effects

Perifollicular hyperkeratosis (phrynoderma, toad skin)

Increased mortality[1,7]

Increased susceptibility to infections,[8] e.g., measles.[9]

Urolithiasis?

Impaired mental development?[10]

REFERENCES

1. **Oomen, H. A. P. C.,** An outline of xerophthalmia, *Int. Rev. Trop. Med.,* 1, 131–213, 1961.
2. **McLaren, D. S.,** *Malnutrition and the Eye,* Academic Press, New York, 1963.
3. **Hume, E. M. and Krebs, H. A.,** *Med. Res. Counc. G.B. Spec. Rep. Ser.,* No. 264, His Majesty's Stationery Office, London, 1949.
4. **Dhanda, R. P.,** Electroretinography in night blindness and other vitamin A deficiencies, *Arch. Opthalmol.,* 54, 841–849, 1955.
5. **McLaren, D. S., Oomen, H. A. P. C., and Escapini, H.,** Ocular manifestations of vitamin-A deficiency in man, *Bull. WHO,* 34, 357–361, 1966.
6. **Teng-Khoen-Hing,** Fundus changes in hypovitaminosis A, *Ophthalmologica,* 137, 81–85, 1959.
7. **McLaren, D. S., Shirajian, E., Tchalian, M., and Khoury, G.,** Xerophthalmia in Jordan, *Am. J. Clin. Nutr.,* 17, 117–130, 1965.
8. **Scrimshaw, N. S., Taylor, C. E., and Gordon, J. E.,** Interactions of nutrition and infection, *WHO Monogr. Ser.,* No. 57, Geneva, 1968.
9. **Franken, S.,** Measles and xerophthalmia in East Africa, *Trop. Geogr. Med.,* 26, 39–44, 1974.
10. **Pek Hien Liang, Tjiook Tiauw Hie, Oey Henk Jan, and Lauw Tjin Giok,** Evaluation of mental development in relation to early malnutrition, *Am. J. Clin. Nutr.,* 20, 1290–1294, 1967.

Table 3
CHARACTERISTICS OF BITOT'S SPOTS RESPONSIVE
AND UNRESPONSIVE TO VITAMIN A

Responsive	Unresponsive
Preschool age group	Older children and adults
Other signs of deficiency present	Isolated lesion, no other signs
Plasma vitamin A low	Plasma vitamin A normal
Usually quite large	Frequently small
Often multiple	Usually single
Relatively short duration	Prolonged duration

Table 4
XEROPHTHALMIA CLASSIFICATION

Primary Signs

1A	Conjunctival xerosis
1B	Bitot's spot with conjunctival xerosis
2	Corneal xerosis
3A	Corneal ulceration with xerosis
3B	Keratomalacia

Secondary Signs

N	Night blindness
F	Xerophthalmia fundus
S	Scars

Note: These signs are descriptive rather than diagnostic. All signs seen at the time of examination are recorded. In general, a progression of severity is reflected in the classification of primary signs. The classification can be used both in field surveys and the routine recording of findings in patients in hospitals and clinics. When tabulating the frequency of these signs each child should be included only one, under his or her most severe sign. Only those Bitot's spots accompanied by conjunctival xerosis, usually in the 0 to 5 years age group, are indicative of vitamin A deficiency. This xerosis may be hidden by the overlying foam of the Bitot's spot, and only revealed when this is rubbed away. Secondary signs often occur in association with, or result from, vitamin A deficiency, and should be noted separately.

From Report of a joint WHO/USAID meeting, *Vitamin A Deficiency and Xerophthalmia,* WHO Tech. Rep. Ser., No. 590, 1976. With permission.

Table 5
PROPOSED CRITERIA FOR COMMUNITY DIAGNOSIS OF XEROPHTHALMIA AND VITAMIN A DEFICIENCY

Clinical		
X1B	$\geqslant$	2.0%
X2 + X3A + X3B	$\geqslant$	0.01%
XS	$\geqslant$	0.1%
Biochemical		
Plasma vitamin A less than 10 μg/100 ml	$\geqslant$	5.0%

Note: The above point prevalence rates are suggested. In a vulnerable population the presence of any one or more of the three clinical criteria should be considered as evidence of a xerophthalmia problem. The biochemical criterion is only indicative of significant vitamin A deficiency and may be used alone if the objective is to improve vitamin A nutriture. The value of the corroborative evidence of more than one indicator has to be stressed.

From Report of a joint WHO/USAID meeting, *Vitamin A Deficiency and Xerophthalmia,* WHO Tech. Rep. Ser., No. 590, 1976. With permission.

Table 6
BIOCHEMICAL ABNORMALITIES
IN VITAMIN A DEFICIENCY

Liver stores of vitamin A (mainly retinyl palmitate) depleted

Plasma levels of vitamin A [mainly retinol as holo-retinol binding protein (RBP)] less than 20 μg/100 ml

Plasma levels of retinol binding protein (apo-RBP) lowered[1]

Liver concentration of apo-RBP two to three times normal[1]

Rhodopsin and opsin levels in retina reduced[2]

Lysosomal arylsulfatases A and B increased[3]

Glycoprotein synthesis impaired[4]

Glycolipid synthesis impaired[5]

REFERENCES

1. **Muto, Y., Smith, J. E., Milch, P. O., and Goodman, D. S.,** Regulation of retinol-binding protein metabolism by vitamin A status in the rat, *J. Biol. Chem.,* 247, 2542—2550, 1972.
2. **Dowling, J. E. and Wald, G.,** The role of vitamin A acid, *Vitam. Horm.,* 18, 515—541, 1960.
3. **Guha, A. and Roels, O. A.,** The influence of α-tocopherol on arylsulfatases A and B in the liver of vitamin A-deficient rats, *Biochim. Biophys. Acta,* 111, 364—374, 1965.
4. **DeLuca, L., Schumacher, M., and Wolf, G.,** Biosynthesis of a fucose-containing glycopeptide from rat small intestine in normal and vitamin A-deficient conditions, *J. Biol. Chem.,* 245, 4551—4558, 1970.
5. **DeLuca, L., Maestri, N., Rosso, G., and Wolf, G.,** Retinol glycopeptides, *J. Biol. Chem.,* 248, 641—648, 1973.

Table 7
HUMAN DISORDERS IN WHICH SECONDARY (ENDOGENOUS) VITAMIN A DEFICIENCY HAS BEEN REPORTED

Celiac disease

Cystic fibrosis

Gastrectomy[1]

Carotene-conversion enzyme deficiency[2]

A-beta-lipoproteinemia

Hyperthyroidism[3]

Protein-energy malnutrition (especially kwashiorkor, usually with primary deficiency)

Liver cirrhosis

Viral hepatitis[3]

REFERENCES

1. **Adams, J. F., Johnstone, J. M., and Hunter, R. D.,** Vitamin-A deficiency following total gastrectomy, *Lancet,* 1, 415—417, 1960.
2. **McLaren, D. S. and Zekian, B.,** Failure of enzymatic cleavage of β-carotene, *Am. J. Dis. Child.,* 121, 278—280, 1971.
3. **Smith, F. R. and Goodman, D. S.,** The effects of diseases of the liver, thyroid, and kidneys on the transport of vitamin A in human plasma, *J. Clin. Invest.,* 50, 2426—2436, 1971.

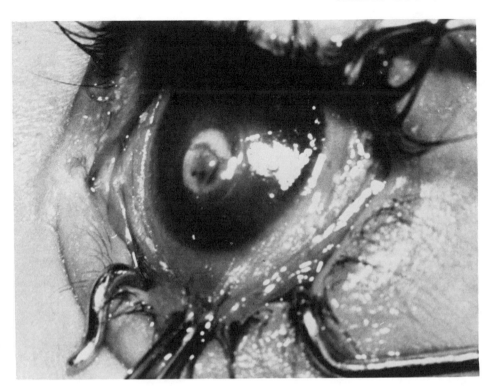

FIGURE 1. Xerosis of the conjunctiva (X1A) with early central keratomalacia (X3B). The bulbar conjunctiva is markedly thickened, wrinkled, and dry. (From McLaren, D. S., Shirajlan, E., Tchalian, M., and Khoury, G., *Am. J. Clin. Nutr.*, 17, 117, 1965. With permission.)

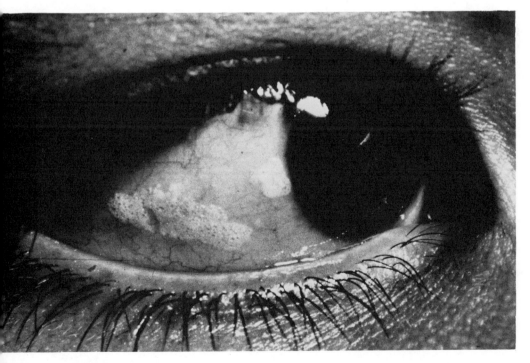

URE 2. Bitot's spots with xerosis of the conjunctiva (X1B). There are several patches of superficial foamy erial on the dry bulbar conjunctiva. (From McLaren, D. S., Shirajian, E., Tchalian, M., and Khoury, G., *Am. J. Clin. :., 17, 117, 1965. With permission.)

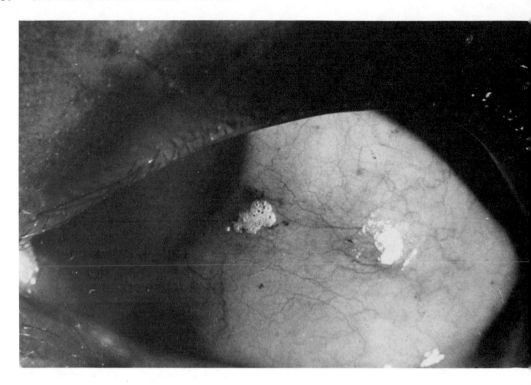

FIGURE 3.   Bitot's spot without xerosis of the conjunctiva. The conjunctiva is normal, except for slight hypere, and focal pigmentation related strictly to the spot, suggesting local conjunctival pathology and *not* systemic (du vitamin A deficiency). (From McLaren, D. S., *Malnutrition and the Eye,* Academic Press, New York, 1963. W permission.)

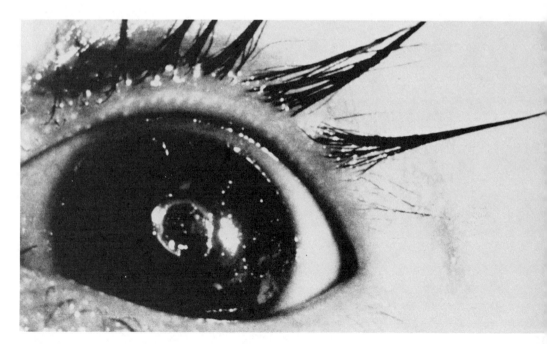

FIGURE 4.   Corneal ulceration with xerosis (X3A).   Early colliquative necrosis characteristically, as here, affec central area of the cornea where nutrient supply from the periphery is poorest.   Ulceration due to infection pro a marked inflammatory reaction, whereas here the eye otherwise is remarkably quiet. (From McLaren, D. S., Oom A. P. C., and Escapini, H., *Bull. WHO,* 34, 357, 1966.   With permission.)

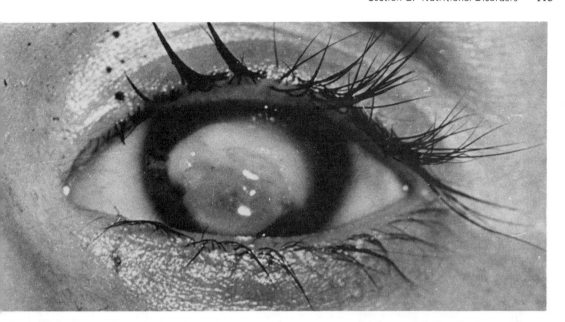

FIGURE 5.   Keratomalacia (X3B). In this case the central area of colliquative necrosis has affected the entire thickness of the cornea and the lens has prolapsed. (From McLaren, D. S., Shirajian, E., Tchalian, M., and Khoury, G., *Am. J. Clin. Nutr.*, 17, 117, 1965. With permission.)

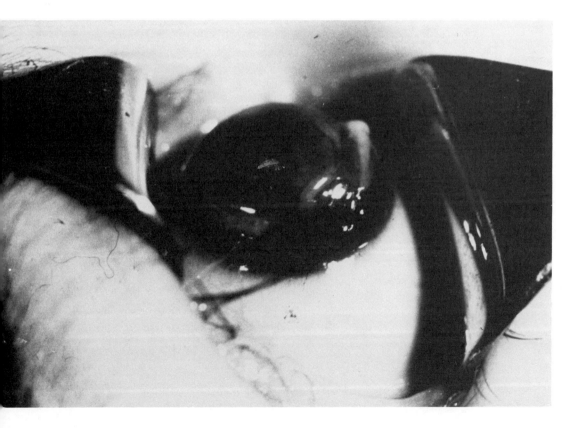

FIGURE 6.   Keratomalacia (X3B). The entire cornea is undergoing liquefaction. The conjunctiva is normal, as it frequently is when the corneal changes are most advanced and develop most rapidly. (From McLaren, D. S., *Malnutrition and the Eye*, Academic Press, New York, 1963. With permission.)

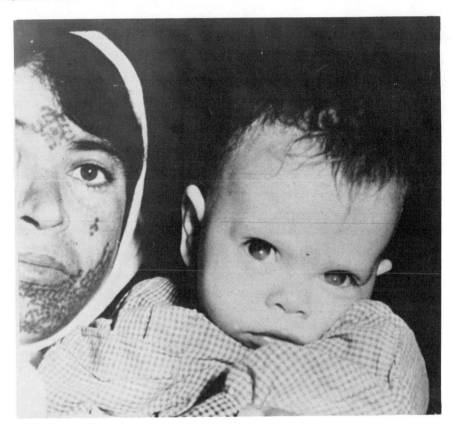

FIGURE 7. Scars (XS). The corneal sequelae of subtotal keratomalacia are frequently bilateral leucomata occupying a lower central position, as in this child. (From Report of a WHO/USAID meeting, WHO Tech. Rep. Ser., No. 590, Geneva, 1976. With permission.)

EFFECT OF NUTRIENT DEFICIENCIES IN MAN: VITAMIN D

H. E. Harrison

Vitamin D is an atypical nutrient in that it is not available in appreciable quantities in most of the foods usually eaten by infants and children. Most importantly, the concentration of vitamin D in human milk is so low that the food which may provide all of the nutrition for infants during the first months of life is deficient in this essential factor (Table 1), unless the mothers are receiving a high intake of vitamin D. The answer lies in the fact that vitamin D is essentially a hormone rather than a nutrient. Biologically, the source of vitamin D for man should be the conversion of 7-dehydrocholesterol in the skin to cholecalciferol. vitamin D_3, by absorption of the energy of ultraviolet radiation in the wavelengths betwwen 250 and 305 nm (Figure 1). When man changed from a hunting and gathering culture to an agricultural one, he also moved from tropical and subtropical climates to temperate zones in which protection from the cold by houses and clothing was necessary for a considerable part of the year. This limited the production of vitamin D in the skin, particularly of infants during those months when exposure of large areas of skin to sunshine was not possible. When a further change to urban living occurred, infants and children in crowded city slums had little opportunity for exposure to sunshine. This was accentuated by the industrial revolution which resulted from the use of fossil fuels (originally coal and then oil) for energy for the driving of machinery as well as for heating. The atmosphere, contaminated by the smoke of burning coal, absorbed most of the shortwave UV energy of sunshine so that little reached the surface of the earth. As a result, rickets, the disease of children resulting from vitamin D deficiency, became widespread in the cities of the northern latitudes in the British Isles, Europe, the U.S., and Canada. Solution to the problem came following development of an animal model for rickets by Mellanby in 1918.[1] This resulted in the clear-cut demonstration of a factor in cod liver oil which prevented or cured rickets in the puppy. Subsequently, McCollum and colleagues[2] were able to produce rickets in the albino rat and show that the antirachitic factor in cod-liver oil was separate from the previously discovered vitamin A, in which cod-liver oil is also rich. They called the antirachitic substance vitamin D.

The role of sunshine in the prevention of rickets had already been determined as the result of epidemiologic studies of the geographic distribution of rickets. The development of the quartz lamp for heliotherapy made it possible to study the effects of artificially

Table 1
AMOUNTS OF VITAMIN D
AVAILABLE IN FOODS
INCLUDING FISH LIVER OIL

Human milk	0–10 U/100 ml
Cow's milk	0.3–4 U/100 ml
Butter	35 U/100 g
Egg yolk	25 U/average yolk
Calf liver	15 U/100 g
Herring	1500 U/100 g
Mackerel	1800 U/100 g
Canned salmon	300 U/100 g
Canned tuna	250 U/100 g
Canned sardines	600 U/100 g
Cod-liver oil	175 U/g

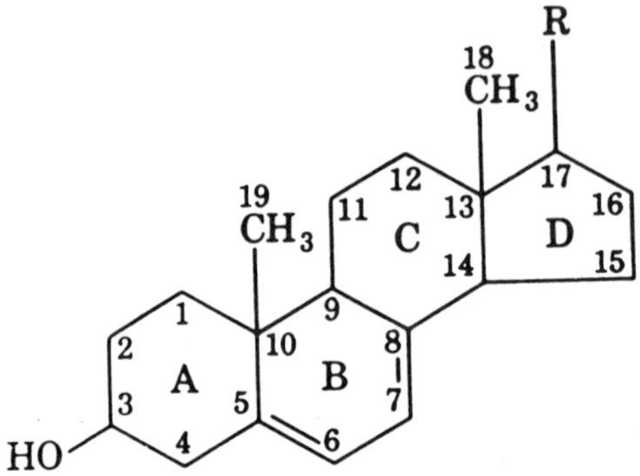

FIGURE 1.   (A) Ring structure of provitamin D. The various forms of provitamin D differ in the side chain R as pictured in Figure 2. (B) Structure of vitamin D produced by absorption of shortwave UV energy by provitamin D with opening of B ring.

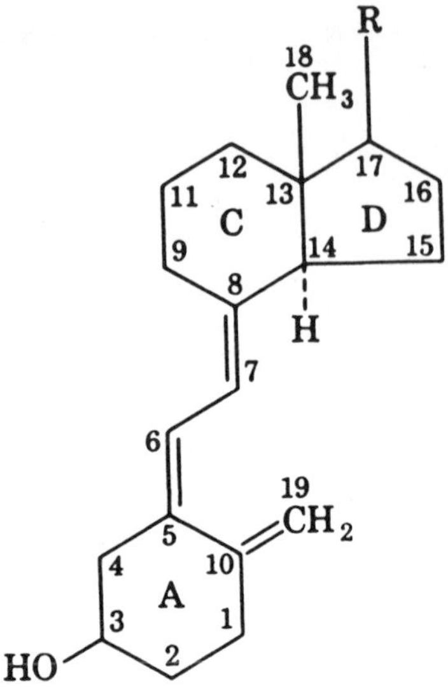

produced shortwave UV radiation on the course of rickets, and it was found that exposure of rachitic infants to such irradiation was curative.[3] These observations were the basis of the further studies of Hess et al.[4] and Steenbock et al.[5] which proved that substances in the nonsaponifiable fat fraction of foods could be given vitamin D activity by irradiation with UV light. From this came further information proving that the substances with provitamin D properties were sterols with a particular configuration of the B ring, namely double bonds at the 5,6 and 7,8 positions. Following absorption of UV radiation energy by these sterols, the B ring is opened with the formation of vitamin D. The two important provitamin D sterols are ergosterol, found in considerable concentrations in yeast and other plant cells, and 7-dehydrocholesterol, the animal provitamin D found in the skin of mammals and the secretion of the preen gland of birds. The two major forms of vitamin D are ergocalciferol (vitamin D_2), which is derived from ergosterol, and cholecalciferol (vitamin D_3), which is derived from 7-dehydrocholesterol.[6] Figure 2 shows the various forms of vitamin D which differ only in the structure of the side chain.

Vitamin D activity can be expressed in units based upon a bioassay in vitamin D-deficient rats or vitamin D-deficient chicks. When assayed in the rachitic rat, ergocalciferol and cholecalciferol are equally active with a potency of 40,000 units/mg of pure steroid. However, in the chick and other birds, ergocalciferol has only about one tenth the activity of cholecalciferol. Of the mammals studied, only the New World monkey has been shown to resemble the bird in discriminating between ergocalciferol and cholecalciferol. Man does not show this discrimination so that either ergocalciferol or cholecalciferol can be used in equivalent amounts for the prevention or treatment of rickets and osteomalacia.

In the past decade, important information has been developed concerning the metabolism of vitamin D, which we shall call the vitamin D cycle. Both cholecalciferol, produced in the skin by UV energy, and ergocalciferol, provided in food, go through the same metabolic conversions. These will be described using vitamin D as the generic term for steroids of the calciferol configuration. Vitamin D is not actually the active molecule.

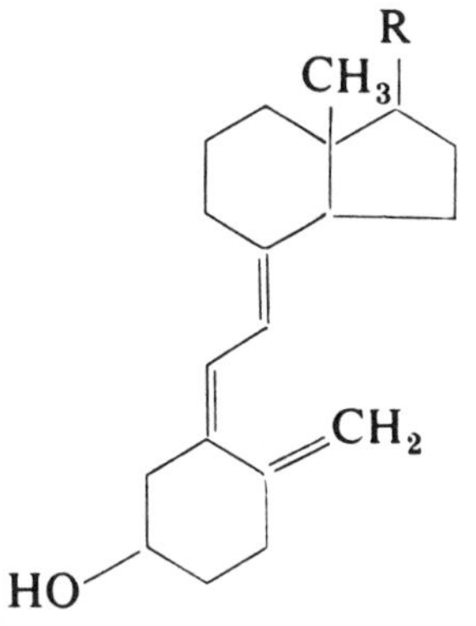

R

CH_3

CH_2

HO

(2)

Vitamin D$_2$ (2a) $R = CHCH=CHCHCH$ with CH_3, CH_3, CH_3, CH_3 substituents

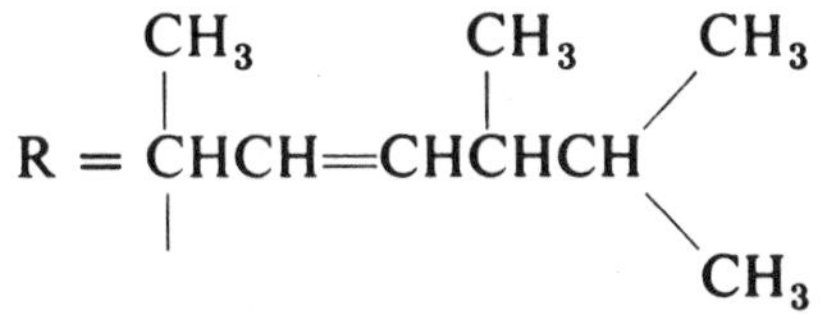

Vitamin D$_3$ (2b) $R = CHCH_2CH_2CH_2CH$ with CH_3, CH_3, CH_3 substituents

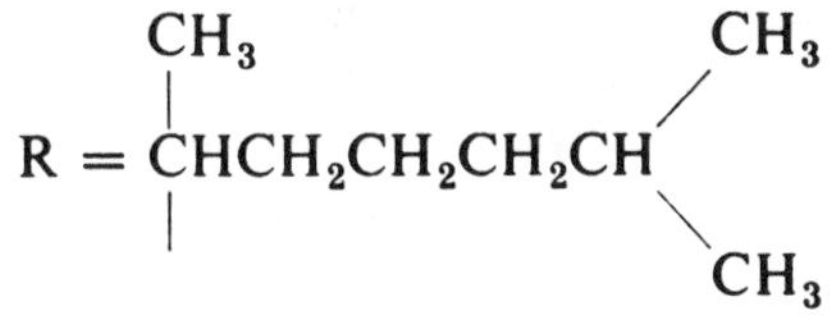

Vitamin D$_4$ (2c) $R = CHCH_2CH_2\overset{*}{C}HCH$ with CH_3, CH_3, CH_3, CH_3 substituents

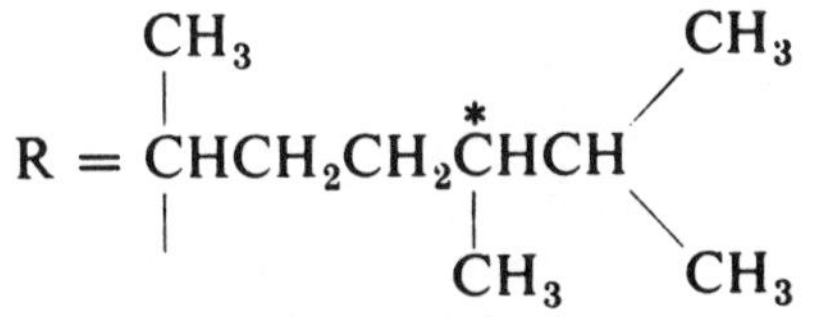

FIGURE 2.   Various forms of vitamin D associated with modifications of side chain R.

The first transformation occurs in the liver, where a microsomal system hydroxylates the 25 carbon in the side chain to produce 25-OH vitamin D[7] (Figure 3). This is probably an active compound but becomes much more active when converted to 1,25-diOH vitamin D (Figure 3) by a mitochondrial system in the kidney. 1,25-diOH vitamin D has been termed the active vitamin D hormone.[8] The liver 25-hydroxylase in vivo does not seem to be under effective negative feedback control since serum 25-OHD concentrations can be determined and values many times the normal concentrations can be obtained when doses of vitamin D of 10 to 20 times the physiologic amount are given. However, in vitro studies have suggested that the liver microsomal vitamin D 25-hydroxylase is subject to product inhibition. The kidney 25-hydroxy vitamin D 1-hydroxylase is controlled since the activity of this system is dependent upon parathyroid hormone and inorganic phosphate concentration.[9] In the absence of parathyroid hormone or in the presence of a high concentration of phosphate, the formation of 1,25-diOH vitamin D is markedly reduced and formation of 24,25-diOH vitamin D is increased. The latter is relatively inactive although it can subsequently be converted to 1,24,25-triOH vitamin D which has physiologic activity. The physiologic control of the kidney 1-hydroxylase enables adaptation of the subject to varying intakes of calcium and phosphate. A low intake of

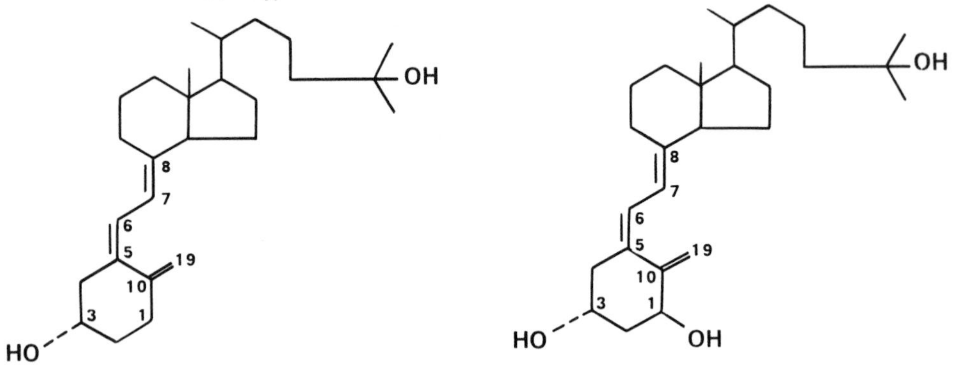

25-OH-cholecalciferol
(3-β-OH; down)

1,25-(OH)₂-cholecalciferol
(3-β-OH; down)
(1-α-OH; up)

FIGURE 3.   Structural configuration of the vitamin D metabolites, 25-OH vitamin D and 1,25-diOH vitamin D.

calcium stimulates parathyroid hormone output since the parathyroid glands are sensitive to a slight reduction in ambient calcium concentration. The increased parathyroid hormone concentration stimulates formation of 1,25-diOH vitamin D and, as will be discussed below, a major physiologic effect of this hormone is to increase the efficiency of calcium absorption by the intestine. This reaction allows adaptation to low calcium diets. Conversely, a high calcium intake suppresses parathyroid hormone output and thus inhibits formation of 1,25-diOH vitamin D, resulting in reduced uptake of calcium from the intestine. The effect of extracellular inorganic phosphate concentration may also be of adaptive significance since formation of 1,25-diOH vitamin D is increased in the hypophosphatemic subject and depressed by hyperphosphatemia. The physiologic actions of 1,25-diOH vitamin D include increase of phosphate transport by the intestine and augmented renal tubular reabsorption of phosphate, both resulting in increase of serum phosphate concentrations so that the physiologic benefit of stimulation of 1,25-diOH vitamin D formation in the hypophosphatemic state is clear.

The physiologic action of 1,25-diOH vitamin D in man requires only 0.5 to 1 μg/day. 25-OH vitamin D may also be physiologically active since its effect can be demonstrated in the anephric subject, but in much higher amounts than 1,25-diOH vitamin D.[10] Vitamin D in high dosage is also active in the anephric subject but presumably only following hepatic conversion to the 25-OH derivative.[11] 25-OH vitamin D circulates in the plasma complexed with a specific protein, 25-OH vitamin D binding protein, which moves electrophoretically with the α1 to α2 globulin fraction. 1,25-diOH vitamin D presumably is also bound by this protein.[12] A purification of this protein has been accomplished and it is a globular protein with α1 mobility and a molecular weight of approximately 50,000.[12] 1,25-diOH vitamin D is taken up by intestinal mucosal cells and is bound by the nuclear chromatin. There is evidence that this initiates the activation of a specific DNA segment with resulting increase of mRNA production and synthesis of specific proteins in the intestinal mucosa through which the intestinal actions of vitamin D are mediated, namely increased intestinal transport of calcium and of phosphate. 25-OH vitamin D and 1,25-diOH vitamin D, as mentioned, also act on renal tubular epithelium to increase renal tubule reabsorption of phosphate. They also affect the bone cells which respond to the vitamin D hormone by increased metabolic activity which causes solubilization of bone mineral and release of calcium and phosphate ions into the extracellular fluid. The net effect of 1,25-diOH vitamin D is to increase the concentration of both calcium and phosphate ions in extracellular fluid which, under conditions of normal intake of calcium and phosphate, results in a maximal rate of deposition of bone mineral in the matrix of cartilage and bone.

Deficiency of vitamin D or specifically of 1,25-diOH vitamin D results in intestinal malabsorption of calcium and phosphate. The resultant calcium deficiency stimulates the parathyroid gland cells so that parathyroid hyperplasia and hypersecretion are secondary features of vitamin D deficiency. Despite the secondary hyperparathyroidism, hypo-calcemia may persist since the calcemic effect of parathyroid hormone is partly dependent on the presence of vitamin D. Hypophosphatemia is also characteristic of vitamin D deficiency due to both impaired absorption of dietary phosphate and reduced renal tubular reabsorption of phosphate. The latter is caused both by deficiency of 1,25-diOH vitamin D and by the secondary excess of parathyroid hormone which inhibits renal tubular reabsorption of phosphate. As the hypophosphatemia becomes more marked, the serum calcium concentration rises toward the normal range, possibly because of diminished precipitation of calcium phosphate in the bone. At this stage the bones are deficient in mineral and show the lesions of rickets and osteomalacia which are an excess of unmineralized matrix, a low mineral content per fat-free dried weight, and deformities due to the lessened structural stability of the poorly mineralized bone particularly at the actively growing metaphyseal ends of the long bones. In the growing bone these lesions are termed rickets; in the adult bone, they are termed osteomalacia.

As the UV energy of sunshine cannot be depended upon to meet the vitamin D requirements of infants and children, vitamin D is ordinarily supplied as a nutrient. This can be either in the form of ergocalciferol or cholecalciferol. In the rapidly growing infant the vitamin D requirement can be assessed in terms of the amount needed to prevent radiological evidences of rickets or changes in serum composition resulting from vitamin D deficiency, namely reduction of serum calcium and phosphate concentrations and elevation of serum alkaline phosphatase. Additional criteria which have been used are calcium and phosphate balances and linear growth. On the basis of many studies, it has been concluded that 400 units of vitamin D, 10 μg daily, will protect practically all infants and children from rickets and promote maximum calcium and phosphate balances and growth.[13] This amount provides a considerable margin of safety and in all likelihood as little as 2.5 μg or 100 units may be adequate for most infants. The amounts produced by skin exposure to sunshine cannot be easily determined but it must provide an additional amount of vitamin D when the infant can be put outdoors and when climatic conditions permit exposure of large areas of skin to sunshine. On the basis of the above requirements the fortification of homogenized cow's milk, evaporated milk and special infant feedings at the level of 400 units per equivalent quart (i.e., the amount providing the calorie intake of a quart of whole milk) was instituted in the U.S. in 1936.[14] Since this coincided with the trend away from prolonged breast feeding of infants, almost all infants fed cow's milk or foods based on cow's milk received their recommended daily allowance in their food. Breast-fed infants or infants unable to ingest cow's milk because of protein hypersensitivity or lactose intolerance require a vitamin D supplement usually given as an irradiated ergosterol concentrate in a dosage of 400 units/day.

The vitamin D requirement of children beyond the first 2 years of life and of adults has been more difficult to determine. It is known that vitamin D is required throughout life, even after growth has ceased. Women living in societies which prohibit married women from freely moving outside the home unless completely covered from head to foot have an appreciable incidence of osteomalacia. Elderly people who also may have restricted activity and diet are known to develop osteomalacia, particularly those living in cities of the northern latitudes as in England and Scotland. The contribution of outdoor exposure to sunshine has been difficult to assess in various populations, but there seems to be no question that active children and adults with ordinary outdoor activities, particularly during the summer months, can obtain their vitamin D needs through endogenous production of cholecalciferol in the skin. A more precise method is now available for measurement of vitamin D requirements through determination of the concentration of the vitamin D metabolite, 25-OH vitamin D, in the serum. This can be done by competitive protein-binding assays using proteins with a high binding capacity for 25-OH vitamin D in kidney cytosol or rat serum. Although normal values differ somewhat in various laboratories, it has been possible to define concentrations of 25-OH vitamin D below which vitamin D deficiency is likely. Utilizing such methods and populations surveys as diet vitamin D estimations, a minimum vitamin D intake of 2.5 μg or 100 units/day has been estimated to be required by the adult.[15] Therefore, there is no relationship between weight, metabolic rate, or surface area and vitamin D requirement since that of the adult is no greater than that of the young infant.

REFERENCES

1. **Mellanby, E.,** Experimental Rickets, Medical Research Council Special Report Series, No. 61, His Majesty's Stationary Office, London, 1921.
2. **McCollum, E. V., Simmonds, N., Becker, J. C., and Shipley, P. G.,** Studies on experimental rickets. XXI. An experimental demonstration of the existence of a vitamin which promotes calcium deposition, *J. Biol. Chem.*, 65, 97, 1925.
3. **Huldschinsky, K.,** Heilung von Rachitis durch künstliche Hohensonne, *Dtsch. Med. Wochenschr.*, 45, 712–713, 1919.
4. **Hess, A. F. and Weinstock, M.,** Antirachitic properties imparted to inert fluids and to green vegetables by ultraviolet irradiation, *J. Biol. Chem.*, 62, 201, 1924.
5. **Steenbock, H.,** The induction of growth promoting and calcifying properties in a ration by exposure to light, *Science*, 60, 224, 1924.
6. **Bills, C. E.,** The chemistry of vitamin D, *JAMA*, 110, 2150–2155, 1938.
7. **Blunt, J. W., Tanaka, Y., and DeLuca, H. F.,** The biological activity of 25-hydroxy cholecalciferol, a metabolite of vitamin D, *Proc. Natl. Acad. Sci. U.S.A.*, 61, 1503–1506, 1968.
8. **Kodicek, E.,** The story of vitamin D from vitamin to hormone, *Lancet*, 1, 325–329, 1974.
9. **DeLuca, H. F.,** The kidney as an endocrine organ for the production of 1,25-dihydroxyvitamin D, a calcium mobilizing hormone, *N. Engl. J. Med.*, 289, 359–365, 1973.
10. **Pavlovitch, H., Garabedian, M., and Balsan, S.,** Calcium-mobilizing effect of large doses of 25-hydroxycholecalciferol in anephric rats, *J. Clin. Invest.*, 52, 2656–2659, 1973.
11. **Harrison, H. E. and Harrison, H. C.,** Dihydrotachysterol: a calcium active steroid not dependent upon kidney metabolism, *J. Clin. Invest.*, 51, 1919–1922, 1972.
12. **Peterson, P. A.,** Isolation and partial characterization of a human vitamin D-binding plasma protein, *J. Biol. Chem.*, 246, 7748–7754, 1971.
13. **Jeans, P. C. and Stearns, G.,** The vitamin D requirement of the child, *Am. J. Dis. Child.*, 54, 189–190, 1937.
14. **Jeans, P. C.,** Vitamin D milk, *JAMA*, 106, 2066–2069, 1936.
15. **Dent, C. E. and Smith, R.,** Nutritional osteomalacia, *Q. J. Med.*, 38, 195–209, 1969.

VITAMIN DEFICIENCY SIGNS IN MAN: VITAMIN E

B. A. Underwood

INTRODUCTION

For a diverse list of man's symptoms, many claims of clinical benefit following supplementation with vitamin E are recorded, but few have stood the test of well-controlled scientific study.[1,2] Clinical and biochemical lesions from induced vitamin E deficiency in animals vary from species to species and have few direct parallels to symptoms seen in man which are proved to be associated with vitamin E depletion. The most universally recognized disease in animals due to vitamin E deficiency is muscular dystrophy, which responds to treatment with the vitamin.[3] Chronic depletion in animals leads to skeletal myopathies with similar gross pathology, irrespective of species but with no parallel consistently seen in skeletal muscles of men who have chemical vitamin E deficiency. Attempts to demonstrate improvement from vitamin E therapy with persons who have various forms of muscular dystrophy have failed.

Depleted animals have red cells that are vulnerable to hemolysis in vitro when exposed to dialuric acid,[4,5] and the red cells of depleted humans show a similar vulnerability to hydrogen peroxide.[6] Correlation of the in vitro test with plasma tocopherol levels is influenced by other factors, and a functional relationship between the in vitro test and shortened red cell survival time in vivo sometimes, but not invariably, occurs.[7]

The closest parallel demonstrable in clinical symptoms of deficiency in animal species and man is deposition of the brown pigment ceroid in smooth muscle tissue following a prolonged period of inadequate levels of vitamin E. Tissue depletion may be the result of absence of vitamin E from the diet or chronic malabsorption.[8] Most believe that ceroid is derived from oxidation products of polyunsaturated fatty acids (PUFA) and copolymerization of the products with protein.[9,10] Only recently, Raychaudhuri and Desai attributed a possible clinical significance to ceroid deposition.[11] They found that rats with ceroid deposited uniformly throughout the uterus and fallopian tubes were infertile because the ceroid interfered with conception and implantation. A parallel in human populations has not been reported. Batten's disease, or neuronal ceroid lipofuscinosis, which is characterized by progressive mental retardation, seizures, and accumulation of large masses of lipopigment (ceroid) in the brain cells, occurs in children who have normal blood levels of vitamin E.[12]

Because vitamin E is present in a number of foods, particularly vegetable oils, whole cereal grains, and similar natural foods,[13,14] and is distributed extensively throughout the body,[15] adults rarely show evidence of inadequate nutriture except as associated with chronic malabsorption. On the other hand, newborns contain minimal levels in tissues (less than 2 to 3 mg/g)[15,16] and blood (less than 0.5 mg/dl),[17,18] possibly due to a placental barrier or to an immature gastrointestinal (G.I.) system for absorbing lipids.[18,19] The low blood level of vitamin E in newborns is related to low lipoprotein levels and is usually within the normal range when expressed as tocopherol per gram blood lipid. Blood and tissue levels of tocopherols rise to normal when the diet includes sources of the vitamin and absorptive efficiency of the G.I. tract matures. Because the primery biochemical role of the vitamin is thought to be that of an antioxidant, requirements are related to the PUFA content of the diet and tissues. The relationship is not direct, since both foods and tissues contain alternate lipid antioxidant and systems capable of scavenging free radicals.[20-22]

The "Elgin Project," 1954 to 1967, is the classical example of a depletion study in which a limited number of adult males were maintained on diets high in PUFA but containing little vitamin E.[23,24] During this period, plasma levels of tocopherol dropped to 0.1 mg/dl or less, but little clinical symptomatology occurred except for a slight decrease in red cell total lifespan.[25]

In some developing countries, dietary intake of vitamin E is low, and a high prevalence of reduced plasma levels (less than 0.5 mg/dl) has been reported.[26] These levels, however, are not usually associated with clinical symptoms or evidence of a compromised physiological function, with the exception of one report of ceroid deposits in the G.I. tract of the Thai-Lao ethnic group.[27] Attempts have been made to correlate low plasma levels of vitamin E (less than 0.5 mg/dl) in children who have protein-energy malnutrition (PEM) with an anemia.[28-30] In these studies, a hemolytic response to vitamin E supplement has not been consistently observed[31,32] and has not been related unequivocably to vitamin E insufficiency.[33]

Although vitamin E deficiency signs can be ameliorated in animals by several compounds that are structurally unrelated to tocopherol, tocopherols are the most commonly encountered, naturally occurring lipid antioxidant in the diet of man.[34] At the tissue level, the selenium-containing glutathione peroxidase and the peroxidase assayed with *p*-phenylenediamine, which is defective in Batten's disease,[35] may be effective physiological alternatives capable of augmenting or partially substituting for tocopherol.[7] There have been no clinical trials in man in which both vitamin E and selenium were intentionally deleted from the diet, to parallel animal studies,[36,37] and few human surveys in which the two nutrients and/or gluthathione peroxidase activity were assayed concurrently.

In view of the distribution of vitamin E in foods and throughout the body, it is not surprising that syndromes for which chemically proved vitamin E deficiency with associated symptomatology is reported are largely confined to low birth weight (LBW) and full-term infants fed formula diets low in vitamin E and rich in PUFA[38-41] and sometimes iron.[42-44] These syndromes are also reported in individuals who chronically malabsorb lipids.[8,45] In addition, a few persons with a rare genetic defect affecting betalipoprotein synthesis (abetalipoproteinemia) are remarkably depleted of vitamin E.[46]

PREMATURE AND FULL-TERM INFANTS

Normal premature and full-term newborns have low plasma tocopherol levels, and some show elevated in vitro hemolysis of red cells[47] but no clinical symptomatology.[17] When such infants are fed formulas high in PUFA, without vitamin E and sometimes containing iron, they may develop a variety of clinical symptoms including irritability, eczema, edema, and an abnormal blood picture that sometimes includes shortened red cell survival time. Efforts to relate symptoms observed in chemically deficient premature children with those seen in deficient animals have not been encouraging. This subject was extensively reviewed in a recent publication.[7] It should be noted that the plasma tocopherol and clinical symptoms in newborns are relieved by feeding vitamin E, but not all investigators have confirmed hematologic benefit in vivo as directly due to supplementation.[38,39] Since the syndrome was recognized as one induced by improperly balanced infant and formula diets,[49] commercial producers have made appropriate adjustments and most now include a vitamin E supplement. In addition, as a precautionary measure, some physicians recommend that supplemental vitamin E be routinely given during the first weeks of life even though it has not been unequivocably proved that tocopherol corrects the hematologic disturbances of prematurity.

ABETALIPOPROTEINEMIA

Abetalipoproteinemia is a genetic defect affecting a limited number of children with its profound effects on vitamin E metabolism. A complete lack of betalipoproteins in plasma occurs due to lack of a particular polypeptide fraction of low-density lipoprotein (LDL).[46] Vitamin E is normally transported from the intestine solubilized in the chylomicron, and it is transported in blood principally in the LDL fraction.[50-52] Children who, due to this genetic defect, are unable to synthesize either the lipoprotein component of the chylomicron carrier in muscosol cells or the LDL carrier in plasma are unable to absorb and transport the vitamin through the usual physiological systems.[46] Fortunately, this is a rare condition in man, described in fewer than 30 patients to date. Serum and red cell tocopherol levels are near exhaustion in affected children,[53-55] and red cells hemolyze spontaneously in vitro.[53,56,57] Autohemolysis is reversible by adding tocopherol in vitro or by supplementation of the diet with a water-miscible preparation of α-tocopheryl succinate. Continued supplementation with oral doses of α-tocopheryl succinate will increase red cells to normal levels but will not correct the abnormal phospholipid, phospholipid fatty acid, or aldehyde distribution[58] and will not maintain normal plasma tocopherol levels.[55] There is neither evidence for a hemolytic anemia nor reports of myopathies, although ceroid deposition in the intestinal tract was reported in one subject.[46] Molenaar et al.[59] described an abnormal cellular ultrastructure of intestinal epithelial cells in two patients with abetalipoproteinemia. These changes paralleled those reported in some vitamin E- and vitamin E and selenium-deficient avian species.[37,60] Additional ultrastructure studies of vitamin E-depleted smooth and striated muscle are warranted in children with this unusual inherited syndrome.

MALABSORPTION SYNDROMES

Young infants and adults who chronically malabsorb lipids for at least 9 months may show biochemical and clinical symptoms resembling some of those seen in vitamin E-deficient animals.[8] Among those most commonly reported are the deposition of ceroid in the intestinal muscularis, low plasma levels, elevated in vitro peroxide hemolysis of red cells, and creatinuria. Occasional lesions in skeletal muscle resembling muscular dystrophy in animals are reported, but these are exceptional and difficult to prove as causally related to vitamin E depletion. By far the largest group of persons in the U.S. with chronic malabsorption and some of the above symptoms that resemble vitamin E deficiency are children with cystic fibrosis (CF).[34]

Cystic Fibrosis and Evidence of Vitamin E Deficiency
Ceroid Deposition

The first clinical report that suggested a tissue deficiency of vitamin E in CF was that of Oppenheimer in 1956.[61] He reported a focal necrosis of striated muscle in 1 child out of 48 studied at necropsy. Blanc et al.[62] were unable to confirm necrosis that could be attributed to vitamin E deficiency in striated muscle of CF patients. They observed, however, that ceroid pigment was deposited extensively in the smooth muscle of the G.I. tract, while none was deposited in striated muscle. Kerner and Goldbloom[63] expanded these studies to other tissues and reported ceroid deposition in the smooth muscle of the portal tract, bladder, and respiratory tract but not in the smooth muscle of vascular tissues. Sung[64] reported neural pathological changes in six patients with prolonged CF that included axonal dystrophy, slight reduction of nerve cells, and an abnormality in the nuclei of nerve cells, all of which he attributed to vitamin E deficiency and possibly an acceleration in cellular aging. In cases of CF and chronic pancreatitis of at least 2 to 4 years duration, Schnitzer and Loesel[65] reported ceroid pigment deposited in smooth muscle in the vessels of almost every organ examined at necropsy. Ceroid is not seen in E-sufficient human populations but has been reported in Thai-Lao ethnic groups whose diets are chronically low in vitamin E.[27] The ultrastructural changes in G.I. epithelial cells described in abetalipoproteinemia[60] are not found in CF.[66,67] Additional quantitative studies are needed to relate ceroid deposition with tissue levels of tocopherol, selenium, glutathione peroxidase, and PUFA in the CF syndrome.

Plasma Tocopherol Levels and Tests on Red Cells

Low plasma levels of tocopherol in CF were first reported in 1949[68] and have subsequently been confirmed by numerous investigators.[8,17,19,67,69-75] An elevated susceptibility in vitro of red cells to hydrogen peroxide, reported first by Gordon et al.,[76] has been a variable finding when plasma levels range from 0.3 to 0.5 mg/dl.[47] These results likely reflect varying levels of tocopherol contained in the red cell per se. Bieri and Poukka[55] reported that the red cells of CF subjects retain tocopherol even at plasma levels less than 0.5 mg/dl, but this was not confirmed by Underwood et al.[77] The latter investigators found that blood levels of 0.1 to 0.4 mg/dl are associated with parallel depletion in red cells, liver, and muscle tissues.[72,73,77] Grimes and Leonard[78] found an almost linear increase in tissue concentrations of vitamin E as plasma concentrations rose to 0.5 mg/dl, and tissue levels in excess of 4 mg/dl were always associated with plasma concentrations above 0.5 mg/dl. Unfortunately, total plasma lipids were not reported in any of these studies. Vitamin E is carried in the betalipoprotein fraction in blood, and the amount of tocopherol distributed between tissues, red cells, and plasma is influenced by the circulating total lipids, particularly by the betalipoprotein fraction,[79] and the level of tissue adiposity.[80] The red cell activity of the selenium-containing enzyme glutathione peroxidase may also contribute to variability in the peroxide hemolysis test.

A measure of the in vivo functioning of tocopherol-depleted red cells of CF patients, although more difficult to assay, is of greater clinical importance than the in vitro hemolysis test. Goldbloom,[17] using ^{51}Cr-RBC to measure the red cell lifespan, reported one CF patient with low plasma tocopherol and a shortened halflife. This finding of a moderately reduced halflife in some unsupplemented patients was recently confirmed by Farrel et al.[67] The latter investigators demonstrated that the shortened survival time was significantly improved, from 19 to 27 days, in six CF patients treated with oral α-tocopherol. Apparently, however, the hematopoietic system in the CF patient compensated adequately for shortened survival, since hematological indices were generally normal.[66]

Absorption Efficiency of Vitamin E

Malabsorption of lipids, including fat-soluble vitamins, is a general finding in about 80% of patients with CF and is caused by intraluminal pancreatic insufficiency secondary to partial acylia of the bile duct and eventual atrophy of acinar activity of the pancreas.[81,82] The mucosal villi appear to be unaffected, suggesting that the steatorrhea is primarily due to alterations occurring in the lumen rather than within mucosal cells. Exogenous enzyme replacement therapy ameliorates but does not fully correct the malabsorption, particularly lipid malabsorption.[83] Attempts to manage malabsorption by other forms of dietary manipulation have led to variable reports of success.[84,85] Reasons for the variability are difficult to pinpoint because of the paucity of studies in vivo of the intraluminal environment during absorption of a regular meal. MacMahon and Neale[86] gave three patients with chronic exocrine pancreatic insufficiency labeled α-tocopherol and found net absorption markedly diminished. In one patient, absorption returned to normal when one capsule of cotazyme (a commercial preparation containing pancreatic enzymes) was given concurrently with vitamin E. In vivo studies that use appropriate intubation techniques are needed to determine the relative contribution of a variety of factors which may be contributing to malabsorption, in addition to incomplete lipolysis due to insufficient pancreatic lipase. For example, some reported a decrease in the total concentration of bile salts,[87,88] inadequate neutralization in the upper jejunum of the acid chyle,[89,90] and an altered ratio of trihydroxy and dihydroxy bile salts.[91] For tocopherols and other highly nonpolar lipids to gain intimate access to the mucosal cell, they must be incorporated into an aqueous phase, and each of the above-mentioned factors can affect this process.[71,92]

The potential significance of bile salt type and concentrations, pH, sodium ion concentration, and the nature of lipolytic products to the efficiency by which polar and nonpolar lipid-soluble vitamins are absorbed is reported in a series of in vivo and in vitro studies[93-96] and summarized, by Underwood, as to how they apply to cystic fibrosis.[91] Incomplete hydrolysis of triglyceride greatly favors partitioning of tocopherol and other nonpolar solutes into the oily rather than micellar phase. This partitioning is further influenced by the nature of the lipolytic products present and of other lipid materials, particularly phospholipid. For example, monoglycerides and phospholipids of long chain fatty acid esters, compared with those of short chain length esters,[95] greatly enhance incorporation of tocopherol into micelles. Concentration of the various products of lipolysis, relative to each other (i.e., fatty acids, monoglycerides and diglycerides, and phospholipids) also influences the partitioning and solubilization into water-miscible micelles, with greater solubilization occurring when both monoglyceride and phospholipid are present concurrently. Water-miscible solutions form only when the concentration of bile salts exceeds the critical micellar concentration, and this in turn is influenced by the relative ratios of trihydroxy and dihydroxy bile salts, taurine and glycine conjugates, pH, and sodium ion concentration.[94] Duodenal aspirates of patients with CF are reported to have reduced concentrations of bile salts,[88] abnormally elevated ratios of trihydroxy to dihydroxy bile salts, a normal to slightly elevated glycine to taurine conjugation ratio,[87,91] abnormally acid pH,[89] and a decreased bicarbonate secretion.[90] How significant these factors are in altering the intraluminal environment at the time a fat-containing meal is present needs to be determined by appropriate intubation studies.[97]

The appearance of clinical evidence of vitamin E deficiency in CF, when subjects rely solely on food sources to meet their tissue needs, is a consequence of malabsorption and not of defective transport or mobilization. The negative factors which limit absorption can be overcome by supplementing diets with either water-miscible or oily forms of vitamin E.[71,98] The level of supplementation advocated by investigators varies considerably,[38,70,73,74,98] but most would agree that 5 mg/kg body weight per day of a water-miscible preparation will, in time, adequately maintain plasma and tissue levels in most afflicted children.

Tissue Levels of Tocopherol and PUFA

Children with CF have lower than normal ratios of linoleic acid in plasma and tissues.[77,99,100] It is not clear whether the decreased linoleic acid (18:2 w 6) is due to a malabsorption from dietary sources or to an increased conversion in tissues to other PUFA of the w 6 series, especially arachidonic acid (20:4 w 6). One study of the fatty acid profiles and α-tocopherol levels of several tissues from a limited number of control and CF subjects supported the hypothesis that conversion occurs at the tissue level to maintain a relatively stable saturated to unsaturated fatty acid ratio and peroxidizable index.[77] Kuo and Huang[84] reported reduced levels of 18:2 w 6 and 20:4 w 6 in plasma lipids and depot fat of children with CF who were fed diets of medium chain triglycerides, but their studies did not report the levels of other PUFA. Similarly, Caren and Corbo[100] reported reduced plasma levels of 18:2 w 6 in CF but did not report levels of other PUFA.

Clearly, there is need for additional studies of fatty acid profiles in CF. There is no doubt that tissues require a lipid-soluble antioxidant to survive and function effectively and that vitamin E is one among several antioxidants occurring naturally that can serve in this role. Tissues containing high levels of PUFA presumably require higher levels of tissue antioxidants. However, it has been difficult to demonstrate in man, including children with CF, that reduced tissue levels of vitamin E significantly alter the peroxidizable index of tissues or lead to an accumulation of oxidation products.[101] Underwood et al.[77] reported that the molar ratio of α-tocopherol to PUFA in plasma and red cells of children with CF was one seventh of that reported by Bieri and Poukka as the minimum needed to prevent 10% hemolysis of red cells in rats[55] and one sixth that found in plasma of non-CF subjects. The clinical significance of this finding is unclear. Supplementing children who had CF with vitamin E did not alter the fatty acid distribution in plasma and red cells but restored to normal the molar ratio of α-tocopherol to PUFA.[67,77] If vitamin E indeed functions in vivo as a tissue antioxidant, the PUFA in tissues of children with CF are highly vulnerable to oxidative damage unless other antioxidants are present at levels sufficient to meet tissue needs. Presumably, this is the case for striated and cardiac muscle, since they do not show the lesions of vitamin E-deficient skeletal muscle described in numerous species of animals. Further, tests of muscle strength and endurance of children with CF and low plasma tocopherol do not differ from test results of normal children. No benefit in muscle function is observed when a vitamin E supplement is provided.[102] The creatinuria generally seen in CF and vitamin E deficiency is not specific for muscle disease, and the more specific enzyme markers for muscular dystrophy of genetic origin, creatine phosphokinase and aldolase,[103] are not consistently elevated in vitamin E-deficient CF patients.[67] On the other hand, smooth muscle tissue that accumulates ceroid and red cells that have a shortened survival time if vitamin E is deficient, do not show these symptoms when the vitamin is sufficient. Bieri and Farrel[7] studied histologically and histochemically four cases of chemically proved vitamin E-deficient CF patients and found only mild histochemical abnormalities. Possibly, these differences are related both to the relative PUFA concentration of the tissue and to its metabolic activity. Hence, the vitamin E and/or total lipid antioxidant requirement of red cell membranes, with their rich content of phospholipid and PUFA relative to that of other tissues, and the requirement of smooth muscle tissue, particularly that of the G.I. tract and lungs, with their rapid rate of metabolism and turnover, may exceed the requirement of other tissue. These tissues (red cell, G.I. tract, and lung) may be the only ones unable to meet metabolic demands for a lipid antioxidant by alternate means.[65] The specific activity of glutathione peroxidase was reported for a variety of rat tissues.[36] Similar studies are needed to delineate the activity of this potentially protective antioxidant system in human tissues from normal and vitamin E-deficient subjects. Little is known of selenium adequacy of tissues in patients with CF, although Underwood et al.[77] found normal levels in plasma of a limited number of children studied.

SUMMARY

Vitamin E is no longer a nutrient in search of a deficiency disease. Although the cases of chemically proved deficiency with physiological consequences are restricted to newborns, rare genetic defects in lipid transport mechanisms, and chronic malabsorption syndromes, other subtle functional consequences may be expected as knowledge is gained of the specific biochemical role of the vitamin at the cellular level. Furthermore, there is a paucity of information in humans on the long-term physiological consequences of chronic, inadequate vitamin E nutriture, such as the deposition of ceroid in the smooth muscle lining of many of the tracts of the body.

REFERENCES

1. *Supplementation of Human Diets with Vitamin E,* National Academy of Sciences, Washington, D.C., 1973.

2. **Farrell, P. M. and Bieri, J. G.,** Megavitamin E supplementation in man, *Am. J. Clin. Nutr.,* 28, 1381–1386, 1975.

3. **Draper, H. H.,** The tocopherols, in *Fat Soluble Vitamins,* Morton, R. A., Ed., Pergamon Press, New York, 1970, 333–393.

4. **Friedman, L., Weiss, W., Wherry, F., and Kline, O. L.,** Bioassay of vitamin E by the dialuric acid hemolysis method, *J. Nutr.,* 65, 143–160, 1958.

5. **Bieri, J. G. and Poukka, R. K. H.,** In vitro hemolysis as related to fat erythrocyte content of α-tocopherol and polyunsaturated fatty acids, *J. Nutr.,* 100, 557–564, 1970.

6. **Rose, C. S. and György, P.,** Specificity of hemolytic reaction in vitamin E deficient erythrocytes, *Am. J. Physiol.,* 168, 414–420, 1952.

7. **Bieri, J. G. and Farrel, P. M.,** Nutritional, metabolic and clinical aspects of vitamin E, in *Vitamins and Hormones,* Academic Press, New York, 1976.

8. **Binder, H. J., Herting, D. C., Hurst, V., Finch, S. C., and Spiro, H. M.,** Tocopherol deficiency in man, *N. Engl. J. Med.,* 273, 1289–1297, 1965.

9. **Pappenheimer, A. M. and Victor, J.,** "Ceroid" pigment in human tissues, *Am. J. Pathol.,* 22, 395–413, 1946.

10. **Tappel, A. L.,** Studies of the mechanism of vitamin E action. III. In vitro copolymerization of oxidized fats with protein, *Arch. Biochem.,* 54, 266–280, 1955.

11. **Raychaudhuri, C. and Desai, I. D.,** Ceroid pigment formation and irreversible sterility in vitamin E deficiency, *Science,* 173, 1028–1029, 1971.

12. **Siakotos, A. N., Koppang, N., Youmans, S., and Bucana, C.,** Blood levels of α-tocopherol in a disorder of lipid peroxidation: Batten's disease, *Am. J. Clin. Nutr.,* 27, 1152–1157, 1974.

13. **Bieri, J. G. and Evarts, R. P.,** Tocopherols and fatty acids in American diets, *J. Am. Diet. Assoc.,* 62, 147–151, 1973.

14. **Thompson, J. N., Beare-Rogers, J. L., Erdödy, P., and Smith, D. C.,** Appraisal of human vitamin E requirement based on examination of individual meals and a composite Canadian diet, *Am. J. Clin. Nutr.,* 26, 1349–1354, 1973.

15. **Dju, M. Y., Mason, K. E., and Filer, L. J., Jr.,** Vitamin E (tocopherol) in human tissues from birth to old age, *Am. J. Clin. Nutr.,* 6, 50–59, 1958.

16. **Filer, L. J., Jr.,** Introductory remarks, *Am. J. Clin. Nutr.,* 21, 3–6, 1968.

17. **Goldbloom, R. B.,** Investigations of tocopherol deficiency in infancy and childhood (studies of serum tocopherol levels and of erythrocyte survival), *Can. Med. Assoc. J.,* 82, 1114–1117, 1960.

18. **Clausen, J. and Friis-Hansen, B.,** Studies on changes in vitamin E and fatty acids of neonatal serum, *Z. Ernaehr.* 10, 264–276, 1971.

19. **Filer, L. J., Wright, S. W., Manning, M. P., and Mason, K. E.,** Absorption of alpha-tocopherol and tocopherol esters by premature and full term infants and children in health and disease, *Pediatrics,* 8, 328–339, 1951.

20. **Jager, F. C.,** Linoleic acid intake and vitamin E requirement in rats and ducklings, *Ann. N.Y. Acad. Sci.,* 203, 199–211, 1972.

21. **Witting, L. A.,** The recommended dietary allowance for vitamin E, *Am. J. Clin. Nutr.,* 25, 257–261, 1972.

22. **Witting, L.,** The role of nutritional factors on free radical reactions, in *Advances in Nutritional Research,* Draper, H. H., Ed., Plenum Press, New York, 1977.

23. **Horwitt, M. K.,** Vitamin E and lipid metabolism in man, *Am. J. Clin. Nutr.,* 8, 451–461, 1960.

24. **Horwitt, M. K.,** Studies of human requirements for vitamin E, *Am. J. Clin. Nutr.,* 27, 1182–1193, 1974.

25. **Horwitt, M. K., Century, B., and Zeman, A. A.,** Erythrocyte survival time and reticulocyte levels after tocopherol depletion in man, *Am. J. Clin. Nutr.,* 12, 99–106, 1968.

26. Nutrition Survey of East Pakistan, Report by the Ministry of Health, Government of Pakistan, U.S. Department of Health, Education, and Welfare, Public Health Service 1966.

27. **Ney, S. W. and Chittayasothorn, K.,** Ceroid in the gastrointestinal smooth muscle of the Thai-Lao ethnic group, *Am. J. Pathol.,* 51, 287–299, 1967.

28. **Majaj, A. S., Dinning, J. S., Azzan, S. A., and Darby, W. J.,** Vitamin E responsive megaloblastic anemia in infants with protein-calorie malnutrition, *Am. J. Clin. Nutr.,* 12, 374–379, 1963.

29. **Majaj, A. S.,** Vitamin E responsive macrocytic anemia in protein-calorie malnutrition. Measurement of vitamin E, folic acid, vitamin C, vitamin B_{12} and iron, *Am. J. Clin. Nutr.,* 18, 362–368, 1966.

30. **Whitaker, J. A., Fort, E. G., Vimokesant, S., and Dinning, J. S.,** Hematologic response to vitamin E in the anemic, *Am. J. Clin. Nutr.,* 20, 783–789, 1967.

31. **Baker, S. J.,** Failure of vitamin E therapy in the treatment of anemia of protein-calorie malnutrition, *Blood,* 32, 717–725, 1968.

32. **Halsted, C. H., Sourial, N., Guindi, S., Mourad, K. A. H., Kattab, A. K., Carter, J. P., and Patwardhan, V. N.,** Anemia of Kwashiorkor in Cairo: deficiencies of protein, iron and folic acid, *Am. J. Clin. Nutr.,* 22, 1371–1382, 1969.

33. **Kulapongs, P.,** The effect of vitamin E on the anemia of protein-calorie malnutrition in northern Thai children, in *Protein-Calorie Malnutrition,* Olson, R. E., Ed., Academic Press, New York, 1975, 263–268.

34. **Bieri, J. G.,** Vitamin E, *Nutr. Rev.,* 33, 161–167, 1975.

35. **Armstrong, D., Dimmitt, S., Grider, L., VanWormer, D., and Austin, J.,** Deficient leucocyte peroxidase activity in the Batten-Spiegelmayer-Vogt syndrome, *Trans. Am. Neurol. Assoc.,* 98, 3–7, 1973.

36. **Tappel, A. L.,** Selenium-glutathione peroxidase and vitamin E, *Am. J. Clin. Nutr.,* 27, 960–965, 1974.

37. **Yarrington, J. T. and Whitehair, C. K.,** Ultrastructure of gastrointestinal smooth muscle in ducks with a vitamin E-selenium deficiency, *J. Nutr.,* 105, 782–790, 1975.

38. **Goldbloom, R. B. and Cameron, D.,** Studies of tocopherol requirements in health and disease, *Pediatrics,* 32, 36–46, 1963.

39. **Hassan, H., Hashim, S. A., Van Itallie, T. B., and Sebrell, W. H.,** Syndrome in premature infants associated with low plasma vitamin E levels and high polyunsaturated fatty acid diet, *Am. J. Clin. Nutr.,* 19, 147–157, 1966.

40. **Hashim, S. A. and Asfour, R. H.,** Tocopherol in infants fed diets rich in polyunsaturated fatty acids, *Am. J. Clin. Nutr.,* 21, 7–14, 1968.

41. **Oski, F. A. and Barness, L. A.,** Vitamin E deficiency: a previously unrecognized cause of hemolytic anemia in premature infants, *J. Pediatr.,* 70, 211–220, 1967.

42. **Ritchie, J. H., Fish, M. B., McMasters, V., and Grossman, M.,** Edema and hemolytic anemia in premature infants: a vitamin E deficiency syndrome, *N. Engl. J. Med.,* 279, 1185–1190, 1968.

43. **Melhorn, D. K., Gross, S., and Childers, G.,** Vitamin E-dependent anemia in the premature infant. I. Effects of large doses of medicinal iron, *J. Pediat.* 79, 569–588, 1971.

44. **Gross, S. and Melhorn, D. K.,** Vitamin E, red cell lipids and red cell stability in prematurity, *Ann. N.Y. Acad. Sci.,* 203, 141–162, 1972.

45. **Binder, H. J. and Spiro, H. M.,** Tocopherol deficiency in man, *Am. J. Clin. Nutr.,* 20, 594–601, 1967.

46. **Kayden, H. J.,** A betalipoproteinemia, *Annu. Rev. Med.,* 23, 285–296, 1972.

47. **MacKenzie, J. B.,** Relation between serum tocopherol and hemolysis in hydrogen peroxide of erythrocytes in premature infants, *Pediatrics,* 13, 346–351, 1954.

48. **Panos, T. C., Stinnett, B., Zapata, G., Eminians, J., Marasigan, B. V., and Beard, A. G.,** Vitamin E and linoleic acid in the feeding of premature infants, *Am. J. Clin. Nutr.,* 21, 15–39, 1968.

49. **Davis, K. C.,** Vitamin E: adequacy of infant diets, *Am. J. Clin. Nutr.,* 25, 933–938, 1972.

50. **McCormick, E. C., Cornwell, D. G., and Brown, J. B.,** Studies on the distribution of tocopherol in human serum lipoproteins, *J. Lipid Res.,* 1, 221–228, 1960.

51. **Silber, R., Winter, R., and Kayden, H. J.,** Tocopherol transport in the rat erythrocyte, *J. Clin. Invest.,* 48, 2089–2095, 1969.

52. **Kayden, H. J. and Bjornson, L.,** The dynamics of vitamin E transport in the human erythrocyte, *Ann. N.Y. Acad. Sci.,* 203, 127–140, 1972.

53. **Kayden, H. J., Silber, R., and Kossman, C. E.,** The role of vitamin E deficiency in the abnormal autohemolysis of acanthocytosis, *Trans. Am. Assoc. Physicians,* 78, 334–342, 1965.

54. **Muller, D. P. R. and Harries, J. J.,** Vitamin E studies in children with malabsorption, *Biochem. J.,* 112, 28, 1969.

55. **Bieri, J. G. and Poukka, R. K. H.,** Red cell content of vitamin E and fatty acids in normal subjects and patients with abnormal lipid metabolism, *Int. J. Vitam. Nutr. Res.,* 40, 344–350, 1970.

56. **Simon, E. R. and Ways, P.,** Incubation hemolysis and red cell metabolism in acanthocytosis, *J. Clin. Invest.,* 43, 1311–1321, 1964.

57. **Dodge, J. T., Cohen, G., Kayden, H. J., and Phillips, G. B.,** Peroxidative hemolysis of red blood cells from patients with abetalipoproteinemia (acanthocytosis), *J. Clin. Invest.,* 46, 357–368, 1967.

58. **Phillips, G. B. and Dodge, J. T.,** Phospholipid and phospholipid fatty acid and aldehyde composition of red cells in patients with abetalipoproteinemia (acanthocytosis), *J. Lab. Clin. Med.,* 71, 629–637, 1968.

59. Molenaar, I., Hommes, F. A., Braams, W. G., and Polman, H. A., Effect of vitamin E on membranes of the intestinal cell, *Proc. Natl. Acad. Sci. U.S.A.*, 61, 982–988, 1968.

60. Molenaar, I., Vos, J., Jager, F. C., and Hommes, F. A., The influence of vitamin E deficiency on biological membranes: an ultrastructural study on the intestinal epithelial cells of ducklings, *Nutr. Metab.*, 12, 358–370, 1970.

61. Oppenheimer, E. H., Focal necrosis of striated muscle in an infant with cystic fibrosis of the pancreas and evidence of lack of absorption of fat-soluble vitamins, *Bull. Johns Hopkins Hosp.*, 98, 353–358, 1956.

62. Blanc, W. A., Reid, J. D., and Andersen, D. H., Avitaminosis E in cystic fibrosis of the pancreas, *Pediatrics*, 22, 494–505, 1958.

63. Kerner, I. and Goldbloom, R. B., Investigations of tocopherol deficiency in infancy and childhood, *Am. J. Dis. Child.*, 99, 597–603, 1960.

64. Sung, J. H., Neuroaxanal dystrophy in mucoviscidosis, *J. Neuropathol. Exp. Neurol.*, 23, 567–583, 1964.

65. Schnitzer, B. and Loesel, L. S., Brown bowel, *Am. J. Clin. Pathol.*, 50, 433–439, 1968.

66. di Sant'Agnese, P. A. and Talamo, R. C., Pathogenesis and physiopathology of cystic fibrosis of the pancreas, *N. Engl. J. Med.*, 277, 1287–1295; 1343–1352; 1399–1408, 1967.

67. Farrel, P. M., Fratantoni, J. C., Bieri, J. G., and di Sant'Agnese, P. A., Effects of vitamin E deficiency in man, *Acta Paediatr. Scand.*, 64, 150–151, 1975.

68. Darby, W. J., Ferguson, M. E., Furman, R. H., Lemley, J. M., Ball, C. T., and Meneely, G. R., Plasma tocopherols in health and disease, *Ann. N.Y. Acad. Sci.*, 52, 328–333, 1949.

69. Nitowsky, H. M., Tildon, J. T., Levin, S., and Gordon, H. H., Studies of tocopherol deficiency in infants and children. VII. The effect of tocopherol on urinary, plasma and muscle creatine, *Am. J. Clin. Nutr.*, 10, 368–378, 1962.

70. Bennett, M. J. and Medwadowski, B. F., Vitamin E and lipids in serum of children with cystic fibrosis or congenital heart defects compared with normal children, *J. Clin. Nutr.*, 20, 415–421, 1967.

71. Harries, J. T. and Muller, D. P. R., Absorption of different doses of fat soluble and water miscible preparations of vitamin E in children with cystic fibrosis, *Arch. Dis. Child.*, 46, 341–344, 1971.

72. Underwood, B. A. and Denning, C. R., Correlations between plasma and liver concentrations of vitamins A and E in children with cystic fibrosis, *Bull. N.Y. Acad. Med.*, 47, 34–39, 1971.

73. Underwood, B. A. and Denning, C. R., Blood and liver concentrations of vitamins A and E in children with cystic fibrosis of the pancreas, *Pediatr. Res.*, 6, 26–31, 1972.

74. Taylor, B. W., Watts, J. L., and Fosbrooke, A. S., Vitamin E therapy in cystic fibrosis, *Arch. Dis. Child.*, 48, 657–658, 1973.

75. McWherter, W. R., Plasma tocopherol in infants and children, *Acta Paediatr. Scand.*, 64, 446–448, 1975.

76. Gordon, H. H., Nitowsky, H. M., and Cornblath, M., Studies of tocopherol deficiency in infants and children. I. Hemolysis of erythrocytes in hydrogen peroxide, *Am. J. Dis. Child.*, 90, 669–681, 1955.

77. Underwood, B. A., Denning, C. R., and Navab, M., Polyunsaturated fatty acids and tocopherol levels in patients with cystic fibrosis, *Ann. N.Y. Acad. Sci.*, 203, 237–247, 1972.

78. Grimes, H. and Leonard, P. J., The relationship between plasma and tissue vitamin E concentrations in man, *Biochem. J.*, 115, 151, 1969.

79. Bieri, J. G. and Evarts, R. P., Effect of plasma lipid level on tissue deposition of α-tocopherol., *Fed. Proc. Fed. Am. Soc. Exp. Biol.*, 34, 913, 1975.

80. Bieri, J. G. and Evarts, R. P., Effect of plasma lipid-levels and obesity on tissue stores of α-tocopherol, *Proc. Soc. Exp. Biol. Med.*, 149, 500–502, 1975.

81. Barbero, G. J., Schwachman, H., Grand, R., and Woodruff, C., Gastrointestinal and nutritional manifestations of cystic fibrosis, in *Cystic Fibrosis: Projections Into the Future*, Mangos, J. A. and Talamo, R. C., Eds., Stratton Intercontinental Medical Book, New York, 1976, 83–111.

82. Johansen, P. G., Anderson, C. M., and Hadorn, B., Cystic fibrosis of the pancreas: a generalized disturbance of water and electrolyte movement in exocrine tissue, *Lancet*, 1, 455–460, 1968.

83. Mullinger, M., The effect of exogenous pancreatic enzymes on fat absorption, *Pediatrics*, 42, 523–525, 1968.

84. Kuo, P. J. and Huang, N. N., The effect of medium chain triglyceride upon fat absorption and plasma lipid and depot fat of children with cystic fibrosis of the pancreas, *J. Clin. Invest.*, 44, 1924–1933, 1965.

85. Weihofen, D. M. and Pringle, D. J., Dietary intake and food tolerance of children with cystic fibrosis, *J. Am. Diet. Assoc.*, 54, 206–209, 1969.

86. MacMahon, M. T. and Neale, G., The absorption of α-tocopherol in control subjects and in patients with intestinal malabsorption, *Clin. Sci.*, 36, 197–210, 1970.

87. **Watkin, J. B., Tercyak, A. M., Szczepanik, P., and Klein, P. D.,** Bile salt kinetics in cystic fibrosis, *Gastroenterology,* 68, 1087, 1975.

88. **Weber, A. M., Roy, C. C., Lepage, G., Chartrand, L., and Lasalle, R.,** Interruption of the entrohepatic circulation (EHC) of bile acids (BA) in cystic fibrosis (CF), *Gastroenterology,* 68, 1066, 1975.

89. **Knauff, R. E. and Adams, J. A.,** Duodenal pH in cystic fibrosis, *Clin. Chem.* (N.Y.), 14, 477–479, 1968.

90. **Hadorn, B., Zoppi, G., Shmerling, D. H., Prader, A., McIntyre, I., and Anderson, C. M.,** Quantitative assesment of exocrine pancreatic function in infants and children, *Pediatrics,* 73, 39–50, 1968.

91. **Underwood, B. A.,** Vitamin A and E nutrition in children with cystic fibrosis, in *Nutrition,* Vol. 2, Proc. 9th Int. Congr. Nutrition, Chavez, A., Bourges, H., and Basta, S., Eds., S. Karger, Basel, 1975, 348–356.

92. **Gallo-Torres, H. E.,** Obligatory role of bile for the intestinal absorption of vitamin E, *Lipids,* 5, 379–384, 1969.

93. **MacMahon, M. T. and Thompson, G. R.,** Comparison of the absorption of a polar lipid, oleic acid, and a non-polar lipid, α-tocopherol from mixed micellar solutions and emulsions, *Eur. J. Clin. Invest.,* 1, 161–166, 1970.

94. **El-Gorab, M. and Underwood, B. A.,** Solubilization of β-carotene and retinol into aqueous solutions of mixed micelles, *Biochim. Biophys. Acta,* 306, 58–66, 1973.

95. **Takahasi, Y. I. and Underwood, B. A.,** Effect of long and medium chain length lipids upon aqueous solubility of α-tocopherol, *Lipids,* 9, 855–859, 1974.

96. **El-Gorab, M., Underwood, B. A., and Loerch, J. P.,** The roles of bile salts on the uptake of β-carotene and retinol by rat everted gut sacs, *Biochim. Biophys. Acta,* 401, 265–277, 1975.

97. "Gap" Conference Report, Nutrition and Cystic Fibrosis, Cystic Fibrosis Foundation, Atlanta, June 23 to 24, 1976.

98. **Harries, J. T., Muller, D. P. R., and Lloyd, J. K.,** Vitamin E absorption in children, *Pediatr. Res.,* 6, 65, 1972.

99. **Kuo, P. T., Huang, N. N., and Bassett, D. R.,** The fatty acid composition of the serum chylomicrons and adipose tissue of children with cystic fibrosis of the pancreas, *J. Pediatr.,* 60, 394–403, 1962.

100. **Caren, R. and Corbo, L.,** Plasma fatty acids in cystic fibrosis and liver disease, *J. Clin. Endocrinol. Metab.,* 26, 470–477, 1966.

101. **Green, J.,** Vitamin E and the biological antioxident theory, *Ann. N. Y. Acad. Sci.,* 203, 29–44, 1972.

102. **Levin, S., Gordon, M. H., Nitowsky, H. M., Goldman, C., di Sant'Agnese, P., and Gordon, H. H.,** Studies of tocopherol deficiency in infants and children. VI. Evaluation of muscle strength and effect of tocopherol administration in children with cystic fibrosis, *Pediatrics,* 27, 578–588, 1961.

103. **Munsat, T. L., Baloh, R., Pearson, C. M., and Fowler, W.,** Serum enzyme alterations in neuromuscular disorders, *JAMA,* 226, 1536–1543, 1973.

EFFECTS OF NUTRIENT DEFICIENCIES IN MAN: VITAMIN K

G. F. Pineo

Vitamin K, a fat-soluble vitamin, is found in nature in two forms: vitamin K_1 or phylloquinone, found naturally in many plants, and vitamin K_2 or menaquinone, synthesized by many Gram-positive bacteria.[1]

Function of vitamin K — The only known biological activity of vitamin K is in the production of the following coagulation factors by the liver: II (prothrombin), VII (proconvertin), IX (Christmas factor), X (Staurt factor), collectively called the vitamin K-dependent factors. The activity of vitamin K at the molecular level has recently been clearly defined. Early studies revealed that vitamin K did not stimulate the production of the polypeptide portion of prothrombin or the other vitamin K-dependent factors[2-4] as evident by the lack of blockage by cyclohexamide of the rapid production of prothrombin seen after the administration of vitamin K.

Several investigators demonstrated the presence of precursor molecules to pro-thrombin and the other vitamin K-dependent factors, either with natural vitamin K deficiency or with the administration of coumarin or indandione, drugs which are antagonists of vitamin K.[5,6] Although the precursor molecules had the same antigenic determinants and were chemically indistinguishable, they did not bind calcium well[7] and were thus inactive in normal coagulation. Detailed studies by Stenflo[7,8] have shown that the precursor molecule of prothrombin lacks γ-carboxyglutamic residues which are necessary for calcium binding and, therefore, the normal physiological function. Similar studies with factor IX and factor X precursor[9] suggested a similar mechanism of action of vitamin K in the production of these factors.

Normal requirements in man — In the absence of an efficient labeling method, the normal requirement in man and the normal amount of vitamin K in body stores remains unknown. On the basis of studies in patients sustained on i.v. fluids who were given antibiotics,[10] a minimal daily requirement of 0.03 $\mu g/kg$ was suggested, although higher requirements have been demonstrated for various experimental animals.[11] A normal diet provides adequate vitamin K, and this is supplemented by endogenous vitamin K produced by bacteria in the distal small bowel and colon.[12] The significance of this contribution is unclear.[13] The main dietary sources of vitamin K are cereals, vegetables, organ meats, and eggs.

Deficiency states in man — Deficiency of vitamin K may result from inadequate intake, absorption, or utilization, or as a result of drugs which interfere with its activity.

Inadequate intake — Although previous reports suggested that vitamin K deficiency only occurred after prolonged antibiotic therapy and inadequate intake,[10,14] a more recent report indicates that vitamin K deficiency can occur rapidly in patients with a poor food intake who are on antibiotics, particularly in the postoperative period.[13]

Mild vitamin K deficiency is seen in all newborn infants and is particularly severe in premature infants. It is due to inadequate liver synthesis and is aggravated by decreased dietary intake and probably inadequate production by intestinal bacterial flora.

Inadequate absorption — Vitamin K is a fat-soluble vitamin which requires the presence of bile salts for its absorption in the upper small intestine; thus, a deficiency state can be caused by intra- or extrahepatic biliary obstruction. Decreased absorption may be caused by any of the malabsorption syndromes such as celiac disease, Crohn's disease, bowel resection, and bacterial overgrowth.

Inadequate utilization — Any form of acute or chronic liver insufficiency may result in decreased utilization of vitamin K for the production of the vitamin K-dependent coagulation factors. Utilization may be competitively blocked by the oral anticoagulants of the coumarin or indandione groups or by large doses of salicylates.[15] The vitamin K antagonistic effect may be enhanced by drugs that displace the anticoagulant from its albumin binding site, e.g., anti-inflammatory drugs, Dilantin®, and sulfonamides.[16] The rapid onset of vitamin K deficiency in patients on multiple antibiotics suggests interference with utilization as one of the mechanisms of action.[13]

Diagnosis of vitamin K deficiency — Vitamin K deficiency leads to a bleeding tendency with ecchymoses, bleeding from mucous membranes, postoperative bleeding, and hematuria being the most common manifestations. Muscle hematomas, intracranial hemorrhages, and submucosal intestinal hemorrhages are seen less commonly.

Prolongation of the prothrombin time in the absence of liver disease suggests the diagnosis which can be confirmed by specific vitamin K-dependent factor assays or by the rapid response to administration of vitamin K. Accidental or surreptitious overdose with oral anticoagulants can be demonstrated by measurement of the specific blood levels.[16]

Treatment — Mild to moderate vitamin K deficiency can be treated with vitamin K (Phenonadione) orally, subcutaneously, or intravenously. Rapid i.v. administration may be accompanied by hypotension, collapse, and, rarely, cardiac arrest; therefore, it should be administered slowly and used only when the oral or subcutaneous route is inadequate. For the hemorrhagic syndrome of the newborn, 1 to 2 mg parenterally is usually adequate.

For overdose of anticoagulants, 25 to 50 mg of vitamin K_1 i.v. should be given. Further therapy may be necessary in 24 to 48 hr as the half-life of vitamin K is less than the half-life of the coumarins, especially when they are given in large doses.[17]

For severe vitamin K deficiency, fresh frozen plasma or a concentrate of the vitamin K-dependent coagulation factors may be required, for example, Konyne-Cutter or prothrombin concentrate-Connaught. These must be used with extreme caution in newborns or in patients with liver disease because of the danger of inducing intravascular thrombosis.[18]

REFERENCES

1. **Isler, O. and Wiss, O.,** Chemistry and biochemistry of the K vitamins, *Vitam. Horm.,* 17, 54–92, 1959.
2. **Shah, D. V. and Suttie, J. W.,** The effect of vitamin K and warfarin on rat liver prothrombin concentrations, *Arch. Biochem. Biophys.,* 150, 91–95, 1972.
3. **Olson, R. E.,** The mode of action of vitamin K, *Nutr. Rev.,* 28, 171–176, 1970.
4. **Suttie, J. W.,** The effect of cyclohexamide administration on vitamin K-stimulated prothrombin formation, *Arch. Biochem. Biophys.,* 141, 571–578, 1970.
5. **Hemker, H. C. and Muller, A. D.,** Kinetic aspects of the interaction of blood-clotting enzymes. VI. Localization of the site of blood-coagulation inhibition by the protein induced by vitamin K absence (PIVKA), *Thromb. Diath. Haemorrh.,* 20, 78–87, 1968.
6. **Bell, R. G. and Matschiner, J. T.,** Synthesis and destruction of prothrombin in the rat, *Arch. Biochem. Biophys.,* 135, 152–159, 1969.
7. **Stenflo, J.,** Vitamin K and the biosynthesis of prothrombin. IV. Isolation of peptides containing prosthetic groups from normal prothombin and the corresponding peptides from dicoumarol-induced prothrombin, *J. Biol. Chem.,* 249, 5527–5535, 1974.
8. **Stenflo, J.,** Vitamin K and the biosynthesis of prothrombin. III. Structural comparison of an NH_2-terminal fragment from normal and dicoumarol-induced bovine prothrombin, *J. Biol. Chem.,* 248, 6325–6632, 1973.
9. **Fujikawa, K., Coan, M. H., Enfield, D. L., Titani, K., Ericsson, L. H., and Davie, W. E.,** A comparison of bovine prothrombin, factor IX (Christmas factor), and factor X (Staurt factor), *Proc. Nat. Acad. Sci. U.S.A.,* 71, 427–430, 1974.
10. **Frick, P. G., Riedler, G., and Brogli, H.,** Dose response and minimal requirement for vitamin K in man, *J. Appl. Physiol.,* 23, 387–389, 1967.

11. Mameesh, M. S. and Johnson, B. C., Production of dietary vitamin K deficiency in the rat, *Proc. Soc. Exp. Biol. Med.,* 101, 467–468, 1959.

12. Hollander, D. and Truscott, T. C., Colonic absorption of vitamin K_3, *J. Lab. Clin. Med.,* 83, 648–656, 1974.

13. Pineo, G. F., Gallus, A. S., and Hirsh, J., Unexpected vitamin K deficiency in hospitalized patients, *Can. Med. Assoc. J.,* 109, 880–883, 1973.

14. Ham, J. M., Hypoprothrombinemia in patients undergoing prolonged intensive care, *Med. J. Aust.,* 11, 716–718, 1971.

15. Seegers, W. H., The influence of certain drugs on blood coagulation and related phenomena, *Pharmacol. Rev.,* 3, 278–340, 1951.

16. Axelrod, J., Cooper, J. R., and Brodie, B. B., Estimation of dicoumarol, 3,3 Methylenebis (4-hydroxycoumarin) in biological fluids, *Proc. Soc. Exp. Biol. Med.,* 70, 693–695, 1949.

17. Koch-Weser, J. and Sellers, E. M., Drug interactions with coumarin anticoagulants, *N. Engl. J. Med.,* 285, 487–498, 547–559, 1971.

18. Gazzard, B. G., Lewis, M. C., Ash, G., Rizza, C. R., Bidwell, E., and Williams, R., Coagulation factor concentrate in the treatment of the hemorrhagic diathesis of fulminant hepatitis, *Gut,* 15, 993–998, 1974.

Minerals

DISORDERS OF CALCIUM METABOLISM

William C. Thomas, Jr. and John Eager Howard

INTRODUCTION

In terms of weight, calcium is one of the more abundant inorganic elements of the body. Most of the calcium is confined to the skeleton, but the small amount that is not participates in a variety of important metabolic events. Information developed in recent years has added greatly to our understanding of calcium metabolism. Although much remains to be learned, application of what is known has permitted development of concepts that have proved useful in delineating the disorders of calcium homeostasis. Before presenting a discussion of specific disorders, however, we shall briefly review certain biochemical and biophysical aspects of calcium metabolism.

CALCIUM HOMEOSTASIS

Role of the Skeleton

There are approximately 1.1 kg of calcium in the body of a man weighing 70 kg, and 99% of this amount is in the skeleton.[1] The extracellular fluids normally contain about 900 mg of calcium. The remaining 10 g of nonskeletal calcium is distributed in the various tissues of the body, with the skin containing proportionately more than other tissues.[1,2] The calcium present in these soft tissues is only partially exchangeable with extracellular calcium.[3-5] Thus, from the quantitative aspects of calcium distribution within the body, it is evident that the skeleton is the only source for replacement of any large losses of this cation through urine or feces. During the course of such losses, normal concentrations of calcium can be maintained in body fluids for prolonged periods of time provided nothing is awry with the responsiveness of the bones to homeostatic forces.[6]

Histologic examination of different parts of the skeleton reveals considerable variation in the functional activity of bone cells in different areas. At one site new bone formation may be in progress, whereas in a nearby area there may be active resorption[6] (Figure 1). Therefore, it is important to conceive of the skeleton as having the capacity to respond locally or generally to stimuli; the response at one site or in one extremity may be quite different from that occurring in other areas. The osteones in compact bone and the trabeculae in cancellous bone are so structurally organized that they are everywhere separated from other compartments of the body by a cellular membrane of osteocytes and osteoblasts.[7] Constituents are transported by way of these cells to or from bone as determined by their metabolic activity. Since calcium and phosphorus in plasma or extracellular fluids are normally present in a supersaturated state with respect to the apatite crystals of bone,[8] the cellular membrane between bone and adjacent environment provides vital protection against the physicochemical forces that might otherwise promote progressive accretion of apatite crystals beyond the confines of the skeleton. There are now significant data to indicate that the concentrations of a number of ions in the extracellular fluid within the confines of the skeleton are quite different from those in ultrafiltrates of sera.[9] Thus, just as the composition of cerebrospinal fluid is unique with respect to that of other circulating fluids, the internal environment of bone is uniquely composed for its special needs.

Within the skeleton, bone mineral is thought to exist in two phases — as amorphous calcium phosphates, probably in colloidal aggregations, and as crystalline apatite.[10] Although the crystalline phase lends the most rigidity to the skeleton, there is evidence that the amorphous phase is the more readily available source of calcium for replacement

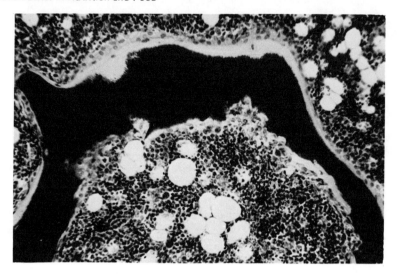

FIGURE 1. Simultaneous bone resorption and formation. Single trabecula (undecalcified) from the tibia of a rat made rachitic by feeding a low-calcium, high-phosphorus diet. (Magnification × 250.) Note the osteoid formation along superior border and evidence of active bone resorption on the inferior border. (From Thomas, W. C., Jr., Howard, J. E., and Connor, T. B., *Bull. Johns Hopkins Hosp.*, 101, 123, 1957. With permission.)

of external loss.[11] The proportion of bone mineral present in the amorphous phase decreases with advancing age, but some is always present.[10,11] Whether in the amorphous or crystalline phase, bone mineral provides a tremendous surface area for the exchange of ions and the maintenance of calcium homeostasis. For example, when large amounts of calcium are quickly removed from the body by exchange transfusions with calcium-free blood or during dialysis against calcium-free solutions, normocalcemia is rapidly restored.[12-14] Conversely, daily infusions of 1 g of calcium for as long as 30 days cause hypercalcemia for only a few hours, despite loss in the urine of only part of the infused calcium.[15]

Such experiments illustrate the capacity of the normally functioning skeleton either to release or to accept the amounts of calcium required to maintain normal concentrations. Because the calcium concentration in body fluids is so vigorously maintained within very narrow limits, McLean and Hastings[16] concluded long ago that the concentration of this ion is "one of nature's physiological constants."

Local factors of which we have little knowledge also affect skeletal metabolism. A bone increases in density and size with use, and disuse is followed by changes of an opposite nature.[17] For example, heavily muscled persons ordinarily have large, dense bones but when an extremity is put in a plaster cast or deprived of its nerve supply, marked atrophy of the bones involved may occur within a few weeks. Finally, other vital forces exist that protect the development or maintenance of the skeleton. Isolated embryonic bone will develop into its predestined form even when cultured in vitro.[18] At the other end of life's spectrum, an aged, markedly atrophic bone still retains the capacity to repair itself when fractured.

The Intestinal Tract

The newborn begins life with 25 to 36 g of total body calcium, which increases to approximately 1100 g in the male by adulthood.[1,2] To achieve this increment in body calcium, there obviously must be adequate dietary provision of calcium and some degree of efficiency in its absorption by the intestines. In normal adult life, probably not more than 250 to 350 mg of calcium are absorbed per day, even when large quantities of this element are provided in the ingested foods.[19] If the individual adult is maintaining calcium equilibrium, the amount of calcium absorbed will be balanced by an equal amount lost in urine, sweat, desquamated skin, and intestinal secretions. However, in times of increased need, as during growth, tissue repair, pregnancy, or lactation, adaptive changes take place in the intestinal mucosa that increase the efficiency for absorption of calcium.[20]

The percentage of ingested calcium that is absorbed generally decreases when the diet contains a large amount of calcium and increases when the dietary intake is restricted. Several weeks may be required for the healthy person to make these adaptive changes to variations in the amount of calcium ingested.[21] Some subjects are more capable of effectively adapting to a restricted intake of calcium than others. For example, young adults may maintain calcium equilibrium when ingesting less than 300 mg of calcium per day,[22] but in some elderly subjects the inefficiency of the intestinal tract may be such that they require a daily intake of 1 g or more of calcium to achieve equilibrium.[23] An age-related decreased deficiency in the intestinal absorption of calcium has also been detected in rats[24] and may occur in all vertebrates.

A small amount of calcium absorption apparently takes place by passive diffusion, but most of the absorption of this element depends on specific processes. Initiated by the observations of Schachter and associates[25] on isolated segments of the intestines, many investigators[26-28] have established in laboratory animals that the duodenum is the most efficient calcium-absorbing segment of the intestines. Its short length is a limiting factor, however, and the majority of the calcium absorbed is by the more distal segments of the small intestine. In addition, Wasserman and Taylor[20,29] have described a calcium-binding protein, which is apparently produced by mucosal cells and translocated to the microvilli, where it possibly exercises a controlling role in the absorption of calcium. The steric configuration or production of this protein is thought to be determined largely by the action of active metabolites of vitamin D.[30] Conceivably, the fundamental role of vitamin D in enhancing the intestinal absorption of calcium is mediated in large part through its effect on this calcium-binding protein.

Considerable attention has been directed to the provision of generous amounts of dietary calcium during the growth years of young people, but curiously little emphasis has been placed on maintaining an optimal intake of this ion during adult years. Animal nutritionists have devoted considerable effort to determining the amount of dietary calcium necessary to achieve maximum growth and development. Interestingly, the currently recommended amounts of dietary calcium for livestock, laboratory animals, and pets constitute a much greater proportion of the diet of these animals than is recommended for man.[31,32] This disparity may be because animal nutritionists have been interested in achieving maximum development, whereas most of the investigations of man have been designed to determine the minimum amount of calcium required to achieve a state in which loss from the body does not exceed intake.[32]

Dairy products remain a major source of calcium for man, and it is probably of teleological relevance that the calcium in milk is absorbed better than that from most other foods.[33] Lysine and lactose are two ingredients of milk that may partially account for the high absorbability of calcium from this source.[34] As would be expected, such dietary constituents as phytic acid or oxalate, which form insoluble compounds with calcium, impair the availability of dietary calcium.[35] The presence of steatorrhea will impair the absorption of fat-soluble vitamin D, and calcium absorption will be additionally affected because of the formation of insoluble calcium soaps. Even though the intestinal tract can be remarkably efficient in absorbing the needed amounts of calcium, these amounts must first be provided in the diet in soluble or potentially soluble form. In old people or in those with disorders of intestinal function, calcium equilibrium may not be easily achieved, but in elderly subjects the skeleton may be spared if generous intakes of calcium are maintained. Although some calcium is present in the feces of subjects fed calcium-free diets, the intestinal secretions apparently are not an adaptable route for excreting excesses of calcium.[19,36]

The Kidneys

The amount of calcium excreted in urine often reflects the internal state of calcium metabolism, but the amount excreted is usually too small to be considered an important controlling factor in calcium homeostasis. Nonetheless, the constant presence of this cation in urine even when the dietary intake is inadequate, may be, in the aggregate, sufficient to cause a significant deficit.

The healthy adult ordinarily excretes between 100 and 300 mg of calcium in the urine per 24 hr. Within this range the amount excreted will usually reflect the dietary intake.[23,27] Because so little of the calcium presented to the kidneys appears in the urine, almost all of that in the glomerular filtrate must be reabsorbed in the renal tubules. Although a transtubular flux may occur, there is no evidence for tubular secretion of this ion.[38]

Several factors modify the excretion of calcium. Metabolic acidosis increases and alkalosis decreases urinary calcium.[39] Similar but less pronounced changes in urinary calcium occur after ingestion of diets that have large amounts of acid or alkaline ash[40] Besides these well-established effects of specific diets, glucose or other rapidly metabolized foods have been observed to cause a transient increase in calcium excretion.[41] Augmenting the urine volume may increase divalent cation excretion to a mild extent, but when an increased volume is accomplished by means of diuretics, a marked increase in divalent ion excretion occurs.[42,43] To some extent, sodium, magnesium, and calcium apparently share common tubular reabsorptive sites; and infusion of one of these ions is followed by increased excretion of all three.[44] Conversely, the healthy person will respond to a markedly restricted intake of sodium by a reduction in urinary sodium, calcium, and magnesium. In certain disorders, however, it can be shown that there are also independent tubular functions that control the excretion of these three cations.[45]

One effect of parathyroid hormone on the kidney is a decrease in the excretion of calcium.[46] However, when parathyroid hormone is present in sufficient excess to cause hypercalcemia, as in hyperparathyroidism, the effect of the hypercalcemia becomes predominant and there is a net increase in urinary calcium. Hypercalciuria in hyperparathyroidism and other hypercalcemic states is thought to occur primarily because of the increase in ultrafilterable calcium presented to the glomeruli. Regardless of the degree of hypercalcemia, we have never seen this derangement induce an excretion of more than 1000 mg of calcium per 24 hr. By contrast, much more marked hypercalciuria can be achieved by inducing a diuresis with hypertonic saline or furosemide.[43]

Other disorders of calcium homeostasis may be accompanied by hypercalciuria. During dissolution of bone as that which may ensue on complete immobilization or after a bone fracture, marked hypercalciuria is usually present.[47] Since hypercalcemia may also occur during such immobilization, this problem will be discussed subsequently in this chapter. Phosphate depletion is another cause of hypercalciuria.[48] Perhaps phosphate depletion accelerates the formation of 1,25-dihydroxycholecalciferol,[49] which in turn may lead to decreased secretion of parathyroid hormone. This latter change might account for this hypercalciuria, but more information is required to support such a hypothesis.

In normocalcemic persons with healthy kidneys, the least excretion of calcium (< 50 mg/24 hr) occurs in those with vitamin D deficiency.[50] Whether or not this minimal excretion is because of increased parathyroid hormone effect is not known, but it is so characteristic that when detected in the normocalcemic patient, it has proved a useful clue to the presence of vitamin D deficiency. As to be expected, urinary calcium is reduced in subjects with hypocalcemia, but returns to the normal range when the cause of the hypocalcemia has been corrected. However, when hypocalcemia is due to hypoparathyroidism and the serum calcium has been restored to the normal range by administration of large doses of vitamin D or calcium salts, urinary calcium will be much greater (at 400 to 800 mg/24 hr) than normal.[51] This hypercalciuria may simply reflect the absence of parathyroid hormone action on the renal tubules.

Vitamin D, Parathyroid Hormone, and Calcitonin

During the past 20 years a series of brilliant investigations have provided a great deal of information on the roles of vitamin D, parathyroid hormone, and the newly discovered calcitonin in regulating calcium metabolism.[52-55] It is beyond the scope of this chapter to review these advances in detail, but in this preamble to disorders of calcium metabolism, certain interrelations should be mentioned.

The parathyroid glands have long been regarded as the primary defense in maintaining calcium homeostasis. Function of these glands has been thought to be controlled primarily by the concentration of ionic calcium in plasma — decreased calcium promoting the release of parathyroid hormone and increased calcium promoting a reduction in hormone secretion. Although seldom encountered clinically except in subjects with advanced renal failure, an increase in serum magnesium also decreases the output of parathyroid hormone.[56] Interestingly, a marked deficiency of magnesium impairs the release of parathyroid hormone and also reduces its effect in mobilizing calcium from bone. Hypocalcemia due to magnesium deficiency is now a well-recognized syndrome.[57,58] In tissue cultures of parathyroid glands, similarly high or low concentrations of either calcium or magnesium cause a similar decrease or increase, respectively, in hormone secretions.[59] Evidence recently has been presented that indicates the most active metabolite of vitamin D, 1,25-dihydroxycholecalciferol, is also capable of reducing secretion of parathyroid hormone from cultured glands.[60]

In experiments conducted some years ago Harrison and Harrison[61] noted that vitamin D-deficient animals were resistant to the hypercalcemic effect of administered parathyroid hormone. Conversely, parathyroid hormone is not a requisite for the hypercalcemia induced by large amounts of vitamin D. The recently defined pathways of vitamin D metabolism and studies on the specific effects of the metabolic products derived from this vitamin provide answers to some of these previous enigmas. After formation of 25-hydroxycholecalciferol from cholecalciferol in the liver, additional hydroxylation to 1,25-dihydroxycholecalciferol in the kidney is stimulated by parathyroid hormone or phosphate depletion.[52,53] This vitamin metabolite not only stimulates formation of intestinal calcium-binding protein, but small amounts can cause rapid mobilization of calcium from bone. The possible interactions of parathyroid hormone and dihydroxycholecalciferol on bone cells has not been well formulated, but parathyroid hormone does appear to act as a modulator of the active metabolites of vitamin D. Variations in magnesium concentration will probably also prove to affect vitamin D metabolism.

The role of calcitonin in human metabolism has not yet been completely defined. This hormone, which retards mobilization of calcium from bone, is demonstratively important to calcium homeostasis in amphibians and marine mammals, but in man the endogenous secretion of calcitonin has little recognized effect on calcium metabolism.[62] Exogenously administered calcitonin, however, has been useful in the clinical management of Paget's disease and in certain hypercalcemic states.[63-65] For example, it is effective in correcting idiopathic hypercalcemia of infancy and hypercalcemia induced by immobilization, and it is sometimes effective in correcting the hypercalcemia associated with certain forms of malignancies.[66] Calcitonin is relatively less effective in modifying the hypercalcemia associated with hyperparathyroidism or other disorders in which a parathyroid hormone-like substance is being produced. The assay of circulating calcitonin has now become important in detecting the presence of medullary carcinoma of the thyroid gland.[67] It may be anticipated that within the near future the true role of this hormone in human metabolism will be established.

Recent investigations have provided a beginning insight into the cellular mechanisms whereby parathyroid hormone, calcitonin, vitamin D, and other substances modify the mobilization of calcium from bone. Parathyroid hormone increases formation of cyclic adenosine monophosphate within bone cells and, by a separate process, increases the permeability of cell membranes to calcium.[68] As a result of this latter process, an initial effect of parathyroid hormone is to increase calcium within the cells of bone and other tissues. The parathyroid hormone stimulation of adenylate cyclase activity is reduced in the absence of vitamin D.[69] Calcitonin has an opposite effect to that of parathyroid hormone on the mobilization of calcium from bone, and by in vitro techniques calcitonin has been shown to retard the egress of calcium from bone by a process that does not interfere with parathyroid hormone activation of adenylate cyclase.[69]

It has been demonstrated recently that various prostaglandins are capable of mobilizing calcium from bone in vitro. The greatest effect was achieved with prostaglandin E_2.[70] These prostaglandins have an effect similar to that of parathyroid hormone in stimulating adenylate cyclase activity of bone cells. This may explain the hypercalcemia occurring with certain malignant tumors, for it has been shown that such tumors may produce prostaglandin E_2 or secrete products that induce formation of prostaglandins in bone.[71,72] In addition to prostaglandins, a secretory product of various white blood cells has been obtained that is also capable of mobilizing calcium from bone.[73-75] This product, currently identified as osteoclast activating factor, may be important in the increased resorption of bone that occurs in a number of clinical disorders.

STATE OF CALCIUM IN BODY FLUIDS

The terms "**hypercalcemia**" and "**hypocalcemia**" are used in this chapter to signify higher or lower concentrations of "biologically active" calcium than normally exist in blood serum. In healthy persons, the nonprotein-bound calcium in serum is ultrafilterable, and most of this latter fraction is considered to be biologically active since only a small amount of the calcium in serum is complexed with citrate and other small ultrafilterable anions.[76,77]

In serum the portion of calcium that is complexed with proteins, primarily albumin, does not pass freely across capillary or other semipermeable membranes. However, this protein-bound calcium exists in a state of equilibrium with that which is nonprotein-bound; thus by dialysis all of the calcium can be removed from the proteins.[78] Depending to some extent on the technique used, 30 to 45% of the total calcium in serum is not ultrafilterable,[76,77,79,80] and the absolute amount is determined primarily by the concentration of albumin. The association constant of calcium proteinate at pH 7.35 and 35°C is such (approximate pK of 2.2) that 1 g of protein binds 0.4 mg of calcium.[81] Thus because the equilibrium between ionic and protein-bound calcium is maintained, a 1-g decrease in the concentration of albumin will be reflected by a concomitant decrease of approximately 0.85 mg in the total calcium concentration in serum.[82] Changes in the pH of serum or plasma also alter the calcium-protein complex by decreasing the binding of calcium in the presence of acidosis and increasing it in alkalotic states. The magnitude of this effect per unit of change in pH is approximately 0.85 mg of calcium per 100 ml of serum.[81]

In clinical medicine there are no recognized disorders characterized by excessive production of albumin sufficient to cause hyperalbuminemia. There are diseases, however, in which large amounts of other calcium-binding proteins may be present in serum, thereby causing an increase in total calcium without a necessary concomitant increase in the concentration of ultrafilterable calcium.

Multiple myeloma is the most frequently occurring entity in which abnormal concentrations of calcium-binding proteins may be present.[83,84] Since hypercalcemia with an increase in ultrafilterable calcium is not unusual in patients with multiple myeloma, special studies are necessary in this disease to determine whether all or, as is usually true, only part of the increase in total calcium is due to abnormal protein binding. We have encountered one seemingly healthy woman with hypercalcemia (13 mg/100 ml) in whom the increase above the normal range for calcium was entirely due to an unexplained increase in the nonultrafilterable fraction — apparently because of the presence of an unidentified calcium-binding protein in her serum. That naturally occurring proteins may have great affinity for calcium is perhaps best exemplified in female birds, whose increase in calcium-binding vitellin during shell formation may cause a twofold to tenfold increase in serum calcium without altering the amount of ultrafilterable calcium.[85]

The foregoing comments are made with the full realization that ion-specific electrodes that are capable of accurately measuring ionic calcium in serum should soon be available. Such electrodes will make it unnecessary to evaluate the effect of various serum proteins in determining the presence or absence of increased concentrations of biologically active calcium. Experimental studies with newly developed electrodes are in progress, but much remains to to learned before these techniques are established as reliable and can be routinely applied in clinical medicine.[81,86]

Saturation

In adults the concentration of calcium in serum or plasma is normally between 8.8 and 10.6 mg/100 ml. The values for women (8.8 to 10.2) are slightly lower than those for men (9.2 to 10.6), and the minimally different normal values reported by several laboratories can be attributed to variations in methodology. Small differences in the concentration of albumin may partially account for the lower value for serum calcium in women, but most of the difference in normal calcium concentrations of men and women may be due to a well-documented but unexplained calcium-reducing effect of estrogens.[87]

Plasma is capable of maintaining in solution much greater concentrations of calcium and phosphorus than are normally present. But if the calcium concentration in plasma is increased to between 15 and 20 mg or the phosphorus concentration to between 12 and 15 mg/100 ml, it can be demonstrated that the usual degree of ultrafilterability of both of these components is reduced.[88] The hypothesis has been formulated that in the presence of extreme hypercalcemia or hyperphosphatemia, a poorly soluble calcium-phosphorus-protein complex is formed.[89-91] The identity of such a complex has not been established, but there is the well-recognized clinical correlation of rapidly deteriorating renal and cerebral function coincident with the advent of sufficient hypercalcemia to alter the usual ultrafilterable proportions of calcium and phosphorus in serum.[92]

CALCIUM DEFICIENCY WITH NORMOCALCEMIA

Rickets and Osteomalacia

The term **rickets** is used to designate those disorders in which mineralization of both proliferative cartilage matrix and osteoid tissue is impaired. Except in the rare

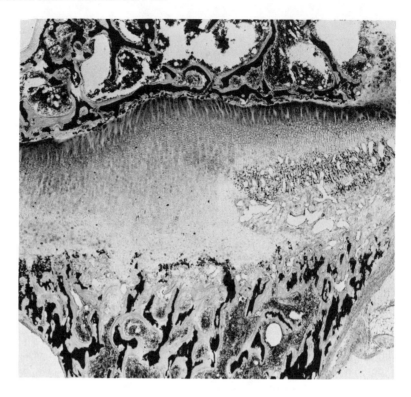

FIGURE 2.   Rachitic bone. Undecalcified section of proximal tibia from a rat fed a high-calcium, low-phosphorus diet. (Magnification × 50) Appearance of the bone section is typical of that seen in human, vitamin D-deficiency rickets. Note the wide epiphysis caused by proliferating cartilage cells and the sparsely mineralized (black), osteoid-encompassed primary trabeculae in the metaphysis. (From Thomas, W. C., Jr., Howard, J. E., and Connor, T. B., *Bull. Johns Hopkins Hosp.*, 101, 123, 1957. With permission)

developmental types of epiphyseal dysplasia,[93],[94] when there is impaired mineralization of cartilage matrix there is also deficient mineralization of bone matrix (Figure 2). As will be mentioned subsequently, the converse is not necessarily true, for in certain experimentally induced pathological states, impaired mineralization may be limited to bone matrix. "Osteomalacia" signifies the presence of increased amounts of unmineralized bone matrix, i.e., osteoid tissue, and occurs clinically in postpuberal persons whose epiphyses are fused and in whom, therefore, proliferative epiphyseal cartilage is absent (Figure 3).

Deficiency of vitamin D is the most common cause of rickets.[95] This deficiency occurs most frequently in inhabitants of northern latitudes where there is limited opportunity for exposure to the ultraviolet portion of the sun's spectrum required for (activation of provitamin D_3 (7-dehydrocholesterol) to cholecalciferol in the skin.) Nutritionally adequate amounts of vitamin D may also be obtained from natural food sources (eggs, butter, cream, liver, and fish) or from foods fortified with vitamin D_2 (ergo-calciferol) such as milk and bread.[53] Vitamins D_2 and D_3 are equally effective in man, but in birds and new-world primates, vitamin D_3 is the more effective antirachitic vitamin.

In addition to deficient mineralization of cartilage matrix and bone, vitamin D deficiency is characterized by markedly impaired intestinal absorption of calcium, hypophosphatemia, and reduced urinary excretion of calcium (often < 50 mg/24 hr).

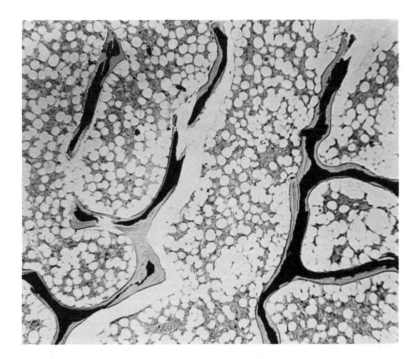

FIGURE 3.   Osteomalacia. Undecalcified section (magnification × 125) of bone from a 52-year-old woman with resistant, hypophosphatemic "rickets." The mineralized portion of the trabeculae is indicated by the black ($AgNO_3$) stain; and the large amount of unmineralized matrix, i.e., osteoid, indicates the severity of the disorder.

Although in this disorder the amount of urinary phosphorus will reflect, as in normals, the dietary intake, the renal tubular reabsorption of phosphorus is reduced. The magnitude of the characteristic increase in serum alkaline phosphatase activity is a reflection of the severity of bone involvement. The serum concentration of calcium is ordinarily maintained within the normal range. However, in Central Europe at the end of World War I, hypocalcemia was present in many poorly nourished rachitic children.[96] Yet it is uncertain whether the hypocalcemia occurred because of more complete deficiency of vitamin D than is usually present or because of a coincident depletion of magnesium.

Occasionally states of vitamin D deficiency are induced by long-term use of drugs that alter metabolism of the vitamin. Phenobarbital and diphenylhydantoin are two such compounds that, by action on liver enzymes, stimulate conversion of 25-hydroxycholecalciferol to inactive products.[53] A supplemental intake of vitamin D is usually adequate to compensate for any deficiency caused by this accelerated catabolism of vitamin D.

In the U.S., familial X-linked, hypophosphatemic rickets occurs more frequently than that due to vitamin D deficiency.[97] Although clinically quite similar to the syndrome caused by deficiency of vitamin D, this familial disorder is not corrected by any of the currently available metabolites of vitamin D, including 1,25-dihydroxycholecalciferol. A clinically similar form of hypophosphatemic rickets or osteomalacia occurs sporadically as an acquired disorder. This nonfamilial variety may develop in either sex at any age. Interestingly, a number of these patients have been cured by removal of localized, small mesenchymal tumors.[98,99] This sequence suggests that the tumors were producing substances that caused the rickets, but to date there is no information on specific secretory products of such tumors.

Recently, another heritable form of rickets has been delineated in which cure can be effected by much larger doses (10,000 to 50,000 IU/day) than the usual physiologic amounts of vitamin D.[100] This vitamin D-dependent disorder differs from vitamin deficiency rickets in that some degree of hypocalcemia is regularly present and the serum concentration of phosphorus may be only slightly reduced. Because afflicted patients have responded to very small doses of 1,25-dihydroxycholecalciferol, the suggestion has been made that patients with this disease, although not azotemic, have impaired production of this specific metabolite.[101]

Hypophosphatasia is a rare developmental disorder characterized by reduced serum concentrations of alkaline phosphatase and rachitic changes in bone.[102] In this disorder the defect is thought to be due to impaired production of alkaline phosphatase by bone and cartilage cells.

Several other unusual osteomalacic states may occur more often than are recognized. There is an acquired osteomalacia of unknown cause that apparently involves primarily the axial skeleton.[103] Another unusual type of osteomalacia, which has been recognized only in adults, is characterized by formation of osteoid tissue that is unique in that it does not refract polarized light.[104] This disorder has been termed "fibrogenesis imperfecta ossium."[104,105] The cause of both of these osteomalacic states is still speculative.

Finally, rachitic-like changes in bone and cartilage can be induced experimentally by administration of a number of trace elements. Fluoride, strontium, beryllium, manganese, and cadmium interfere with the process of mineralization when present in excess.[106-110] Occasionally, clinical instances of rickets or osteomalacia have been recognized after purposeful or accidental ingestion of large amounts of these elements.[111,112]

Steatorrhea

In chronic diarrheal disorders, especially in steatorrhea, large losses of calcium occur by way of the stools, and skeletal rarefaction may ensue. In patients with tropical sprue, hypocalcemia is infrequent and rickets or osteomalacia is only rarely seen.[113-115] However, the nontropical variant of sprue in northern zones is frequently characterized by the presence of osteomalacia and hypocalcemia. These differences in clinical aspects may be due in part to the greater likelihood of vitamin D deficiency in people of the temperate and northern zones where there is often reduced exposure to sunlight, and the marked steatorrhea prevents absorption of compensatory amounts of fat-soluble vitamin D. That deficiency of vitamin D may not be the explanation for the hypocalcemia in nontropical sprue is suggested by the finding of normal concentrations of vitamin D in sera of three hypocalcemic patients with nontropical sprue,[116] and it is now known that coexistence of magnesium deficiency is frequent and may account for the hypocalcemia in many instances.

In recent years there has been a widespread surgical practice of creating a jejunal bypass to cause malabsorption and thereby ameliorate morbid obesity. This diversion of the intestinal tract is regularly followed by marked steatorrhea. Not only has much of the intestine where calcium is normally absorbed been excluded, but the loss of calcium is aggravated by formation of calcium soaps.[117] Hypocalcemia is not rare in these patients and, though there are several possible causes,[118] in our experience it has always been associated with the presence of hypomagnesemia. Furthermore, parenteral administration of magnesium has been followed by correction of the hypocalcemia. It should be anticipated that patients with an extensive intestinal bypass will, unless carefully managed, eventually develop severe skeletal lesions secondary to the excessive loss of calcium and to malabsorption of other essential nutrients.

Acidosis

In the absence of renal insufficiency with marked azotemia, metabolic acidosis is characterized by an increased excretion of calcium in the urine.[39,43,119] Unless this loss is compensated for by an increase in intestinal absorption of calcium, skeletal rarefaction results. Normocalcemia is regularly present. In contrast to metabolic acidosis, respiratory acidosis causes little if any increase in urinary calcium.[119,120] The concept is developing that hypercalciuria induced by acidosis may be accounted for by alteration in renal tubular function.[43,121] The loss of calcium in urine is then balanced by withdrawal of calcium from bone to maintain normal concentration in serum and extracellular fluids.

Acidosis may be induced by medication but most often occurs as a result of deranged renal metabolism. Renal tubular defects in hydrogen ion excretion or bicarbonate wastage may be developmental or acquired. When present, there is hyperchloremia and reduced serum bicarbonate in addition to the hypercalciuria. Serum concentration of phosphate may also be reduced, in which case osteomalacia or rickets may be present. Patients with renal tubular acidosis often develop nephrocalcinosis or renal calculi.[122] In some of the more extensive disorders of renal tubular function, marked aminoaciduria and renal glycosuria may be present.[123,124] Correction of the acidosis by administration of bicarbonate or citrate salts is followed by a decrease in urinary calcium and restoration of the skeleton to normal.

Osteoporosis

In normal healthy persons, bones increase in density and strength until approximately age 25. Then a slow process of atrophy begins, so that in elderly subjects bones may be quite thin and easily broken.[125-127] This decrease in bone mass with age may be due in part to decrease in physical activity, but aging per se has an effect. The normally less dense bones of women often become so fragile by age 50 or 60 that spontaneous fractures are not uncommon. The maximum density of bones is modified by race and sex. For example, in the U.S. the bones of black men are denser and less easily crushed than are bones of similarly aged white men.[125] Black women have bones that are essentially of the same density and strength as those of white men. There is also a correlation between muscular development and bone mass — heavily muscled individuals often develop dense, strong bones.[17,128]

The role of the menopause in aggravating the course of senile atrophy of bone remains somewhat uncertain. Most investigators, however, consider a lack of estrogens to be an additive factor in the development of osteoporosis in women. Interestingly, administration of estrogens in doses too small to cause a detectable change in conventional calcium or phosphorus balance will lessen bone tenderness and possibly the frequency of fractures in osteoporotic women. Although long-term therapy with estrogens appears to be beneficial in elderly osteoporotic women, any estrogen-induced bone changes that might occur have not been sufficient to improve the radiologic density of the skeleton.[127,129]

By using balance techniques, Malm[23] has shown that most elderly people require more than 400 mg of calcium daily and some more than 800 mg to achieve calcium balance. Since many people do not maintain a calcium intake of this magnitude, a chronic deficiency of calcium may contribute importantly to the development of senile osteoporosis.

Emotional illnesses during childhood often impair optimal growth of the skeleton.[130] In addition, there are disorders that may specifically affect the skeleton, e.g., osteogenesis imperfecta. Bones remain slender, and epiphyses do not close in patients with gonadal insufficiency. In certain hematologic disorders the bones are often quite osteoporotic.[131] Idiopathic osteoporosis, a disorder apparently affecting only the skeleton, is characterized by progressive loss of bone mass and, eventually, spontaneous fractures with consequent deformities which may be life-threatening.[132] Idiopathic osteoporosis may begin in children or young adults and is apparently due to an impairment of bone formation rather than increased resorption.

In all of these osteoporotic states, the concentration of calcium in serum remains normal. Urinary calcium is also usually within the normal range, but may be increased during the early phases of an acquired disorder in which there is interference with normal rates of bone formation.

Cushing's Syndrome

Prolonged exposure to an excess of adrenal glucocorticoid hormones, whether from hypersecretion or oral administration, regularly leads to some diminution of skeletal mass. Evidence to date indicates that glucocorticoids do not cause an increase in bone resorption, but do reduce bone formation rates.[133] After removal of the excess hormone in patients with glucocorticoid-induced bone atrophy, bone strength improves; but in adults at least, the radiologic appearance never returns to the density expected for the patient's age and sex.[134]

HYPOCALCEMIC STATES

Hypocalcemia may be present without any overt signs or symptoms. A hyperexcitable state of the nervous system, as evidenced by the presence of a Chvostek or a Trousseau phenomenon, or occasionally a generalized convulsion, is what usually leads the physician to suspect hypocalcemia. These evidences of hyperexcitability of the nervous system, however, may not be present, and this is especially so if the hypocalcemia has been of long duration. It is interesting, though not explained, that in some hypocalcemic patients the Chvostek sign may be absent but the Trousseau phenomenon present or vice versa. Rarely, chronically hypocalcemic patients may present with papilledema and other signs of increased intracranial pressure.[135] Premature cataracts and calcification of the basal ganglia are other findings that may lead the physician to suspect the presence of hypocalcemia.

Hypoparathyroidism

The most important function of parathyroid hormone is to maintain a stable, normal concentration of calcium in serum and extracellular fluids. Removal of the parathyroid glands is promptly followed by development of hypocalcemia, and, unless treatment is instituted, the serum concentration of calcium then becomes stabilized at a lower-than-normal concentration. Besides the increased neuromuscular irritability that occurs with abrupt reduction in calcium concentration, chronic hypocalcemia may cause peripheral cataracts and "trophic" changes in skin and hair.[136] If hypoparathyroidism is present during the formative stage of tooth development, dental changes may also occur.[137] These trophic changes, however, are not specific for hypoparathyroidism and may appear during persistent hypocalcemia of any cause. Rarely, the bones become unusually dense when hypoparathyroidism has been untreated for many years.[136] Although hypoparathyroidism is most commonly due to inadvertent removal of the parathyroid glands or their blood supply during operations on the thyroid gland, this endocrine deficiency may develop spontaneously.[138-140]

Hyperphosphatemia is regularly present in untreated hypoparathyroidism. As with calcium, the serum concentrations of phosphorous then become stabilized at a new increased concentration, and although the renal clearance of phosphorus is decreased, the amount in the urine of the stabilized patient is similar to that of patients with normal parathyroid function.

Pseudohypoparathyroidism is an interesting syndrome characterized by short stature, facial rounding, developmental defects of bone, and changes in the concentrations of calcium and phosphorus in serum identical to those present in idiopathic or postoperative hypoparathyroidism. Surgical exploration of a few individuals with this syndrome has revealed histologically normal or hyperplastic parathyroid glands.[141] Increased serum concentrations of immunoassayable parathyroid hormone are present, and the current concept is that afflicted patients do not have the cellular capacity to respond normally to parathyroid hormone.[142] Initially, it was thought that patients with pseudohypoparathyroidism were completely unresponsive to parathyroid hormone, but it is often possible to obtain a partial response to administered hormone.[143,144]

Magnesium Deficiency

During recent years, hypomagnesemia has been clearly delineated as a cause of hypocalcemia.[57,59,145] In this hypocalcemic state, parathyroid hormone concentrations in serum are mildly increased but are disproportionately low with respect to the degree of hypocalcemia. There is also an impaired calcemic response to injected parathyroid hormone. Interestingly, magnesium deficiency also impairs the calcemic response to vitamin D_2.[146] Although the mechanism is unknown, it appears that nearly normal serum concentrations of magnesium are necessary for parathyroid hormone-induced mobilization of calcium from bone. The phosphaturic effect of parathyroid hormone is less affected by magnesium depletion than is the effect on calcium homeostasis.

Malabsorption states, including steatorrhea, are the most common causes of magnesium deficiency. In hospitalized patients, loss through exudative wounds coupled with inadequate replacement of magnesium is not an infrequent cause of magnesium depletion. Hypomagnesemia has also been recognized in infants with persistent hypocalcemia, and the serum calcium was restored to normal concentration when supplementary magnesium was provided.[147,148]

Posthyperparathyroidism

Hypocalcemia sometimes occurs after the removal of a functioning parathyroid adenoma, especially if the patients have detectable bone disease (osteitis fibrosa).[149] In such patients the phosphorus concentration in serum remains low, and the low phosphorus is a useful means of differentiating this hypocalcemic state from that due to postoperative hypoparathyroidism. The hypocalcemia and hypophosphatemia present after removal of a parathyroid adenoma is thought to be caused by the avidity of the skeleton for these minerals during their vigorous remineralization after relief from the influence of excessive parathyroid hormone. The hypocalcemia may persist for weeks or months, and the presence of tetany may necessitate treatment.

Acute Pancreatitis

This disorder is sometimes accompanied by an abrupt reduction in serum calcium concentration.[150,151] Postmortem analyses of the pancreas and adjacent tissues from patients with pancreatitis have revealed the presence of large quantities of calcium as calcium soaps. It seems possible that the rapid deposition of grams of calcium could place so great a drain on extracellular calcium that homeostatic forces could not immediately replace the large losses. However, processes other than calcium deposition as soaps may participate in causing hypocalcemia in patients with pancreatitis. Hypomagnesemia may occur, presumably because of precipitation with fatty acids, and thereby interfere with parathyroid hormone mobilization of calcium from bone.[151] Hyperglucagonemia is often present in patients with pancreatitis, and this hormone increases the release of calcitonin.[152] Although an excess of calcitonin would not be expected to reduce serum concentrations of calcium to any great degree in healthy adults, the effects may be exaggerated in patients with acute pancreatitis and certainly so if responsiveness to parathyroid hormone is impaired.

Miscellaneous Disorders

Several unusual traumatic events may induce acute hypocalcemia. Overloading the circulating fluids quickly with inorganic phosphorus, administered either parenterally or orally, causes a marked increase in serum phosphorus concentration and a rapid reduction in calcium concentration.[153,154] Although it is difficult to determine from these reports, in such circumstances serum phosphorus probably has been increased to concentrations at which ultrafilterability of calcium and phosphorus are reduced.[88]

Acute anuria, such as may follow transfusion of incompatible blood, crush injuries, or septic shock, is another clinical state in which marked hyperphosphatemia and hypocalcemia may develop. [155,156] In all of these acute hyperphosphatemic situations, there is often deposition of calcium phosphate aggregations in soft tissues, and it is thought that the increased phosphorus concentration interferes with the usual homeostatic forces at the skeletal-extracellular fluid boundary to prevent the release of calcium.

In chronic azotemic renal disease, increased secretion of parathyroid hormone is usually sufficient to maintain normocalcemia despite the persistence of appreciable hyperphosphatemia. Occasionally, however, patients with chronic azotemic renal disease become hypocalcemic,[157,158] and it is now thought that marked deficiency of the dihydroxy derivative of vitamin D may account for the inability of these patients to maintain a normal concentration of calcium.[52] In a few patients with chronic renal disease, hypermagnesemia may be of such magnitude as to reduce parathyroid hormone secretion enough to cause hypocalcemia. In our experience, this situation occurs when the azotemic patients have been ingesting large amounts of absorbable magnesium.

Transient hypoparathyroidism may explain the hypocalcemia that occurs frequently during the neonatal period of babies, and this hypocalcemia is more apt to occur in those babies receiving cow's milk.[159-161] A more protracted neonatal and infantile hypocalcemia due to functional hypoparathyroidism is sometimes seen when the mothers are hypercalcemic (e.g., because of hyperparathyroidism) at the time of delivery.[162] Occasionally, magnesium deficiency with hypomagnesemia may be the cause of hypocalcemia in the neonatal period or during early infancy.[147,148]

The homeostatic forces for maintenance of normocalcemia are apparently more easily overwhelmed in infants and children than in adults. Transient hypocalcemia may occur in infants with hypernatremia,[163] in children with acute infections,[6] and, rarely, in children with renal tubular disorders.[164,165]

In people of all ages, ingestion of sufficient quantities of substances, such as fluoride or oxalate which form insoluble compounds with calcium, may induce hypocalcemia.[166,167] Similarly, chronic ingestion of cation-binding exchange resins may lead to the development of hypocalcemia.[168] These resins were originally thought to remove calcium at a rate which exceeded that of the replacement from skeletal stores, but it is quite likely that magnesium depletion could have been a more important factor in causing this hypocalcemia.

In a few instances, hypocalcemia has been noted in patients with rapidly developing osteoblastic metastases or during mineralization of previously lytic skeletal metastases after hypophysectomy or effective hormonal treatment.[169-171] It appears that the explanation for the reduced calcium concentration in these situations is rapid mineralization of the skeletal lesions with removal of calcium from extracellular fluid at a rate which exceeds that of replacement.[149] This is analogous to Albright and Reifenstein's[172] hypothesis regarding the hypocalcemia that follows removal of a parathyroid tumor.

Finally, there are a few other unusual causes of hypocalcemia. The administration of several antituberculous drugs may cause renal tubule dysfunction with marked electrolyte abnormalities and hypocalcemia.[173,174] Hypomagnesemia may be the particular abnormality responsible for the hypocalcemia in such patients,[174] but other factors may also be operative. Totally unexplained at present is the acute, transient hypocalcemia that may develop early in the postoperative period after lung resection.[175] A suggestion that hormones of the anterior pituitary gland may assist in maintaining normocalcemia is gleaned from the rare occurrences of hypocalcemia in patients with pituitary insufficiency.[176]

HYPERCALCEMIC STATES

Hypercalcemia is of itself a potentially dangerous derangement of calcium homeostasis. Minimal degrees of hypercalcemia may be tolerated for months or years with little evidence of ill effects, but when the serum concentration of calcium is increased by 2 to 3 mg/100 ml or more, adverse effects usually occur. Hypercalcemia impairs renal function, and this impaired function may persist after correction of the hypercalcemia. Nausea, vomiting, pruritis, and polyuria are frequent symptoms in patients with modest hypercalcemia; but in markedly (Ca > 15 mg/100ml) hypercalcemic patients, drowsiness, stupor, coma, and anuria may develop in rapid progression to a fatal outcome. Occipital lobe blindness is not a rare event in patients with extreme degrees of hypercalcemia (personal observations). Thus, once detected, the cause of hypercalcemia should be determined, and corrective measures should be instituted.

Exogenously Induced Hypercalcemia

It must be exceptionally rare for the intestinal tract to absorb so much ingested calcium that homeostatic mechanisms are overcome. In those instances when ingested calcium can be shown to induce significant hypercalcemia, however, there is usually something detectably awry with the patient to make him susceptible to this derangement. For example, patients with severe hypothyroidism readily become hypercalcemic after ingesting several grams of calcium, as in calcium chloride, per day.[177] In this situation a slowed disposition of absorbed calcium appears to be the explanation. Patients ingesting large amounts of vitamin D (20,000 to 50,000 IU/day) may become overtly hypercalcemic when receiving supplementary calcium. A susceptibility to becoming hypercalcemic upon ingesting calcium may persist for months in patients who have developed the milk-alkali syndrome (see following). A similarly persistent susceptibility to become recurrently hypercalcemic has been noted in children with idiopathic hypercalcemia.[178] We have observed one patient, however, with acute postoperative hypoparathyroidism who became markedly hypercalcemic within 24 hr after beginning treatment with calcium chloride, and in this patient we were unable to detect any reason for her sensitivity to the oral calcium.

Hyperparathyroidism

A regular accompaniment of hyperparathyroidism is an increased serum concentration of calcium. Except when renal insufficiency coexists, hypercalciuria is nearly always present and the inorganic phosphate in serum is reduced. Although increased intestinal absorption of calcium can be demonstrated in patients with hyperparathyroidism, the major source of the excess calcium in serum and extracellular fluids is from the skeleton. In most cases, bone resorption is not sufficient to cause roentgenographically evident lesions, but histologic examination of bone will reveal evidence of active resorption long before development of the gross changes that are required for radiologic demonstration. Hyperparathyroidism is much less common in children than in adults, but when present in the young, detectable bone lesions occur with great frequency.[179] Hyperparathyroidism may occur as a familial disorder, either as the only endocrinopathy or as one part of multiple endocrine adenopathies.[180] Most often a single benign adenoma will be the cause of hyperparathyroidism, but occasionally, there is diffuse hyperplasia of all the glands or there are two or more adenomas.[181] Rarely, the parathyroid tumor will be malignant.[182] Surgical removal of overactive tissue is the only curative treatment.

Milk-Alkali Syndrome

It is now well known that some patients taking calcium orally in conjunction with an absorbable alkali (usually as therapy for a peptic ulcer) may become hypercalcemic.[183-185] Apparently, the absorbable alkali is the major determining factor in the classic examples of this syndrome and there is reason to believe that extracellular alkalosis, however induced, may render patients unusually liable to develop exogenously induced hypercalcemia. That the serum half-life of radioactive calcium is significantly prolonged in mildly alkalotic subjects[186] possibly explains the role of absorbable alkali in the pathogenesis of this syndrome.

A common misconception is that hypercalciuria does not occur in patients with the milk-alkali syndrome. Although such patients are usually not hypercalciuric, this is because of the attendant impaired renal function, which occurs quickly in alkalotic hypercalcemic patients.[187] Hypercalciuria is demonstrable in such patients if they are evaluated before they develop azotemia.

Hypervitaminosis D

Dosages of vitamin D (10,000 to 20,000 IU/day) larger than those required to prevent rickets (400 to 800 IU/day) enhance the intestinal absorption of calcium.[188] Larger dosages ($>$ 100,000 IU/day) may cause hypercalcemia. This hypercalcemic effect is accentuated by oral administration of calcium salts.[189-192] However, the hypercalcemic effect of large doses of vitamin D is not necessarily dependent on calcium intake for greater amounts (500,000 IU/day) may induce hypercalcemia even though the diet is free of calcium. In experimental animals with hypervitaminosis D, evidence of increased bone resorption can be detected histologically.[193-195] Paradoxically, such animals also develop an excessive amount of osteoid tissue, or osteomalacia.[196-198]

In the absence of azotemia, patients receiving large amounts of vitamin D are almost invariably hypercalciuric even in the absence of hypercalcemia.[51] Today, hypoparathyroidism is the disorder in which large doses of vitamin D are most often used. This vitamin, supplemented by calcium salts, is effective in correcting the hypocalcemia in such patients. However, the magnitude of the calcemic response to vitamin D is inexplicably variable,[51] so the calcium concentration in serum must be monitored from time to time to avoid hypercalcemia. Interestingly, the hypercalcemia due to hypervitaminosis D is often accompanied by an extracellular alkalosis.[51,197]

Neoplasms With and Without Overt Metastases in Bone

It is remarkable how cells from certain types of cancers can induce local resorption of bone when they metastasize and multiply in skeletal tissue. Carcinoma of the breast has a special tendency to spread to bone and cause such resorption. When prostatic cancer metastasizes to bone, however, the usual effect is the opposite, and there is stimulation of new bone formation and a radiologic appearance of increased bone density.

The hypercalcemia that sometimes occurs in patients with osteolytic metastases was initially attributed to the release of calcium from the sites of the osteolytic lesions.[199] This concept is inadequate to explain the observations that despite the presence of osteolytic lesions of similar magnitude, only a minority of these patients are hypercalcemic. Thus, products of the tumor cells must alter bone cell metabolism differently in the patients who become hypercalcemic than in those who remain normocalcemic. If the neoplasm in the hypercalcemic patients with metastases is responsive to treatment, the hypercalcemia will subside before, or coincident with, histologic or radiologic evidence that the activity of the cancer cells has been altered. This

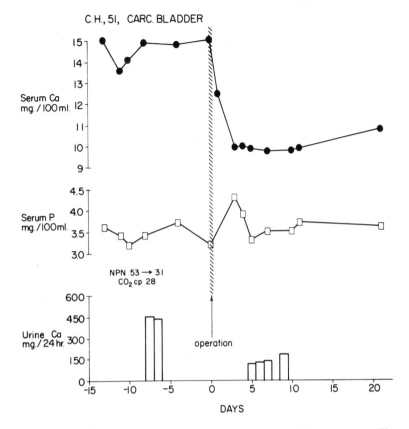

C.H., 5l, CARC. BLADDER

Serum Ca
mg./100ml.

Serum P
mg./100ml.

NPN 53 → 31
CO_2 cp 28

Urine Ca
mg./24 hr.

operation

DAYS

FIGURE 4.   Correction of hypercalcemia after removal of bladder cancer. The patient was a 51-year-old man with marked hypercalcemia and a carcinoma in the dome of the bladder. The changes in serum and urine calcium after operative resection of the tumor were dramatic and quite like those seen after removal of a parathyroid tumor.

sequence of events is not unusual in patients with metastatic breast cancer that is responsive to therapy with androgen or estrogen.[172,200-202] Conversely, if the metastases are inadvertently stimulated by specific agents, hypercalcemia may be much worse or induced if not present.[203]

Hypercalcemia may occur in patients with leukemia and various types of lymphoma. In these disorders, increased resorption of bone may be histologically evident.[204-206] There is evidence to suggest that the previously mentioned osteoclast activating factor may participate in the augmented bone resorption that sometimes occurs in leukemia and lymphoma.[73,74]

Hypercalcemia has often occurred in patients with discrete neoplasms that have not metastasized to bone.[207-209] In some of these patients it can be demonstrated that the tumor is producing a parathyroid hormone-like substance capable of promoting the release of calcium from the skeleton.[210,211] In other patients whose tumors cannot be shown to be producing such a substance, there is speculation, based largely on limited experimental studies, that the tumors are secreting either large amounts of a prostaglandin or a substance that induces increased formation of prostaglandin E by bone cells, with a consequent increase in skeletal resorption.[71,72,212] Regardless of what such tumors may be secreting to cause increased resorption of bone, local removal of the tumor or effective treatment by radiation or other means has repeatedly corrected the hypercalcemia just as promptly as would removal of a parathyroid adenoma in a patient with hyperparathyroidism (Figure 4).[208,213]

Neoplasms originating in many tissues (lung, esophagus, colon, ovary, kidney, etc.) have been shown to be capable of inducing hypercalcemia by a remote effect.[214,215] Such tumors need not be large, and afflicted patients are often mistakenly thought to have hyperparathyroidism. There are, however, subtle differences between the usual clinical presentation of patients who have malignant neoplasms causing hypercalcemia and that of patients with hyperparathyroidism. Extracellular alkalosis, hypokalemia, anorexia, and a malaise out of keeping with the degree of hypercalcemia are some of the manifestations which suggest that a nonparathyroid neoplasm, rather than hyperparathyroidism, is causing the hypercalcemia.

Hyperthyroidism

An excess of thyroid hormones, whether occurring in clinical hyperthyroidism or as a result of exogenous administration, causes a marked increase in urinary and fecal calcium.[216] In chronic severe hyperthyroidism, radiologic and histologic evidence of bone resorption may be detectable.[217] Apparently as a result of accelerated bone resorption, hypercalcemia may also be present.[218] Usually the magnitude of the hypercalcemia is not great but occurs rather frequently in patients with hyperthyroidism.[219-222] The hypercalcemia is promptly corrected when the patient is restored to a euthyroid state, but the repair of induced bone changes may require a number of months. How thyroid hormones induce these changes in calcium and bone metabolism is not known, but presumably it is by a direct effect upon bone cells.[223]

Disuse Atrophy

When a large area of the body is immobilized (e.g., as in a body cast or after spinal cord injury), large amounts of calcium are introduced into circulating body fluids from the rapidly atrophying bones. In such circumstances hypercalcemia often occurs, especially in young people. As might be expected, hypercalcemia is more likely to occur during immobilization if this stress has been compounded by an injury to bone, such as a fracture.[224-227]

It is not adequately appreciated that confining a patient who already has increased bone resorption and hypercalcemia to bed, e.g., patients with hyperparathyroidism, hypervitaminosis D, or Paget's disease, may induce a sufficient increase in the hypercalcemia to be lethal within a few days.[92,228,229] In a few instances, patients with resistant rickets who were receiving large amounts of vitamin D as therapy became hypercalcemic only after confinement to bed.[230,231]

Sarcoidosis

Hypercalcemia occurs frequently in sarcoidosis, and as many as one third of the patients with this disorder may develop hypercalcemia at some time during the course of the disease.[232,233] At any one time, however, considerably less than one third of a series of such patients will be hypercalcemic.[234,235] Radiologically detectable bone lesions of sarcoid are not rare,[236] but there is no correlation between the presence of such lesions and the existence of hypercalcemia.[237] Interestingly, the abnormally increased concentrations of calcium in this disorder may persist for only a few weeks and then subside spontaneously. In the unusual patient, however, hypercalcemia may persist for years unless treated.

The concentration of various proteins in serum is frequently altered in patients with sarcoidosis, but the presence of these protein abnormalities does not correlate with the occurrence of hypercalcemia.[207,232] Renal damage may occur as a consequence of the hypercalcemia or because of sarcoid lesions within the kidneys, and nephrocalcinosis or renal calculi are occasionally present.[238,239]

The cause of hypercalcemia in sarcoidosis is not known. The similarities in the altered calcium metabolism of patients with sarcoidosis to those of subjects receiving large amounts of vitamin D are intriguing.[240,241] Hypercalciuria and increased intestinal absorption of calcium are often present — with much greater frequency than is hypercalcemia. Assay of serum from affected patients for vitamin D-like activity has not revealed any evidence of increased amounts of vitamin D,[116] but this report does not negate the possibility that patients with sarcoidosis might have an enhanced conversion of vitamin D to active metabolites. Evidence was cited in recent reports to show that the metabolism of vitamin D may be accelerated in patients with sarcoidosis,[242] but there was no increase in the formation rate of dihydroxycholecalciferol.[243]

An increased sensitivity to exogenous vitamin D has been repeatedly demonstrated in patients who have sarcoidosis.[232,240,241,244,245] Yet a similar sensitivity may exist in other hypercalcemic states. Interestingly, patients with sarcoidosis also have an increased sensitivity to parathyroid hormone.[246] Deprivation of vitamin D improves the disordered calcium metabolism in sarcoidosis patients,[247] but this response might occur in patients with other causes of hypercalcemia. Because adrenal corticosteroids regularly correct the hypercalcemia and ameliorate the other aspects of sarcoidosis, it is possibly the particular type of tissue inflammation characteristic of this disease that induces changes responsible for the altered calcium metabolism.

Idiopathic Hypercalcemia of Infants

Hypercalcemia may occur during infancy and early childhood in association with renal tubular dysfunction,[248] after ingestion of apparently excessive amounts of vitamin D,[249-251] or as one aspect of a syndrome in which the infants have a peculiar facies and developmental anomalies of the aortic valve.[252] In these hypercalcemic states, either an excess of or unusual sensitivity to vitamin D is thought to be the cause of the hypercalcemia.[253,254] Assay of serum for vitamin D from a number of hypercalcemic infants in the U.S. revealed increased concentrations of vitamin D in some but not in others.[116,255] In England, during a period after World War II when milk powder was fortified with vitamin D, infantile hypercalcemia occurred frequently; the reduced incidence of recognized hypercalcemia after cessation of milk fortification provided support for the hypothesis that an excess of or unusual sensitivity to vitamin D was responsible for many of these instances.

In other hypercalcemic states in infancy there is little reason to suspect the cause to be undue sensitivity to vitamin D. Two such syndromes in which hypercalcemia has been noted are the "blue diaper syndrome," which is characterized by an abnormality in tryptophan metabolism, and subcutaneous fat necrosis.[256,257]

Adrenal corticosteroids, and also calcitonin, have been quite effective in correcting the hypercalcemia in a number of afflicted infants,[64,66,258,259] but the response to these agents does not provide additional insight into cause.

Rare Causes of Hypercalcemia

There are certain other unusual circumstances in which hypercalcemia may occur. High concentrations of calcium have been noted in beryllium intoxication.[260] The similarity of the histological features in berylliosis and sarcoidosis suggests that the pathogenesis of the hypercalcemia in these two disorders may also be similar. Patients with acute adrenal insufficiency developing spontaneously or after removal of hyperfunctioning adrenal tissue may become hypercalcemic.[261,262] This disorder and other aspects of the importance of adrenal corticosteroids to calcium metabolism will be discussed in the subsequent section. Mild hypercalcemia has been noted in some patients with acromegaly but has been corrected when this endocrinopathy was controlled.[263,264] Although hypercalciuria and osteoporosis are commonly present in acromegalic patients,[265] there is relatively little specific information on the effect of an excess of growth hormone on calcium and bone metabolism.

An increased incidence of hypercalcemia has been reported in patients with tuberculosis.[266,267] Reviewing these reports reveals that factors other than tuberculosis may have contributed to or accounted for the hypercalcemia: The patients were confined to bed and some of them ingested excessive amounts of alkali and calcium. These reports, however, provide reason for speculating that tuberculosis, like sarcoidosis, is conceivably accompanied by an increased susceptibility to hypercalcemia.

Another acute hypercalcemic state of unknown cause occurs in children and adolescents. We have seen only two such patients, but Harrison[268] has encountered similar patients and the true incidence is unknown. In all of these young patients abrupt onset of anorexia, nausea, and vomiting was followed by lethargy. Marked hypercalcemia (approx. 15 mg/100 ml) has been present and has promptly responded to adrenal corticosteroids with no sequellae and no recurrence of the hypercalcemia. In none of these patients could the hypercalcemia be adequately explained. Several instances of marked hypercalcemia due to ingestion of large amounts of vitamin A have been reported.[269,270] Although hypervitaminosis A is known to cause periosteal calcification,[271] there is not enough information to permit any conjecture as to the cause of the hypercalcemia.

During recent years it has been demonstrated that thiazide diuretics may induce a mild but significant increase in the serum concentration of calcium in patients with hyperparathyroidism, those receiving large amounts of vitamin D, and possibly others in whom there is increased resorption of bone.[272] The cause of this thiazide-induced change has not been defined, but is is possible that the hypokalemia and extracellular alkalosis may interfere with the uptake of calcium by bone.

CALCIUM METABOLISM AND ADRENAL CORTICOIDS

There is abundant evidence that the corticosteroids of the adrenal cortex have an effect upon skeletal tissues. General rarefaction of bones, especially of the spine and ribs, is a regular feature of chronic hyperadrenocorticism. However, in acute adrenal insufficiency, whether occurring in a patient with recognized Addison's disease[261] or in the early postoperative period after removal of hyperfunctioning adrenal tissue,[262] hypercalcemia may occur. In this circumstance, administration of cortisol or one of its analogues promptly restores serum calcium to normal values.[262]

Initially, adrenocorticotrophic hormone[273] and subsequently cortisone, cortisol, and various synthetic corticosteroids were demonstrated to ameliorate or correct the hypercalcemia in a variety of disorders.[258,274] In certain of these disorders, among which are sarcoidosis, multiple myeloma, lymphomas, immobilization hypercalcemia, and idiopathic hypercalcemia of infancy, the hypercalcemia responds dramatically to steroid therapy. In some disorders, however, corticosteroids induce only a modest reduction in calcium concentration, e.g., in hypervitaminosis D.

Initially, it was thought that corticosteroid administration did not reduce the serum calcium in patients with hyperparathyroidism,[275] but instances have now been recorded in which the calcium was decreased by 1 to 2 mg/100 ml.[276,277] The hypercalcemia associated with malignant tumors of soft tissues, especially carcinoma of the lung, often does not change on treatment with steroids.

Just how adrenal glucocorticoids modify calcium metabolism has not been clearly elucidated. In kinetic studies in dogs it has been demonstrated that chronic steroid administration impairs bone formation rates.[133] These observations support the morphologic studies of Sissons[278] on the paucity of bone and decreased osteoblastic activity in man and animals chronically exposed to an excess of corticosteroids.

In contrast to the effects of long-term dosage with steroids, it can be shown in acute experiments with young animals that administration of corticosteroids impairs bone resorption and this may be sufficiently marked to result in increased density of bones.[279] Such a decrease in bone resorption rates is probably one means whereby steroids acutely modify the degree of hypercalcemia. In experimental hypervitaminosis D, the effect of the vitamin on bone resorption is not prevented by steroids; but the hypercalcemia is lessened, and the histologic effects of both vitamin D and steroids are detectable in bone.[192,195,280] In other hypercalcemic states, such as multiple myeloma or lymphoma, it is conceivable that corticosteroids reduce the cellular release or formation of the osteoclast activating factor or other substances that promote bone resorption.

REFERENCES

1. Mitchell, H. H., Hamilton, T. S., Steggerda, F. R., and Bean, H. W., The chemical composition of the adult human body and its bearing on the biochemistry of growth, *J. Biol. Chem.*, 158, 625, 1945.
2. Widdowson, E. M. and Dickerson, J. W. T., Chemical composition of the body, in *Mineral Metabolism; an Advanced Treatise, Part A*, Vol, 2, Comar, C. L. and Bronner, F., Eds., Academic Press, New York, 1964, 2.
3. Lienke, R. I., Cullen, G., and Armstrong, W. D., Studies on the excretion and distribution of radioactive calcium, in *Metabolic Interrelations: Transactions of the First Conference*, Reifenstein, E. C., Jr., Ed., Josiah Macy, Jr. Foundation, New York, 1949, 73.
4. Armstrong, W. D., Johnson, J. A., Singer, L., Lienke, R. I., and Premer, M. L., Rates of transcapillary movement of calcium and sodium and of calcium exchange by the skeleton, *Am. J. Physiol.*, 171, 641, 1952.
5. Gilbert, D. L. and Fenn, W. O., Calcium equilibrium in muscle, *J. Gen. Physiol.*, 40. 393, 1957.
6. Park, E. A., Bone growth in health and disease, *Arch. Dis. Child.*, 29, 269, 369, 1954.
7. Howard, J. E., Calcium metabolism, bones and calcium homeostasis: a review of certain current concepts, *J. Clin. Endocrinol. Metab.*, 17, 1105, 1957.
8. Neuman, W. F. and Neuman, M. W., *The Chemical Dynamics of Bone Mineral*, University of Chicago Press, Chicago, 1958, 34.
9. Neuman, W. F., The milieu interieur of bone: Claude Bernard revisited, *Fed. Proc.*, 28, 1846, 1969.
10. Posner, A. S., Crystal chemistry of bone mineral, *Physiol. Rev.*, 49, 760, 1969.
11. Posner, A. S., Bone mineral on the molecular level, *Fed. Proc.*, 32, 1933, 1973.
12. Hastings, A. B. and Huggins, C. B., Experimental hypocalcemia, *Proc. Soc. Exp. Biol. Med.*, 30, 458, 1933.
13. Hastings, A. B., Studies on the effect of alteration in the concentration of calcium in circulating fluids on the metabolism of calcium, in *Metabolic Interrelations; Transactions of the Third Conference*, Reifenstein, E. C., Jr., Ed., Josiah Macy, Jr. Foundation, New York, 1951, 38.
14. Looney, W. B., Maletskos, C. J., Helmick, M., Reardan, J., Cohen, J., and Guild, W. R., A study of the dynamics of strontium and calcium metabolism and radioelement removal, *J. Clin. Invest.*, 37, 913, 1958.
15. Baylor, C. H., Van Alstine, H. E., Keutmann, E. H., and Bassett, S. H., The fate of intravenously administered calcium: effect of urinary calcium and phosphorus and calcium-phosphorus balance, *J. Clin. Invest.*, 29, 1167, 1950.
16. McLean, F. C. and Hastings, A. B., Clinical estimation and significance of calcium ion in the blood, *Am. J. Med. Sci.*, 189, 601, 1935.
17. Albright, F. and Reifenstein, E. C., Jr., *The Parathyroid Glands and Metabolic Bone Disease; Selected Studies*, Williams & Wilkins, Baltimore, 1948, 135.
18. Fell, H. B., Skeletal development in tissue culture, in *The Biochemistry and Physiology of Bone*, Bourne, G. H., Ed., Academic Press, New York, 1956, 401.
19. Howard, J. E., Normal calcium and phosphorus transport and body fluid homeostasis, in *Metabolic Interrelations; Transactions of the Fifth Conference*, Reifenstein, E. C., Jr., Ed., Josiah Macy, Jr. Foundation, New York, 1953, 11.
20. Wasserman, R. H. and Taylor, A. N., Some aspects of the intestinal absorption of calcium with special reference to vitamin D, in *Mineral Metabolism; an Advanced Treatise*, Vol. 3, Comar, C. L. and Bronner, F., Eds., Academic Press, New York, 1969, 321.
21. Whedon, G. D., The combined use of balance and isotope studies in the study of calcium metabolism, in *Proceedings of the Sixth International Congress of Nutrition*, E. and S. Livingstone, Edinburgh, Scotland, 1964, 425.

22. Nicolaysen, R., Eeg-Larsen, N., and Malm, O. J., Physiology of calcium metabolism, *Physiol. Rev.,* 33, 424, 1953.

23. Malm, O. J., Adaptations to alterations in calcium intake, in *The Transfer of Calcium and Strontium Across Biological Membranes,* Wasserman, R. H., Ed., Academic Press, New York, 1963, 143.

24. Schachter, D., Dowdle, E. B., and Schenker, H., Accumulation of Ca^{45} by slices of the small intestine, *Am. J. Physiol.,* 198, 275, 1960.

25. Schachter, D. and Rosen, S. M., Active transport of Ca^{45} by the small intestine and its dependence on vitamin D, *Am. J. Physiol.,* 196, 357, 1959.

26. Harrison, H. E. and Harrison, H. C., Transfer of Ca^{45} across intestinal wall in vitro in relation to action of vitamin D and cortisol, *Am. J. Physiol.,* 199, 265, 1960.

27. Rasmussen, H., The influence of parathyroid function upon the transport of calcium in isolated sacs of rat small intestine, *Endocrinology,* 65, 517, 1959.

28. Wasserman, R. H., Metabolic basis of calcium and strontium discrimination: studies with surviving intestinal segments, *Proc. Soc. Exp. Biol. Med.,* 104, 92, 1960.

29. Wasserman, R. H. and Taylor, A. N., Vitamin D_3-induced calcium-binding protein in chick intestinal mucosa, *Science,* 1952, 791, 1966.

30. Wasserman, R. H. and Taylor, A. N., Evidence for a Vitamin D_3-induced calcium-binding protein in new world primates, *Proc. Soc. Exp. Biol. Med.,* 136, 25, 1970.

31. National Academy of Sciences, *Nutrient Requirements of Domestic Animals,* No. 1–10, National Academy of Sciences, Washington, D. C., 1972.

32. National Academy of Sciences, *Recommended Dietary Alowances,* 8th ed., National Academy of Sciences, Washington, D. C., 1974, 82.

33. Lengemann, F. W., Comar, C. L., and Wasserman, R. H., Absorption of calcium and phosphorus from milk and nonmilk diets, *J. Nutr.,* 61, 571, 1957.

34. Wasserman, R. H., Comar, C. L., and Schooley, J. C., Interrelated effects of L-lysine and other dietary factors on the gastrointestinal absorption of calcium-45 in the rat and chick, *J. Nutr.,* 62, 367, 1957.

35. Gontzea, I. and Sutzescu, P., *Natural Antinutritive Substances in Foodstuffs and Forages,* S. Karger, New York, 1968.

36. McCance, R. A. and Widdowson, E. M., The fate of calcium and magnesium after intravenous administration to normal persons, *Biochem. J.,* 33, 523, 1939.

37. Nordin, B. E. C., Peacock, M., and Wilkinson, R., Hypercalciuria and calcium stone disease, in *Clinics in Endocrinology and Metabolism,* Vol. 1, MacIntyre, I., Ed., W. B. Saunders Ltd., London, 1972, 169.

38. Bronner, F. and Thompson, D. D., Renal transtubular flux of electrolytes in dogs with special reference to calcium, *J. Physiol.* (London), 157, 232, 1961.

39. Albright, F. and Reifenstein, E. C., Jr., *The Parathyroid Glands and Metabolic Bone Disease; Selected Studies,* Williams & Wilkins, Baltimore, 1948, 234.

40. Farquharson, R. F., Salter, W. T., Tibbetts, D. M., and Aub, J. C., Studies on calcium and phosphorus metabolism. XII. The effect of acid-producing substances, *J. Clin. Invest.,* 10, 221, 1931.

41. Lindeman, R. D., Adler, S., Yiengst, M. J., and Beard, E. S., Influence of various nutrients on urinary divalent cation excretion, *J. Lab. Clin. Med.,* 70, 236, 1967.

42. Walser, M., Calcium clearance as a function of sodium clearance in the dog, *Am. J. Physiol.,* 200, 1099, 1961.

43. Walser, M., Renal excretion of alkaline earths, in *Mineral Metabolism; an Advanced Treatise,* Vol. 3, Comar, C. L. and Bronner, F., Eds., Academic Press, New York, 1969, 235.

44. Massry, S. G., Coburn, J. W., Chapman, L. W., and Kleeman, C. R., Effect of NaCl infusion on urinary Ca^{++} and Mg^{++} during reduction in their filtered loads, *Am. J. Physiol.,* 213, 1218, 1967.

45. Massry, S. G., Coburn, J. W., Chapman, L. W., and Kleeman, C. R., The effect of long term desoxycorticosterone acetate administration on the renal excretion of calcium and magnesium, *J. Lab. Clin. Med.,* 71, 212, 1968.

46. Kleeman, C. R., Bernstein, D., Rockney, R., Dowling, J. T., and Maxwell, M. H., Studies on the renal clearance of diffusible calcium and the role of the parathyroid glands in its regulation, in *The Parathyroids,* Greep, R. O. and Talmage, R. V., Eds., Charles C Thomas, Springfield, Ill., 1961, 353.

47. Whedon, G. D., Osteoporosis: atrophy of disuse, in *Bone as a Tissue,* Rodahl, K., Nicholson, J. T., and Brown, E. M., Jr., Eds., McGraw-Hill, New York, 1960, 67.

48. Coburn, J. W. and Massry, S. G., Changes in serum and urinary calcium during phosphate depletion: studies on mechanism, *J. Clin. Invest.,* 49, 1973, 1970.

49. DeLuca, H. F., Tanaka, Y., and Castillo, L., Interrelationships between vitamin D and phosphate metabolism, in *Calcium-Regulating Hormones,* Talmage, R. V., Owen, M., and Parsons, J. A., Eds., Elsevier, New York, 1975, 305.

50. Nicolaysen, R. and Eeg-Larsen, N., The biochemistry and physiology of vitamin D, *Vitam. Horm.* (N.Y.), 11, 29, 1953.

51. Howard, J. E. and Connor, T. B., Some experiences with the use of vitamin D in the treatment of hypoparathyroidism, *Trans. Assoc. Am. Physicians,* 67, 199, 1954.

52. Omdahl, J. L. and DeLuca, H. F., Relation of vitamin D metabolism and function, *Physiol. Rev.,* 53, 327, 1973.

53. Coburn, J. W., Hartenbower, D. L., and Norman, A. W., Metabolism and action of the hormone vitamin D: its relation to diseases of calcium homeostasis, *West. J. Med.,* 121, 22, 1974.

54. Talmage, R. V., Owen, M., and Parsons, J. A., *Calcium-Regulating Hormones,* Elsevier, New York, 1975.

55. Talmage, R. V. and Munson, P. L., *Calcium Parathyroid Hormone and the Calcitonins,* Excerpta Medica, Amsterdam, 1972.

56. Massry, S. G., Coburn, J. W., and Kleeman, C. R., Evidence for suppression of parathyroid gland activity by hypermagnesemia, *J. Clin. Invest.,* 49, 1619, 1970.

57. Shils, M. E., Experimental human magnesium depletion, *J. Clin. Invest.,* 48, 61, 1969.

58. Levy, J., Massry, S. G., Coburn, J. W., Llach, F., and Kleeman, C. R., Hypocalcemia in magnesium-depleted dogs: evidence for reduced responsiveness to parathyroid hormone and relative failure of parathyroid gland function, *Metabolism,* 23, 323, 1974.

59. Sherwood, L. M., Herrmann, I., and Bassett, C. A., In vitro studies of normal and abnormal parathyroid tissue, *Arch. Intern. Med.,* 124, 426, 1969.

60. Au, W. Y. W. and Bukowski, A., Inhibition of PTH secretion by vitamin D metabolites in organ culture of rat parathyroids, *Fed. Proc.,* 35, 301, 1976.

61. Harrison, H. E. and Harrison, H. C., Physiology of vitamin D, in *Bone as a Tissue,* Rodahl, K., Nicholson, J. T., and Brown, E. M., Jr., Eds., McGraw-Hill, New York, 1960, 300.

62. Copp, D. H., Parathormone, calcitonin and calcium homeostasis, in *Mineral Metabolism; an Advanced Treatise,* Vol. 3, Comar, C. L. and Bronner, F., Eds., Academic Press, New York, 1969, 453.

63. Krane, S. M., Harris, E. D., Jr., Singer, F. R., and Potts, J. T., Jr., Acute effects of calcitonin on bone formation in man, *Metabolism,* 22, 51, 1973.

64. Milhaud, G. and Job, J., Thyrocalcitonin: effect on idiopathic hypercalcemia, *Science,* 154, 794, 1966.

65. Neer, R. M., Parsons, J. A., Krane, S. M., Deftos, L. J., Shields, C. L., Copp, D. H., and Potts, J. T., Jr., Pharmacology of calcitonin: human studies, in *Calcitonin 1969; Proceedings of the Second International Symposium,* Springer-Verlag, New York, 1970, 547.

66. Silva, O. L. and Becker, K. L., Salmon calcitonin in the treatment of hypercalcemia, *Arch. Intern. Med.,* 132, 337, 1973.

67. Goltzman, D., Potts, J. T., Jr., Ridgway, E. C., and Maloof, F., Calcitonin as a tumor marker: use of the radioimmunoassay for calcitonin in the postoperative evaluation of patients with medullary thyroid carcinoma, *N. Engl. J. Med.,* 290, 1035, 1974.

68. Rasmussen, H., Bordier, P., Kurokawa, K., Nagata, N., and Ogata, E., Hormonal control of skeletal and mineral homeostasis, *Am. J. Med.,* 56, 751, 1974.

69. Forte, L. R., Nickols, G. A., and Anast, C. S., Renal adenylate cyclase and the interrelationship between parathyroid hormone and vitamin D in the regulation of urinary phosphate and adenosine cyclic 3'5'-monophosphate excretion, *J. Clin. Invest.,* 57, 559, 1976.

70. Dietrich, J. W. and Raisz, L. G., Prostaglandin in calcium and bone metabolism, *Clin. Orthop. Relat. Res.,* 111, 228, 1975.

71. Tashjian, A. H., Voelkel, E. F., Goldhaber, P., and Levine, L., Prostaglandins, calcium metabolism and cancer, *Fed. Proc.,* 33, 81, 1974.

72. Seyberth, H. W., Segre, G. V., Morgan, J. L., Sweetman, B. J., Potts, J. T., Jr., and Oates, J. A., Prostaglandins as mediators of hypercalcemia associated with certain types of cancer, *N. Engl. J. Med.,* 293, 1278, 1975.

73. Horton, J. E., Raisz, L. G., Simmons, H. A., Oppenheim, J. J., and Mergenhagen, S. E., Bone-resorbing activity in supernatant fluid from cultured human peripheral blood leukocytes, *Science,* 177, 793, 1972.

74. Mundy, G. R., Luben, R. A., Raisz, L. G., Oppenheim, J. J., and Buell, D. N., Bone-resorbing activity in supernatants from lymphoid cell lines, *N. Engl. J. Med.,* 290, 867, 1974.

75. Raisz, L. G., Luben, R. A., Mundy, G. R., Dietrich, J. W., Horton, J. E., and Trummel, C. L., Effect of osteoclast activating factor from human leukocytes on bone metabolism, *J. Clin. Invest.,* 56, 408, 1975.

76. **Hopkins, T., Howard, J. E., and Eisenberg, H.,** Ultrafiltration studies on calcium and phosphorus in human serum, *Bull. Johns Hopkins Hosp.,* 91, 1, 1952.

77. **Toribara, T. Y., Terepka, A. R., and Dewey, P. A.,** The ultrafiltrable calcium of human serum. I. Ultrafiltration methods and normal values. *J. Clin. Invest.,* 36, 738, 1957.

78. **Armstrong, W. D., Johnson, J. A., Singer, L., Lienke, R. I., and Premer, M. L.,** Rates of transcapillary movement of calcium and sodium and of calcium exchange by the skeleton, *Am. J. Physiol.,* 171, 641, 1952.

79. **Neuman, W. F. and Neuman, M. W.,** *The Chemical Dynamics of Bone Mineral,* University of Chicago Press, Chicago, 1958, 11.

80. **Loken, H. F., Havel, R. J., Gordan, G. S., and Whittington, S. L.,** Ultracentrifugal analysis of protein-bound and free calcium in human serum, *J. Biol. Chem.,* 235, 3654, 1960.

81. **Moore, E. W.,** Ionized calcium in normal serum, ultrafiltrates, and whole blood determined by ion-exchange electrodes, *J. Clin. Invest.,* 49, 318, 1970.

82. **Root, A. W. and Harrison, H. E.,** Recent advances in calcium metabolism. I. Mechanisms of calcium homeostasis. *J. Pediatr.,* 88, 1, 1976.

83. **Kyle, R. A.,** Multiple myeloma: review of 869 cases, *Mayo Clin. Proc.,* 50, 29, 1975.

84. **Lingärde, F. and Zettervall, O.,** Hypercalcemia and normal ionized serum calcium in a case of myelomatosis, *Ann. Intern. Med.,* 78, 396, 1973.

85. **Urist, M. R., Schjeide, O. A., and McLean, F. C.,** The partition and binding of calcium in the serum of the laying hen and of the estrogenized rooster, *Endocrinology,* 63, 570, 1958.

86. **Schwartz, H. D.,** New techniques for ion-selective measurements of ionized calcium in serum after pH adjustment of aerobically handled sera, *Clin. Chem.,* 22, 461, 1976.

87. **Gallagher, J. C. and Nordin, B. E. C.,** Oestrogens and calcium metabolism, *Front. Horm. Res.,* 2, 98, 1973.

88. **Hopkins, T. R., Connor, T. B., and Howard, J. E.,** Ultrafiltration studies on calcium and phosphorus in pathological human serum, *Bull. Johns Hopkins Hosp.,* 93, 249, 1953.

89. **McLean, F. C. and Hinrichs, M. A.,** The formation of colloidal calcium phosphate in blood, *Am. J. Physiol.,* 121, 580, 1938.

90. **Gersh, I.,** The fate of colloidal calcium phosphate in the dog, *Am. J. Physiol.,* 121, 589, 1938.

91. **Hebert, L. A., LeMann, J., Jr., Petersen, J. R., and Lennon, E. J.,** Studies of the mechanism by which phosphate infusion lowers serum calcium concentration, *J. Clin. Invest.,* 45, 1886, 1966.

92. **Thomas, W. C., Jr., Wiswell, J. G., Connor, T. B., and Howard, J. E.,** Hypercalcemic crisis due to hyperparathyroidism, *Am. J. Med.,* 24, 229, 1958.

93. **Fairbank, H. A. T.,** Dysplasia epiphysialis multiplex, *Br. J. Surg.,* 34, 225, 1947.

94. **Spranger, J. W., Langer, L. O., and Weidemann, H. R.,** *Bone Dysplasias. An Atlas of Constitutional Disorders of Skeletal Development,* W. B. Saunders, Philadelphia, 1974.

95. **Harrison, H. E.,** Rickets, in *Practice of Pediatrics,* Brenneman, J., Ed., W. F. Prior, Hagerstown, Maryland, 1963, 1.

96. **Shipley, P. G.,** The relation of the diet to rickets, *Trans. Congr. Am. Phys. Surg.,* 12, 31, 1922.

97. **Yendt, E. R.,** Disorders of calcium phosphorus and magnesium metabolism, in *Clinical Disorders of Fluid and Electrolyte Metabolism,* Maxwell, M. H. and Kleeman, C. R., Eds., McGraw-Hill, New York, 1972, 401.

98. **Salassa, R. M., Jowsey, J., and Arnaud, C. D.,** Hypophosphatemic osteomalacia associated with "nonendocrine" tumors, *N. Engl. J. Med.,* 283, 65, 1970.

99. **Olefsky, J., Kempson, R., Jones, H., and Reaven, G.,** 'Tertiary' hyperparathyroidism and apparent 'cure' of vitamin D-resistant rickets after removal of an ossifying mesenchymal tumor of the pharynx, *N. Engl. J. Med.,* 286, 740, 1972.

100. **Scriver, C. R.,** Vitamin D dependency, *Pediatrics,* 45, 361, 1970.

101. **Fraser, D., Kooh, S. W., Kind, H. P., Holick, M. F., Tanaka, Y., and DeLuca, H. F.,** Pathogenesis of hereditary vitamin D-dependent rickets, *N. Engl. J. Med.,* 289, 817, 1973.

102. **Fraser, D.,** Hypophosphatasia, *Am. J. Med.,* 22, 730, 1957.

103. **Frame, B., Frost, H. M., Ormond, R. S., and Hunter, R. B.,** Atypical osteomalacia involving the axial skeleton, *Ann. Intern. Med.,* 55, 632, 1961.

104. **Baker, S. L.,** Fibrogenesis imperfecta ossium, *J. Bone J. Surg.,* 38(B), 378, 1956.

105. **Thomas, W. C., Jr. and Moore, T. H.,** Fibrogenesis imperfecta ossium, *Trans. Am. Clin. Climatol. Assoc.,* 80, 54, 1968.

106. **Gutman, A. B. and Yü, T. F.,** A further consideration of the effects of beryllium salts on in vitro calcification of cartilage, in *Metabolic Interrelations; Transactions of the Third Conference,* Reifenstein, E. C., Jr., Ed., Josiah Macy, Jr. Foundation, New York, 1951, 90.

107. **Hiatt, H. H. and Shorr, E.,** Inhibition of endochondral calcification in vitro by beryllium and L-histidine, in *Metabolic Interrelations; Transactions of the Third Conference,* Reifenstein, E. C., Jr., Ed., Josiah Macy, Jr. Foundation, New York, 1951, 105.

108. **Sobel, A. E.,** Studies on the "local factor" of calcification, in *Metabolic Interrelations; Transactions of the Fourth Conference,* Reifenstein, E. C., Jr., Ed., Josiah Macy, Jr. Foundation, New York, 1952, 113.
109. **Bird, E. D. and Thomas, W. C., Jr.,** Effect of various metals on mineralization in vitro, *Proc. Soc. Exp. Biol. Med.,* 112, 640, 1963.
110. **Follis, R. H.,** Bone changes resulting from parenteral strontium administration, *Fed. Proc.,* 14, 403, 1955.
111. **Cass, R. M., Croft, J. D., Perkins, P., Nye, W., Waterhouse, C., and Terry, R.,** New bone formation in osteoporosis following treatment with sodium fluoride, *Arch. Intern. Med.,* 118, 111, 1966.
112. **Adams, R. G., Harrison, J. F., and Scott, P.,** The development of cadmium-induced proteinuria, impaired renal function, and osteomalacia in alkaline battery workers, *Q. J. Med.,* 38, 425, 1969.
113. **Fairley, N. H.,** Tropical sprue with special reference to intestinal absorption, *Trans. R. Soc. Trop. Med. Hyg.,* 30, 9, 1936.
114. **Snell, A. M.,** Tropical and nontropical sprue (chronic idiopathic steatorrhea): their probable interrelationship, *Ann. Intern. Med.,* 12, 1632, 1939.
115. **Barker, W. H.,** Sprue, *Med. Clin. North Am.,* 27, 451, 1943.
116. **Thomas, W. C., Jr., Morgan, H. G., Connor, T. B., Haddock, L., Bills, C. E., and Howard, J. E.,** Studies of antirachitic activity in sera from patients with disorders of calcium metabolism and preliminary observations on the mode of transport of vitamin D in human serum, *J. Clin. Invest.,* 38, 1078, 1959.
117. **Earnest, D. L., Johnson, G., Williams, H. E., and Admirand, W. H.,** Hyperoxaluria in patients with ileal resection: an abnormality in dietary oxalate absorption, *Gastroenterology,* 66, 1114, 1974.
118. **Johnson, C. D. and Bynum, T. E.,** Hypocalcemia as a complication of jejunoileal bypass for morbid obesity, *South. Med. J.,* 69, 616, 1976.
119. **Epstein, F. H.,** Calcium and the kidney, *J. Chronic Dis.,* 11, 255, 1960.
120. **Silberg, B. N., Calder, D., Carter, N., and Seldin, D.,** Urinary calcium excretion in parathyroidectomized rats during metabolic and respiratory acidosis, *Clin. Res.,* 12, 50, 1964.
121. **Lennon, E. J. and Piering, W. F.,** A comparison of the effects of glucose ingestion and NH_4Cl acidosis on urinary calcium and magnesium excretion in man, *J. Clin. Invest.,* 49, 1458, 1970.
122. **Buckalew, V. M., Jr., Purvis, M. L., Shulman, M. G., Herndon, C. N., and Rudman, D.,** Hereditary renal tublar acidosis: report of a 64 member kindred with variable clinical expression including idiopathic hypercalciuria, *Medicine,* 53, 229, 1974.
123. **Dent, C. E.,** Rickets and osteomalacia from renal tubular defects, *J. Bone J. Surg.,* 34(B), 266, 1952.
124. **Seldin, D. W. and Wilson, J. D.,** Renal tubular acidosis, in *The Metabolic Basis of Inherited Disease,* 3rd ed., Stanbury, J. B., Wyngaarden, J. B., and Fredrickson, D. S., Eds., McGraw-Hill, New York, 1972, 1548.
125. **Trotter, M., Broman, G. E., and Peterson, R. R.,** Densities of bones of white and negro skeletons, *J. Bone J. Surg.,* 42(A), 50, 1960.
126. **Garn, S. M., Rohman, C. G., and Wagner, B.,** Bone loss as a general phenomenon in man, *Fed. Proc.,* 26, 1729, 1967.
127. **Urist, M. R., Gurvey, M. S., and Fareed, D. O.,** Long-term observations on aged women with pathologic osteoporosis, in *Osteoporosis,* Barzel, U. S., Ed., Grune & Stratton, New York, 1970, 3.
128. **Ascenzi, A. and Bell, G. H.,** Bone as a mechanical engineering problem, in *The Biochemistry and Physiology of Bone; Structure,* Vol. 1, 2nd ed., Bourne, G. H., Ed., Academic Press, New York, 1972, 311.
129. **Henneman, P. H. and Wallach, S.,** A review of the prolonged use of estrogens and androgens in postmenopausal and senile osteoporosis, *A.M.A. Arch. Intern. Med.,* 100, 715, 1957.
130. **Powell, G. F., Brasel, J. A., Raiti, S., and Blizzard, R. M.,** Emotional deprivation and growth retardations simulating idiopathic hypopituitarism, *N. Engl. J. Med.,* 276, 1271, 1279, 1967.
131. **Moseley, J. E.,** *Bone Changes in Hematologic Disorders (Roentgen Aspects),* Grune & Stratton, New York, 1963, 261.
132. **Hioco, D., Miravet, L., and Bordier, P.,** Physiopathology and treatment of osteoporosis in younger men, in *Bone and Tooth; Proceedings of the First European Symposium,* Blackwood, H. J. J., Ed., Macmillan, New York, 1964, 365.
133. **Collins, E. J., Garrett, E. R., and Johnston, R. L.,** Effect of adrenal steroids on radio-calcium metabolism in dogs, *Metabolism,* 11, 716, 1962.
134. **Iannaccone, A., Gabrilove, J. L., Brahms, S. A., and Soffer, L. J.,** Osteoporosis in Cushing's syndrome, *Ann. Intern. Med.,* 52, 570, 1960.

135. Grant, D. K., Papilloedema and fits in hypoparathyroidism, *Q. J. Med.*, 22, 243, 1953.
136. Emerson, K., Jr., Walsh, F. B., and Howard, J. E., Idiopathic hypoparathyroidism: a report of two cases, *Ann. Intern. Med.*, 14, 1256, 1941.
137. Albright, F. and Reifenstein, E. C., Jr., *The Parathyroid Glands and Metabolic Bone Disease: Selected Studies*, Williams & Wilkins, Baltimore, 1948, 27.
138. Drake, T. G., Albright, F., Bauer, W., and Castleman, B., Chronic idiopathic hypoparathyroidism: report of six cases with autopsy findings in one, *Ann. Intern. Med.*, 12, 1751, 1939.
139. Bronsky, D., Kushner, D. S., Dubin, A., and Snapper, I., Idiopathic hypoparathyroidism and pseudo-hypoparathyroidism: case reports and review of the literature, *Medicine*, 37, 317, 1958.
140. Murley, R. S. and Peters, P. M., Inadvertent parathyroidectomy, *Proc. R. Soc. Med.*, 54, 487, 1961.
141. Elrick, H., Albright, F., Bartter, F. C., Forbes, A. P., and Reeves, J. D., Further studies on pseudohypoparathyroidism: report on four new cases, *Acta Endocrinol.* (Copenhagen), 5, 199, 1950.
142. Greenberg, S. R., Karabel, S., and Saade, G. A., Pseudohypoparathyroidism: a disease of the second messenger, *Arch. Intern. Med.*, 129, 633, 1972.
143. Marcus, R., Wilbur, J. F., and Aurbach, G. D., Parathyroid hormone-sensitive adenyl cyclase from the renal cortex of a patient with pseudohyperparathyroidism, *J. Clin. Endocrinol. Metab.*, 33, 537, 1971.
144. Drezner, M. K., Neelon, F. A., McPherson, H. T., and Lebovitz, H. E., 1,25-dihydroxychole-calciferol (DHCC) deficiency: cause of hypocalcemia in pseudohypoparathyroidism, *Clin. Res.*, 23, 42(A), 1975.
145. Chase, L. R. and Slatopolsky, E., Secretion and metabolic efficacy of parathyroid hormone in patients with severe hypomagnesemia, *J. Clin. Endocrinol. Metab.*, 38, 363, 1974.
146. Medalle, R., Waterhouse, C., and Hahn, T. J., Vitamin D resistance in magnesium deficiency, *Am. J. Clin. Nutr.*, 29, 854, 1976.
147. Friedman, M., Hatcher, G., and Watson, L., Primary hypomagnesaemia with secondary hypocalcaemia in an infant, *Lancet*, 1, 703, 1967.
148. Paunier, L., Radde, I. C., Kooh, S. W., Conen, P. E., and Fraser, D., Primary hypomagnesemia with secondary hypocalcemia in an infant, *Pediatrics*, 41, 385, 1968.
149. Thomas, W. C., Jr. and Howard, J. E., Disturbances of calcium metabolism, in *Mineral Metabolism; an Advanced Treatise*, Vol. 2A, Comar, C. L. and Bronner, F., Eds., Academic Press, New York, 1969, 445.
150. Edmondson, H. A. and Berne, C. J., Calcium changes in acute pancreatic necrosis, *Surg. Gynecol. Obstet.*, 79, 240, 1944.
151. Edmondson, H. A., Berne, C. J., Homann, R. E., Jr., and Wertman, M., Calcium, potassium, magnesium and amylase disturbances in acute pancreatitis, *Am. J. Med.*, 12, 34, 1952.
152. Paloyan, E., Paloyan, D., and Harper, P. V., The role of glucagon hypersecretion in the relationship of pancreatitis and hyperparathyroidism, *Surgery*, 62, 167, 1967.
153. Binger, C., Toxicity of phosphates in relation to blood calcium and tetany, *J. Pharmacol. Exp. Ther.*, 10, 105, 1917.
154. Selvensen, H. A., Hastings, A. B., and McIntosh, J. F., Blood changes and clinical symptoms following oral administration of phosphates, *J. Biol. Chem.*, 60, 311, 1924.
155. Burnett, C. H., Shapiro, S. L., Simeone, F. A., Beecher, H. K., Mallory, T. B., and Sullivan, E. R., Post-traumatic renal insufficiency, *Surgery*, 22, 994, 1947.
156. Meroney, W. H. and Herndon, R. F., The management of acute renal insufficiency, *J.A.M.A.*, 155, 877, 1954.
157. Liu, S. H. and Chu, H. I., Studies of calcium and phosphorus metabolism with special reference to pathogenesis and effects of dihydrotachysterol (A.T. 10) and iron, *Medicine*, 22, 103, 1943.
158. Dent, C. E., Harper, C. M., and Philpot, G. R., The treatment of renal-glomerular osteodystrophy, *Q. J. Med.*, 30, 1, 1961.
159. Guild, H. G., Tetany of the newborn, *Int. Clin.*, 3, 47, 1932.
160. Gardner, L. I., Tetany and parathyroid hyperplasia in the newborn infant: influence of dietary phosphate load, *Pediatrics*, 9, 534, 1952.
161. Root, A. W. and Harrison, H. E., Recent advances in calcium metabolism. II. Disorders of calcium homeostasis, *J. Pediatr.*, 88, 177, 1976.
162. Wagner, G., Transbøl, I., and Melchior, J. C., Hyperparathyroidism and pregnancy, *Acta Endocrinol.* (Copenhagen), 47, 549, 1964.
163. Finberg, L. and Harrison, H. E., Hypernatremia in infants, *Pediatrics*, 16, 1, 1955.
164. Tegelaers, W. H. H. and Tiddens, H. W., Nephrotic-glucosuricaminoaciduric dwarfism and electrolyte metabolism, *Helv. Paediatr. Acta*, 10, 269, 1955.

165. Stanbury, S. W. and Macaulay, D., Defects of renal tubular function in the nephrotic syndrome, *Q. J. Med.,* 26, 7, 1957.

166. Peters, J. H., Therapy of acute fluoride poisoning, *Am. J. Med. Sci.,* 216, 278, 1948.

167. Zarembski, P. M. and Hodgkinson, A., Plasma oxalic acid and calcium levels in oxalate poisoning, *J. Clin. Pathol.,* 20, 283, 1967.

168. Greenman, L., Shaler, J. B., and Danowski, T. S., Biochemical disturbances and clinical symptoms during prolonged exchange resin therapy in congestive heart failure, *Am. J. Med.,* 14, 391, 1953.

169. Ray, B. S. and Pearson, O. H., Hypophysectomy in the treatment of advanced cancer of breast, *Ann. Surg.,* 144, 394, 1956.

170. Hall, T. C., Griffiths, C. T., and Petranek, J. R., Hypocalcemia — an unusual metabolic complication of breast cancer, *N. Engl. J. Med.,* 275, 1474, 1966.

171. Raskin, P., McClain, C. J., and Medsger, T. A., Jr., Hypocalcemia associated with metastatic bone disease: a retrospective study, *Arch. Intern. Med.,* 132, 539, 1973.

172. Albright, F. and Reifenstein, E. C., Jr., *The Parathyroid Glands and Metabolic Bone Disease; Selected Studies,* Williams & Wilkins, Baltimore, 1948, 46.

173. Werner, C. A., Tompsett, R., Muschenheim, C., and McDermott, W., The toxicity of viomycin in humans, *Am. Rev. Tuberc.,* 63, 49, 1951.

174. Holmes, A. M., Hesling, C. M., and Wilson, T. M., Drug-induced secondary hyperaldosteronism in patients with pulmonary tuberculosis, *Q. J. Med.,* 39, 299, 1970.

175. Unpublished observations by the authors.

176. Rupp, J. J. and Paschkis, K. E., Panhypopituitarism and hypocalcemic tetany in a male: case presentation, *Ann. Intern. Med..* 39, 1103, 1953.

177. Lowe, C. E., Bird, E. D., and Thomas, W. C., Jr., Hypercalcemia in myxedema, *J. Clin. Endocrinol. Metab.,* 22, 261, 1962.

178. Smith, D. W., Blizzard, R. M., and Harrison, H. E., Idiopathic hypercalcemia: a case report with assays of vitamin D in the serum, *Pediatrics,* 24, 258, 1959.

179. Nolan, R. B., Hayles, A. B., and Woolner, L. B., Adenoma of the parathyroid gland in children, *Am. J. Dis. Child.,* 99, 622, 1960.

180. Steiner, A. L., Goodman, A. D., and Powers, S. R., Study of a kindred with pheochromacytoma, medullary thyroid carcinoma, hyperparathyroidism and Cushing's disease: multiple endocrine neoplasia, Type 2, *Medicine,* 47, 371, 1968.

181. Straus, F. H. and Paloyan, E., The pathology of hyperparathyroidism, *Surg. Clin. North Am.,* 49, 27, 1969.

182. Holmes, E. C., Morton, D. L., and Ketcham, A. S., Parathyroid carcinoma: a collective review, *Ann. Surg.,* 169, 631, 1962.

183. Cope, C. L., Base changes in the alkalosis produced by treatment of gastric ulcer with alkalies, *Clin. Sci.,* 2, 287, 1936.

184. Burnett, C. H., Commons, R. R., Albright, F., and Howard, J. E., Hypercalcemia without hypercalciuria or hypophosphatemia, calcinosis and renal insufficiency, *N. Engl. J. Med.,* 240, 787, 1949.

185. Scholz, D. A. and Keating, F. R., Milk-alkali syndrome, *A.M.A. Arch. Intern. Med.,* 95, 460, 1955.

186. Thomas, W. C., Jr., Lewis, A. M., and Bird, E. D., Effect of alkali administration on calcium metabolism, *J. Clin. Endocrinol. Metab.,* 27, 1328, 1967.

187. Epstein, F. H., Calcium nephropathy, in *Diseases of the Kidney,* Vol. II, 2nd ed., Strauss, M. B. and Welt, L. G., Eds., Little, Brown, Boston, 1971, 918.

188. Ackermann, P. C. and Toro, G., Effect of added vitamin D on the calcium balance in elderly males, *J. Gerontol.,* 8, 451, 1953.

189. Jones, J. H. and Rapoport, M., Further observations on regulation of calcium and phosphorus intake to hypercalcemia and hyperphosphatemia induced by irradiated ergosterol, *J. Biol. Chem.,* 93, 153, 1931.

190. Bauer, W., Marble, A., and Claflin, F., Studies on the mode of action of irradiated ergosterol, *J. Clin. Invest.,* 11, 47, 1932.

191. Shelling, D. H. and Asher, D. E., Calcium and phosphorus studies. IV. The relation of calcium and phosphorus of the diet to the toxicity of viosterol, *Bull. Johns Hopkins Hosp.,* 50, 318, 1932.

192. Cruickshank, E. M. and Kodicek, E., The antagonism between cortisone and vitamin D: experiments on hypervitaminosis D in rats, *J. Endocrinol.,* 17, 35, 1958.

193. Hess, A. F., Benjamin, H. R., and Gross, J., Sources of excess calcium in hypercalcemia induced by irradiated ergosterol, *J. Biol. Chem.,* 94, 1, 1931.

194. McLean, F. C., Activated sterols in the treatment of parathyroid insufficiency, *J.A.M.A.* 117, 609, 1941.

195. Thomas, W. C., Jr. and Morgan, H. G., The effect of cortisone in experimental hypervitaminosis D, *Endocrinology,* 63, 57, 1958.

196. Ham, A. W. and Lewis, M. D., Hypervitaminosis D rickets, *Br. J. Exp. Pathol.,* 15, 228, 1934.

197. Follis, R. H., Jr., Studies on hypervitaminosis D, *Am. J. Pathol.,* 51, 568, 1955.

198. Thomas, W. C., Jr., Comparative studies on bone matrix and osteoid by histochemical techniques, *J. Bone Joint Surg.,* 43(A), 419, 1961.

199. Gutman, A. B., Tyson, T. L., and Gutman, E. B., Serum calcium, inorganic phosphorus and phosphatase activity in hyperparathyroidism, Paget's disease, multiple myeloma and neoplastic disease of bone, *AMA Arch. Intern. Med.,* 57, 379, 1936.

200. Kennedy, B. J., Tibbets, D. M., Nathanson, I. T., and Aub, J. C., Hypercalcemia, a complication of hormone therapy of advanced breast cancer, *Cancer Res.,* 13, 445, 1953.

201. Pearson, O. H., West, C. D., Hollander, V. P., and Treves, N. E., Evaluation of endocrine therapy for advanced breast cancer, *J.A.M.A.* 154, 234, 1954.

202. Kennedy, B. J., Nathanson, I. T., Tibbets, D. M., and Aub, J. C., Biochemical alterations during steroid hormone therapy for advanced breast cancer, *Am. J. Med.,* 19, 337, 1955.

203. Herrman, J. B., Kirsten, E., and Krakauer, J. S., Hypercalcemic syndrome associated with androgenic and estrogenic therapy, *J. Clin. Endocrinol.,* 9, 1, 1949.

204. Mawdsley, C. and Holman, R. L., Hypercalcemia in acute leukemia, *Lancet,* 1, 78, 1957.

205. Kabakow, B., Mines, M. F., and King, F. H., Hypercalcemia in Hodgkin's disease, *N. Engl. J. Med.,* 256, 59, 1957.

206. Moses, A. M. and Spencer, H., Hypercalcemia in patients with malignant lymphoma, *Ann. Intern. Med.,* 59, 531, 1963.

207. Howard, J. E., Carey, R. A., Rubin, P. S., and Levine, M. D., Diagnostic problems in patients with hypercalcemia, *Trans. Assoc. Am. Physicians,* 62, 264, 1949.

208. Connor, T. B., Thomas, W. C., Jr., and Howard, J. E., The etiology of hypercalcemia associated with lung carcinoma, *J. Clin. Invest.,* 35, 697, 1956.

209. Plimpton, C. H. and Gellhorn, A., Hypercalcemia in malignant diseases without evidence of bone destruction, *Am. J. Med.,* 21, 750, 1956.

210. Tashjian, A. H., Jr., Levine, L., and Munson, P. L., Immunochemical identification of parathyroid hormone in nonparathyroid neoplasms, *J. Exp. Med.,* 119, 467, 1964.

211. Sherwood, L. M., O'Riordan, J. L. H., Aurbach, G. D., and Potts, J. T., Jr., Production of parathyroid hormone by nonparathyroid tumor, *J. Clin. Endocroinol. Metab.,* 27, 140, 1967.

212. Brereton, H. D., Halushka, P. V., Alexander, R. W., Mason, D. M., Keiser, H. R., and DeVita, V. T., Jr., Indomethacin-responsive hypercalcemia in a patient with renal-cell adenocarcinoma, *N. Engl. J. Med.,* 291, 83, 1974.

213. Connor, T. B., Thomas, W. C., Jr., Stodghill, W. B., and Howard, J. E., Malignant Tumors Mimicking Hyperparathyroidism, Paper presented at 43rd Annu. Mtg. of the Endocrine Soc., New York, 1961.

214. Lipsett, M. B., Odell, W. D., Rosenberg, L. E., and Waldmann, T. A., Humoral syndromes associated with nonendocrine tumors, *Ann. Intern. Med.,* 61, 733, 1964.

215. Bower, B. F. and Gordan, G. S., Hormonal effects of nonendocrine tumors, *Annu. Rev. Med.,* 16, 83, 1965.

216. Aub, J. C., Bauer, W., Heath, C., and Ropes, M., Studies of calcium and phosphorus metabolism. III. The effects of the thyroid hormone and thyroid disease, *J. Clin. Invest.,* 7, 97, 1929.

217. Follis, R. H., Jr., Skeletal changes associated with hyperthyroidism, *Bull. Johns Hopkins Hosp.,* 92, 405, 1953.

218. Krane, S. M., Brownell, G. L., Stanbury, J. B., and Corrigan, H., The effect of thyroid disease on calcium metabolism in man, *J. Clin. Invest.,* 35, 874, 1956.

219. Stanley, M. M. and Fazekas, J., Thyrotoxicosis simulating hyperparathyroidism, *Am. J. Med.,* 7, 262, 1949.

220. Rose, E. and Boles, R. S., Jr., Hypercalcemia in thyrotoxicosis, *Med. Clin. North Am.,* 37, 1715, 1953.

221. Epstein, F. H., Freedman, L. R., and Levitin, H., Hypercalcemia, nephrocalcinosis and reversible renal insufficiency associated with hyperthyroidism, *N. Engl. J. Med.,* 258, 782, 1958.

222. Parfitt, A. M. and Dent, C. E., Hyperthyroidism and hypercalcemia, *Q. J. Med.,* 39, 171, 1970.

223. Kleeman, C. R., Tuttle, S., and Bassett, S. H., Metabolic observations in a case of thyrotoxicosis with hypercalcemia, *J. Clin. Endocrinol. Metab.,* 18, 477, 1958.

224. Albright, F., Burnett, C. H., Cope, O., and Parson, W., Acute atrophy of bone (osteoporosis) simulating hyperparathyroidism, *J. Clin. Endocrinol.,* 1, 711, 1941.

225. Dodd, K., Graubarth, H., and Rapoport, S., Hypercalcemia nephropathy and encephalopathy following immobilization: case report, *Pediatrics,* 6, 124, 1950.

226. Mason, A. S., Acute osteoporosis with hypercalcemia, *Lancet,* 1, 911, 1957.

227. Lawrence, G. D., Loeffler, R. G., Martin, L. G., and Connor, T. B., Immobilization hypercalcemia, *J. Bone J. Surg.,* 55(A), 87, 1973.

228. Albright, F. and Reifenstein, E. C., Jr., *The Parathyroid Glands and Metabolic Bone Disease; Selected Studies,* Williams & Wilkins, Baltimore, 1948, 295.

229. James, P. R. and Richards, P. G., Parathyroid crisis: treatment of emergency parathyroidectomy, *Arch. Surg.,* 72, 553, 1956.

230. Guild, H. G. and Howard, J. E., unpublished observations, 1941.

231. Howard, J. E., Hypercalcemia and renal injury, *Ann. Intern. Med.,* 16, 176, 1942.

232. Harrell, G. T. and Fisher, S., Blood chemical changes in Boeck's sarcoid with particular reference to protein, calcium and phosphatase values, *J. Clin. Invest.,* 18, 687, 1939.

233. Longcope, W. T. and Freiman, D. G., A study of sarcoidosis, *Medicine,* 31, 1, 1952.

234. Taylor, R. L., Lynch, H. J., Jr., and Wysor, W. G., Jr., Seasonal influence of sunlight on the hypercalcemia of sarcoidosis, *Am. J. Med.,* 34, 221, 1963.

235. James, D. G., Siltzbach, L. E., Sharma, O. P., and Carstairs, L. D., A tale of two cities: a comparison of sarcoidosis in London and New York, *Arch. Intern. Med.,* 123, 187, 1969.

236. Mather, G., Calcium metabolism and bone changes in sarcoidosis, *Br. Med. J.,* 1, 248, 1957.

237. Winnacker, J. L., Becker, K. L., and Katz, S., Endocrine aspects of sarcoidosis, *N. Engl. J. Med.,* 278, 427, 1968.

238. Davidson, C. N., Dennis, J. M., McNinch, E. R., Willson, J. K. V., and Brown, W. H., Nephrocalcinosis associated with sarcoidosis: a presentation and discussion of seven cases, *Radiology,* 62, 203, 1954.

239. Murphy, G. P. and Schirmer, H. K., Nephrocalcinosis, urolithiasis and renal insufficiency in sarcoidosis, *J. Urol.,* 86, 702, 1961.

240. Anderson, J., Dent, C. E., Harper, C., and Philpot, G. R., Effect of cortisone on calcium metabolism in sarcoidosis with hypercalcemia: possible antagonistic actions of cortisone and vitamin D, *Lancet,* 2, 720, 1954.

241. Henneman, P. H., Dempsey, E. F., Carroll, E. L., and Albright, F., The cause of hypercalciuria in sarcoid and its treatment with cortisone and sodium phytate, *J. Clin. Invest.,* 35, 1229, 1956.

242. Avioli, L. V., McDonald, J. E., Lund, J., and DeLuca, H., The fate of intravenously administered tritiated vitamin D_3 in normal subjects and in patients, *J. Clin. Invest.,* 44, 1026, 1965.

243. Mawer, E. B., Backhouse, J., Lumb, G. A., and Stanbury, S. W., Evidence for formation of 1,25-dihydroxycholecalciferol during metabolism of vitamin D in man, *Nature, New Biol.,* 232, 188, 1971.

244. Bell, N. H., Gill, J. R., Jr., and Bartter, F. C., An abnormal calcium absorption in sarcoidosis: evidence for increased sensitivity to vitamin D, *Am. J. Med.,* 36, 500, 1964.

245. Scadding, J. G., Sarcoidosis with special reference to lung changes, *Br. Med. J.,* 1, 745, 1950.

246. Rhodes, J., Reynolds, E. H., Fitzgerald, J. D., and Fourman, P., Exaggerated response to parathyroid extract in sarcoidosis, *Lancet,* 2, 598, 1963.

247. Hendrix, J. Z., Abnormal skeleton mineral metabolism in sarcoidosis, *Ann. Intern. Med.,* 64, 797, 1966.

248. Fanconi, G., Girardet, P., Schlesinger, B., Butler, N., and Black, J., Cronische Hypercalcäemie, krombiniert mit Osteosklerose, Hyperazotämie, Minderwuchs und kongenitalen Missbildungen, *Helv. Paediatr. Acta,* 4, 314, 1952.

249. Lightwood, R., Idiopathic hypercalcemia with failure to thrive: nephrocalcinosis, *Proc. R. Soc. Med.,* 45, 401, 1952.

250. Lowe, K. G., Henderson, J. L., Park, W. W., and McGreal, D. A., The idiopathic hypercalcemic syndromes of infancy, *Lancet,* 2, 101, 1954.

251. Creery, R. D. G. and Neill, D. W., Idiopathic hypercalcemia in infants with failure to thrive, *Lancet,* 2, 110, 1954.

252. Friedman, W. F. and Mills, L. F., The relationship between vitamin D and the craniofacial and dental anomalies of the supravalvular aortic stenosis syndrome, *Pediatrics,* 43, 12, 1969.

253. Graham, S., Idiopathic hypercalciuria, *Postgrad. Med.,* 26, 67, 1959.

254. American Academy of Pediatrics Committee on Nutrition, The relation between infantile hypercalcemia and vitamin D — public health implications in North America, *Pediatrics,* 40, 1050, 1967.

255. Fellers, F. X. and Schwartz, R., Etiology of the severe form of idiopathic hypercalcemia of infancy: a defect in vitamin D metabolism, *N. Engl. J. Med.,* 259, 1050, 1958.
256. Drummond, K. N., Michael, A. F., Ulstrom, R. A., and Good, R. A., The blue diaper syndrome: familial hypercalcemia with nephrocalcinosis and indicanuria, *Am. J. Med.,* 37, 928, 1964.
257. Michael, A. F., Hong, R., and West, C. D., Hypercalcemia in infancy associated with subcutaneous fat necrosis and calcification, *Am. J. Dis. Child.,* 104, 235, 1962.
258. Connor, T. B., Hopkins, T. R., Thomas, W. C., Jr., Carey, R. A., and Howard, J. E., The use of cortisone and ACTH in hypercalcemic states, *J. Clin. Endocrinol. Metab.,* 16, 945, 1956.
259. Fraser, D., Kidd, B. S. L., Kooh, S. W., and Paunier, L., A new look at infantile hypercalcemia, *Pediatr. Clin. North Am.,* 13, 503, 1966.
260. Hardy, H. L., Proc. Int. Conf. on Sarcoidosis, *Am. Rev. Respir. Dis.,* 84, 53, 1961.
261. Leeksma, C. H. W., de Graeff, J., and de Cock, J., Hypercalcemia in adrenal insufficiency, *Acta Med. Scand.,* 156, 455, 1957.
262. Sprague, R. G., Kvale, W. F., and Priestley, J. T., Management of certain hyperfunctioning lesions of the adrenal cortex and medulla, *J.A.M.A.,* 151, 629, 1953.
263. Nadarajah, A., Hartog, M., Redfern, B., Thalassinos, N., Wright, A. D., Joplin, G. F., and Fraser, T. R., Calcium metabolism in acromegaly, *Br. Med. J.,* 4, 797, 1968.
264. Brown, J. and Singer, F. R., Calcium metabolism in acromegaly, *Br. Med. J.,* 1, 50, 1969.
265. Albright, F. and Reifenstein, E. C., Jr., *The Parathyroid Glands and Metabolic Bone Disease; Selected Studies,* Williams & Wilkins, Baltimore, 1948, 188.
266. Shai, F., Baker, R. K., Addrizzo, J. R., and Wallach, S., Hypercalcemia in myobacterial infection, *J. Clin. Endocrinol. Metab.,* 34, 251, 1972.
267. Braman, S. S., Goldman, A. L., and Schwarz, M. I., Steroid-responsive hypercalcemia in disseminated bone tuberculosis, *Arch. Intern. Med.,* 132, 269, 1973.
268. Harrison, H. E., Personal communication.
269. Katz, C. M. and Tzagournis, M., Chronic adult hypervitaminosis A with hypercalcemia, *Metabolism,* 21, 1171, 1972.
270. Frame, B., Jackson, C. E., Reynolds, W. A., and Umphrey, J. E., Hypercalcemia and skeletal effects in chronic hypervitaminosis A, *Ann. Intern. Med.,* 80, 44, 1974.
271. Caffey, J., Chronic poisoning due to an excess of vitamin A, *Am. J. Roentgenol. Radium Ther. Nucl. Med.,* 65, 12, 1951.
272. Brickman, A. S., Massry, S. G., and Coburn, J. W., Changes in serum and urinary calcium during treatment with hydrochlorothiazide, *J. Clin. Invest.,* 51, 945, 1972.
273. Shulman, L. E., Schoenrich, E. H., and Harvey, A. M., The effects of adrenocorticotropic hormone (ACTH) and cortisone on sarcoidosis, *Bull. Johns Hopkins Hosp.,* 91, 371, 1952.
274. Thomas, W. C., Jr., Connor, T. B., and Morgan, H. G., Diagnostic considerations in hypercalcemia, *N. Engl. J. Med.,* 260, 591, 1959.
275. Dent, C. E., Cortisone test for hyperparathyroidism, *Br. Med. J.,* 1, 230, 1956.
276. Gwinup, G. and Sayle, B., Cortisone responsive hypercalcemia in proved hyperparathyroidism, *Ann. Intern. Med.,* 55, 1001, 1961.
277. Garcia, D. A. and Yendt, E. R., Temporary remission of hypercalcemia in hyperparathyroidism induced by corticosteroids, *Can. Med. Assoc. J.,* 99, 1047, 1968.
278. Sissons, H. A., Osteoporosis of Cushing's syndrome, in *Bone as a Tissue,* Rodahl, K., Nicholson, J. T., and Brown, E. M., Jr., Eds., McGraw-Hill, New York, 1960, 3.
279. Follis, R. H., Jr., Effect of cortisone on the growing bones of the rat, *Proc. Soc. Exp. Biol. Med.,* 76, 722, 1951.
280. Harrison, H. C., Harrison, H. E., and Park, E. A., Vitamin D and citrate metabolism: inhibition of vitamin D effect by cortisol, *Proc. Soc. Exp. Biol. Med.,* 96, 768, 1957.

EFFECT OF SPECIFIC NUTRIENT DEFICIENCIES IN MAN: MAGNESIUM

J. K. Aikawa

THE CLINICAL SYNDROME

Although there was doubt for many years about the existence of a pure magnesium-deficiency state in man, it is now established that there is such a condition.[1-3] It is characterized by the following features: (a) spasmophilia, gross muscular tremor, choreiform movements, ataxia, tetany, and in some instances, predisposition to epileptiform convulsions; (b) hallucinations, agitation, confusion, tremulousness, delirium, depression, vertigo, and muscular weakness; (c) a low serum magnesium concentration associated with a normal serum calcium concentration and a normal blood pH; (d) a low voltage T-wave in the electrocardiogram; (e) a positive Chvostek and Trousseau sign; and (f) prompt relief of the tetany when the serum magnesium concentration is restored to normal. Some investigators recognize the presence of other manifestations of clinical magnesium deficiency, such as phlebothrombosis, constitutional thrombasthenia and hemolytic anemia, an allergic or osseous form of the deficiency, and oxalate lithiasis.[4]

EXPERIMENTAL PRODUCTION OF A PURE MAGNESIUM DEFICIENCY

It is difficult to achieve a significant magnesium depletion in normal individuals by simple dietary restriction because of the exceedingly efficient renal and gastrointestinal mechanisms for conservation. The urinary magnesium in normal individuals falls to trivial amounts within 4 to 6 days of magnesium restriction. In spite of these conservatory mechanisms, investigators have successfully induced deficits approaching 10% of the total body content of magnesium by infusing sodium sulfate and adding calcium supplements to the magnesium-deficient diet.[5] The concentration of magnesium in plasma and erythrocytes fell moderately. Because the muscle magnesium content remained normal, the presumption was that bone was the source of the loss. No untoward clinical effects were noted. Other clinical data[6] suggest that total body depletion of magnesium may result in psychiatric and neuromuscular symptoms. Administration of magnesium by the parenteral route or in the diet was associated with clinical improvement which occasionally was dramatic.

The best study to date of magnesium deficiency in man is that reported by Shils.[7] Seven subjects were placed on a magnesium-deficient diet containing 0.7 mEq of magnesium per day. The concentration of magnesium in plasma declined perceptibly in all subjects within 7 to 10 days. Urinary and fecal magnesium decreased markedly, as did urinary calcium. At the height of the deficiency, the plasma magnesium concentration fell to a range of 10 to 30% of the control values, while the red cell magnesium declined more slowly and to a smaller degree. All male subjects developed hypocalcemia; one female did not. Marked and persistent symptoms developed only in the presence of hypocalcemia. The serum potassium concentration decreased and in four of the five subjects in whom the measurement was made, the ^{42}K space was decreased. The serum sodium concentration was not altered significantly. Three of the four subjects with the most severe symptoms also had metabolic alkalosis.

A positive Trousseau sign, which occurred in five of the seven subjects, was the most common neurologic sign observed. Electromyographic changes, which were characterized by the development of myopathic potentials, occurred in all five of the patients tested. Anorexia, nausea, and vomiting were frequently experienced. When magnesium was added to the experimental diet, all clinical and biochemical abnormalities were corrected.

NORMAL DISTRIBUTION AND TURNOVER OF MAGNESIUM IN MAN[8]

Anatomy

Analysis of human carcasses indicates that the magnesium content of the human body ranges between 23 and 35 mEq/kg wet weight of tissue. The body content of magnesium for a man weighing 70 kg would be approximately 2,000 mEq (24 g).

Of all the magnesium in the body, 89% resides in bone and muscle (Figure 1). Bone contains about 60% of the total body content of magnesium at a concentration of about 90 mEq/kg wet weight. Most of the remaining magnesium is distributed equally between muscle and the nonmuscular soft tissues. Of the nonosseus tissues, liver and striated muscle contain the highest concentrations, 14 to 16 mEq/kg. Approximately 1% of the total body content of magnesium is extracellular. The levels of magnesium in serum of healthy people are remarkably constant, averaging 1.7 mEq/l, and varying less than 15% from this mean value. Approximately one third of the extracellular magnesium is bound nonspecifically to plasma proteins. The remaining 65%, which is diffusible or ionized, appears to be the biologically active component.

Intake

The average American ingests daily between 20 and 40 mEq of magnesium; magnesium intakes of from 0.30 to 0.35 mEq/kg per day are thought to be adequate to maintain magnesium balance in normal adults (Figure 2). A daily intake of 17 mEq (0.25 mEq/kg) may meet nutritive requirements, provided that the individual remains in positive nitrogen balance. The estimated daily requirement for a child is 12.5 mEq (150 mg). The greater importance of magnesium in childhood is suggested by the relative ease with which deficiency states are produced experimentally in young animals as compared with adult animals.

Absorption

Tracer studies with ^{28}Mg administered orally to normal subjects revealed fecal excretion within 120 hr of 60 to 88% of the administered dose. When ^{28}Mg was injected intravenously, only 1.8% of the radioactivity was recovered in the stools within 72 hr.[9] The fecal magnesium appears to be primarily magnesium that is not absorbed by the body rather than magnesium secreted by the intestine. Ingested magnesium appears to be absorbed mainly by the small intestine. The factors controlling the gastrointestinal absorption of magnesium are poorly understood.

Secretion

There undoubtedly is considerable secretion of magnesium into the intestinal tract from bile, pancreatic, and intestinal juices. This secretion is followed by almost complete reabsorption. The observation that hypomagnesemia can occur in patients suffering from large losses of intestinal fluids suggests that intestinal juices contain enough magnesium to deplete the serum when magnesium is not reabsorbed by the colon.

Excretion

Most of that portion of the magnesium which is absorbed into the body is excreted by the kidney; fecal magnesium represents largely the unabsorbed fraction. In subjects on a normal diet, one third or less of the ingested magnesium (5 to 17 mEq) is excreted by the kidney. The maximal renal capacity for excretion is not known, but it is probably quite high, perhaps greater than 164 mEq/day. Magnesium deficiency does not occur in human beings with healthy kidneys. When the dietary intake of magnesium is low, renal mechanisms are efficient enough to conserve all but 1 mEq of magnesium per day. Both the mercurial and the thiazide diuretics increase renal excretion of magnesium, calcium, potassium, and sodium.

Turnover

Studies with the radioactive isotope of magnesium, ^{28}Mg, in human subjects have shown that the labile or "exchangeable" pool of magnesium is on the order of 135 to 400 mEq (2.6 to 5.3 mEq/kg) of body weight.[9] Since the total body content of magnesium is estimated to be 30 mEq/kg, it appears that less than 16% of the total body content of magnesium is labile. This labile pool is contained primarily in connective tissue, skin, and the soft tissues of the abdominal cavity (such as the liver and intestine); the magnesium in bone, muscle, and red cells exchanges very slowly.

Multicompartmental analysis indicates that in man there are at least three exchangeable magnesium pools with varied rates of turnover: compartments 1 and 2, exemplifying pools with a relatively fast turnover, together approximating extracellular fluid in distribution; compartment 3, an intracellular pool containing over 80% of the exchanging magnesium with a turnover rate one half that of the most rapid pool; and compartment 4, which probably accounts for most of the whole-body magnesium.[10] Only 15% of whole body magnesium is accounted for by relatively rapid exchange processes.[11]

Homeostatis

We do not yet understand clearly the physiologic mechanisms which are responsible for maintaining the plasma magnesium concentration at a constant level. The absorption of magnesium by the gut and its excretion by the kidney appear to be the main factors determining plasma magnesium levels. Because parathyroid hormone markedly affects the handling of magnesium by the gut and kidney, this hormone appears to be the main regulator of magnesium. The following hypothetical scheme of magnesium control is suggested.[12] The absorption of magnesium by the gut is largely controlled by parathyroid hormone. When plasma magnesium tends to fall, parathyroid hormone secretion increases due to a direct effect on the parathyroid gland. This increases magnesium absorption from the gastrointestinal tract and diminishes magnesium excretion by the kidney. The effect of the skeleton is mainly to smooth out any rapid changes by a passive exchange process of plasma magnesium with magnesium in the bone crystals.

CLINICAL CONDITIONS ASSOCIATED WITH DEPLETION OF MAGNESIUM

Fasting

Prolonged fasting is associated with a continued renal excretion of magnesium.[13] After 2 months of fasting, the deficit in some subjects may amount to 20% of the total body content of magnesium. Despite evidence for depletion of magnesium in muscle, the concentration of magnesium in plasma remains unchanged. The excess acid load presented for excretion to the kidney and the absence of intake of carbohydrate might be factors contributing to the persistent loss of magnesium. The magnitude of the excretion of magnesium parallels the severity of the acidosis. The ingestion of glucose decreases the urinary loss of magnesium.

Excessive Loss from the Gastrointestinal Tract

Persistent vomiting or prolonged removal of intestinal secretions by mechanical suction coupled with the administration of magnesium-free intravenous infusions can induce clinical magnesium deficiency.

Surgery

There are postoperative changes in magnesium metabolism in patients undergoing a variety of operations involving a moderate degree of trauma.[14] A lowered serum magnesium concentration is observed on the day after operation in 56% of the patients, but it is usually corrected by the second or third postoperative day. Surgery is followed by a negative magnesium balance of 3 days' duration, and similar changes are observed after dietary restriction in normal subjects. Usually the magnitude of the magnesium loss following surgery is minimal and does not result in symptomatic magnesium deficiency. However, maximum daily losses from the gastrointestinal tract may exceed 100 mEq per day and may require a similar rate of replacement.[15]

Gastrointestinal Disorders

The intestinal tract plays a major role in magnesium homeostasis. The rate of transport of magnesium across the intestine appears to be slower than that of calcium and directly proportional to intestinal transit time. Malabsorption of magnesium, therefore, occurs in conditions in which intestinal transit is abnormally rapid[12] or in which the major absorbing site, the distal small intestine, has been resected.

Malabsorption

Hypomagnesemia is associated frequently with malabsorption due to a variety of causes. In general, there appears to be a correlation between the degree of hypomagnesemia and the severity of the underlying disease. The increased fecal loss of magnesium that has been demonstrated in this disorder may be due to steatorrhea.[16]

Idiopathic Steatorrhea

A case of idiopathic steatorrhea with hypomagnesemia was reported by MacIntyre et al.[17], whose careful studies utilized balance techniques, bone and muscle biopsies, and measurements of the exchangeable magnesium content. The amount of magnesium in muscle was reduced to two thirds of the normal value, but the quantity in bone was affected little. The presence of magnesium depletion was further confirmed by the finding that the 24-hr exchangeable magnesium content was markedly reduced. Balance studies showed that fecal loss of magnesium was the cause of the deficiency and that this loss was intensified by a high intake of calcium. The results of this study appear to refute the concept that the bone magnesium always acts as a reservoir. The suggestion that bone magnesium in the adult man may be largely unavailable in disease state appears to correlate with the evidence that very little intracellular magnesium is mobilized during magnesium deficiency.

Malabsorption Syndrome

The magnesium deficiency of severe malabsorption syndrome is associated with metabolic abnormalities which affect the renal and intestinal transport of several other electrolytes.[18] Negative balances of magnesium, calcium, phosphorus, and potassium have been reported. Total exchangeable potassium content is reduced by one third during magnesium deficiency. The total 24-hr exchangeable magnesium content is reduced by more than 50%, due mainly to a decrease in the size of the slow pool which is believed to include skeletal muscle magnesium.

In patients with steatorrhea and tetany associated with hypomagnesemia, hypocalcemia, and absence of demonstrable bone disease, the replenishment of magnesium results in a very rapid restoration of serum calcium towards normal. This rapid change suggests that secretion of parathyroid hormone had been insufficient during magnesium deficiency and that replenishment of magnesium in some way stimulates the parathyroid glands.

Regional Enteritis

Patients with inflammatory bowel disease may develop symptomatic hypomagnesemia with symptoms and signs comparable to those induced in experimental animals placed on a diet deficient in magnesium. Although not a common complication, patients with this disease are uniquely disposed to the development of magnesium deficiency, either as a result of the underlying disease or as a consequence of therapy. The symptoms of hypomagnesemia are readily reversible. Unrecognized magnesium depletion may interfere with the complete restoration of deficits of other ions such as calcium and potassium.

Celiac Disease

Hypomagnesemia may occur in this disease in which the fecal loss of magnesium is increased, renal excretion of magnesium is decreased, and sodium and potassium is retained in the body. All mineral balances are restored to normal when gluten is eliminated from the diet.

Ulcerative Colitis

Patients with severe ulcerative colitis may develop hypomagnesemia along with subnormal levels of serum sodium, potassium, and chloride.

In *acute pancreatitis,* magnesium apparently forms fatty soaps in the necrotic omental fat and causes hypomagnesemia.[19]

Mastocytosis

A patient with the telangiectatic type of urticaria pigmentosa had clinical evidence of intestinal malabsorption and increased amounts of mast cells and histamine in the skin, stomach, and gut.[20] This patient had daily attacks of tetany; the serum magnesium concentration was 1.1 mEq/l. There was an excessive fecal loss of calcium and magnesium.

Another patient with systemic mastocytosis had attacks of facial flushing, abdominal pain, tachycardia, and explosive diarrhea.[21] These clinical symptoms and signs occurred coincidentally with a decrease in serum magnesium concentration.

MAGNESIUM DEFICIENCY IN CHILDHOOD

Alterations in the physiology of the gastrointestinal tract by surgical procedures, such as resection or creation of short circuits and fistulae, can result in neonatal magnesium deficiency.

Convulsions or tremors have been reported recently in children with hypomagnesemia but without associated hypocalcemia. Upon recovery, either spontaneously or after administration of magnesium, the serum magnesium concentration returned to normal levels. Convulsions have also been reported in association with both hypomagnesemia and hypocalcemia. Hypomagnesemic convulsions have been reported in a well-nourished infant in association with gastroenteritis, in a newborn infant from a hypophosphatemic mother, and in association with maternal hyperparathyroidism. Caddell[22] has advanced the hypothesis that sudden unexpected death in infancy is a preventable condition which results from the magnesium-deprivation syndrome of growth.

Kwashiorkor and Protein-calorie Malnutrition (PCM).

Protein-calorie malnutrition is the major nutritional deficiency syndrome of the world. Magnesium supplementation improves the rate and extent of recovery from severe PCM.[23]

A characteristic feature of diets which give rise to kwashiorkor is often the complete lack of milk. A pint of milk contains 6 to 7 mEq of magnesium; a satisfactory curative diet for kwashiorkor contains 13 mEq, of which 11 are derived from milk. A kwashiorkor-producing diet contains about one half of the magnesium in a curative diet.

The first controlled clinical studies of the relationship between magnesium deficiency and protein-calorie malnutrition were reported from Nigeria by Caddell and her associates.[23] Children with PCM have several clinical signs or symptoms compatible with magnesium deficiency. These include weakness, emaciation, anorexia, sleeplessness, trophic changes, and hyperirritability.

It is now well documented that magnesium deficiency occurs in PCM. This is evidenced by the low magnesium content of muscle, by a very low level of urinary magnesium, and by a prolonged positive balance of magnesium during recovery. This positive balance is much greater than could be accounted for by the corresponding retention of nitrogen. The magnesium content of the serum and red blood cells could be within normal range. There is impaired gastrointestinal absorption of magnesium in kwashiorkor, as well as a loss of tissue potassium, a retention of sodium, and osteoporosis. Children with kwashiorkor have an increased concentration of sodium and calcium and a decreased concentration of magnesium in the nails when compared with a control group.

Pregnancy

There may exist an occult deficiency of magnesium even in normal pregnancy. In 87 normal subjects, the average value for serum magnesium was 1.6 ± 0.14 mEq/l. The value for erythrocytes in 56 specimens was 5.2 ± 0.81 mEq/l. For normal pregnant women in the third trimester, the mean value for 105 serum specimens was 1.41 ± 0.04 mEq/l. The mean value for 104 erythrocyte specimens was 4.35 ± 0.41 mEq/l. These respective differences between the two groups are statistically significant.[24]

Magnesium deficiency may be a cause of the spasmophilic gravid uterus. Administration of magnesium reduces uterine hyperexcitability, insomnia, anxiety, and asthenia. In 24 of 80 observations, the serum magnesium level was less than 1.48 mEq/l.[25]

Excessive lactation may result in tetany attributable to hypomagnesemia. In one case, the intravenous administration of 12.5 mEq of magnesium resulted in the retention of some 78% of the administered dose, indicating the prior existence of a depletion of this ion.[26] Studies are in progress to determine by means of a parenteral magnesium load in postpartum women the nutritional status for this element.[27]

Acute Alcoholism

The mean serum magnesium value in patients with delirium tremens in one study was 1.53 ± 0.27 mEq/l. In alcoholics without delirium tremens, it was 1.89 ± 0.22 mEq/l. In the control group of 157 nonalcoholics, the mean serum magnesium value was 1.84 ± 0.18 mEq/l. There was a tendency for the lowest serum magnesium levels to coincide with the highest values for serum glutamic oxaloacetic transaminase.[28] Hypomagnesemia occurs frequently in patients with chronic alcoholism with and without delirium tremens.

Patients exhibiting alcohol withdrawal signs and symptoms have low serum and cerebral spinal fluid levels of magnesium, low exchangeable magnesium levels, a lowered muscle content of magnesium, and conservation of magnesium following intravenous loading. A transient decrease in serum magnesium may occur during the withdrawal state, even though prewithdrawal levels are normal. An ethanol-induced increase of magnesium in the urine occurs only when the blood alcohol level is rising. It does not persist once the subject has established high blood alcohol levels. However, in the presence of hypomagnesemia and delirium tremens, sudden death can occur as a result of cardiovascular collapse, infection, and hyperthermia. The red cell concentration of magnesium is abnormally low in all patients with delirium tremens, whereas the plasma concentration is abnormally low in only 58% of them. Intracellular fluid levels of magnesium, as reflected in the erythrocyte, correlate better with clinical symptoms and signs than do extracellular fluid levels. The predominant factor accounting for magnesium depletion in acute alcoholism is most likely an inadequate intake of magnesium, but another factor may be increased excretion of magnesium in the urine and feces.[29,30]

Independent of the phenomena described above, an abrupt and significant fall in serum magnesium level may occur following cessation of drinking. This acute fall in serum magnesium level is associated with a transient decrease in concentration of other serum electrolytes and with respiratory alkalosis and coincides with the onset of neuromuscular hyperexcitability that characterizes the withdrawal state. In most patients there is little correlation between serum magnesium levels and such clinical findings as hallucination, tremor, and tremulous handwriting. However, the combination of hypomagnesemia with delirium tremens or predelirium tremens may be of etiological significance for the occurrence of alcoholic encephalopathy.[31] A kinetic analysis of radiomagnesium turnover was performed in a group of partially repleted alcoholic subjects. Despite the continued presence of hypomagnesemia and of decreased urinary excretion of magnesium there was little evidence of continued depletion of magnesium in the extracellular space or in the tissue pools.[32]

Cirrhosis

The magnesium content of the liver tissue per unit weight is decreased in cirrhosis. This decrease appears to be due mainly to the substitution of parenchymal tissue of high magnesium content with connective tissue of low magnesium content. There is good relationship between histological changes (extent of fibrosis and degree of infiltration of inflammatory cells) and decrease of the magnesium concentration per number of cells. The actual changes in the concentration of magnesium in the parenchymal cells of the cirrhotic liver appear to be negligible.[33]

Cirrhotic patients have diminished skeletal muscle magnesium content but normal values for serum, bone, and erythrocyte magnesium; the clinical features of the magnesium deficiency are corrected with adequate replacement therapy.[34]

Diseases of the Parathyroid Gland

There have been references to abnormalities of magnesium metabolism in hyperparathyroidism for years. In clinical hypoparathyroidism, serum magnesium concentration tends to be lower than the average for normal individuals. Patients with primary hyperparathyroidism prior to treatment tend to develop a negative magnesium balance; this tendency appears to be related to the degree of hypercalcemia, the presence of advanced secondary renal disease, and inadequate intake of magnesium in the diet.[35] These patients retain an abnormally large portion of an injected dose of magnesium. Individuals reverting from hyperparathyroidism to normal or to the hypoparathyroid state store magnesium. The probable causes of the hypomagnesemia are an increased excretion of magnesium due to impaired renal conservation and an impaired absorption of magnesium from the gut. Severe hypomagnesemia may result in an end-organ unresponsiveness to the physiological effects of parathyroid hormone.[36] Parathyroid hormone responsiveness in hypomagnesemic patients may, at least in part, be dependent upon the adequacy of intracellular magnesium stores.

The effect of parathyroid hormone on magnesium metabolism is qualitatively similar to its influence on calcium metabolism but clearly less prominent. Significant and sustained hypomagnesemia after parathyroidectomy occurs only in patients with generalized bone disease. Its development appears to be dependent upon the inadequacy of the ordinary diet to meet the combined requirements for magnesium for formation of new soft tissue and for deposition of mineralizing bone.

There are reports of patients with an acute parathyroid crisis treated by surgical removal of a parathyroid adenoma who postoperatively developed evidence of magnesium deficiency; this cleared dramatically following the parenteral administration of magnesium.[37,38]

Burns

Patients with extensive burns may develop the magnesium-deficiency syndrome.[39] Serum magnesium may decrease significantly and clinical symptoms of magnesium deficiency may develop. Some of the psychiatric symptoms exhibited by many patients with burns may be due to, or may be aggravated by, magnesium deficiency.

Renal Disease

Hypomagnesemia has been observed in patients with glomerulonephritis, hydronephrosis, pyelonephritis, and renal tubular acidosis. Although this reduction is presumably due to a defective tubular reabsorptive mechanism, its occurrence is unpredictable. Frequent hemodialysis may result in severe muscle cramps due to acute hypomagnesemia if the dialysis fluid contains an inadequate amount of magnesium.[40]

Hypomagnesemia due to renal disease of unknown etiology may be associated with tetanic convulsions, arthritic pains, hypermagnesuria, hypokalemia, hypercalciuria, progressive nephrocalcinosis, and chrondrocalcinosis.[41] Children may have hypomagnesemia with carpopedal spasm and fail to thrive due to a renal tubular defect in the reabsorption of magnesium.[42]

Cardiovascular Disease

Some patients with hypertension and others with primary aldosteronism have subnormal magnesium levels. The exchangeable magnesium content per kilogram of body weight is decreased in hypertensive men as compared to controls. These changes have not been found in hypertensive women.[43]

The administration of hydrochlorothiazide to patients with hypertension results in a decrease in serum magnesium concentration; despite this, the erythrocyte magnesium concentration increases significantly. The change in diastolic pressure induced by hydrochlorothiazide correlates with the ratio of the difference of intracellular potassium to intracellular magnesium. These results suggest that changes in magnesium metabolism influence vascular tone and play an integral role in the control of blood pressure.[44]

Congestive Heart Failure

Although the serum magnesium levels in patients with congestive heart failure are similar to those in healthy individuals, the magnesium content of erythrocytes may be decreased in patients with long-standing failure. Also reduced are the magnesium concentrations in heart and skeletal muscles. There is a correlation between the intracellular deficiency of magnesium and that of potassium, yet the deficiency of potassium is greater.[45]

Digitalis Toxicity

Many patients who are receiving digitalis have a significantly lowered magnesium level which predisposes them to digitalis toxicity.[46] This toxicity can be terminated promptly by the administration of magnesium sulfate. It is recommended that levels of serum magnesium as well as those of serum potassium be determined routinely in patients suspected of digitalis toxicity.

Paroxysmal Ventricular Fibrillation

Paroxysmal ventricular fibrillation, which is not associated with heart block, may occur in patients with hypomagnesemia.[47]

Familial Disorder of Magnesium Metabolism

Since 1966 there have appeared reports in the literature of chronic congenital hypomagnesemia which seemed to be a familial disorder with autosomal recessive transmission. Such patients may have repeated episodes of hypomagnesemia, symptoms of intermittent weakness and abnormal personality behavior, and a history of dermatographism and bronchial asthma.

There have been several other recent reports of a familial hypomagnesemia. In a Scandinavian report,[48] two brothers at the ages of 15 and 23 days, developed repeated tetanic convulsions. The tetany, hypomagnesemia, hypocalcemia, and hypophosphatemia all subsided when magnesium was given as the only form of therapy. Histologic examination by electron microscopy revealed hepatic cellular necrosis, dilatation of the endoplasmic reticulum, and mitochondrial swelling. Net intestinal absorption of magnesium as measured by ^{28}Mg was abnormally low, whereas its renal handling appeared normal; no definite signs of generalized malabsorption were found. Therefore, the defect appeared to be one related specifically to the intestinal absorption of magnesium.

DIAGNOSIS OF MAGNESIUM DEFICIENCY

If the clinical situation is suggestive and the serum magnesium value is subnormal, magnesium deficiency is present. If the serum values are normal, erythrocyte magnesium content and 24 hr urinary excretion should be measured. When the magnesium content of erythrocytes or the urinary excretion of magnesium is normal, magnesium deficiency is very unlikely.

REFERENCES

1. Wacker, W. E. C., Moore, F. D., Ulmer, D. D., and Vallee, B. L., Normocalcemic magnesium deficiency tetany, *J.A.M.A.,* 180, 161–163, 1962.
2. Flink, E. B., Magnesium deficiency syndrome in man, *J.A.M.A.,* 160, 1406-1409, 1956.
3. Wacker, W. E. C. and Parisi, A. F., Magnesium metabolism, *N. Engl. J. Med.,* 278, 658–662, 712–717, 772–776, 1968.
4. Durlach, J., Le Magnésium en pathologie humaine. Problèmes practiques et incidences diététiques, *Gaz. Med. Fr.,* 74, 3303–3320, 1967.
5. Dunn, M. J. and Walser, M., Magnesium depletion in normal man, *Metabolism,* 15, 884–895, 1966.
6. Randall, R. E., Rossmeisl, E. C., and Bleifer, K. H., Magnesium depletion in man, *Ann. Intern. Med.,* 50, 257–287, 1959.
7. Shils, M. E., Experimental human magnesium depletion, *Medicine,* 48, 61–85, 1969.
8. Aikawa, J. K., *The Relationship of Magnesium to Disease in Domestic Animals and in Humans,* Charles C Thomas, Springfield, Ill., 1971.
9. Aikawa, J. K., Gordon, G. S., and Rhoades, E. L., Magnesium metabolism in human beings: Studies with Mg^{28}, *J. Appl. Physiol.,* 15, 503–507, 1960.
10. Wallach, S., Rizek, S. E., Dimich, A., Prasad, N., and Siler, W., Magnesium transport in normal and uremic patients, *J. Clin. Endocrinol.,* 26, 1069–1080, 1966.
11. Avioli, L. V. and Berman, M., Mg^{28} kinetics in man, *J. Appl. Physiol.,* 21, 1688–1694, 1966.
12. MacIntyre, I. and Robinson, C. J., Magnesium in the gut: Experimental and clinical observations, *Ann. N.Y. Acad. Sci.,* 162, 865–873, 1969.
13. Drenick, E. J., Hunt, I. F., and Swendseid, M. E., Magnesium depletion during prolonged fasting of obese males, *J. Clin. Endocrinol.,* 29, 1341–1348, 1969.
14. Heaton, F. W., Magnesium metabolism in surgical patients, *Clin. Chim. Acta,* 9, 327–333, 1964.
15. Barnes, B. A., Magnesium conservation: A study of surgical patients, *Ann. N.Y. Acad. Sci.,* 162, 786–801, 1969.
16. Gitelman, H. J. and Welt, L. G., Magnesium deficiency, *Ann. Rev. Med.,* 20, 233–242, 1969.
17. MacIntyre, I., Hanna, S., Booth, C. C., and Read, A. E., Intracellular magnesium deficiency in man, *Clin. Sci.,* 20, 297–305, 1961.
18. Petersen, V. P., Potassium and magnesium turnover in magnesium deficiency, *Acta Med. Scand.,* 174, 595–604, 1963.
19. Martin, H. E., Mehl, J., and Wertman, M., Clinical studies of magnesium metabolism, *Med. Clin. North Am.,* 36, 1157–1171, 1952.

20. **Jarnum, S. and Zachariae, H.,** Mastocytosis (urticaria pigmentosa) of skin, stomach, and gut with malabsorption, *Gut,* 8, 64–68, 1967.

21. **Broitman, S. A., McCray, R. S., May, J. C., Deren, J. J., Ackroyd, F., Gottlieb, L. S., McDermott, W., and Zamcheck, N.,** Mastocytosis and intestinal malabsorption, *Am. J. Med.,* 48, 382–389, 1970.

22. **Caddell, J. L.,** Magnesium deprivation in sudden unexpected infant death, *Lancet,* 258–262, August 5, 1972.

23. **Caddell, J. L.,** Magnesium deficiency in protein-calorie malnutrition: A follow-up study, *Ann. N.Y. Acad. Sci.,* 162, 874–890, 1969.

24. **Lim, P., Jacob, E., Dong, S., and Khoo, O. T.,** Values for tissue magnesium as a guide in detecting magnesium deficiency, *J. Clin. Pathol.,* 22, 417–421, 1969.

25. **Dumont, M. and Bernard, P.,** Carence en magnésium et thérapeutique magnésiée au cours de l'état gravido-puerpéral, *Lyon Med.,* 216, 307–358, 1966.

26. **Greenwald, J. H., Dubin, A., and Cardon, L.,** Hypomagneseinic tetany due to excessive lactation, *Am. J. Med.,* 35, 854–860, 1963.

27. **Caddell, J. L., Saier, F. L., and Thomason, C. A.,** Parenteral magnesium load tests in post partum American women, *Am. J. Clin. Nutr.,* 28, 1099–1104, 1975.

28. **Nielsen, J.,** Magnesium metabolism in acute alcoholics, *Dan. Med. Bull.,* 10, 225–233, 1963.

29. **Frankushen, D., Raskin, D., Dimich, A., and Wallach, S.,** The significance of hypomagnesemia in alcoholic patients, *Am. J. Med.,* 37, 802–814, 1964.

30. **McCollister, R. J., Flink, E. B., and Lewis, M. D.,** Urinary excretion of magnesium in man following the ingestion of ethanol, *Am. J. Clin. Nutr.,* 12, 415–420, 1963.

31. **Stendig-Lindberg, G.,** Hypomagnesaemia in alcohol encephalopathies, *Acta Psychiatr. Scand.,* 50, 465–480, 1974.

32. **Wallach, S. and Dimich, A.,** Radiomagnesium turnover studies in hypomagnesemic states, *Ann. N.Y. Acad. Sci.,* 162, 963–972, 1969.

33. **Wilke, H. and Spielmann, H.,** Untersuchungen über den Magnesiumgehalt der Leber bei der Cirrhose, *Klin. Wochenschr.,* 46, 1162–1164, 1968.

34. **Lim, P. and Jacob, E.,** Magnesium deficiency in liver cirrhosis, *Q. J. Med.,* 163, 291–300, 1972.

35. **King, R. G. and Stanbury, S. W.,** Magnesium metabolism in primary hyperparathyroidism, *Clin. Sci.,* 39, 281–303, 1970.

36. **MacManus, J., Heaton, F. W., and Lucas, P. W.,** A decreased response to parathyroid hormone in magnesium deficiency, *J. Endocrinol.,* 49, 253–258, 1971.

37. **Jacobs, J. K. and Merritt, C. R.,** Magnesium deficiency in hyperparathyroidism. Case report of toxic psychosis, *Ann. Surg.,* 163, 260–262, 1966.

38. **Jones, C. T. A., Sellwood, R. A., and Evanson, J. M.,** Symptomatic hypomagnesemia after parathyroidectomy, *Br. Med. J.,* 391–392, August 18, 1973.

39. **Broughton, A., Anderson, I. R. M., and Bowden, C. H.,** Magnesium-deficiency syndrome in burns, *Lancet,* 2, 1156–1158, 1968.

40. **Triger, D. R. and Jokes, A. M.,** Severe muscle cramps due to acute hypomagnesaemia in haemodialysis, *Br. Med. J.,* 2, 804–805, 1969.

41. **Runeberg, L., Collan, Y., Jokinen, E. J., Lähdevirta, J., and Aro, A.,** Hypomagnesemia due to renal disease of unknown etiology, *Am. J. Med.,* 59, 873–881, 1975.

42. **Booth, B. E. and Johanson, A.,** Hypomagnesemia due to renal tubular defect in reabsorption of magnesium, *J. Pediatr.,* 84, 350–354, 1974.

43. **Bauer, F. K., Martin, H. E., and Mickey, M. R.,** Exchangeable magnesium in hypertension, *Proc. Soc. Exp. Biol. Med.,* 120, 466–468, 1965.

44. **Seller, R. H., Ramirez-Muxo, O., Brest, A., and Moyer, J. H.,** Magnesium metabolism in hypertension, *J.A.M.A.,* 191, 654–656, 1965.

45. **Szczepański, L.,** Cellular magnesium deficiency in congestive heart failure, *Pol. Med. J.,* 6, 7–12, 1967.

46. **Beller, G. A., Hood, W. B., Smith, T. W., Abelmann, W. H., and Wacker, W. E. C.,** Correlation of serum magnesium levels and cardiac digitalis intoxication, *Am. J. Cardiol.,* 33, 225–229, 1974.

47. **Iseri, L. T., Freed, J., and Bures, A. R.,** Magnesium deficiency and cardiac disorders, *Am. J. Med.,* 58, 837–846, 1975.

48. **Strømme, J. H., Nesbakken, R., Normann, T., Skorten, F., Skyberg, D., and Johannessen, R.,** Familial hypomagnesemia, *Acta Paediatr. Scand.,* 58, 434–444, 1969.

EFFECT OF NUTRIENT DEFICIENCIES IN MAN — CHLORINE

B. G. Shah

INTRODUCTION

Chlorine occurs in living organisms mostly as the chloride ion. The total body chloride in a human adult is about 1 g (28 meq)/kg body weight as determined by neutron activation analysis. This value is about 15% less than the results previously obtained by chemical methods.[1] Approximately 90% of it is exchangeable, and the ratio between extracellular and intracellular chloride is about 2:1. The chloride ion is found in low concentration in bone and connective tissues but there are no appreciable reserves of this anion in the body.[2-3] As the affinity of chloride for binding with protein is very weak, it is a major contributor to the acid-base balance and the osmotic equilibrium in body fluids. Normally, the efficient handling of chloride by the small intestine and the colon and by the kidney is responsible for chloride homeostasis. The serum chloride is maintained at about 100 meq/l. Chloride is not under direct hormonal or central control but homeostasis is achieved through the movement of cations which are subject to such mechanisms.[4]

Dietary chloride is chiefly derived from salt. The daily salt intake in the Western countries varies from 5 to 20 g (80 to 350 meq), although many primarily herbivorous peoples living in isolated areas of the world where salt is not used have survived on a daily intake of as little as 10 meq of salt. Thus, a simple dietary deficiency of chloride in normal individuals has not been reported.[5-6] Excessive loss of chloride or genetic defect in the intestinal absorption of chloride can, however, lead to deficiency and its sequelae.

CONGENITAL CHLORIDE DIARRHEA

About a dozen cases of congenital chloride diarrhea have been reported since it was first described in 1945,[7-8] indicating that it is a rare disorder. The disease is characterized by diarrhea from birth. Fecal chloride concentrations are in excess of the combined concentrations of sodium and potassium, usually more than 150 mM, indicating that there is a primary defect in secretion or reabsorption of chloride in the ileum and colon. The excessive fecal loss of chloride is associated with hypochloremia, hypokalemia, metabolic alkalosis, and an almost complete absence of chloride in the urine.[9-11] Kidney biopsies of children suffering from the disease revealed hypertrophy of juxtaglomerular apparatus, hyperplastic cortical arteries, pericapsular fibrosis, vacuolization of epithelium in the proximal convoluted tubules, and calcification of distal convoluted tubules. Treatment of a case with sodium chloride, potassium bicarbonate, a combination of sodium chloride and potassium chloride, and potassium chloride alone, showed that serum chloride progressively increased with increasing intake of potassium chloride with or without sodium chloride but not with increasing intake of sodium chloride alone. Stool concentrations of chloride remained elevated and urinary excretion remained low despite high chloride intake, confirming the defect in the intestinal absorption.[11-12]

CHLORIDE DEPLETION ALKALOSIS

Chloride depletion can result from frequent vomiting or repeated gastric drainage. Vomiting may be caused by a chronic gastric ulcer followed by pyloric stenosis. Gastric drainage is often recommended in gastric ulcer patients. Chloride loss may occur via the rectum in patients with adenoma of the sigmoid colon. Excessive loss of chloride through urine can also result from chronic hypercapnia and prolonged treatment with diuretics.

As chloride depletion alkalosis is associated with loss of body potassium and hypokalemia, excess of mineralocorticoids caused by various factors also leads to metabolic alkalosis and potassium depletion. Although chloride depletion metabolic alkalosis can be corrected partially by the provision of chloride in the form of sodium chloride, administration of potassium chloride is the therapy of choice, since potassium loss usually coexists with the loss of chloride.[13-14] Azotemia accompanies hypochloremia only when coincident dehydration is rapid.[15]

The symptoms of metabolic alkalosis are indistinguishable from those of hypokalemia. The major effect involves the overexcitability of the nervous system. Usually, the peripheral nerves are affected before the central nervous system. Tetany appears first in the muscles of the forearm followed by those of the face; it then spreads throughout the body. Cardiac arrhythmias have also been noted. Alkalotic patients may die from tetany of the respiratory muscles. The symptoms of overexcitability of the central nervous system (viz., extreme nervousness or convulsions) are only rarely seen.[13,16]

Calcification of the kidneys was reported in six fatal cases of pyloric stenosis and was attributed to chloride depletion alkalosis. The tubules showed gross degeneration, lipid changes, and deposits of calcium phosphate in every case. The congestion found in the kidney sections suggested slowing of the circulation. These functional and structural changes in the kidneys were similar to those seen in experimental nephritis in animals due to uranium nitrate.[17]

REFERENCES

1. **Ellis, K. J., Vaswani, A., Zanzi, I., and Cohn, S. H.,** Total body sodium and chloride in normal adults, *Metab. Clin. Exp.,* 25, 645–654, 1976.
2. **Randall, H. T.,** Water, electrolytes and acid-base balance, in *Modern Nutrition in Health and Disease: Dietotherapy,* Goodhart, R. S. and Shils, M. E., Eds., Lea & Febiger, Philadelphia, 1974, 324–361.
3. **Deane, N., Ziff, M., and Smith, H. W.,** The distribution of total body chloride in man, *J. Clin. Invest.,* 31, 200–203, 1952.
4. **Cotlove, E. and Hogben, C. A. M.,** Chloride, in *Mineral Metabolism: an Advanced Treatise,* Vol. 2B, Comar, C. L. and Bronner, F., Eds., Academic Press, New York, 1962, 109–173.
5. **Davidson, S., Passmore, R., Brock, J. F., and Truswell, A. S.,** *Human Nutrition and Dietetics,* Churchill, Livingstone, New York, 1975, 94–106.
6. **Dahl, L. K.,** Salt and hypertension, *Am. J. Clin. Nutr.,* 25, 231–244, 1972.
7. **Gamble, J. L., Fahey, K. R., Appleton, J., and MacLachlan, E.,** Congenital alkalosis with diarrhea, *J. Pediatr.,* 26, 509–518, 1945.
8. **Darrow, D. C.,** Congenital alkalosis with diarrhea, *J. Pediatr.,* 26, 519–532, 1945.
9. **Schedl, H. P.,** Water and electrolyte transport — clinical aspects, *Med. Clin. North Am.,* 58, 1429–1448, 1974.
10. **Turnberg, L. A.,** Absorption and secretion of salt and water by the small intestine, *Digestion,* 9, 357–381, 1973.
11. **Pasternack, A., Percheentupa, J., Launiala, K., and Hallman, N.,** Kidney biopsy findings in familial chloride diarrhea, *Acta Endocrinol.* (Copenhagen), 55, 1–9, 1967.
12. **McReynolds, E. W., Shane, R., III, and Etteldorf, J. N.,** Congenital chloride diarrhea, *Am. J. Dis. Child.,* 127, 566–570, 1974.
13. **Makoff, D. L.,** Acid-base metabolism, in *Clinical Disorders of Fluid and Electrolyte Metabolism,* Maxwell, M. H. and Kleeman, C. R., Eds., McGraw-Hill, New York, 1972, 297–346.
14. **Fein, H. D.,** Nutrition in diseases of the gastrointestinal tract, in *Modern Nutrition in Health and Disease: Dietotherapy,* Goodhart R. S. and Shils, M. E., Eds., Lea & Febiger, Philadelphia, 1974, 770–784.
15. **Kirsner, J. B., Palmer, W. L., and Knowlton, K.,** Studies on experimental and clinical hypochloremia in man, *J. Clin. Invest.,* 22, 95–102, 1943.
16. **Guyton, A. C.,** *Textbook of Medical Physiology,* W. B. Saunders, Philadelphia, 1971, 439.
17. **Cooke, A. M.,** Calcification of the kidneys in pyloric stenosis, *Q. J. of Med.,* 26, 539–548, 1933.

NUTRIENT DEFICIENCIES IN MAN: SULFUR

Marian E. Swendseid and Marian Wang

Humans require sulfur for the synthesis of proteins, mucopolysaccharides, and a number of low-molecular weight, sulfur-containing compounds with important catalytic and homeostatic roles. Present evidence indicates that in humans, the requirement for synthesis of all these compounds can be met by a single, sulfur-containing amino acid, methionine. It is possible that in infants there is an additional requirement for taurine, an amino sulfonic acid. Breast-fed infants have higher plasma levels of taurine than infants fed casein-based formulas.[1] Metabolic effects of sulfur deficiencies are presumed to be the same as those of methionine deficiency. (See chapter on Effects of Nutrient Deficiencies in Man: Specific Amino Acids.)

REFERENCES

1. Sturman, J. A., Rassin, D. K., and Gaull, G. E., Taurine in development: is it essential in the neonate?, *Pediatr. Res.,* 10, 415, 1976.

NUTRIENT DEFICIENCIES IN MAN: IRON

V. F. Fairbanks

INTRODUCTION

Iron is an element that is abundant in the soil of most of the world, and it is present in vast quantities in the environment of civilized man. Iron is, furthermore, essential to all forms of life. In the absence of iron, photosynthesis would not take place and aerobic glycolysis, which requires iron in numerous enzymatic reactions, would cease. In view of these facts, the high prevalence of iron deficiency throughout the world is a striking paradox.

Clinical manifestations of iron deficiency have long been recognized.[1] As "chlorosis," iron deficiency anemia was considered epidemic in Europe and in the European settlements of the western hemisphere more than a century ago. Only near the turn of the century did it become clear that chlorosis was, in fact, iron deficiency anemia.

Perhaps the earliest epidemiologic study of iron deficiency was that of MacKay,[2] who, in the years 1925 to 1930, surveyed "nutritional anaemia" in 1100 infants and 168 expectant or nursing mothers in London, England. This survey was based on hemoglobin measurements. She found that, of breast-fed infants, 45% were anemic, and that of formula-fed infants, 51% were anemic. Anemia responded to iron therapy.

Subsequent studies have confirmed the high prevalence of iron deficiency anemia, not only in infants and young children but also in adults. This remains true of "affluent" western societies but is particularly characteristic of developing countries, in which the greater part of the population may be iron deficient.

DEFINITIONS

Iron deficiency is said to exist when total body iron content is less than that expected in healthy persons. However, this definition cannot be applied clinically, as no method exists for measuring total body iron in living persons. Therefore, the practical or clinical definition of iron deficiency is based on sampling of one or more iron compartments. These compartments are (1) hemoglobin iron; (2) storage iron, comprising ferritin and hemosiderin; (3) transport iron, in the form of transferrin-bound iron in plasma; (4) "tissue iron," comprising iron enzymes, coenzymes, and cytochromes; (5) myoglobin iron, contained in muscle cells; and (6) a "labile iron pool," a concept derived from iron kinetic studies. These iron compartments will be further discussed in the following section entitled "Iron Compartments and Iron Metabolism."

For clinical purposes, it is important to recognize three stages of iron deficiency.[3] These are (1) iron depletion, (2) iron deficiency anemia without anemia, and (3) iron deficiency anemia. *Iron depletion* represents minimal iron deficiency and is characterized by absent storage iron, normal transport (serum) iron, and normal hemoglobin iron. *Iron deficiency without anemia* is the intermediate phase, characterized by absent storage iron, reduced transport (serum) iron, but normal hemoglobin iron. In *iron deficiency anemia,* storage iron, transport iron, and hemoglobin iron are all characteristically diminished. In practice these distinctions are often blurred; for example, it is not rare to observe frank iron deficiency anemia in persons whose serum iron concentrations are normal.[4]

IRON COMPARTMENTS AND IRON METABOLISM

The approximate iron content of some of the iron compartments for a healthy adult male of about 70 kg weight is as follows:

Hemoglobin iron	2500 mg
Storage iron	800 mg
Transport iron	3 mg
Myoglobin iron	120 mg
Tissue iron	8 mg
Labile iron	80 mg

The iron compartments of women and children are usually smaller in proportion to weight. In rapidly growing children or in women of child-bearing years, iron stores are commonly depleted.

Hemoglobin is a protein with a molecular weight of 64,456 daltons. It contains 0.34% iron by weight. It is responsible for the transport and exchange of oxygen and carbon dioxide in the blood. Ferritin is a complex of iron with apoferritin. Each molecule of ferritin may incorporate as many as 2000 FeOOH molecules into a semicrystalline lattice. The molecular weight of apoferritin is 450,000 daltons. Ferritin is approximately 20% iron by weight. Ferritin occurs in almost all cells of the body, but particularly in hepatic parenchymal cells (hepatocytes) and in cells of the reticuloendothelial system. In these cells it may coexist with the other major storage form, hemosiderin. In addition, ferritin is found in minute concentration in normal plasma. Whether it subserves a transport function in plasma is uncertain. However, plasma ferritin concentration appears to reflect total body iron stores. Hemosiderin is the other major storage form of iron. It appears to be principally a large aggregate of many ferritin molecules. In contrast to ferritin, hemosiderin is water insoluble and thus is easily seen microscopically in bone marrow smears or tissue sections. It contains about 25 to 30% iron by weight. The major iron transport protein is transferrin, a plasma β-1 globulin that can bind two atoms of trivalent iron per transferrin molecule. It functions to transport iron from storage sites to developing red cell precursors in the bone marrow. Myoglobin is closely related to hemoglobin structurally but is monomeric, whereas hemoglobin is tetrameric. Thus, the molecular weight of myoglobin is approximately one fourth that of hemoglobin. It occurs in striated muscle (skeletal muscle and myocardium), where it may serve as an oxygen reservoir under circumstances of extreme oxygen deprivation. In terrestrial mammals such as man, the myoglobin content of striated muscle is very small, and myoglobin does not appear to have a major physiologic role or contribute in a major way to iron metabolism. "Tissue iron" is a term used to include the large number of enzymes, coenzymes, and cytochromes that contain or require iron for their metabolic role. The amount of iron in this compartment is very small, but this is a very active compartment that is involved in the metabolism of most of the cells of the body. Some of the iron enzymes and cytochromes are very sensitive to changes in total body iron and exhibit decreased activity very early in the course of iron depletion. The concept of a "labile iron pool" is a hypothetic one derived from iron kinetic studies and will not be further considered here.

An average adult American male ingests approximately 10 to 15 mg of iron daily in his diet, or about 6 mg/1000 calories.[5] Approximately half the dietary iron is inorganic; most of the remainder is heme iron, in the form of hemoglobin, myoglobin, and other heme proteins. Heme appears to be well absorbed as such and may be catabolized in the intestinal mucosal cells. Inorganic iron is most readily absorbed in the divalent state by the duodenal mucosa. Altogether, approximately 1 mg of iron is absorbed daily. If iron requirement increases, more dietary iron can be absorbed. Iron absorption is balanced by iron excretion, mostly in the form of iron contained in mucosal cells lost by desquamation into the lumen of the gastrointestinal tract.

Of each milligram of iron absorbed from the diet, approximately 0.8 mg is directly incorporated into hemoglobin. However, approximately 25 mg of iron is required each day for hemoglobin synthesis. Of this, most is derived from hemoglobin catabolism. Phagocytic reticuloendothelial cells engulf and digest effete erythrocytes (the red cells that have been in circulation for nearly 4 months) and release iron into plasma. In the plasma, iron binds to transferrin and is transported to developing erythrocytic precursor cells of the bone marrow. The synthesis of new hemoglobin, requiring about 25 mg of iron daily, is balanced by release of about 25 mg of iron as a result of hemoglobin catabolism. There is also a slight exchange with iron in storage and tissue iron compartments.

IRON REQUIREMENTS

Physiologic iron loss in adult males is approximately 0.5 to 1 mg daily, almost entirely in the form of iron contained in exfoliated mucosal cells of the intestinal tract. Normally there is scant iron loss in sweat or urine, irrespective of climatologic conditions.[6] Adult women, during the reproductive years, lose an average of about 20 mg of iron monthly as a result of menstruation and approximately 800 mg of iron for each pregnancy. These losses are in addition to an approximately 1-mg daily iron loss from exfoliation of intestinal epithelial cells. To balance these losses, adult males should ingest approximately 10 to 15 mg of iron daily, and women during the reproductive years should ingest nearly twice this amount of iron exclusive of the amount required to replace the iron cost of pregnancy. These iron needs seem generally to be met in adult American males, but they commonly are not met by women, whose mean iron intake is about 10 mg/day.[7]

Rapidly growing children have, in proportion to size, a greater iron requirement than adults. This is particularly so in infants and during the pubertal growth spurt. The iron need of infants has been estimated at about 1 mg daily, which should be met by about 1.5 mg of iron per kilogram daily, to a maximum of 15 mg/day. Prepubertal children require 0.5 to 1 mg daily, necessitating about 5 to 10 mg of dietary iron. During the pubertal growth spurt, iron need is about 1 to 2 mg daily, requiring 10 to 20 mg of iron in dietary sources.[3,8]

ETIOLOGY OF IRON DEFICIENCY

Iron deficiency arises when, over a sufficient period of time, the need for iron exceeds the availability of iron. Iron need is increased when there is either rapid growth or accelerated iron loss — for example, from bleeding, hemoglobinuria, pregnancy, or lactation. Iron availability is diminished (1) when there is a low dietary iron content, (2) when the dietary iron is quantitatively sufficient but is in a form not readily absorbed, or (3) when the mucosa of the small intestine is unable to absorb iron efficiently.

An iron-poor diet and rapid growth are prime causes of iron deficiency in infants and preschool children. However, in many such children, intestinal bleeding (from hookworm infestation or from bovine milk sensitivity)[9] may also be a factor. Similarly, the suboptimal iron content of the diet of young women contributes to the high prevalence of iron deficiency in this group, although blood loss is the major factor.

In agricultural societies in many rural areas of Central America, the major dietary constituents are black beans and maize, and meat is virtually lacking. Black beans have a high iron content.[10,11] However, phytic acid in maize chelates iron and markedly reduces its absorption.[11] In consequence, iron deficiency is very prevalent in such societies, although the iron content of the bean-maize diet is quantitatively adequate.

In adults, the demonstration of iron deficiency must be taken as evidence of iron loss, which is usually through bleeding. It is very rare for this condition to arise simply from an iron-poor diet, since dietary deficiency would, in the absence of blood loss, require for its genesis at least 10 years of a severely restricted diet. In males or postmenopausal women, this is most commonly gastrointestinal bleeding. In women of reproductive age, iron loss is usually the result of menstrual bleeding and the iron cost of pregnancy and lactation.

Table 1 lists the causes of iron deficiency; the recognized causes of gastrointestinal bleeding are also tabulated. In adults the most frequent causes of gastrointestinal blood loss are hemorrhoids, peptic ulcer, diaphragmatic hiatus hernia, and gastritis. Hemorrhoids are often overlooked as a cause of blood loss and iron deficiency anemia, both by the patient and by the physician. There may be very severe iron deficiency anemia resulting from recurrent hemorrhoidal bleeding in persons who fail to recognize hemorrhoidal bleeding as a significant source of blood loss, probably because of social and personal mores. Diaphragmatic esophageal hiatus hernia is more likely to be a cause of bleeding and iron deficiency if it is of a large sliding type. Gastritis due to aspirin ingestion is a common cause of chronic low-grade bleeding that leads to iron deficiency. This often is not detectable by guaiac-based methods for testing for occult fecal blood, because the guaiac test is usually insensitive to quantities of blood loss of the order of 5 ml/day — an amount sufficient, however, if long persistent, to result in iron deficiency. Other drugs known to cause bleeding are adrenocortical steroids, phenylbutazone, indomethacin, ethacrynic acid, and thiazide diuretics combined with potassium chloride in enteric-coated tablets.

Approximately 2% of adults with gastrointestinal bleeding have an underlying malignancy of the gastrointestinal tract, such as carcinoma of the stomach or colon. In such cases, iron deficiency may be the earliest clue to the presence of an otherwise asymptomatic malignancy.

Intestinal malabsorption leading to iron deficiency may be the result of gluten sensitivity (celiac disease and nontropical sprue), or infiltration of bowel by carcinoma or lymphoma, or it may result from Whipple's disease or follow gastrointestinal surgery. Subtotal gastrectomy, especially of the types employing gastrojejunal anastomosis (Billroth II), is often followed by iron deficiency anemia, which may develop several years after surgery.[12] This occurs because of reduced absorption of dietary iron. Low-grade bleeding from the site of anastomosis may also be contributory in some cases.

Table 1
CAUSES OF IRON DEFICIENCY

I.   Dietary deficiency of iron
    A. In infants and young children
    B. In adults, rarely, except as contributory cause in women
II. Intestinal disorders
    Malabsorption
        Gluten sensitivity (nontropical sprue)
        Postsurgical (gastrectomy with gastroenterostomy)
        Other causes of malabsorption
III. Blood loss from anatomic lesions
    A. Respiratory tract
        Epistaxis
        Hemoptysis
          Idiopathic pulmonary hemosiderosis
          Goodpasture's syndrome
          Cardiac valvulopathy with pulmonary hypertension
          Bronchiectasis
          Vascular anomalies

Table 1 (continued)
CAUSES OF IRON DEFICIENCY

  B. Alimentary tract
      Esophagus
         Varices
         Hiatus hernia
      Stomach
         Varices
         Ulcer
         Carcinoma
         Gastritis
         Leiomyoma
      Small bowel
         Ulcer
         Aberrant pancreas
         Telangiectasia
         Polyp
         Carcinoma
         Regional enteritis
         Helminthiasis
         Vascular occlusion
         Intussusception
         Volvulus
         Leiomyoma
         Aberrant gastric mucosa
         Meckel's diverticulum
      Biliary tract
         Trauma
         Cholelithiasis
         Neoplasm
         Aberrant pancreas or gastric mucosa
         Ruptured aneurysm
         Intrahepatic bleeding
      Colon
         Ulcerative colitis
         Amebiasis
         Carcinoma
         Telangiectasia
         Diverticulum
         Benign polyps
         Vascular occlusion
      Rectum
         Hemorrhoids
         Ulceration
         Carcinoma
  C. Genitourinary tract
      Menorrhagia
      Malignancy of uterus or bladder
      Renal neoplasms or tuberculosis
  D. Trauma or surgical wounds
  E. Hemodialysis (trapping of blood in apparatus)
IV. Pregnancy
V. Intravascular hemolysis and hemoglobinuria
  A. Cardiac prostheses (faulty valves or septal patches)
  B. Paroxysmal nocturnal hemoglobinuria
  C. Paroxysmal cold hemoglobinuria
  D. March hemoglobinuria
  E. Microangiopathy due to pathologic deposition of fibrin

Iron deficiency has commonly resulted from hemodialysis in patients with chronic renal disease. This has resulted from trapping of blood in the dialyzer. The problem may be largely averted by returning all blood to the patient.

Iron may be lost as a result of hemoglobinuria. When there has been intravascular hemolysis, the hemoglobin in excess of that bound by haptoglobin is excreted in the urine. The hemoglobinuria is also accompanied by hemosiderinuria and ferritinuria. The causes of hemoglobinuria are indicated in Table 1 and include, besides paroxysmal nocturnal hemoglobinuria, faulty cardiac septal patches and valvular prostheses. The intracardiac causes of hemoglobinuria have usually been related to a stream of high pressure blood flow — for example, in the ventricular outflow path — and are due to excess shear forces acting on erythrocytes. Variance (that is, loss of complete spheroidicity) of a ball valve has often been cited as a cause. Erythrocytes may also be sheared by abnormally present fibrin strands that traverse the lumens of small blood vessels, with resultant hemolysis, hemoglobinuria, and iron deficiency in the "microangiopathic hemolytic anemia," that may attend chronic renal disease or metastatic disease.

Many drugs may cause intravascular hemolysis. Classic among these are phenylhydrazine, which is not currently used therapeutically, and dapsone, a sulfone commonly used in dermatologic practice for dermatitis herpetiformis. Dapsone appears to cause low-grade intravascular hemolysis in virtually all persons receiving this drug. Such antimalarials as primaquine may cause acute intravascular hemolysis and anemia in susceptible persons, but they do not lead to chronic hemoglobinuria and iron deficiency.

CLINICAL AND LABORATORY MANIFESTATIONS

In common with anemias of other cause, moderate to severe iron deficiency anemia may give rise to easy fatigue, pallor, dyspnea, palpitation, edema, tinnitus, and headaches. However, since iron deficiency usually develops insidiously, symptoms related to anemia may be minimal or absent altogether. Soreness of the mouth or tongue or dysphagia may occur in the absence of frank anemia. Pica also may occur in persons who have iron deficiency with or without anemia. This may take various forms of perversion of appetite, such as compulsive eating of ice, snow, dirt, or laundry starch. Very rarely, iron deficiency may masquerade with manifestations of increased intracranial pressure — headache, visual blurring, and optic papillitis. Whether atrophic rhinitis or excess hair loss can be due to iron deficiency is doubtful, although these disorders have been ascribed to iron deficiency.

Common physical findings in iron deficiency include atrophy and erythema at the corners of the mouth (cheilitis), atrophy of the mucosa of the tongue, pallor (if anemia is moderate to severe), and tachycardia (if anemia is moderate to severe). The spleen may be palpable (as much as a few centimeters below the costal margin) if anemia is severe and long-standing. In the author's experience, no more than 5% of patients have mild splenomegaly. Koilonychia (spoon nail) is very rarely observed even in severely iron-deficient persons in the United States. As with anemia of other etiologies, severe iron deficiency anemia (hemoglobin concentration less than 5 g/dl) may be accompanied by retinal hemorrhage.

Rare clinical manifestations attributed to iron deficiency, either alone or in combination with other deficiency states, have been reviewed in detail elsewhere[1] and will not be recapitulated here. These include iron deficiency with hypocupremia and hypoproteinemia, iron deficiency with protein-calorie malnutrition, iron deficiency with idiopathic pulmonary hemosiderosis, and a syndrome of iron deficiency anemia, dwarfism, hypogonadism, and geophagia which has been attributed to combined iron and zinc deficiency.

Anemia occurs when iron deficiency is severe. When anemia is mild (hemoglobin

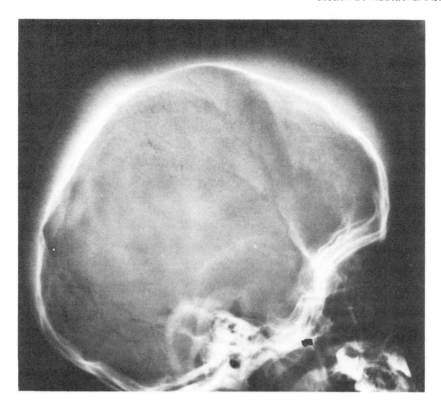

FIGURE 1.   Roentgenographic examination of the skull of a child with severe iron deficiency anemia, showing pronounced widening of the diploic spaces due to expansion of bone marrow. A fine vertical striation ("hair on end") is noted when anemia is very severe and of long duration. This roentgenographic finding is associated with observation of prominence of frontal, temporal, and parietal portions of the cranium, or "bossing," on physical examination and corresponds to "spongy bones" or porotic hyperostosis found in archeologic specimens. Very similar roentgenographic changes are seen in thalassemia major and other severe anemias of early childhood. However, development and pneumatization of maxillae are normal in iron deficiency anemia, whereas there is decreased pneumatization and increased prominence of maxillae in thalassemia major. [From Moseley, J.E., *Semin. Roentgenol.*, 9, 169–184, 1974. By permission of Grune & Stratton.]

concentration greater than 10 g/dl), it is usually normochromic and normocytic. As anemia worsens, it becomes microcytic and finally hypochromic. At this stage, erythrocytes exhibit much pleomorphism: there may be elliptocytes, elongated cigar-shaped erythrocytes ("pencil cells"), schistocytes, and target cells. Leukocytes are usually normal, but in slightly more than 10% of iron-deficient persons, there may be mild leukopenia, with a white blood cell count rarely less than 3000/μl and a normal differential count. Special tests for iron deficiency are discussed in a subsequent section.

Esophagoscopy in patients with dysphagia may reveal a pharyngeal web. This is characteristic of the Patterson-Kelly or Plummer-Vinson syndrome, which is generally ascribed to long-standing iron deficiency anemia.

Roentgenographic studies may reveal pronounced changes in the skull of young children with severe iron deficiency anemia, due to expansion of the erythropoietic marrow[13] (Figure 1). This is not specific for iron deficiency but is seen also in other severe chronic anemias such as thalassemia major. Roentgenograms may also demonstrate a postcricoid esophageal web in those with Plummer-Vinson syndrome.[14]

BIOCHEMICAL EFFECTS

It was once believed that all the effects of iron deficiency are due to decrease in hemoglobin concentration in the peripheral blood. However, this view is clearly erroneous. Hemoglobin, although quantitatively the major iron compound of the body, is in fact only one of a large number of iron-containing or iron-dependent proteins. These include myoglobin, ferritin, hemosiderin, cytochromes, peroxidases, catalase, aconitase, and nearly half the enzymes of Kreb's tricarboxylic acid cycle. These iron compounds are present in minute amounts in almost all cells of the body, and aerobic energy metabolism is, thus, strongly dependent on a sufficient supply of iron.

Numerous studies (which have been summarized elsewhere[1]) have demonstrated unequivocally that the activities of many of these enzymes are diminished in tissues of persons with iron deficiency and that this effect occurs even before the onset of anemia. These biochemical changes appear to be responsible for many of the clinical features and histologic changes observed in iron deficiency, particularly those due to atrophy of epithelial tissues with a high rate of cell regeneration: atrophy of circumoral epithelium and epithelium of the tongue, pharynx, stomach, and small intestine, and koilonychia.

PATHOLOGY

Intestinal mucosal biopsies in children with iron deficiency anemia have shown atrophy and fusion of villi[15,16] (Figure 2). Sequential biopsies have also shown that this lesion is reversible with iron therapy. Gastric mucosal atrophy has also been demonstrated in severe iron deficiency anemia. Changes in pharyngeal mucosal epithelium have been observed which are thought to be premalignant. Carcinoma of the pharynx appears to be a sequel of postcricoid web formation, although this has been disputed.

Extraordinary malformations of the skull have frequently been observed in archeologic specimens from the southwestern United States, from Mesoamerica, and from Peru[17,18] (Figure 3). These malformations, termed "porotic hyperostosis," are believed to be an expression of severe iron deficiency anemia resulting from a predominantly maize diet. (In canyon bottom sites, where this agricultural system predominated, 34% of skulls exhibited this malformation.) Similar skeletal malformations have been described in bones of modern American Indians subsisting on this diet and in an African tribe whose members consume an iron-poor diet.

TREATMENT OF IRON DEFICIENCY

Treatment of iron deficiency has been extensively reviewed by the author elsewhere[1] and will be only briefly reviewed here.

Once a diagnosis of iron deficiency has been established, it is incumbent on the physician to determine the cause. In very young children, it may be evident that dietary deficiency (for example, diet restricted to milk) is the cause. However, it is important to exclude gastrointestinal bleeding (for example, from Meckel's diverticulum) as a cause. In tropical or semitropical areas, an examination must be made for hookworm infestation of an iron-deficient person. In young women, it may be evident that menorrhagia is responsible for iron deficiency. For postmenopausal women or for men, it is necessary that gastrointestinal bleeding be evaluated as a cause of iron deficiency and, if bleeding is confirmed, that careful search be made for an anatomic lesion that may be amenable to surgery.

While the ancillary examinations are being made, treatment with iron should be instituted. There are more than 100 preparations commercially available for oral treatment of iron deficiency. Most of these are quite satisfactory. To be avoided are

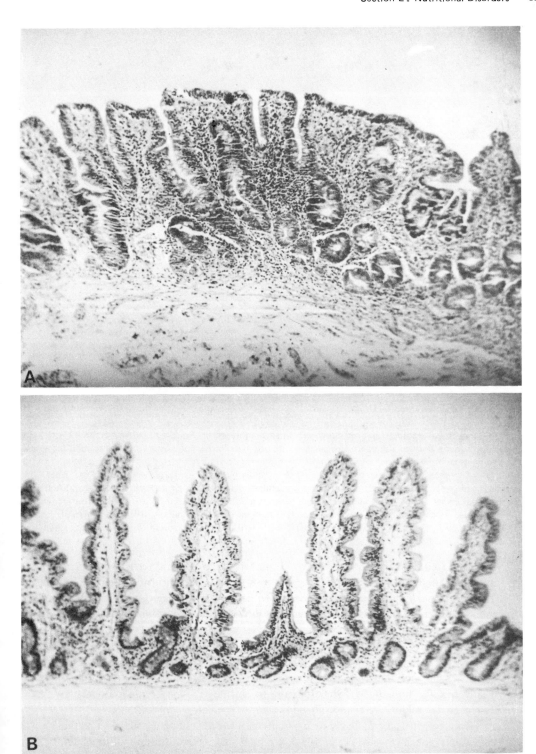

FIGURE 2.   Marked epithelial atrophy in an intestinal mucosa biopsy specimen from a child with iron deficiency anemia. A. Before treatment, showing shortening, blunting, and fusion of villi. B. After treatment with iron, showing restoration of normal histologic features. [From Guha, D.K., Walia, B.N., Tandon, B.N., Deo, M.G., and Ghai, O.P., *Arch. Dis. Child.,* 43, 239–244, 1968. By permission of the American Medical Association.]

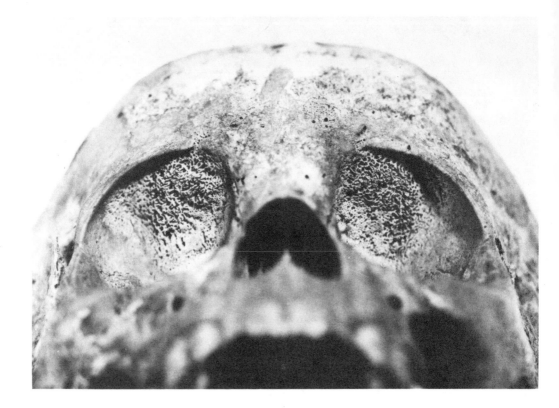

FIGURE 3.   Paleopathology of iron deficiency anemia in Anasazi (American Indian, ca. 1000 A.D.) child's skull from canyon bottom site in southwestern United States. Porotic hyperostosis is due to multiple apertures on the external skull surface, associated with enlargement of diploic spaces, particularly evident in orbital, frontal, and parietal bones. Roentgenograms of such skulls disclose the features illustrated in Figure 1. Note normal maxillary structure, which contrasts with maxillary deformity in thalassemia major. In addition to skull changes, malformations of pelvis and long bones have been noted. These changes are ascribed to iron deficiency, the only cause of severe anemia known to be endemic among American Indians. (From El-Najjar, M.Y., Lozoff, B., and Ryan, D.J., *Am. J. Roentgenol.*, 125, 918–924, 1975. By permission of Charles C Thomas, Publisher.]

enteric-coated, "prolonged release," or "sustained release" forms; ferric salts; ferrous salts in combination with other hematinic; or those forms in which iron content is miniscule. The author has rarely found any occasion to use an oral preparation other than ferrous sulfate, which is inexpensive and generally well tolerated. This contains 65 mg of elemental iron in either a 0.3-g tablet of ferrous sulfate, USP, or a 0.2-g tablet of ferrous sulfate exsiccated, USP. Such tablets should ordinarily be taken three times daily between meals until the anemia has been corrected (usually for about 2 months), and for 6 months thereafter in order to replete iron stores. Generic prescriptions for ferrous sulfate should clearly specify "nonenteric," since pharmacists in the U.S. may erroneously assume that enteric-coated tablets are generically equivalent to nonenteric-coated tablets.

Treatment of iron deficiency by a parenteral route should be restricted to certain clear indications, the most obvious of which are intestinal malabsorption and unreliability of the patient. Parenteral administration of iron is never justified simply for the convenience of the physician or simply because a patient with iron deficiency anemia has failed to respond to oral iron therapy. Failure to respond to oral therapy should alert the physician to the likelihood that (1) the diagnosis is wrong, (2) the patient's blood loss has not been adequately controlled, (3) the iron preparation dispensed was in enteric-coated or a prolonged-release form, (4) the patient did not take the medication as prescribed, or (5) there is intestinal malabsorption. When these possibilities have been evaluated and ruled out, parenteral administration of iron may be contemplated. It should be remembered, however, that parenteral iron therapy always poses an increased frequency of side effects, some of which are dangerous and may be lethal, a greater inconvenience, and a greater cost than does oral therapy. Also, there is no demonstrable difference in the rate of response to oral compared with that to parenteral iron therapy. Furthermore, some persons utilize parenteral iron poorly and may continue to exhibit characteristic clinical and laboratory features of iron deficiency anemia, even though storage iron (in bone marrow) may appear adequate. Such persons have subsequently responded satisfactorily to oral iron therapy.

PREVENTION OF IRON DEFICIENCY

Measures that have been aimed at preventing iron deficiency in entire populations are summarized in a subsequent section. To be considered here are measures that may be applied for two high-risk groups: namely, infants and women of reproductive age.

The iron content of the diet of infants may be supplemented in two ways. The first of these is by early weaning. It is a common practice in the United States to begin semisolid foods (cereals) at about 4 weeks of age. This ensures a better iron intake, since infant cereals contain about 14 mg of iron per ounce. Although phytates may partially bind the iron, this practice does appear to reduce the frequency of subsequent anemia. Furthermore, the introduction of cereals early may make easier an early transition to other foods that provide iron in a better-absorbed form. A second approach is the direct addition of iron to formula or fruit juice, or even by dropper into the infant's mouth. Ferrous sulfate "drops" are well tolerated and may be given in a dose of 0.3 ml (8 mg) daily for infants of age up to 3 months and 0.6 ml daily for older infants.

For prophylaxis of iron deficiency in menstruating women, a ferrous sulfate tablet (65 mg elemental iron) can be taken two or three times weekly. For prophylaxis of iron deficiency in pregnancy, the same dose of iron should be taken at least daily throughout pregnancy. It is also a common, and reasonable, practice to administer both iron and folic acid throughout pregnancy, since deficiency of both substances commonly occurs.

Young women may have young children in their home. It is important to stress that iron preparations must be kept out of reach of little children, as severe and even fatal iron poisoning may follow ingestion of a child's handful of iron tablets.

METHODS FOR ASCERTAINMENT OF IRON DEFICIENCY IN POPULATION SURVEYS

Hemoglobin Concentration

Many large-scale surveys for iron deficiency have been based primarily on measurement of hemoglobin concentration of venous or capillary blood specimens. These surveys have, in fact, determined the prevalence of anemia. Since iron deficiency is, in all populations, by far the commonest cause of anemia, the frequency of anemia has served as a useful index to the frequency of iron deficiency. It must be remembered, however, that other causes of anemia are also common in many populations. These include thalassemias and hemoglobinopathies, chronic infectious diseases, and other nutritional deficiencies. On the other hand, since anemia is a manifestation of severe iron deficiency, it is virtually certain that the frequency of anemia underestimates the frequency of iron deficiency (including iron depletion and iron deficiency anemia as defined above).

A World Health Organization Scientific Group on Nutritional Anemias[19] recommended that surveys use the following lower-normal limits for defining anemia for the age and sex groups indicated. (No sex difference is recognized before puberty.) These limits assume residence at sea level.

	Hemoglobin concentration (g/dl)
Age 6 months–6 years	11
Age 6–14 years	12
Adult men	13
Adult women, pregnant	11
Adult women, nonpregnant	12

These standards are inappropriate for surveys of populations living at high altitudes; standards for such groups must be developed independently of the WHO criteria. For example, in Mexico City (elevation 2240 meters), in a group of 115 healthy adult males (mostly physicians), none had a hemoglobin concentration of less than 15 g/dl.[20]

Hypochromia and Microcytosis

These are morphologic characteristics of the erythrocytes in advanced iron deficiency anemia. Although microscopic examination of the blood has been of some use in some surveys to help substantiate the diagnosis of iron deficiency anemia, it must be stressed that morphologic changes in erythrocytes are both very nonspecific and very insensitive as criteria of iron deficiency. Hypochromia and microcytosis commonly are evident when anemia is pronounced but not when anemia is mild.[21,22] Furthermore, thalassemia minor and certain hemoglobinopathies cannot reliably be differentiated from iron deficiency anemia on the basis of erythrocyte morphology. This would be particularly a problem in the Mediterranean littoral or in Southeast Asia but also in many areas of the United States.

Serum Iron, Iron Binding Capacity, and Transferrin Saturation

A low serum iron concentration is regarded as a defining characteristic of "iron deficiency without anemia," an intermediate phase in the progression of iron deficiency. However, even a significant proportion of persons with early iron deficiency anemia have normal values for serum iron concentration.[4] Thus, serum iron concentration, if used as a criterion of iron deficiency, may underestimate the true frequency of this disorder. Furthermore, the serum iron concentration is extremely labile and exhibits large diurnal and day-to-day variations in healthy persons.[23-25] It is also markedly diminished as a result of other processes in addition to iron deficiency, including acute or chronic inflammatory processes,[26] onset of menstruation,[27-29] ethanol withdrawal,[30,31] and myocardial infarction.[32,33] Some of these conditions also lower the transferrin concentration, which is measured by the total iron-binding capacity of serum. On the other hand, in iron deficiency, transferrin concentration is not diminished. Transferrin saturation is an index widely used to differentiate iron deficiency from other causes of sideropenia. This is calculated from $\frac{100 \times \text{serum iron concentration}}{\text{total iron binding capacity}}$. In healthy adults who are not iron deficient, transferrin saturation is generally greater than 16%. However, we and others have encountered lower values (as low as 12%) in healthy young women who have no evidence of iron deficiency. Conversely, low transferrin saturation has been found in persons with chronic inflammatory disorders who are not iron deficient.[34] Furthermore, use of transferrin saturation as a criterion of iron deficiency appears to underestimate the true frequency of iron deficiency.[4]

The WHO Scientific Group on Nutritional Anemias[19] accepted as evidence of iron deficiency in adults serum iron concentration of less than 50 μg/dl or transferrin saturation of less than 15%.

The author's experience indicates that, for diagnostic purposes, and with recognition of the limitations of these procedures, realistic lower limits should be as follows:

	Serum iron concentration (μg/dl)	Transferrin saturation (%)
Adult males	70	16
Adult women	60	12
Children	50	12

Erythrocyte Indices

Of the erythrocyte corpuscular indices, only the mean corpuscular volume (MCV) merits consideration for the identification of iron deficiency anemia. The MCV was originally derived from the ratio of hematocrit to erythrocyte count, but now it is more commonly measured directly by automated instruments. It is clearly a more sensitive indicator of iron deficiency than are other erythrocyte indices. MCV of less than 80 μm³ indicates microcytosis, which is characteristic of iron deficiency anemia. However, low values of MCV are also found in thalassemias and in chronic disorders such as infections, inflammation, and malignancy. Thus, the MCV is also too nonspecific to be reliable for diagnosis of iron deficiency anemia. Measurement of the two other erythrocyte indices — mean corpuscular hemoglobin and mean corpuscular hemoglobin concentration — should be abandoned because these measurements are of little diagnostic value, particularly when obtained with modern equipment.[35]

Estimation of Iron Stores by Serial Phlebotomy

Storage iron is mobilized and utilized in new hemoglobin synthesis after removal of blood by phlebotomy, usually in 500-ml units. If, for example, the hemoglobin concentration were 14.7 g/dl, 73.5 g of hemoglobin would be removed in a 500-ml phlebotomy, equivalent to 250 mg of iron (from 14.7 g/dl $\times$ 5 dl $\times$ 0.0034). For determination of the size of iron stores experimentally, phlebotomy may be performed once or twice weekly until the hemoglobin concentration of blood exhibits a decline and thereby signifies the inability of bone marrow to regenerate hemoglobin, as a consequence of iron depletion. By measuring the volume and hemoglobin concentration of each unit of blood removed, it is easy to calculate the total quantity of hemoglobin removed and hence the quantity of iron mobilized from the storage compartment. This procedure is the most direct means of assessing storage iron. Because the size of hemoglobin iron and transport iron compartments is easily ascertained, and because other compartments are relatively minute, the measurement of storage iron by serial phlebotomy permits a nearly complete quantitation of total body iron. This technique is obviously not practical for population surveys. Yet it has been employed in the study of some population groups, as indicated in Table 2.

Measurement of Hepatic Iron Content

A few studies have assessed geographic regional differences in iron stores by assaying the iron content of liver specimens obtained at autopsy from persons who have been killed in accidents. Such studies assume that victims of accidental death reflect the larger populations of which they were members. Collaborative international studies using this approach have, indeed, implied that there is substantial geographic variability in liver iron content.[110]

Table 2
EPIDEMIOLOGIC SURVEYS OF IRON DEFICIENCY[a]

Area of world, country	Investigators (reference)	Group studied	Criterion for anemia	Criterion for iron deficiency	Frequency of anemia (%)	Frequency of iron deficiency (%)	Frequency of iron deficiency anemia (%)
Western Hemisphere							
Canada	Valberg et al.[36]	Infants and children		Serum ferritin			
		0—1 years					
		1—4 years				48	
		Children and adults		<20 ng/ml, ≦12 years M or F			
		5—9 years				39	
		10—19 M				35	
		10—19 F		<30 ng/ml, >12 years M or F		40	
		20—64 M				12	
		20—64 F				30	
		Pregnant women				60	
	de Leeuw et al.[37]	Healthy pregnant women		Marrow iron (histochemical)			
		No iron supplement				100	
		Iron supplement				90	
United States	Haddy et al.[38]	Children 4 months—5 years	Hb <11 g/dl	Transferrin saturation <17%	22	42	20
	Scott and Pritchard[39]	Healthy nulliparous young women		Marrow iron (histochemical) or phlebotomy		66	

Table 2 (continued)
EPIDEMIOLOGIC SURVEYS OF IRON DEFICIENCY[a]

Area of world, country	Investigators (reference)	Group studied	Criterion for anemia	Criterion for iron deficiency	Frequency of anemia (%)	Frequency of iron deficiency (%)	Frequency of iron deficiency anemia (%)
Tennessee	Hutcheson and Hutcheson[40]	White 0–2 years	Hct <31%		21		
		3–5 years			7		
		Black 0–2 years			29		
		3–5 years			13		
Ohio	Schubert and Hunter[41]	Infants 0–18 months No-iron-supplement formula	Hb <10 g/dl	Transferrin saturation <15% on two consecutive visits	28	44	28
		Homogenized milk			33	59	33
Maryland	Stine et al.[42]	White 4 years	Hct <33%		5		
		Nonwhite 4 years			20		
Chicago, Ill. Gainesville, Fla. Jacksonville, Fla. Augusta, Ga. Houston, Tex.	Pearson et al.[43]	Children 4–6 years, low socio-economic status	Hct <31%		4 2 3 8 1		
New York City	Christakis et al.[44]	Children 10–13 years	Hb <10 g/dl		3		
Iowa	Kripke and Sanders[45]	Children 6 months–3 years	Hb <10 g/dl	Hypochromic erythrocytes	4		4

Table 2 (continued)
EPIDEMIOLOGIC SURVEYS OF IRON DEFICIENCY[a]

Area of world, country	Investigators (reference)	Group studied	Criterion for anemia	Criterion for iron deficiency	Frequency of anemia (%)	Frequency of iron deficiency (%)	Frequency of iron deficiency anemia (%)
California	Fuerth[46]	Infants seen in private practice	Hb <10 g/dl		3		
Washington	Cook[47]	Men Women	Distribution analysis		2 9		
Massachusetts	Edozien[48]	Pregnant women	Hb <11 g/dl		16		
Major cities of U.S. (all regions)	Fomon and Weckwerth[49]	Children 0–6 months 6–12 months 1–2 years 3–4 years 4–9 years	Hb <10 g/dl		12 20 26 3 2		
Mississippi	Owen et al.[50-52]	Preschool children			24		
		Home income <$500/year per capita		Serum iron <40 µg/dl		30	
		Income >$500/year per capita	Hb <10 g/dl		12	20	
	Johnson and Futrell[53]	Black preschool children	Hb <10 g/dl	Serum iron <40 µg/dl	9	5	

Table 2 (continued)
EPIDEMIOLOGIC SURVEYS OF IRON DEFICIENCY[a]

Area of world, country	Investigators (reference)	Group studied	Criterion for anemia	Criterion for iron deficiency	Frequency of anemia (%)	Frequency of iron deficiency (%)	Frequency of iron deficiency anemia (%)
New York City	Haughton[54]	Children of low income families >3 years Black / Puerto Rican / White	Hb <10 g/dl		33 / 26 / 29		
		Black / Puerto Rican / White			4 / 4 / 0		
Washington, D.C.	Gutelius[55]	Black preschool children	Hb <10 g/dl	Response to oral iron therapy	29		29
United States (all areas)	U.S. Public Health Service[56]	Children 1–5 years White / Black	Hb >10 g/dl	Transferrin saturation <15%	2 / 8	9 / 19	
		6–11 years White / Black	Hb <11.5 g/dl	Transferrin saturation <20%	9	12 / 10	
		12–17 years White / Black	Hb <13 g/dl (M) Hb <11.5 g/dl (F)	Transferrin saturation <20% (M) Transferrin saturation <15% (F)	2 / 18	6 / 10	

Table 2 (continued)
EPIDEMIOLOGIC SURVEYS OF IRON DEFICIENCY[a]

Area of world, country	Investigators (reference)	Group studied	Criterion for anemia	Criterion for iron deficiency	Frequency of anemia (%)	Frequency of iron deficiency (%)	Frequency of iron deficiency anemia (%)
United States (all areas) (continued)		18—44 years M Black White }	Hb <14 g/dl	Transferrin saturation <20%	11 3	4 5	
		18—44 years F Black White }	Hb <12 g/dl	Transferrin saturation <15%	16 5	5 5	
		45—59 years Black White }	Hb <14 g/dl (M) Hb <12 g/dl (F)	Transferrin saturation <20% (M) Transferrin saturation <15% (F)	8 5	6 5	
Mexico City	Loría et al.[57]	Healthy children 0—3 years, high socio-economic status	Hb<11.5 g/dl	Serum iron <60 µg/dl			
		0—4 months			16		20
		4 months—1 year			6		33
		1—2 years			3		21
		2—3 years			3		20

Table 2 (continued)
EPIDEMIOLOGIC SURVEYS OF IRON DEFICIENCY[a]

Area of world, country	Investigators (reference)	Group studied	Criterion for anemia	Criterion for iron deficiency	Frequency of anemia (%)	Frequency of iron deficiency (%)	Frequency of iron deficiency anemia (%)
Mexico City (continued)	Loría et al.[58]	Children 0–7 years, low socio-economic status	Hb <11.5 g/dl	Serum iron <60 μg/dl			
		0–4 months			21		35
		4 months– 1 year			25		73
		1–2 years			55		70
		2–3 years			9		46
		3–7 years			<1		19
Mexico	W.H.O.[19]	Women		Serum iron <50 μg/dl			
		Pregnant	Hb <12 g/dl		27	30	
		Nonpregnant	Hb <13 g/dl		12	22	
		Men	Hb <14 g/dl		1	6	
Central America	Viteri et al.[59]	Adult men	Hb <12 g/dl				
		With hookworm			32		
		Without hookworm			7		
		Adult women					
		With hookworm			18		
		Without hookworm			11		
		Children 5–12 years					
		With hookworm			16		
		Without hookworm			5		
		Children 1–4 years					
		With hookworm			33		
		without hookworm			20		

Table 2 (continued)
EPIDEMIOLOGIC SURVEYS OF IRON DEFICIENCY[a]

Area of world, country	Investigators (reference)	Group studied	Criterion for anemia	Criterion for iron deficiency	Frequency of anemia (%)	Frequency of iron deficiency (%)	Frequency of iron deficiency anemia (%)
South America Venezuela	Molina et al.[60]	Pregnant women	Hb <11 g/dl	Serum iron <50 µg/dl	25	18	
	W.H.O[19]	Women Pregnant	Hb <11 g/dl	Serum iron <50 µg/dl	37	57	
		Nonpregnant	Hb <12 g/dl		15	14	
		Men	Hb <13 g/dl		2	8	
	Layrisse and Roche[61]	Rural villages in zone free of hookworm infestation					
		Adult men	Hb <13 g/dl		4		
		Adult women	Hb <11.5 g/dl		15		
		Children 2–6 years	Hb <11.5 g/dl		14		
		7–14 years			16		
		Rural villages in hookworm infested zone					
		Adult men	Hb <13 g/dl		30		
		Adult women	Hb <11.5 g/dl		33		
		Children 2–6 years	Hb <11.5 g/dl		46		
		7–14 years			45		
Brazil	Szarfarc[62]	Newborn	Hb <13.6 g/dl		21		
		Postpartum women	Hb <12 g/dl		52		
	Roncada and Szarfarc[63]	Pregnant women	Hb <11 g/dl		30		
		Nonpregnant women	Hb <11 g/dl		8		

Table 2 (continued)
EPIDEMIOLOGIC SURVEYS OF IRON DEFICIENCY[a]

Area of world, country	Investigators (reference)	Group studied	Criterion for anemia	Criterion for iron deficiency	Frequency of anemia (%)	Frequency of iron deficiency (%)	Frequency of iron deficiency anemia (%)
Latin America (Guatemala, Mexico, Peru, Venezuela, Colombia, Panama)	Cook et al.[64]	Adult men	Hb ≤ 13 g/dl	Transferrin saturation <15%	4	3	
		Nonpregnant women	Hb ≤ 12 g/dl		17	21	
		Pregnant women	Hb ≤ 11 g/dl		38	48	
Trinidad and Tobago	Chopra and Byam[65]	Children 5–9 years	Hb <12 g/dl	Serum iron <50 µg/dl	40	11	
		10–13 years			18	5	
		14–17 years			16	6	
		Adults Men	Hb <12 g/dl	Serum iron <50 µg/dl	4	9	
		Nonpregnant women	Hb <12 g/dl		21	23	
		Pregnant women	Hb <10 g/dl		31	38	
Europe United Kingdom	McFarlane et al.[66]	Adult women	Hb <12 g/dl	Transferrin saturation <16%	2	7	
	Burman[67]	Healthy infants, no iron supplement	Hb <10 g/dl		17		
	Powell et al.[68]	Hospitalized adults >50 years old, high socioeconomic status		Serum iron <50 µg/dl, transferrin saturation <16%		40	

Table 2 (continued)
EPIDEMIOLOGIC SURVEYS OF IRON DEFICIENCY[a]

Area of world, country	Investigators (reference)	Group studied	Criterion for anemia	Criterion for iron deficiency	Frequency of anemia (%)	Frequency of iron deficiency (%)	Frequency of iron deficiency anemia (%)
United Kingdom (continued)	Kilpatrick and Hardisty[69]	Welsh Adults					
		Men	Hb <12.6 g/dl	Hypochromic erythrocytes	3	3	13
		Women	Hb <12.1 g/dl		14		
	Rees et al.[70]	Adults					
		Men	Hb <12 g/dl		6		8
		Women	Hb <11 g/dl		10		
		Children					
		0—9 years	Hb <11 g/dl		72		44
		10—14 years	Hb <12 g/dl		10		4
	Elwood et al.[71]	Adult nonpregnant women	Hb <12 g/dl		11		
Norway	Natvig[72]	Men 15—21 years	Hb <13 g/dl		4		
	Natvig et al.[73]	Children					
		7—10 years	Hb <11 g/dl	Response to oral iron therapy			1
		11—13 years	Hb <11.5 g/dl				2
		14—16 years M	Hb <13 g/dl				3
		14—16 years F	Hb <12 g/dl				3
	Natvig and Vellar[74]	Adult men					
		15—39 years	Hb <14 g/dl	Response to oral iron therapy			2
		40—59 years					5
		>60 years					6
		Adult women					
		15—29 years	Hb <12.5 g/dl				3
		30—39 years					12
		40—49 years					4
		>50 years					7

EPIDEMIOLOGIC SURVEYS OF IRON DEFICIENCY[a]

Area of world, country	Investigators (reference)	Group studied	Criterion for anemia	Criterion for iron deficiency	Frequency of anemia (%)	Frequency of iron deficiency (%)	Frequency of iron deficiency anemia (%)
Sweden	Garby et al.[75,76]	Women 18–48 years		Response of Hb to iron therapy			17
Denmark	Marner[77]	Children 3–6 years		Response of Hb to iron therapy			0
Italy	Dughera and Naldi[78]	Adult inpatients >65 years	Hb <11.2 g/dl	Serum iron <60 µg/dl	35		8
	Piancino et al.[79]	Adult inpatients Men	Hb <12 g/dl				9
		Women	Hb <11 g/dl				15
Greece	Kattamis et al.[80]	Adult men	Hb <14 g/dl	Serum iron <60 µg/dl, transferrin saturation <16%		26	10
		Children ½–5 years	Hb <11 g/dl			44	34
Yugoslavia	Hibšer et al.[81]	Secondary school children Boys	Hb <13 g/dl	Transferrin saturation <15%		21	13
		Girls	Hb <12 g/dl			23	12
Hungary	István[82]	Boys 10–14 years	Hb <12 g/dl		23		
		19–24 years	Hb <13.5 g/dl		9		
		Girls 10–14 years	Hb <12 g/dl		22		
		19–24 years			17		
		Pregnant women	Hb <12 g/dl		62		
			Hb <10.5 g/dl		14		

Table 2 (continued)
EPIDEMIOLOGIC SURVEYS OF IRON DEFICIENCY[a]

Area of world, country	Investigators (reference)	Group studied	Criterion for anemia	Criterion for iron deficiency	Frequency of anemia (%)	Frequency of iron deficiency (%)	Frequency of iron deficiency anemia (%)
Poland	W.H.O.[19]	Pregnant women	Hb <11 g/dl	Serum iron <50 μg/dl	11		
Finland	Antila[83]	Adults Male	Hb <14 g/dl	Serum iron <80 μg/dl	55	70	
		Female	Hb <12 g/dl	Serum iron <60 μg/dl	28	71	
	Takkunen and Seppänen[84]	Adults Male	Hb <13 g/dl		2		
		Female	Hb <12 g/dl		7		
	Hirsjärvi[85]	Elderly inpatients (84 F, 6 M)	Hb <12 g/dl		76		17
Germany	Göltner and Schlunk[86]	Pregnant women	Hb <12 g/dl	Serum iron <80 μg/dl	35	73	
Africa Egypt	Abdel-Fattah et al.[87]	Infants 0—6 months	Hb <11 g/dl		77		
Union of South Africa	Mayet et al.[88]	Indian females	Hb <12 g/dl	Transferrin saturation <16%	48	33	
		Indian males (ages not specified)	Hb <13 g/dl		55	20	

Table 2 (continued)
EPIDEMIOLOGIC SURVEYS OF IRON DEFICIENCY[a]

Area of world, country	Investigators (reference)	Group studied	Criterion for anemia	Criterion for iron deficiency	Frequency of anemia (%)	Frequency of iron deficiency (%)	Frequency of iron deficiency anemia (%)
Ethiopia[b]	Hofvander[8 9]	Urban adult men	Hb <15 g/dl		20		0
		Urban adult women	Hb <13 g/dl	Serum iron <50 µg/dl	16		10
		"Priveleged children"					
		½–4 years	Hb <11.6 g/dl		5		
		5–9 years	Hb <11.9 g/dl		6		3.4
		10–14 years	Hb <12.4 g/dl		7		
		Rural adult men	Hb <15 g/dl		45		8
		Rural adult women	Hb <13 g/dl	MCHC <27%	54		8
		Rural children					
		½–1 years	Hb <11.6 g/dl		80		35
		1–4 years	Hb <11.9 g/dl		48		10
		5–9 years	Hb <11.9 g/dl		69		7
		10–14 years	Hb <12.4 g/dl		45		6
Middle East Israel	W.H.O[19]	Women	Hb <11 g/dl		47	36	
		Women Pregnant	Hb <12 g/dl	Serum Fe <50 µg/dl	29	15	
		Nonpregnant Men	Hb <13 g/dl		14	12	
Indian Ocean Mauritius	Stott[90]	Children					
		0–3 years			55		
		3–5 years	Hb <10.4 g/dl		32		
		5–9 years			18		
		9–15 years	Hb <11 g/dl		18		

[b] Elevation greater than 2 000 meters. Hookworm and other parasitic infestations are common in rural population. Selection of mean corpuscular hemoglobin concentration (MCHC) of less than 27% as criterion grossly underestimates iron deficiency anemia.

Table 2 (continued)
EPIDEMIOLOGIC SURVEYS OF IRON DEFICIENCY[a]

Area of world, country	Investigators (reference)	Group studied	Criterion for anemia	Criterion for iron deficiency	Frequency of anemia (%)	Frequency of iron deficiency (%)	Frequency of iron deficiency anemia (%)
Mauritius (continued)		Adults					
		Men	Hb <14 g/dl		30		
		Women	Hb <12 g/dl		34		
		Pregnant women	Hb <10 g/dl		60		
Asia India Sri Lanka (Ceylon)	De Silva and Fernando[91]	Healthy children					
		1–5 years, Low economic status			99		
		1–5 years, Affluent status	Hb <10.6 g/dl		66		
		5–10 years, Affluent status			71		
	Manchanda et al.[92]	Children 10 days to 14 years Healthy					
		1–9 months	Hb <9 g/dl				
		9 months–6 years	Hb <10 g/dl		56		
		6–14 years	Hb <11 g/dl				
		Malnutrition			78	85	
		Chronic infection			56	83	
		Chronic systemic disease (noninfectious)	Hb <8 g/dl		54		
	Banerji et al.[93]	Accidental		Liver histo-		60	

Table 2 (continued)
EPIDEMIOLOGIC SURVEYS OF IRON DEFICIENCY[a]

Area of world, country	Investigators (reference)	Group studied	Criterion for anemia	Criterion for iron deficiency	Frequency of anemia (%)	Frequency of iron deficiency (%)	Frequency of iron deficiency anemia (%)
India (continued)	Sood et al.[94]	Healthy nonanemic adult Indian males		Repeated phlebotomy			
Delhi	W.H.O.[19]	Women Pregnant Nonpregnant	Hb <12 g/dl Hb <11 g/dl	Serum iron <50 µg/dl	80 64		
Vellore	W.H.O.[19]	Women Pregnant Nonpregnant Men	Hb <12 g/dl Hb <11 g/dl Hb <13 g/dl	Serum iron <50 µg/dl	56 35 6		
	Jadhav and Baker[95]	Children, outpatient and inpatient	Hb ≤ 10 g/dl	Hypochromia	22		12
	Dhar et al.[96]	Children <14 yr	Hb <12 g/dl	Serum iron "low"	Not stated		91% of anemic children
Japan	Takaku[97]	Medical outpatients 1955–1960 and in 1966–1970	Not stated	Not stated	79	45 36	
		Healthy pregnant women	Not stated		91 31		
Malaysia	Sindhu[98]	Pregnant women Malayan Chinese Indian	Hb <11 g/dl	Serum iron <20 µg/dl	58 33 65		16 18 47

Table 2 (continued)
EPIDEMIOLOGIC SURVEYS OF IRON DEFICIENCY[a]

Area of world, country	Investigators (reference)	Group studied	Criterion for anemia	Criterion for iron deficiency	Frequency of anemia (%)	Frequency of iron deficiency (%)	Frequency of iron deficiency anemia (%)
Malaysia (continued)	Ong[99]	Pregnant women Aborigine (Orang Asli)	Hb <10 g/dl	Microcytic hypochromic erythrocytes	26		Approximately 60% of anemic persons were thought to have iron deficiency alone (6%) or in combination with folic acid deficiency (55%)
Pacific Philippines	Marzan et al.[100]	Infants and preschool children		Serum iron <50 µg/dl, transferrin saturation <15%			
		0–4 years	Hb <10.8 g/dl		62		62
		5–7 years	Hb <11.5 g/dl		22		18
		All ages			42		40
	Tantengco et al.[101]	Children		Serum iron <60 µg/dl			
		6–14 years	Hb <12 g/dl		47	21	
		10–13 years M			21	23	
		10–13 years F			33	21	
Indonesia	Kho and Halimoen[102]	Hospitalized children	Hb <10 g/dl	Hypochromia	34		27
Australia	McEwin et al.[103]	Women 6–16 years	Hb <12 g/dl		4		
	Lovric[104,105]	Well children 6–36 months	Hb <10 g/dl	Hypochromic microcytic erythrocytes	3	3	
	Kamien and Cameron[106]	Hospitalized children <5 years	Hb <10 g/dl				
		White			7		
		Aborigine			36		

EPIDEMIOLOGIC SURVEYS OF IRON DEFICIENCY[a]

Area of world, country	Investigators (reference)	Group studied	Criterion for anemia	Criterion for iron deficiency	Frequency of anemia (%)	Frequency of iron deficiency (%)	Frequency of iron deficiency anemia (%)
Australia (continued)	Fleming et al.[107]	Young adult women		Hypochromic microcytic erythrocytes			
		Nonpregnant	Hb <12 g/dl		3	16	
		Pregnant	Hb <11 g/dl		14	24	
New Zealand	Anyon and Desmond[108]	Healthy Caucasian infants followed serially					
		4 months			3		
		9 months	Hb <10.5 g/dl		7		
		14 months			17		
		2 years			10		
	Neave et al.[109]	Maori					"Majority" of anemias were iron-deficiency type on basis of hypochromia
		Adult men	Hb <12.5 g/dl		0		
		Adult women	Hb <12 g/dl		18		
		Infants	Hb <12 g/dl	Hypochromic microcytic erythrocytes	75		
		Children 5 years	Hb <12 g/dl		26		
		Children 5–14 years	Hb <12 g/dl		10		

Other Procedures

Several other procedures have been considered as being potentially useful for population surveys but have not yet been widely used for this purpose. These include tests for free erythrocyte protoporphyrin,[111,112] [57]Co excretion test,[113-118] and serum ferritin assay.[119-124]

An elevation in free erythrocyte protoporphyrin occurs in a variety of anemias, including iron deficiency anemia, in the presence of infection, and, most strikingly, in lead poisoning. Thus, the nonspecificity of this procedure appears to be a major drawback.

An absorption test with $^{57}CoCl_2$ has been proposed, based on the evidence that Co^{++} appears to be absorbed by the same intestinal mechanism as is Fe^{++} but that Co^{++} is not retained and is promptly excreted in the urine. Since there is enhanced absorption of a dose of radioiron in iron-deficient persons, there is a parallel enhancement in Co^{++} absorption. Iron-deficient persons thus exhibit enhanced urinary excretion of an oral dose of ^{57}Co compared with normal subjects. This procedure appears to have promise but has not yet been widely applied. It is the only completely noninvasive method that might be employed for screening or surveys for iron deficiency. It requires ingestion of 0.5 μCi of $^{57}Co^{++}$, a nearly infinitesimal amount in terms of radiation dosage. Adults who receive this dose of $^{57}CoCl_2$ would absorb 2×10^{-4} rads, whole body. (This is approximately $\frac{1}{250}$ the radiation dose to the chest received from a diagnostic chest roentgenogram.) Thus, the radiation that would be absorbed from this dose of $^{57}Co^{++}$ is low enough that there should be little reluctance to use this procedure as a screening or survey method for iron deficiency, should further experience substantiate its sensitivity and specificity.

Plasma or serum ferritin concentration appears in general to correlate well with iron stores, and since ferritin can be readily assayed by the immunoradiometric method, this measurement may well be used in future surveys of iron deficiency. It is the author's impression, based on experience to date with ferritin assays, that this measurement does not give a sharp distinction between normal and iron-deficient persons. Furthermore, persons with inflammatory diseases may have serum ferritin concentration in the normal range despite advanced iron deficiency.[125] Surveys for iron deficiency using low serum ferritin as the method of ascertainment have confirmed a high frequency of iron deficiency among Canadians of various age and sex groups.[36] However, in this survey, the limits may have been too generously set.

Response to Iron Therapy

Garby et al.[75,76] compared several criteria for iron deficiency and found that of serum iron concentration, transferrin saturation, and mean corpuscular hemoglobin, none of these was more sensitive in identifying iron deficiency than was measurement of hemoglobin concentration. These workers found the most sensitive and most specific test of iron deficiency to be the response, measured by change in hemoglobin concentration, to a trial of oral iron therapy. This criterion has not been widely adopted and may prove of limited practical usefulness for population screening. However, results of its use in a small number of surveys are indicated in Table 2.

PREVALENCE OF ANEMIA AND IRON DEFICIENCY

The literature of iron deficiency leaves no doubt that this condition has long been recognized as being widely prevalent in Europe and elsewhere. However, it is only in the past few decades that detailed epidemiologic surveys of iron deficiency have been carried out on a large scale throughout the world. These studies are summarized in Table 2. The studies there summarized are predominantly, although not exclusively, those reported since 1967. Whenever possible, the most recent data have been included, in some cases at the expense of omitting citation of earlier similar studies. An effort has been made to include results of surveys in all regions of the world. Unfortunately, Africa is underrepresented in Table 2, and few data are accessible for the Soviet Union, China, Pakistan, or the Antilles (except for an older study for Trinidad and Tobago). Thus, epidemiologic data are missing for a large proportion of the earth's population. Whether such surveys have not been undertaken or whether they are inaccessible because of language barriers is not known. It is reasonable to assume, however, that in these densely populated areas, the prevalence of iron deficiency will mirror that of other surveyed countries with similar dietary, economic, and climatologic circumstances.

A careful examination of the studies tabulated leads to the following conclusions concerning the prevalence of iron deficiency.

1. It is widely prevalent in all population groups that have been studied.

2. It is particularly frequent in preschool children of economically deprived families irrespective of race and in women of childbearing age or pregnant women.

3. It has a higher prevalence in persons of all ages and both sexes in tropical areas, due both to the frequency of intestinal infestation with hookworm and to consumption of a meat-poor diet, common in many populations of tropical areas.

A composite estimate of the prevalence of iron deficiency in high-risk population subgroups of the U.S. would be about as follows:

	Iron deficiency (%)	Iron deficiency anemia (%)
Infants	50	20
Preschool children		
Good economic status	5	2
Poor economic status	30	20
Young women, nonpregnant	33	10
Pregnant women (no iron supplement)	100	30

In the U.S., the population groups most often affected by iron deficiency are children or young women of black, Puerto Rican, or poor white families and multiparous women of all socioeconomic strata.

Although dietary iron deficiency may be a serious problem for a small proportion of the population of the U.S., in many areas of the world this problem is far more pervasive. Thus, in India, where for much of the population both protein intake and caloric consumption are marginal, iron deficiency appears to affect a very large proportion of the population. In Sri Lanka the majority of children, even those regarded as of affluent status, have been found to be anemic.[114]

SIGNIFICANCE OF SURVEY RESULTS

The results of these surveys reveal that iron deficiency is a worldwide problem of extraordinary magnitude. It is less clear what the consequences of iron deficiency may be in terms of increased morbidity or reduced function of persons so affected. De Silva and Fernando[91] have suggested that criteria for anemia based on studies in Caucasians of European origin may not be appropriate for other ethnic groups, because even in their healthy affluent group of Ceylonese (Sri Lanka) children, 77% were anemic by generally accepted criteria. There appears to be no other evidence to support this point of view, and investigators in countries with populations of similar ethnic composition (for example, India) have not adopted standards differing from those generally accepted. Wood and Elwood[126] were unable to find any adverse effect of anemia on health. However, their study was largely limited to persons with very mild anemia. It has been shown that the longitudinal growth rate in iron-deficient children is in the lowest percentiles[127] and that iron-deficient children perform in the lower percentiles in school work.[128,129] It must be stressed that these associations do not necessarily imply that iron deficiency causes slower growth or poorer school performance. It may be that other socioeconomic factors are the root cause of poor diet, poor growth, and poor school performance. However, this is clearly an important area for further investigation. Were such effects as growth retardation and impaired intellectual performance ascribable to iron deficiency, this would amply justify large-scale, worldwide efforts to increase the dietary iron content.

Several studies have been made of work tolerance in iron-deficient subjects.[130-134] Some of these studies seem to indicate no adverse effect of iron deficiency on exercise performance. Others have demonstrated decreased performance. In experimental animals, exercise performance is diminished in iron deficiency with or without anemia.[135,136] In man, although results of various studies have been contradictory, on balance, the evidence tends to support the contention that function of skeletal muscle is reduced in iron deficiency.

IRON SUPPLEMENTATION OF FOOD

The close association between high prevalence of iron deficiency and poverty would seem to imply that rising standards of living should be accompanied by a better diet and the disappearance of anemia and iron deficiency.

Indeed, this may already have begun to be evident in Japan[97] and Denmark.[77] From 1955 to 1970, among outpatients in a major medical center in Tokyo, iron deficiency had steadily declined as a cause of anemia, although it remained the single most frequent cause of anemia.[97] However, such trends have not yet become evident elsewhere. It is, furthermore, possible that the state of nutrition worldwide may deteriorate rather than improve, as populations continue to expand in the face of declining food production. Alleviation of anemia and iron deficiency should not await a hypothetical global rise in affluence. Furthermore, because of religious convictions or dietary customs prevalent in some countries, it is not to be expected that an increase in standard of living will necessarily be accompanied by an increase in protein and iron content of food consumed.

In the United States, a wheat-flour enrichment program was begun in 1941, and since that time "enriched flour" has had iron added at a level of 13.0 to 16.5 mg/lb, most commonly in the form of reduced iron — that is, powdered metallic iron.[137] Similarly, "enriched bread, rolls, or buns" contain 8.0 to 12.5 mg of iron per pound. Some bakeries add ferrous sulfate at the time of preparation of bread dough to attain this level of iron enrichment. However, since unacceptable odors and flavors may develop in ferrous sulfate-enriched flour if it is not promptly used by the bakery, ferrous salts cannot be added at the mill; thus, reduced iron is used almost exclusively by millers. In practice, many bakers use flour enriched at the mill. In the U.S. 36 states currently require by statute the enrichment of wheat flour.* Because of the patchwork of state laws, and

---

* These states are Alabama, Alaska, Arizona, Arkansas, California, Colorado, Connecticut, Florida, Georgia, Hawaii, Idaho, Indiana, Kansas, Kentucky, Louisiana, Maine, Massachusetts, Mississippi, Montana, Nebraska, New Hampshire, New Jersey, New Mexico, New York, North Carolina, North Dakota, Ohio, Oklahoma, Oregon, Rhode Island, South Dakota, Texas, Utah, Washington, West Virginia, and Wyoming. In addition, California, Oregon, and Utah require that in all foods containing 25% or more flour, enriched flour must be used.

uncertainty over the final destination of wheat products, most flour mills enrich all their flour. Thus, enrichment of flour in the U.S. is very nearly universal.

For some years, the Millers' National Federation had urged enrichment at a higher level, and new standards, tentatively set for February 1974, were to have been 40 mg of iron per pound of enriched flour and 25 mg per pound of enriched bread, rolls, or buns.[138] These standards have not been adopted, however, as of September 1976, and it is believed that they will most likely be abandoned, so that iron enrichment will remain at the current level.

The practice of iron enrichment of flour and bread may be criticized on several points. First, the ferric oxides and hydroxides that readily form from reduced (metallic) iron are inherently poorly absorbed. Secondly, iron is chelated by phytates naturally present in grain. For example, in Ethiopia, where bread contains large amounts of iron, the dietary iron intake may be as high as 500 mg daily, yet in rural areas of that country the frequency of iron deficiency anemia in children is no less than in economically deprived groups in the U.S., and clinical disorders of iron overload, such as hemochromatosis, are not unduly common.[89,137] Thus, the ingestion of iron mixed with bread may not result in an increase in the quantity of iron absorbed. Thirdly, the attempt to increase dietary iron content by this means appears to assume that all segments of the population will have a nearly proportional consumption of bread. It is not known whether this is true and whether this practice is likely to provide a significant iron supplement for those portions of the population especially at risk. However, since a person eating seven slices of enriched bread daily would thereby have only a 2-mg/day increment in iron ingested, the resultant increment in iron absorbed daily would be small even if iron in bread were absorbed as readily as that in other foods. Fourthly, as indicated in several nutrition surveys conducted in the U.S. within the past several years, and which are summarized in Table 2, iron deficiency remains widely prevalent in the U.S. despite a 35-year-old program of iron enrichment of bread and flour.

Lastly, there is at least one group that may be adversely affected by iron fortification of bread or other foodstuffs: those with hemochromatosis. Idiopathic hemochromatosis is believed to occur with a frequency of approximately 1:10,000 population. Thus, in the U.S. there are estimated to be approximately 20,000 persons genetically at risk of developing hemochromatosis. In addition, many persons with chronic anemias experience secondary hemochromatosis at present levels of dietary iron intake, in part as a result of multiple transfusions. This is particularly true of those with thalassemia major and those with sideroblastic anemia. Undoubtedly, for those persons who are subject to iron overload, hemochromatosis and its complications will develop many years earlier if the iron content of their diet is substantially increased. Whether iron enrichment of bread and flour will adversely affect this group is not presently known. Experience in Ethiopia with iron-rich flour suggests that even persons predisposed to hemochromatosis may not absorb appreciable iron from bread. However, any effort to improve the iron content of the diet must not overlook potential jeopardy to the health of those who inherently accumulate iron more rapidly than others.

More ingenuity seems needed in the future to direct iron supplementation programs to those most likely to become iron depleted. This might mean iron fortification of milk or of infant formulas and a more widespread practice of routine use of iron supplements by women during the reproductive years.

Finally, in order to permit a clearer judgment of the potential benefit versus cost of iron supplementation programs, it is necessary that better delineation be achieved of the adverse effects of iron deficiency in human subjects in terms of growth, development, and intellectual and motor performance. At present, this appears to be the issue that is least resolved, although it is one around which revolve all the programs of iron enrichment and iron therapy.

SUMMARY

Iron deficiency and iron deficiency anemia are prevalent in all areas of the world, and they affect especially infants, young children, and women of the reproductive years. Poverty, meat-poor diets, and hookworm infestation are major contributing factors. Improvement in nutritional standards has reduced the frequency of iron deficiency in Denmark and Japan, but this has not yet been observed elsewhere. Further efforts are needed to increase the iron intake of susceptible groups while avoiding an increment in iron intake for those in whom hemochromatosis may develop. Iron enrichment of flour does not appear to have been successful, because of the form and small quantity of iron added and because of chelation by phytates. Impairment in growth, development, and motor functions in iron-deficient persons has been demonstrated in some studies but not in others and requires additional documentation.

ACKNOWLEDGMENTS

Dr. Haruo Okazaki kindly translated an article from the Japanese.[97] Mr. William Mailhot, of General Mills, Inc., was helpful in providing information concerning present practices of the milling industry in iron enrichment of flour.

REFERENCES

1. **Fairbanks, V. F., Fahey, J. L., and Beutler, E.,** *Clinical Disorders of Iron Metabolism,* 2nd ed., Grune & Stratton, New York, 1971.
2. **MacKay, H. M. M.,** Nutritional anaemia in infancy: With special reference to iron deficiency, *Med. Res. Counc. (G.B.), Spec. Rep. Ser.,* No. 157, pp. 1–125, 1931.
3. Committee on Iron Deficiency, Iron Deficiency in the United States, *J.A.M.A.,* 203, 407–412, 1968.
4. **Beutler, E., Robson, M. J., and Buttenwieser, E.,** A comparison of the plasma iron, iron-binding capacity, sternal marrow iron and other methods in the clinical evaluation of iron stores, *Ann. Intern. Med.,* 48, 60–82, 1958.
5. **Monsen, E. R., Kuhn, I. N., and Finch, C. A.,** Iron status of menstruating women, *Am. J. Clin. Nutr.,* 20, 842–849, 1967.
6. **Green, R., Charlton, R., Seftel, H., Bothwell, T., Mayet, F., Adams, B., Finch, C., and Layrisse, M.,** Body iron excretion in man: A collaborative study, *Am. J. Med.,* 45, 336–353, 1968.
7. **White, H. S.,** Iron nutriture of girls and women: A review. I. Dietary iron and hemoglobin concentrations, *J. Am. Diet. Assoc.,* 53, 563–569, 1968.
8. World Health Organization, Nutritional Anaemias, Report of a WHO Group of Experts, *W.H.O. Tech. Rep. Ser.,* No. 503, 5–29, 1972.
9. **Wilson, J. F., Heiner, D. C., and Lahey, M. E.,** Milk-induced gastrointestinal bleeding in infants with hypochromic microcytic anemia, *J.A.M.A.,* 189, 568–572, 1964.
10. **Layrisse, M. and Martinez-Torres, C.,** Food iron absorption: Iron supplementation of food, *Prog. Hematol.,* 7, 137–160, 1971.
11. **Maisterrena, J. A., Tovar Zamora, E., and Murphy, C. A.,** Absorción de hierro en dos tipos diferentes de alimentacion, *Rev. Invest. Clin.,* 25, 1–7, 1973.
12. **Hobbs, J. R.,** Iron deficiency after partial gastrectomy, *Gut,* 2, 141–149, 1961.
13. **Moseley, J. E.,** Skeletal changes in the anemias, *Semin. Roentgenol.,* 9, 169–184, 1974.
14. **Hutton, C. F.,** Plummer Vinson syndrome, *Br. J. Radiol.,* 29, 81–85, 1956.
15. **Naiman, J. L., Oski, F. A., Diamond, L. K., Vawter, G. F., and Shwachman, H.,** The gastrointestinal effects of iron-deficiency anemia, *Pediatrics,* 33, 83–99, 1964.
16. **Guha, D. K., Walia, B. N., Tandon, B. N., Deo, M. G., and Ghai, O. P.,** Small bowel changes in iron-deficiency anaemia of childhood, *Arch. Dis. Child.,* 43, 239–244, 1968.
17. **El-Najjar, M. Y., Lozoff, B., and Ryan, D. J.,** The paleoepidemiology of porotic hyperostosis in the American southwest: Radiological and ecological considerations, *Am. J. Roentgenol.,* 125, 918–924, 1975.
18. **El-Najjar, M. Y. and Robertson, A. L., Jr.,** Spongy bones in prehistoric America, *Science,* 193, 141–143, 1976.

19. World Health Organization, Nutritional Anaemias: Report of a WHO Scientific Group, *W.H.O. Tech. Rep. Ser.,* 405, 5–37, 1968.

20. **Loría, A., Piedras, J., Labardini, J., and Sánchez-Medal, L.,** Anemia nutricional. I. Valores de serie roja en varones adultos sanos residentes a 2 240 metros sobre el nivel del mar, *Rev. Invest. Clin.,* 23, 3–9, 1971.

21. **Beutler, E.,** The red cell indices in the diagnosis of iron-deficiency anemia, *Ann. Intern. Med.,* 50, 313–322, 1959.

22. **Fairbanks, V. F.,** Is the peripheral blood film reliable for the diagnosis of iron deficiency anemia? *Am. J. Clin. Pathol.,* 55, 447–451, 1971.

23. **Hamilton, L. D., Gubler, C. J., Cartwright, G. E., and Wintrobe, M. M.,** Diurnal variation in the plasma iron level of man, *Proc. Soc. Exp. Biol. Med.,* 75, 65–68, 1950.

24. **Høyer, K.,** Physiologic variations in the iron content of human blood serum. I. The variations from week to week, from day to day and through twenty-four hours. II. Further studies of the intra diem variations, *Acta Med. Scand.,* 119, 562–576; 577–585, 1944.

25. **Speck, B.,** Diurnal variation of serum iron and the latent iron-binding in normal adults, *Helv. Med. Acta,* 34, 231–238, 1967.

26. **Cartwright, G. E.,** The anemia of chronic disorders, *Semin. Hematol.,* 3, 351–375, 1966.

27. **Zilva, J. F. and Patson, V. J.,** Variations in serum-iron in healthy women, *Lancet,* 1, 459–462, 1966.

28. **Fujino, M., Dawson, E. B., Holeman, T., and McGanity, W. J.,** Interrelationships between estrogenic activity, serum iron and ascorbic acid levels during the menstrual cycle, *Am. J. Clin. Nutr.,* 18, 256–260, 1966.

29. **Mardell, M. and Zilva, J. F.,** Effect of oral contraceptives on the variations in serum-iron during the menstrual cycle, *Lancet,* 2, 1323–1325, 1967.

30. **Waters, A. H., Morley, A. A., and Rankin, J. G.,** Effect of alcohol on haemopoiesis, *Br. Med. J.,* 2, 1565–1568, 1966.

31. **Lindenbaum, J. and Lieber, C. S.,** Hematologic effects of alcohol in man in the absence of nutritional deficiency, *N. Engl. J. Med.,* 281, 333–338, 1969.

32. **Handjani, A. M., Banihashemi, A., Rafiee, R., and Tolou, H.,** Serum iron in acute myocardial infarction, *Blut,* 23, 363–366, 1971.

33. **Syrkis, I. and Machtey, I.,** Hypoferremia in acute myocardial infarction, *J. Am. Geriat. Soc.,* 21, 28–30, 1973.

34. **Bainton, D. F. and Finch, C. A.,** The diagnosis of iron deficiency anemia, *Am. J. Med.,* 37, 62–70, 1964.

35. **Klee, G. G., Fairbanks, V. F., Pierre, R. V., and O'Sullivan, M. B.,** Use of routine erythrocyte measurements in diagnosis of iron-deficiency anemia and thalassemia minor, *Am. J. Clin. Pathol.,* 66, 875–877, 1976.

36. **Valberg, L. S., Sorbie, J., Ludwig, J., and Pelletier, O.,** Serum ferritin and the iron status of Canadians, *Can. Med. Assoc. J.,* 114, 417–421, 1976.

37. **De Leeuw, N. K. M., Lowenstein, L., and Hsieh, Y.-S.,** Iron deficiency and hydremia in normal pregnancy, *Medicine* (Baltimore), 45, 291–315, 1966.

38. **Haddy, T. B., Jurkowski, C., Brody, H., Kallen, D. J., Czajka-Narins, D. M.,** Iron deficiency with and without anemia in infants and children, *Am. J. Dis. Child.,* 128, 787–793, 1974.

39. **Scott, D. E. and Pritchard, J. A.,** Iron deficiency in healthy young college women, *JAMA,* 199, 897–900, 1967.

40. **Hutcheson, R. H. and Hutcheson, J. K.,** Iron and vitamin C and D deficiencies in a large population of children, *Health Serv. Rep.,* 87, 232–235, 1972.

41. **Schubert, W. K. and Hunter, R. E.,** Iron nutrition in infancy, *Rep. Ross Conf. Pediatr. Res.,* 62, 20–23, 1970.

42. **Stine, O. C., Saratsiotic, J. B., and Furno, O. F.,** Appraising the health of culturally deprived children, *Am. J. Clin. Nutr.,* 20, 1084–1095, 1967.

43. **Pearson, H. A., Abrams, I., Fernbach, D. J., Gyland, S. P., and Hahn, D. A.,** Anemia in preschool children in the United States of America, *Pediatr. Res.,* 1, 169–172, 1967.

44. **Christakis, G., Miridjanian, A., Nath, L., Khurana, H. S., Cowell, C., Archer, M., Frank, O., Ziffer, H., Baker, H., and James, G.,** A nutritional epidemiologic investigation of 642 New York City children, *Am. J. Clin. Nutr.,* 21, 107–126, 1968.

45. **Kripke, S. S. and Sanders, E.,** Prevalence of iron-deficiency anemia among infants and young children seen at rural ambulatory clinics, *Am. J. Clin. Nutr.,* 23, 716–724, 1970.

46. **Fuerth, J. H.,** Incidence of anemia in full-term infants seen in private practice, *J. Pediatr.,* 79, 560–562, 1971.

47. **Cook, J. D.,** Approaches to analysis of survey data on iron nutritional deficiency, *Extent and Meanings of Iron Deficiency in the U.S. Summary Proceedings of a Workshop,* Committee on Iron Nutritional Deficiencies, Food and Nutrition Board, National Academy of Sciences, National Research Council, Washington, D.C., 1971.

48. **Edozien, J.,** Prevalence of iron deficiency in the United States, with or without anemia – national survey data, *Extent and Meanings of Iron Deficiency in the U.S. Summary Proceedings of a Workshop,* Committee on Iron Nutritional Deficiencies, Food and Nutrition Board, National Academy of Sciences, National Research Council, Washington, D.C., 1971.

49. **Fomon, S. J. and Weckwerth, V. E.,** Hemoglobin concentrations of children registered for care in C & Y projects during March and April 1968, *Extent and Meanings of Iron Deficiency in the U.S. Summary Proceedings of a Workshop,* Committee on Iron Nutritional Deficiencies, Food and Nutrition Board, National Academy of Sciences, National Research Council, Washington, D.C., 1971.

50. **Owen, G. M., Garry, P. J., Kram, K. M., Nelsen, C. E., and Montalvo, J. M.,** Nutritional status of Mississippi preschool children: a pilot study, *Am. J. Clin. Nutr.,* 22, 1444–1458, 1969.

51. **Owen, G. M., Nelsen, C. E., and Garry, P. J.,** Nutritional status of preschool children: Hemoglobin, hematocrit, and plasma iron values, *J. Pediatr.,* 76, 761–763, 1970.

52. **Owen, G. M., Lubin, A. H., and Garry, P. J.,** Preschool children in the United States: Who has iron deficiency? *J. Pediatr.,* 79, 563–568, 1971.

53. **Johnson, C. C. and Futrell, M. F.,** Anemia in black preschool children in Mississippi, *J. Am. Diet. Assoc.,* 65, 536–541, 1974.

54. **Haughton, J. G.,** Nutritional anemia of infancy and childhood, *Am. J. Public Health,* 53, 1121–1126, 1963.

55. **Gutelius, M. F.,** The problem of iron deficiency anemia in preschool Negro children, *Am. J. Public Health,* 59, 290–295, 1969.

56. Preliminary Findings of the First Health and Nutrition Examination Survey, United States, 1971–1972: Dietary Intake and Biochemical Findings, Department of Health, Education and Welfare, Washington, D.C., 1974.

57. **Loría, A., García Viveros, J., Sánchez Medal, L., Hoffs, M. D., Shein, M., and Berger, I.,** Anemia nutricional. II. Deficiencia de hierro en niños de 0 a 36 meses de edad y de buena condición socioeconómica, *Bol. Med. Hosp. Infant. Mex.* (Span. Ed.), 27, 251–260, 1970.

58. **Loría, A., Sánchez Medal, L., García Viveros, J., and Piedras, J.,** Anemia nutricional. III. Deficiencia de hierro en niños menores de 7 años de edad y de baja condición socioeconómica, *Rev. Invest. Clin.,* 23, 11–19, 1971.

59. **Viteri, F. E., Guzmán, M. A., Mata, L. J.,** Anemias nutricionales en Centro América influencia de infección por uncinaria, *Arch. Latinoam. Nutr.,* 23, 33–53, 1973.

60. **Molina, V. R. A., Diez-Ewald, M., Fernández, G., and Velázquez, N.,** Nutritional anaemia during pregnancy: A comparative study of two socio-economic classes, *J. Obstet. Gynaecol. Br. Commonw.,* 81, 454–458, 1974.

61. **Layrisse, M. and Roche, M.,** The relationship between anemia and hookworm infection: Results of surveys of rural Venezuelan population, *Am. J. Hyg.,* 79, 279–301, 1964.

62. **Szarfarc, S. C.,** Anemia ferropriva em parturientes e recém-nascidos, *Rev. Saude Publica.,* 8, 369–374, 1974.

63. **Roncada, M. J. and Szarfarc, S. C.,** Hipovitaminose a e anemia ferropriva em gestantes de duas comunidades do vale do ribeira (estado de São Palo, Brazil), *Rev. Saude Publica.,* 9, 99–106, 1975.

64. **Cook, J. D., Alvarado, J., Gutnisky, A., Jamra, M., Labardini, J., Layrisse, M., Linares, J., Loría, A., Maspes, V., Restrepo, A., Reynafarje, C., Sánchez-Medal, L., Vélez, H., and Viteri, F.,** Nutritional deficiency and anemia in Latin America: A collaborative study, *Blood,* 38, 591–603, 1971.

65. **Chopra, J. G. and Byam, N. T. A.,** Anemia survey in Trinidad and Tobago, *Am. J. Public Health,* 58, 1922–1936, 1968.

66. **McFarlane, D. B., Pinkerton, P. H., Dagg, J. H., and Goldberg, A.,** Incidence of iron deficiency, with and without anaemia, in women in general practice, *Br. J. Haematol.,* 13, 790–796, 1967.

67. **Burman, D.,** Haemoglobin levels in normal infants aged 3 to 24 months, and the effect of iron, *Arch. Dis. Child.,* 47, 261–271, 1972.

68. **Powell, D. E. B., Thomas, J. H., and Mills, P.,** Serum iron in elderly hospital patients, *Gerontol. Clin.,* 10, 21–29, 1968.

69. **Kilpatrick, G. S. and Hardisty, R. M.,** The prevalence of anaemia in the community: A survey of a random sample of the population, *Br. Med. J.,* 1, 778–782, 1961.

70. **Rees, E. G., Moore, R. M. A., and Wycherley, P. A.,** Unsuspected anaemia: The case for population screening, *J. R. Coll. Gen. Pract.,* 17, 155–161, 1969.

71. **Elwood, P. C., Waters, W. E., Greene, W. J., and Wood, M. M.,** Evaluation of a screening survey for anaemia in adult non-pregnant women, *Br. Med. J.,* 4, 714–717, 1967.

72. **Natvig, K.,** Studies on hemoglobin values in Norway. V. Hemoglobin concentration and hematocrit in men aged 15–21 years, *Acta Med. Scand.,* 180, 613–620, 1966.

73. **Natvig, H., Vellar, O. D., and Andersen, J.,** Studies on hemoglobin values in Norway, *Acta Med. Scand.,* 182, 183–191, 1967.

74. **Natvig, H. and Vellar, O. D.,** Studies on hemoglobin values in Norway. VIII. Hemoglobin, hematocrit and MCHC values in adult men and women, *Acta Med. Scand.,* 182, 193–205, 1967.

75. **Garby, L., Irnell, L., and Werner, I.,** Iron deficiency in women of fertile age in a Swedish community. II. Efficiency of several laboratory tests to predict the response to iron supplementation, *Acta Med. Scand.,* 185, 107–111, 1969.

76. **Garby, L., Irnell, L., and Werner, I.,** Iron deficiency in women of fertile age in a Swedish community. III. Estimation of prevalence based on response to iron supplementation, *Acta Med. Scand.,* 185, 113–117, 1969.

77. **Marner, T.,** Haemoglobin, erythrocytes and serum iron values in normal children 3–6 years of age, *Acta Paediatr. Scand.,* 58, 363–368, 1969.

78. **Dughera, L. and Naldi, R.,** Considerazioni cliniche sull'anemia sideropenica in età senile, *Minerva Med.,* 66, 1942–1948, 1975.

79. **Piancino, G., Battistini, V., and Fontana, F.,** Incidenza dell'anemia sideropenica nei malati di medicina generale, *Minerva Med.,* 66, 1949–1952, 1975.

80. **Kattamis, C., Metaxotou-Mavromnati, A., Konidaris, C., Touliatos, N., Constantsas, N., and Matsaniotis, N.,** Iron deficiency in Greece: Epidemiologic and hematologic studies, *J. Pediatr.,* 84, 666–671, 1974.

81. **Hibšer, M., Hreljac-Hibšer, L., and Pamuković, Z.,** Učestalost manifestnog i latentnog nedostatka željeza kod učenika koji se upisuju u srednju školu, *Lijec. Vjesn.,* 97, 205–207, 1975.

82. **Istvan, S.,** Epidemiology of anaemia, *Santé Publique* (Bucharest), 2, 173–185, 1970.

83. **Antila, V.,** Iron deficiency in the Finnish population, *Acta Med. Scand. Suppl.,* 393, 1–71, 1962.

84. **Takkunen, H. and Seppänen, R.,** Iron deficiency and dietary factors in Finland, *Am. J. Clin. Nutr.,* 28, 1141–1147, 1975.

85. **Hirsjärvi, E.,** Anemia and iron deficiency in aged hospital patients, *Geron.,* No. 20, 72–78, 1974–1975.

86. **Göltner, E. and Schlunk, T.,** Warum noch Eisenmangel in der Schwangerschaft? *Munch. Med. Wochenschr.,* 116, 841–844, 1974.

87. **Abdel-Fattah, M., Shalaby, S., and El-ashmawi, S.,** An epidemiologic study of iron deficiency anaemia in early infancy, *Gaz. Egypt. Paediatr. Assoc.,* 22, 144–147, 1974.

88. **Mayet, F. G. H., Adams, E. B., Moodley, T., Kleber, E. E., and Cooper, S. K.,** Dietary iron and anaemia in an Indian community in Natal, *S. Afr. Med. J.,* pp. 1427–1430, 1972.

89. **Hofvander, Y.,** Hematological investigations in Ethiopia: with special reference to a high iron intake, *Acta Med. Scand. Suppl.,* 494, 11–74, 1968.

90. **Stott, G.,** Anaemia in Mauritius, *Bull. W.H.O.,* 23, 781–791, 1960.

91. **De Silva, C. C. and Fernando, R. P.,** Anemias of Ceylonese children, *Isr. J. Med. Sci.,* 2, 499–505, 1966.

92. **Manchanda, S. S., Lal, H., and Khanna, S.,** Iron stores in health and disease: Bone-marrow studies in 1134 children in Punjab, India, *Arch. Dis. Child.,* 44, 580–584, 1969.

93. **Banerji, L., Sood, S. K., and Ramalingaswami, V.,** Geographic pathology of iron deficiency with special reference to India. I. Histochemical quantitation of iron stores in population groups, *Am. J. Clin. Nutr.,* 21, 1139–1148, 1968.

94. **Sood, S. K., Banerji, L., and Ramalingaswami, V.,** Geographic pathology of iron deficiency with special reference to India. II. Quantitation of iron stores by repeated phlebotomy in Indian volunteers, *Am. J. Clin. Nutr.,* 21, 1149–1155, 1968.

95. **Jadhav, M. and Baker, S. J.,** A study of etiology of severe anemia in 50 children, *Indian J. Child. Health,* 10, 235–243, 1961.

96. **Dhar, N., Agarwal, K. N., Taneja, P. N., and Gupta, S.,** Iron deficiency anemias in children, *Indian J. Pediatr.,* 36, 436–441, 1969.

97. **Takaku, F.,** Hypochromic anemia – lists of statistics pertinent to diagnosis, *Jpn. J. Clin. Med.,* 32 (special no.), 1756–1764, 1974.

98. **Sinhu, S. S.,** Serum levels of iron, folic acid and vitamin B_{12} in the maternal and cord blood in the three major ethnic groups in Malaysia, *Indian Pediatr.,* 11, 775–780, 1974.

99. **Ong, H. C.,** Anaemia in pregnancy in an aboriginal population, *J. Trop. Med. Hyg.,* 77, 22–26, 1974.

100. **Marzan, A. M., Tantengco, V. O., Caviles, A. P., and Villanueva, L.,** Nutritional anaemia in Filipino infants and preschoolers, *Southeast Asian J. Trop. Med. Public Health,* 5, 90–95, 1974.

101. **Tantengco, V. O., Marzan, A. M., Rapanot, N., Villanueva, L., and DeCastro, C.,** Nutritional anaemia in Filipino school children, *Southeast Asian J. Trop. Med. Public Health,* 4, 524–533, 1973.

102. **Kho, L. K. and Halimoen, E. M.,** Pediatric hematology in the tropics, *Ann. Paediatr.,* 206, 276–286, 1966.

103. **McEwin, R., Hinton, J., and Sills, I.,** Iron deficiency anaemia in Australian women, *Med. J. Aust.,* 1, 293–298, 1974.

104. **Lovric, V. A.,** Iron deficiency in children in Sydney, *Paediatr. Indones.,* 15, 34–54, 1975.

105. **Lovric, V. A.,** Normal haematological values in children aged 6 to 36 months and socio-medical implications, *Med. J. Aust.,* 2, 366–370, 1970.

106. **Kamien, M. and Cameron, P.,** An analysis of white and aboriginal children under five years of age admitted to the Bourke district hospital from September 1971 to August 1972, *Aust. Paediatr. J.,* 10, 343–349, 1974.

107. **Fleming, A. F., Martin, J. D.. and Stenhouse, N. S.,** Pregnancy anaemia, iron and folate deficiency in Western Australia, *Med. J. Aust.,* 2, 479–484, 1974.

108. **Anyon, C. P. and Desmond, F. B.,** Normal haematological values in New Zealand European infants, *N.Z. Med. J.,* 80, 383–387, 1974.

109. **Neave, M., Prior, I. A. M., and Toms, V.,** The prevalence of anaemia in two Maori rural communities, *N.Z. Med. J.,* 62, 20–28, 1963.

110. **Charlton, R. W., Hawkins, D. M., Mavor, W. O., and Bothwell, T. H.,** Hepatic storage iron concentrations in different population groups, *Am. J. Clin. Nutr.,* 23, 358–370, 1970.

111. **Piomelli, S.,** A micromethod for free erythrocyte porphyrins: The FEP test, *J. Lab. Clin. Med.,* 81, 932–940, 1973.

112. **Stockman, J. A., III, Weiner, L. S., Simon, G. E., Stuart, M. J., and Oski, F. A.,** The measurement of free erythrocyte porphyrin (FEP) as a simple means of distinguishing iron deficiency from beta-thalassemia trait in subjects with microcytosis, *J. Lab. Clin. Med.,* 85, 113–119, 1975.

113. **Pollack, S., George, J. N., Reba, R. C., Kaufman, R. M., and Crosby, W. H.,** The absorption of nonferrous metals in iron deficiency, *J. Clin. Invest.,* 44, 1470–1473, 1965.

114. **Valberg, L. S., Ludwig, J., and Olatunbosun, D.,** Alteration in cobalt absorption in patients with disorders of iron metabolism, *Gastroenterology,* 56, 241–251, 1969.

115. **Sorbie, J., Olatunbosun, D., Corbett, W. E. N., and Valberg, L. S.,** Cobalt excretion test for the assessment of body iron stores, *Can. Med. Assoc. J.,* 104, 777–782, 1971.

116. **Valberg, L. S., Sorbie, J., Corbett, W. E. N., and Ludwig, J.,** Cobalt test for the detection of iron deficiency anemia, *Ann. Intern. Med.,* 77, 181–187, 1972.

117. **Olatunbosun, D., Corbett, W. E. N., Ludwig, J., and Valberg, L. S.,** Alteration of cobalt absorption in portal cirrhosis and idiopathic hemochromatosis, *J. Lab. Clin. Med.,* 75, 754–762, 1970.

118. **Wahner-Roedler, D. L., Fairbanks, V. F., and Linman, J. W.,** Cobalt excretion test as index of iron absorption and diagnostic test for iron deficiency, *J. Lab. Clin. Med.,* 85, 253–259, 1975.

119. **Jacobs, A., Miller, F., Worwood, M., Beamish, M. R., and Wardrop, C. A.,** Ferritin in the serum of normal subjects and patients with iron deficiency and iron overload, *Br. Med. J.,* 4, 206–208, 1972.

120. **Siimes, M. A., Addiego, J. E., Jr., and Dallman, P. R.,** Ferritin in serum: Diagnosis of iron deficiency and iron overload in infants and children, *Blood,* 43, 581–590, 1974.

121. **Lipschitz, D. A., Cook, J. D., and Finch, C. A.,** A clinical evaluation of serum ferritin as an index of iron stores, *N. Engl. J. Med.,* 290, 1213–1216, 1974.

122. **Miles, L. E. M., Lipschitz, D. A., Bieber, C. P., and Cook, J. D.,** Measurement of serum ferritin by a 2-site immunoradiometric assay, *Anal. Biochem.,* 61, 209–224, 1974.

123. **Addison, G. M., Beamish, M. R., Hales, C. N., Hodgkins, M., Jacobs, A., and Llewellin, P.,** An immunoradiometric assay for ferritin in the serum of normal subjects and patients with iron deficiency and iron overload, *J. Clin. Pathol.,* 25, 326–329, 1972.

124. **Jacobs, A. and Worwood, M.,** Ferritin in serum: Clinical biochemical implications, *N. Engl. J. Med.,* 292, 951–956, 1975.

125. **Walters, G. O., Miller. F. M., and Worwood, M.,** Serum ferritin concentration and iron stores in normal subjects, *J. Clin. Pathol.,* 26, 770–772, 1973.

126. **Wood, M. M. and Elwood, P. C.,** Symptoms of iron deficiency anaemia: A community survey, *Br. J. Prev. Soc. Med.,* 20, 117–121, 1966.

127. **Judisch, J. M., Naiman, J. L., and Oski, F. A.,** The fallacy of the fat iron-deficient child, *Pediatrics,* 37, 987–990, 1966.

128. Webb, T. E. and Oski, F. A.,Iron deficiency anemia and scholastic achievement in young adolescents, *J. Pediatr.*, 82, 827–829, 1973.

129. Sandstead, H. H., Carter, J. P., House, F. R., McConnell, F., Horton, K. B., and Vander Zwaag, R., Nutritional deficiencies in disadvantaged preschool children, *Am. J. Dis. Child.*, 121, 455–463, 1971.

130. Davies, C. T., The physiological effects of iron deficiency anaemia and malnutrition on exercise performance in East African school children, *Acta Paediatr. Belg.*, 28 Suppl., 253–256, 1974.

131. Liedén, G. and Adolfsson, L., Physical work capacity in blood donors, *Scand. J. Clin. Lab. Invest.*, 34, 37–42, 1974.

132. Vellar, O. D. and Hermansen, L., Physical performance and hematological parameters: With special reference to hemoglobin and maximal oxygen uptake, *Acta Med. Scand. Suppl.*, 522, 11–40, 1971.

133. Andersen, H. T. and Barkve, H., Iron deficiency and muscular work performance: An evaluation of the cardio-respiratory function of iron deficient subjects with and without anaemia, *Scand. J. Clin. Lab. Invest Suppl.*, 114, 6–72, 1970.

134. Ericsson, P., The effect of iron supplementation on the physical work capacity in the elderly, *Acta Med. Scand.*, 188, 361–374, 1970.

135. Edgerton, V. R., Bryant, S. L., Gillespie, C. A., and Gardner, G. W., Iron deficiency anemia and physical performance and activity of rats, *J. Nutr.*, 102, 381–399, 1972.

136. Finch, C. A. and Mackler, B., Striate muscle dysfunction in iron deficiency, *Trans. Assoc. Am. Physicians*, 89, 116–119, 1976.

137. American Wheat Institute, *From Wheat to Flour; the Story of Man in a Grain of Wheat*, Chicago, Ill., 1965.

138. Fine, S. D., Wheat flour and related products and baked products: Proposed improvement of nutrient levels of enriched foods, *Fed. Regis.*, 36, 23074–23076, 1971.

Trace Elements

NUTRITION DEFICIENCIES IN MAN: COPPER

Angel Cordano

INTRODUCTION AND REVIEW

After many years of controversy, and long after pathological observations in man were first described, copper deficiency in man is now a recognized clinical condition. Interest in the biological role of copper has increased significantly in the last decade. Many publications and reviews have appeared on human copper metabolism,[1-8] and many points have been clarified regarding the specific sites of action of copper and their role in certain signs of deficiency.

Copper Deficiency Associated with Dietary Deficiencies

The existence of copper deficiency in man has been debated since 1931, when Josephs[9] first suggested that copper deficiency could account for incidences of resistant iron-deficiency anemia in milk-fed infants. His careful descriptions were subsequently questioned, mostly on the basis of unrelated observations.[10] In the 1950s, a series of infants were described with anemia, hypoproteinemia, hypoferremia, and hypocupremia believed to be of dietary origin.[11-15] Some may have had true dietary deficiency, but the majority probably suffered intestinal losses of the copper carrier protein metalloprotein ceruloplasmin due to enteropathies.

Attempts by Bush[10] and Wilson and Lahey[16] to produce copper deficiency experimentally in human infants were unsuccessful. The failure to induce copper deficiency in a relatively short period of time in a group of seven prematures (average weight 1238 g) led the latter investigators[16] to conclude that the daily requirements of copper in early infancy may be satisfied by as little as 15 μg of copper per kilogram of body weight per day. This estimate of the infant's copper requirement was much lower than requirements of other mammals.

In 1964, Cordano et al.[17] reported the development of striking neutropenia and severe demineralization of bone followed by intractable anemia in four infants recovering from malnutrition on a diet of modified cow's milk, with apparently adequate intake of vitamins and iron. These clinical manifestations are now recognized as the main signs of copper deficiency in man. In the four infants, there was absolutely no evidence of malabsorption, and total serum proteins and albumin were normal. This was the first report in humans showing many of the same characteristics of copper deficiency produced experimentally in pigs.[18]

In 1966, Cordano and Graham[19] described severe copper deficiency in a 6-year-old girl with a long history of intestinal malabsorption and an iatrogenically restricted diet. She responded to copper therapy with a striking improvement in the malabsorption syndrome, followed by a dramatic growth spurt, suggesting that chronic copper deficiency can also affect intestinal enzyme activity. In 1968, Cordano[20] expressed concern that premature infants have a higher need for copper than full-term infants. The following publications on copper deficiency in prematures[21-24] support those views.

In 1971, Griscom et al.[21] described a peculiar generalized bone disease in three very small premature infants about 2 months old. Because of similarities with copper deficiency syndrome, these investigators concluded that copper depletion secondary to placental insufficiency and a diet chronically low in copper remained a possible diagnosis.

Oral presentation of these cases to groups of pediatric radiologists led to the realization that similar cases have been observed in many other centers in the U.S. In 1971, Al Rashid and Spangler[22] described a small premature with sepsis, who later developed signs of copper deficiency as well as hypocupremia and low ceruloplasmin. After administration of copper, all symptoms related to copper deficiency disappeared. They speculated that the copper content of the liver was probably too low to maintain the infant while receiving a low copper diet and that he may also have lost some of his body stores of copper due to sepsis. Some intrauterine mechanism may also have contributed to the reduction in the total copper content of this premature.

In 1972 Seely et al.[23] found similar signs of copper deficiency in a premature infant (birth weight of 1100 g) at 3 months of age who had been fed an iron-fortified formula; he responded to a 10-day course of oral copper.

In 1973, Ashkenazi et al.[24] presented another case of copper deficiency in a 6-month-old premature infant (birth weight 1140 g). Along with the usual signs of deficiency, the authors found depigmentation of skin and hair, distended blood vessels thought to be related to changes in the elastin of the vessel walls, and central nervous system abnormalities including hypotonia, psychomotor retardation, and apparent visual impairment. Treatment with oral 1% $CuSO_4$ solution (1 to 3 mg/day for 8 weeks) dramatically cured the infant.

Copper Deficiency Associated with Total Parenteral Nutritions

With the advent of new methods of total parenteral nutrition (TPN) by Dudrick et al.,[25-29] it was recommended that weekly or biweekly plasma transfusions be given in order to avoid essential fatty acids and trace mineral deficiencies.[30,31] However, many believed transfusions were not necessary if TPN was administered for less than 1 month.

In 1970, James and MacMahon[32] called attention to the need for adding some trace minerals, including copper, to solutions intended for TPN. He calculated that solutions then used provided a daily intake of only 10.3 μg of copper for a 3- to 4-kg infant and only 164 μg for a 55-kg adult. Greene et al.[33] also found the trace elements content of such nutrient solutions to be deficient in copper.

In 1972, Karpel and Peden[34] reported the first case of copper deficiency in an infant after 7½ months on TPN with intermittent oral feedings during the course. The signs of copper deficiency responded promptly to oral copper supplementation. Another similar case was described by Ashkenazi in 1973.[25]

In 1974 Dunlap et al.[35] described the occurrence of copper deficiency in an adolescent and an adult patient with short bowel syndrome subsequent to TPN. Anemia, neutropenia, and hypocupremia were present in both cases and responded to treatment with copper. Vilter et al.[36] and Fleming et al.[37] also reported copper deficiencies subsequent to TPN in four adult patients.

Solomons et al.[38] and Fleming et al.[39] demonstrated a marked decrease in plasma copper concentrations in 21 patients with active gastrointestinal disease undergoing TPN.

Copper Deficiency Associated with Genetic Disorders

Menkes' kinky hair syndrome (KHS), first described by Menkes et al.[40] in 1962, appears to be inherited as a recessive X-linked trait. It now appears that all the clinical features of KHS can be explained by a defect in copper-dependent enzymes,[41,46,47] and that a similar array of abnormalities is found in the mottled mutants of mice.[48] Symptoms of KHS generally begin between birth and 3 months of age, and the main characteristics are pili torti, signs of progressive cerebral degeneration with frequent seizures, coarse facies, temperature instability (hypothermia), abnormalities of the metaphyses of long bone, widespread arterial tortuosity and variation in the lumen of arteries with fragmentation of internal elastic lamina and thickening of the intima, low plasma copper and ceruloplasmin, impaired T-cell function, and increased susceptibility to infection with death at an early age.[41-44] French et al.[45] noted the virtual absence of cytochrome oxidase in KHS.

These nearly identical inherited disorders of copper metabolism in men and mice have opened new horizons for research. It has been found that intestinal absorption of copper is markedly decreased,[49,50] but that copper isotopes given intravenously are cleared from the serum at the normal rate.[42,49] Debakan et al.[50] found that the biological half-life of [67]Cu in KHS increases by a factor of two to three over normal controls; it seems that the copper is retained by the patient's liver, while in the control subjects there is a more rapid movement of the copper to circulating ceruloplasmin. Danks[47] assumes that the transport and uptake by the liver are probably normal since ceruloplasmin appears quickly and has normal characteristics. Copper excretion in bile has not been determined.

There is generally no clinical improvement in KHS following oral and parenteral administration of copper, even though liver copper and plasma ceruloplasmin return to normal.[41,42,51] However, one infant 3 days old diagnosed (with copper treatment started at age 28 days) had an encouraging response. There were no seizures and some developmental progress.[52] It is important to know the follow-up on this case, since at 5 months of age this infant was developing better than his older brother who was not treated early.

The poor results thus far could simply mean that irreversible lesions develop very early. However, it is also possible that copper supplements get to the serum but do not gain access to intracellular sites of action. This view is supported by observations in heterozygous females and in cultured cells.[47]

The placental transfer of copper was also considered to be controlled by the chromosomal genes, because some symptoms of KHS are present from birth in babies with Menkes' syndrome.[41,51] However, in a fetus suspected of having KHS, the liver was the only tissue containing less copper than controls, while in kidney, spleen, pancreas, and placenta the levels were significantly higher than controls. These findings suggest that the existence of a defective placental transport of copper is unlikely, but confirm the supposition of inadequate storage of the element in the liver.[53] In spite of poor intestinal absorption, the gut mucosa of patients with KHS has a higher copper content than controls.[46] This is also noted in the mutant mice.[48] However, copper in the mucosal cell could not be differentiated from copper bound to the surface of the cells, and the subcellular localization of the element has yet to be determined.[47]

In an infant with Menkes' syndrome, Lott et al.[54] found that although less than 1% of orally administered ^{64}Cu was absorbed, the infant was able to absorb enough copper to triple the serum copper level when given an oral dose about ten times the normal daily requirements. However, synthesis of ceruloplasmin was not observed until copper was given intravenously. They concluded that the dose of oral copper needed to circumvent the intestinal block may be variable in KHS.

Holtzman[55] recently described the individual involvement of copper-dependent enzymes (ceruloplasmin, connective tissue, amine oxidases, cytochrome oxidase, dopamine-beta-hydroxylase, superoxide dismutase, and tyrosinase) in the pathogenesis of KHS. He discusses normal erythropoietic and neutrophil production in KHS in spite of low serum ceruloplasmin and copper values and states that "while there may be more than one transport system for copper (only one of which is affected in KHS) it is also possible that the hematopoietic tissue in KHS, like the intestinal cells, has an abnormally high affinity for copper." He adds that "the presence of multiple alleles at the KHS locus (and at other genetic loci) in man, which cause different degrees of reduction in copper transport, could account for variations in the susceptibility to copper deficiency observed in infant populations."

Goka et al.[56] analyzed the copper content of cultured skin fibroblasts from patients with KHS and found consistent elevation of copper concentration. This finding is considered a genetic marker and should prove valuable in the diagnosis and perhaps in the antenatal detection of KHS.

DISCUSSION

Copper in Prematurity

Copper, along with other trace elements, must enter the fetus early in gestation. The mechanism of this placental transfer is not well understood. Henkin[57] found that the level of free serum copper in pregnant women was not significantly different from that in nonpregnant women and corresponded to 8% (18 μg/100 ml) of the total serum copper. The proportion of free copper in fetal serum was 16% (5 μg/100 ml). This gives a concentration gradient of 13 μg/100 ml and presumably allows copper to move across the placenta by passive transfer.

While serum values of ceruloplasmin and copper are lowest at birth, the copper content of some fetal and neonatal tissues (like liver, muscle, adrenal glands, thyroid, testes, uterus, and skin) is higher than in adults,[58,59] so that the body concentration of copper in the newborn infant is about three times that of adults. For example, dried infant liver tissue contains between 11 to 32.5 mg of copper per 100 g, while the adult liver has only 5 mg/100 g. The liver is the main depot of copper in fetal life and sometimes contains more than half the total body copper. This copper is located in the mitochondria as neonatal hepatic mitochondrocuprein, which is peculiar to the newborn and the fetus. This compound may contain ten or more times as much copper as any other known copper protein. Soon after birth there is decrease of tissue copper levels and a corresponding increase in concentration of copper and ceruloplasmin in plasma.[60,61] The mitochondrocuprein in the liver begins to disappear and the copper set free is presumably utilized for the needs of other growing tissues. Widdowson et al.[62] state that "since it is in the liver that copper is incorporated into ceruloplasmin, it seems likely that some of the copper set free is used for this purpose."

Very small prematures (750 to 1250 g) are more prone to develop copper deficiency than full-term infants because they have very small livers (39 to 47 g) compared to those of 1500-g prematures (65 g) and normal term babies (151 g). Since the majority of copper deposition occurs in the last trimester of fetal life, it seems evident that the copper content in the liver of a small premature is not sufficient to meet the needs after birth, unless an adequate supplement is provided.

Balance studies on 1-week-old breast-fed infants have shown that some were in negative copper balance. Although very small amounts of copper were excreted in the urine, the major losses were through the intestines.[63] In this review, no studies in prematures were found, but an assumption may be made that losses may persist longer due to immaturity of some enzyme system, as Widdowson[64] postulated in full-term babies.

"Premature or low birth weight babies requiring an exclusive milk diet for more than four months may also become copper deficient. While the copper reserve in the liver of a normal newborn can carry him through the first four months of life with almost no dietary copper, the premature infant has a much reduced reserve. Copper should be added to the milk diet of these infants."[20]

Copper in Malnutrition

During our early experience in Peru prior to 1964, the dietary regimen in the recovery of malnourished infants was mainly based on modified cow's milk formula. This diet was used for 2 to 8 weeks, and under some circumstances it was continued for several months. In the presence of rapid growth and normal serum proteins, four infants developed copper deficiency characterized by a marked neutropenia followed by intractable anemia. Both symptoms responded dramatically to copper administration.

Since hematological values were regularly obtained, it was possible to identify those infants who developed neutropenia below 1500 neutrophils per cubic millimeter. Determinations of serum copper, ceruloplasmin,[65] and serum iron and total iron-binding capacity were determined retrospectively in stored samples and revealed 47 cases of copper deficiency in patients who consumed a similar diet for 1 month or more, as well as 20 other infants who developed no evidence of copper deficiency. Copper deficiency was observed most frequently in infants just below 1 year of age. This suggested that beyond this age, the diet generally provides adequate copper intakes. Nevertheless, if losses continue and the diet is restricted, copper depletion can and does occur.

Spontaneous remission was seen in many cases when milk was replaced by other diets. A few patients that remained deficient were promptly treated and evaluated. Subsequently, if a malnourished infant were given a modified cow's milk formula, copper sulfate was added routinely at some point during the first 2 weeks of rehabilitation.

The earliest clinically detectable evidence of copper deficiency is persistent neutropenia, while the earliest detectable manifestation of copper depletion was found to be a fall in serum copper and ceruloplasmin, with the latter reaching undetectable levels when serum copper was still measurable.[66]

The modified cow's milk formula administered was iron fortified, and many infants received an additional 15 to 60 mg of iron. Some still developed anemia and received intramuscular iron. Most of them, however, did not develop significant anemia or had it corrected by oral iron before copper deficiency was noted, suggesting that iron was utilized despite a low copper diet and that dietary copper was probably not essential for iron absorption. When copper deficiency was present and intramuscular iron was not given, anemia was more likely to develop earlier and to respond poorly to copper unless iron was given simultaneously. This suggested that the iron given in the copper-deficient state was not well absorbed or stored. If intramuscular iron was given earlier in the course of copper deficiency, anemia was prevented. This suggested that although iron was not well absorbed, it was well utilized. The longer the duration of the copper-deficient state, the more likely the development of an intractable, progressive anemia that no longer responded to intramuscular iron. In a few cases, after intramuscular iron had been given without response, nonheme iron was still detectable histochemically in the bone marrow.[67] Similar findings on early impairment of iron absorption and later development of impaired erythropoiesis in experimental copper-deficient pigs were described by Underwood.[68]

Late in the course of copper deficiency, advance bone changes were observed,[17,19] suggestive of scurvy by lacking the hemorrhagic component and occasional pathological fractures.

In the cases described originally,[17] a modest reticulocyte response was noted after high intakes of vitamin C or folic acid, which suggests an increased need for these vitamins or, more likely, a pharmacological "mass action" effect, attempting to correct the inadequate erythropoiesis produced by copper deficiency. In two of the four initial cases of severe copper deficiency, partial neutrophil and good reticulocyte and hemoglobin responses were observed when a pharmacological dose of vitamin B_{12} was given parenterally. Supplemental copper produced further and complete responses.

In deficient infants in whom serum copper was repeatedly analyzed following administration of oral copper sulfate, a further fall in serum copper level was noticed in the first few hours. This was followed by a rapid or marked increase that preceded the elevation of ceruloplasmin levels and was associated with an early rise in ferroxidase activity.[69]

Graham and Cordano[67] analyzed hemoglobin values in 90 malnourished infants receiving modified cow's milk and iron on admission and at approximately 30 and 60 days later in order to determine if inadequate copper intake and early copper depletion had any effect on iron absorption. All patients received oral iron supplements. Results are summarized in Table 1. On admission and after 30 days, hemoglobin levels were quite similar. After 60 days, the hemoglobin levels of those receiving supplementary copper were significantly (about 1 g/100 ml) higher than those at 30 days (by paired t-test) or those at 60 days of the infants not receiving copper supplements who eventually became deficient (by unmatched t-test). For 11 infants who did not become deficient after 60 days on the copper-poor diet, the hemoglobin levels on admission and at 30 and 60 days were very similar to those of the infants who received copper supplements. These results suggest that some impairment of iron metabolism may have occurred relatively early (30 to 60 days) in copper-deficient infants.

Table 2 summarizes hemoglobin, serum iron, and TIBC of infants who were copper deficient as evidenced by hypocupremia and neutropenia. All four subjects who were

Table 1
EFFECT OF SUPPLEMENTAL COPPER ON HEMOGLOBIN OF MALNOURISHED INFANTS RECEIVING MODIFIED COW'S MILK AND IRON[a]

Copper supplement	Copper deficiency	n	Age (months)	Hemoglobin (g/100 ml)		
				Admission	30 days	60 days
Yes	No	23	15.6 ± 8.1	9.30 ± 1.22	9.78 ± 1.10	10.74 ± 0.75
No	No	11	11.5 ± 9.9	9.49 ± 1.60	10.04 ± 0.97	10.52 ± 0.73
No	No	9		9.36 ± 1.17	9.84 ± 1.04	—
No	Yes	47	12.0 ± 7.8	9.25 ± 1.65	9.68 ± 1.03	9.70 ± 1.27

[a] Mean ±SD.

[b] Mean of all 20 cases receiving copper-deficient diet and not developing signs of copper deficiency.

From Graham, G. G. and Cardano, A., in *Trace Elements in Human Health and Disease,* Prasad, A. S., Ed., Academic Press, New York, in press. With permission.

Table 2
HEMOGLOBIN, SERUM IRON, AND TIBC OF MALNOURISHED INFANTS WHO RECEIVED MODIFIED COW'S MILK AND IRON AND DEVELOPED COPPER DEFICIENCY

Condition and duration of neutropenia	n	Hemoglobin (g/100 ml) Mean ±SD	Range	Serum iron (µg/100 ml) Mean ±SD	Range	TIBC (µg/100 ml) Mean ±SD	Ran
Deficient on admission	4	8.7 ± 0.3	8.3–8.9	34 ± 9	25–46	177 ± 73	90–
Hypocupremia only	5	10.1 ± 1.5	8.3–11.7	78 ± 47	20–135	181 ± 123	30–
Neutropenia <30 days	11	9.6 ± 0.8	8.3–10.7	65 ± 40	21–130	123 ± 62	30–
Neutropenia 31–148 days	12	10.0 ± 1.3	7.8–12.5	50 ± 24[a]	18–115	170 ± 93	54–
Following copper therapy (7–247 days)	12	9.8 ± 1.0	7.4–11.0	77 ± 25[a]	35–119	189 ± 75	105–

[a] These two groups were significantly different: $t = 2.80$, $p < 0.02$.

From Graham, G. G. and Cardano, A., in *Trace Elements in Human Health and Disease,* Prasad, A. S., Ed., Academic P New York, in press. With permission.

copper deficient on admission had hemoglobin and serum iron (less than 50 mg/100 ml) levels suggestive of iron deficiency. The TIBC was often strikingly low. In five infants with hypocupremia and a low to undetectable ceruloplasmin level but no neutropenia, the mean serum iron was in the normal range, with only one low value, and anemia was relatively mild. When neutropenia had been present for up to 30 days, mean serum iron was lower, but not significantly so, and hemoglobin levels were not different. When neutropenia was of considerably longer duration, the mean serum iron was significantly lower, with most values consistent with iron deficiency.

After copper was given, serum iron levels returned to normal, except for a few which were in the deficient range. The differences in the results were due to the variation of status of iron stores at the time copper deficiency developed or when supplementary copper was administered.

In four copper-deficient infants, the response of serum copper and iron levels to a single intravenous infusion of cupric acetate at 50 μg Cu per kilogram body weight was evaluated. While there was little or no effect on serum copper, a positive response in the serum iron of three out of four cases was suggested or clearly evident.[67] Holtzman et al.[69] found that this response was preceded by increase of ferroxidase activity, followed by a reticulocyte response. Neutrophil response was also noted on the third day after copper infusion.

Copper In Total Parenteral Nutrition

It is clearly evident that intravenous solutions utilized for total parenteral alimentation are deficient in copper and should be supplemented. While it takes only a few weeks to notice the decreased serum copper levels,[38,39] it may take several months before some patients present the characteristic signs of copper deficiency including leukopenia, neutropenia, intractable anemia, and development of bone lesions.[24,34-37] Little information is available concerning the Cu requirements of adult patients or small prematures while on total parenteral nutrition.

Green[70] states that "what appears to be nutritionally unimportant or of minor importance in the older child or adult may be extremely significant in an infant who may increase his body mass by 50 to 75% during several weeks of total intravenous nutrition."

An accepted procedure to supply the need for essential fatty acids and trace minerals, including copper, has been to infuse 10 to 20 ml of plasma twice weekly.[30,31] However, according to Shaw,[71] supplementing the infant on total intravenous feeds with trace metals by a twice weekly infusion of 20 ml plasma was totally inadequate as it supplied only 6 μg per/kg/day.

The recommendation for copper in formulas for full-term infants is 60 μg/100 kcal or about 72 μg/kg/day (Committee on Nutrition: American Academy of Pediatrics)[72] and is probably higher (90 μg/100 kcal), or about 110 μ/kg/day, for prematures. Using an estimated 33% absorption as suggested by Shils[73] would imply a requirement of about 20 μg copper per kilogram per day for full-term infants on TPN and about 35 μg/kg/day for prematures on TPN.

Franklin et al.[74] stated that their current parenteral alimentation routine is to give small premature infants approximately 150 ml/kg of a solution to which Cu has been added in the amout of 400 μg/l. This will deliver about 60 μg/kg/day. Older children receive first a liter of fluid each day, which contains the 400 μg of Cu; if an additional liter is needed, no Cu is added.

Others[75] administer $CuSo_4$ per os whenever possible to infants or children who receive TPN. Greene[76] is currently attempting to determine the exact requirements of copper given intravenously in prematures and infants.

A number of trace elements are currently being studied in adults on long-term TPN. Jeejeeboy[77] maintains normal levels of copper in patients on TPN at home for up to 5 years of observation by giving copper in the amount of 1.6 mg copper per day.

Shils[73] states that in adults one third of the daily requirement of 2 mg daily should be sufficient when given intravenously. He proposes between 0.5 to 1 mg/day, and according to his experience 0.4 mg/day has been sufficient to keep normal copper and ceruloplasmin serum levels.

It is hoped that the various studies on the use of copper in TPN will soon result in agreement on supplying copper in intravenous solutions.

REMARKS

Copper is an element that, up to now, has not been included in the recommended dietary allowance. With the recent recommendation for zinc, it is possible that many foods will be enriched with this element, and there is concern about the possible change in the ratio of intake of zinc to copper. Recently, Klevay[78] postulated that a metabolic imbalance in regard to zinc and copper is a major factor in the etiology of coronary heart disease. He added that this imbalance is either a relative or an absolute deficiency of copper characterized by a high ratio of zinc to copper. If recommended dietary allowances are established for other elements, it will be essential that these be made in terms of given amounts per unit of energy for any new nontraditional product that may become an important source of calories, in order to avoid alterations that may interfere with the interactions of the elements.

Based on the recommendations of the Committee on Nutrition of the American Academy of Pediatrics recommendations,[72] the FDA in December 1971 published regulations which set the minimum level of copper in infant formula at 60 μg/100 kcal. Shaw[71] has compiled figures from the literature showing that the rate of accumulation of Cu between 24 and 36 weeks of gestation is from 78 to 92 μg/kg/day. There are various estimates of the daily requirement for copper in the diet of infants. Wilson and Lahey[16] estimated that it may be as low as 15 μg/kg/day for prematures; Bush[10] suggested 80 μg/kg/day, and Cartwright and Wintrobe[2] recommend between 50 and 100 μg/kg/day or less for infants. Ashkenazi et al.[24] recommended supplemental copper therapy in the amount of 100 to 500 μg/day for several months to all small premature infants likely to be born with insufficient copper stores. Alexander et al.[79] estimated that the infant's requirement is between 200 to 500 μg/day. Cordano and Graham[19] suggested that the copper requirement of rapidly growing infants with poor stores was between 42 and 135 μg/kg/day.

Most manufacturers of infant formulas now include copper at about 60 μg/100 kcal. However, this level is probably not adequate for small premature infants. Formulas for these infants should contain no less than 90 μg/100 kcal (0.6 mg/l); this will provide the estimated requirement (100 μg/kg/day) when fed at the usual 110 to 120 kcal/kg/day.[80]

Breast milk is highly variable in its copper content, ranging from 90 to 603 μg/l.[62,81-83] According to calculations of Picciano and Guthrie,[82] fully breast-fed infants under 3 months of age receive approximately 60 μg Cu per kilogram per day. The advantage of breast milk is that copper is probably absorbed more efficiently than from cow's milk because of a better zinc:copper ratio.[60] The zink:copper ratio in human milk is about 4:6, but that in cow's milk may be as high as 38,[83] and a high zinc:copper ratio is known to depress the absorption of copper.[52] Copper as well as other minerals in human milk may be inadequate for growing premature infants. It is not suggested to abandon such feedings, but chemical analysis should be done in order to know what is being fed.

The observations by Graham and Cordano,[65] although fragmentary and largely retrospective, support the concept of two effects of copper deficiency on iron metabolism in man. The first, which occurs early, is an adverse effect of copper deficiency on iron absorption. It is suggested that the impairment of iron metabolism may be due to the loss of ferroxidase activity, not necessarily ceruloplasmin. The second and later effect of copper deficiency on iron metabolism is inadequate erythropoiesis, even in the presence of abundant iron stores. In this condition, anemia responds promptly to copper supplementation alone.

With the recent discovery by Goka et al.[56] of a genetic marker (increase copper concentration) in the cultured skin fibroblasts from patients with KHS, a unique tool has been provided for the investigation of the fundamentals of Menkes' disease, as well as a better understanding of normal copper metabolism. Continuation of studies in the mottled mice will also contribute to this.

It seems that the field of trace elements and their metabolism is on the verge of a period of explosive growth. Hopefully, the application of this knowledge will eliminate trace mineral deficiency as a clinical significant condition.

REFERENCES

1. Cartwright, G. E. and Wintrobe, M. M., Copper metabolism in normal subjects, *Am. J. Clin. Nutr.*, 14, 224–232, 1964.
2. Cartwright, G. E. and Wintrobe, M. M., The question of copper deficiency in man, *Am. J. Clin. Nutr.*, 15, 94–110, 1964.
3. Dowdy, R. P., Copper metabolism, *Am. J. Clin. Nutr.*, 22, 887–892, 1969.
4. Scheinberg, I. H. and Sternlieb, I., *Metabolism of Trace Metals. Diseases of Metabolism,* 6th ed., W. B. Saunders, Philadelphia, 1969.
5. Evans, G. W., Copper homeostasis in the mammalian system, *Physiol. Rev.*, 53, 535–562, 1973.
6. Reinhold, J. G., Trace elements – a selective survey, *Clin. Chem.*, 21, 485–487, 1975.
7. Alexander, F. W., Copper metabolism in children, *Arch. Dis. Child.*, 49, 589–590, 1974.
8. Burch, R. E., Hahn, H. K. J., and Sullivan, J. F., Newer aspects of the roles of zinc, manganese, and copper in human nutrition, *Clin. Chem.*, 21, 501–517, 1975.
9. Josephs, H. W., Treatment of anemia in infants with iron and copper, *Bull. Johns Hopkins Hosp.*, 49, 246, 1931.
10. Bush, J. A., The role of trace elements in hemopoiesis and in the therapy of anemia, *Pediatrics,* 17, 586–595, 1956.
11. Lahey, M. E. and Schubert, W. K., New deficiency syndrome occurring in infancy, *Am. J. Dis. Child.*, 93, 31, 1957.
12. Sturgeon, P. and Brubaker, C., Copper deficiency in infants; a syndrome characterized by hypocupremia, iron deficiency anemia, and hypoproteinemia, *Am. J. Dis. Child.*, 92, 254–265, 1956.
13. Ulstrom, R. A., Smith, N. J., and Heimlich, E. M., Transient dyspotreinemia in infants, a new syndrome; clinical studies, *Am. J. Dis. Child.*, 92, 219–265, 1956.
14. Zipursky, A., Dempsey, H., Markowitz, H., Cartwright, G. E., and Wintrobe, M. M., Studies on copper metabolism. XXIV. Hypocupremia in infancy, *Am. J. Dis. Child.*, 96, 148–158, 1958.
15. Schubert, K. W. and Lahey, M. E., Copper and protein depletion complicating hypoferremia in infancy, *Pedatrics,* 24, 710–733, 1959.
16. Wilson, J. F. and Lahey, M. E., Failure to induce dietary deficiency of copper in premature infants, *Pediatrics,* 25, 40–49, 1960.
17. Cordano, A., Baertl, J. M., and Graham, G. G., Copper deficiency in infancy, *Pediatrics,* 34, 324–336, 1964.
18. Lahey, M. E., Gubler, C. J., Chase, M. S., Cartwright, G. E., and Wintrobe, M. M., Studies on copper metabolism. II. Hematologic manifestations of copper deficiency in swine, *Blood,* 7, 1053–1074, 1952.
19. Cordano, A. and Graham, G. G., Copper deficiency complicating severe chronic intestinal malabsorption, *Pediatrics,* 38, 596–604, 1966.
20. Cordano, A., How important are trace elements in diet?, *JAMA,* p. 206: September 30, 1968.
21. Griscom, N. T., Craig, J. N., and Neuhauser, E. B. D., Systemic bone disease developing in small premature infants, *Pediatrics,* 48, 883–895, 1971.
22. Al Rashid, R. A. and Spangler, J., Neonatal copper deficiency, *N. Engl. J. Med.*, 285, 841–843, 1971.
23. Seely, J. R., Humphrey, G. B., and Matter, B. J., Copper deficiency in a premature infant fed an iron-fortified formula, *N. Engl. J. Med.*, 286, 109, 1972.

24. Ashkenazi, A., Levin, S., Djaldetti, M., Fishel, E., and Benvenisti, D., The syndrome of neonatal copper deficiency, *Pediatrics,* 52, 525–533, 1973.
25. Dudrick, S. J., Wilmore, D. W., and Vars, H. M., Long-term total parenteral nutrition with growth in puppies and positive nitrogen balance in patients, *Surg. Forum,* 18, 356–357, 1967.
26. Dudrick, S. J., Wilmore, D. W., Vars, H. M., and Rhoads, J. E., Can intravenous feeding as the sole means of nutrition support growth in the child and restore weight loss in the adult? An affirmative answer, *Ann. Surg.,* 169, 974–984, 1969.
27. Dudrick, S. J., Wilmore, D. W., Vars, H. M., and Rhoades, J. E., Long-term total parenteral nutriton with growth, development, and positive nitrogen balance, *Surgery,* 64, 134–142, 1968.
28. Wilmore, D. W., Groff, D. B., Bishop, H. C., and Dudrick, S. J., Total parenteral nutrition in infants with catastrophic gastrointestinal anomalies, *J. Pediatr. Surg.,* 4, 181–189, 1969.
29. Wilmore, D. W. and Dudrick, S. J., Growth and development of an infant receiving all nutrients exclusively by vein, *JAMA,* 203, 860–864, 1968.
30. Filler, R. M., Eraklis, A. J., Rubin, V. G., and Das, J. B., Long-term total parenteral nutrition in infants, *N. Engl. J. Med.,* 281, 589–594, 1969.
31. Lloyd-Still, J. D., Schwachman, H., and Filler, R. M., Intravenous hyperalimentation in pediatrics, *Am. J. Digest. Dis.,* 17, 1043–1052, 1972.
32. James, B. E. and MacMahon, R. A., Trace elements in intravenous fluids, *Med. J. Aust.,* 2, 1161–1163, 1970.
33. Greene, H. L., Hambidge, M., and Herman, Y. F., Trace elements and vitamins, *Adv. Exp. Med. Biol.,* 46, 131–145, 1974.
34. Karpel, J. T. and Peden, V., Copper deficiency in long-term parenteral nutrition, *J. Pediatr.,* 80, 32–36, 1972.
35. Dunlap, W. M., James, J. C., and Hume, D. M., Anemia and neutropenia caused by copper deficiency, *Ann. Intern. Med.,* 80, 470–476, 1974.
36. Vilter, R. W., Bozian, R. C., and Hess, E. V., Manifestations of copper deficiency in a patient with systemic sclerosis on intravenous hyperalimentation, *N. Engl. J. Med.,* 291, 188–191, 1974.
37. Fleming, C. R., Hodges, R. E., Smith, L. M., and Hurley, L. S., Abstr. 4329, in *10th Int. Congr. of Nutrition,* Science Council of Japan, Kyoto, 1975.
38. Solomons, N. W., Layden, T. S., Rosenberg, I. H., Khactu, K. V., and Sandstead, H. H., Plasma trace metals during total parenteral alimentation, *Gastroenterology,* 70, 1022–1025, 1976.
39. Fleming, C. R., Hodges, R. E., and Hurley, L. S., A prospective study of serum copper and zinc levels in patient receiving total parenteral nutrition, *Am. J. Clin. Nutr.,* 29, 70–77, 1976.
40. Menkes, J. H., Alter, A., Steigleder, G. K., Weakley, D. R., and Sung, J. H., A sex linked recessive disorder with retardation of growth, peculiar hair, and focal cerebral and cerebeller degeneration, *Pediatrics,* 29, 764–779, 1962.
41. Danks, D. M., Campbell, P. E., Stevens, B. J., Mayne, V., and Cartwright, G. E., Menkes' kinky hair syndrome. An inherited defect in copper absorption with widespread effects, *Pediatrics,* 50, 188–201, 1972.
42. Bucknall, W. E., Haslam, R. H. A., and Holtzman, N. A., Kinky hair syndrome: response to copper therapy, *Pediatrics,* 52, 653–657, 1973.
43. Wesenberg, R. L., Gwinn, J. L., and Barnes, G. R., Radiologic findings in kinky-hair syndrome, *Radiology,* 92, 500–506, 1969.
44. Pedroni, E., Bianchi, E., Ugazio, A. G., and Burgio, G. R., Letter: immunodeficiency and steely hair, *Lancet,* 1, 1303–1304, 1975.
45. French, J. H., Sherard, E. S., Lubell, H., Brotz, M., and Moore, C. L., Trichopoliodistrophy. I. Report of a case and biochemical studies, *Arch. Neurol.,* 26, 229–244, 1972.
46. Danks, D. M., Cartwright, G. E., and Stevens, B. J., Menkes' kinky hair disease. Further definition of the defect in copper transport, *Science,* 179, 1140–1142, 1973.
47. Danks, D. M., Steely-hair mottled mice and copper metabolism, *N. Engl. J. Med.,* 293, 1147–1148, 1975.
48. Hunt, D. M., Primary defect in copper transport underlies mottled mutant in the mouse, *Nature,* 249, 852–854, 1974.
49. Danks, D. M., Stevens, B. J., Campbell, P. E., Gillespie, J. M., Walker-Smith, J. A., Blomfield, J., and Turner, B., Menkes' kinky hair syndrome, *Lancet,* 1, 1100–1103, 1972.
50. Debakan, A., Aamodt, R., Rumble, W. F., Johnston, G. E., and O'Reilly, S., Kinky-hair disease: study of copper metabolism with use of ^{67}Copper, *Arch. Neurol.,* 32, 672–675, 1975.
51. Anon., Copper and steely hair, *Lancet,* 1, 902–903, 1975.
52. Grover, W. D. and Scrutton, M. C., Copper infusion therapy in trichopoliodystrophy, *J. Pediatr.,* 86, 216–220, 1975.

53. Horn, N., Mikkelsen, M., Heydorn, K., Damsgaard, E., and Tygstrup, I., Letter: copper and steely hair, *Lancet,* 1, 1236, 1975.

54. Lott, I., DiPaolo, R., Schwartz, D., Janoswska, S., and Kanfer, J. N., Copper metabolism in the steely hair syndrome, *N. Engl. J. Med.,* 292, 197–199, 1975.

55. Holtzman, N. A., *Fed. Am. Soc. Exp. Biol.,* in press.

56. Goka, T. J., Stevenson, R. E., Hefferan, P. M., and Howell, R. D., Menkes' disease: a biochemical abnormality in cultured human fibroblast, *Proc. Natl. Acad. Sci. U.S.A.,* 73, 604–606, 1976.

57. Henkin, R. I., *Newer Trace Elements in Nutrition,* Mertz, W. and Cornatzer, W. F., Eds., Marcel Dekker, New York, 1971, 255–312.

58. Fazekas, I. G., Regei, B., and Romhanyi, I., Copper content of fetal organs, *Kiserl. Orvostud.,* 15, 230, 1963.

59. Tipton, I. H., Cook, M. J., Steiner, C. A., Cook, M. J., Steiner, R. L., Boye, C. A., Perry, H. M., Jr., and Schroeder, H. A., Trace elements in human tissue, *Health Phys.,* 9, 89–101, 1963.

60. Nusbaum, R. E., Alexander, G. V., Butt, E. M., Gilmour, T. C., and Didio, S. L., Some spectrographic studies of trace elements storage in human tissues, *Soil Sci.,* 85, 95–99, 1958.

61. Briickmann, G. and Zondek, S. G., Iron, copper, manganese in human organs at various ages, *Biochem. J.,* 33, 1845–1857, 1939.

62. Widdowson, E. M., Dauncey, J., and Shaw, J. C. L., Trace element in foetal and early postnatal development, *Proc. Nutr. Soc.,* 33, 275–284, 1974.

63. Cavell, P. A. and Widdowson, E. M., Intakes and excretions of iron, copper and manganese in the neonatal period, *Arch. Dis. Child.,* 39, 496–501, 1964.

64. Widdowson, E. M., *Trace Elements and Early Development,* Kasek, J., Osancova, K., and Cuthberson, D. P., Eds., Excerpta Medica, Amsterdam, 1970, 179–181.

65. Graham, G. G. and Cordano, A., Copper depletion and deficiency in the malnourished infant, *Johns Hopkins Med. J.,* 124, 139–150, 1969.

66. Holtzman, N. A., Charache, P., Cordano, A., and Graham, G. G., Distribution of serum copper in copper deficiency, *Johns Hopkins Med. J.,* 126, 34–42, 1970.

67. Graham, G. G. and Cordano, A., *Trace Elements in Human Health and Disease,* Prasad, A. S., Ed., Academic Press, New York, in press.

68. Underwood, E. J., Ed., *Trace Elements in Human and Animal Nutrition,* 3rd ed., Academic Press, New York, 1971, 57–115.

69. Holtzman, N. A., Graham, G. G., and Bucknall, W. E., Role of copper and copper dependent ferroxidases in hematopoesis, in *Proc. 3rd IIIEME Symp. Int. Maladie de Wilson,* Paris, 1973.

70. Greene, H. L., *Intravenous Nutrition in the High Risk Infant,* Winters, R. W. and Hasselmeyer, E. G., Eds., John Wiley & Sons, New York, 1975, 237–284.

71. Shaw, J. C. L., Parenteral nutrition in the management of sick and low birthweight infants, *Pediatr. Clin. North Am.,* 20, 333–358, 1973.

72. Committee on Nutrition, Proposed changes in FDA regulations concerning formula products and vitamin-mineral dietary supplements for infants, *Pediatrics,* 40, 916, 1967.

73. Shils, M. E., *Modern Nutrition in Health and Disease: Total Parenteral Nutrition,* 5th ed., Goodhart, R. S. and Shils, M. E., Eds., Lea & Febiger, Philadelphia, 1973, 966–980.

74. Franklin, F., Filler, R., and Watkins, J., personal communications.

75. Nichols, B., personal communication.

76. Greene, H. L., personal communication.

77. Jeejeeboy, K. M., personal communication.

78. Klevay, L. M., Coronary heart disease: the zinc/copper hypothesis, *Am. J. Clin. Nutr.,* 28, 764–774, 1975.

79. Alexander, F. W., Clayton, B. E., and Delves, H. T., Mineral and trace metal balances on children receiving normal and synthetic diets, *J. Med.,* 43, 89–111, 1974.

80. Cordano, A., Copper requirements and actual recommendations per 100 kilocalories of infant formula, *Pediatrics,* 54, 524, 1974.

81. Murthy, G. K. and Rhea, U. S., Cadmium, copper, iron, lead, manganese, and zinc in evaporated milk, infant products and human milk, *Dairy Sci.,* 54, 1001–1005, 1971.

82. Picciano, M. F. and Guthrie, H. A., Copper, iron, and zinc contents of mature human milk, *Am. J. Clin. Nutr.,* 29, 242–254, 1976.

83. Schroeder, H. A. and Kraemer, L. A., Cardiovascular mortality, municipal water and corrosion, *Arch. Environ. Health,* 28, 303–311, 1974.

IODINE DEFICIENCY IN MAN

John T. Dunn

The major consequences of iodine deficiency in man are endemic goiter and cretinism. The World Health Organization estimated in 1960 that perhaps 200 million people were affected by endemic goiter, making it a major obstacle in the road to good health in many parts of the world. Despite the availability for more than 50 years of simple and effective means for correcting goiter, its incidence continues high.

This review begins with a brief summary of iodine metabolism, followed by data on iodine requirements for man, and then by a survey of the clinical features of endemic goiter and its complications, particularly cretinism. It closes with recommendations for its treatment. Several book-length publications[1-6] summarize the bulk of available information on endemic goiter and its therapy. They are quoted frequently in this review and should be consulted for further details.

UTILIZATION OF IODINE BY HUMANS

The only proven importance of iodine in man is as a constituent of the thyroid hormones. Details of its metabolism are available in several texts.[7-9] The major steps are discussed in the following paragraphs.

Absorption and Concentration of Iodide

Ingested iodine-containing material is absorbed as iodide into the circulation by active transport across the gut mucosa. Once in the blood, iodide is actively concentrated by several tissues, of which the thyroid is the most important. Others include the salivary glands, placenta, breast, choroid plexus, and intestinal mucosa, in addition to the gastric mucosa.[10] Iodide concentration by the thyroid takes place against an electrochemical gradient and requires energy. It may also involve a sodium-potassium-ATPase and complexing of iodide with basal membrane phospholipid receptors, but the chemical steps are poorly understood. Other tissues probably concentrate by the same mechanism, since defective concentration by the thyroid in some familial goiters has been shared by other tissues.[11] Several anions can compete with iodide concentration by the thyroid including thiocyanate, perchlorate, bromide, pertechnetate, and perrhenate, and these are of occasional importance in the natural history of endemic goiter and in the diagnosis and therapy of thyroid diseases.

Thyroidal Hormone Synthesis

The thyroid produces thyroglobulin, a large glycoprotein of molecular weight 660,000 which serves as a matrix for hormone synthesis and as a vehicle for subsequent storage in the thyroid follicular lumen. Iodide concentrated by the thyroid is oxidized by a peroxidase to an intermediate, perhaps I_2, and attached to tyrosyl residues within the thyroglobulin molecule to form the hormone precursors, MIT* and DIT. Next, some of these iodotryosines interact to form the two hormones T_4 (from two DIT's) and T_3 (from DIT + MIT). Under conditions of normal iodine intake, the average completed thyroglobulin molecule contains five to ten residues each of DIT and MIT, one to three of T_4 and less than one of T_3. Thus only about one third of the iodine of thyroglobulin is in the form of hormone. The finished molecule is stored in the follicular lumen and represents the so-called colloid. Thyroglobulin shows considerable heterogeneity in

---

* The following abbreviations are used: MIT, monoiodotyrosine; DIT, diiodotyrosine; T_3, 3,5,3'-triiodothyronine; T_4, thyroxine; TSH, thyrotropin; TRH, thyrotropin releasing hormone.

composition of iodine, carbohydrates, and amino acids, and probably in subunit structure as well.[12-14] All of these may be important in adapting to iodine deficiency, as discussed below.

Hormone Release and Circulation

To provide hormone for peripheral tissues, colloid is resorbed by membranous pseudopods from surrounding thyroid cells. Thyroglobulin is digested by proteases and the hormones T_4 and T_3 are released into the circulation. DIT and MIT are not normally released, but instead are deiodinated and their iodine retained within the thyroid for recycling. This important mechanism of iodine conservation depends on a deiodinase, and absence of this enzyme in certain familial goiters leads to iodine deficiency.[11]

The thyroid hormones are bound principally to two serum proteins, thyroxine-binding globulin and thyroxine-binding prealbumin, both synthesized by the liver. Serum halftimes are approximately 6 days for T_4 and 1 to 2 days for T_3.[15] There is an appreciable conversion of T_4 to T_3 in the circulation, probably involving 50% or more of T_4's hormonal effect. This conversion may be decreased in ill and malnourished subjects[16] and the inactive isomer "reverse T_3" $(3,3'\text{-}5'\text{-triiodothyronine})$ formed instead.[17]

Excretion

Thyroid hormones are metabolized principally in the liver, although iodinases are present in other tissues including kidney and muscle. About 90% of excreted iodine appears in the urine[18] chiefly as iodide. Most of the remaining 10% is in the feces. T_4 is conjugated, mostly as the glucuronide, by the liver and secreted into the bowel.[19] Absorption in this form is poor, but is facilitated by hydrolysis in the lower intestine. In rats, perhaps half of biliary T_4 is absorbed. In humans, many factors may affect absorption of oral T_4, and probably its reabsorption from bowel. The variability of absorption does not often lead to clinical problems, but might occasionally be important in areas of endemic goiter, where intestinal parasites, diarrheal diseases, and poor nutrition are frequent additional medical problems. Heavy sweating can result in significant iodine removal[18] and this might also be of occasional importance when coupled with iodine deficiency.

Controlling Factors

TSH enhances thyroid hormone production in a number of ways.[20] It increases iodide concentration, iodination of tyrosyls, conversion of iodotyrosyls to hormone, and resorption of colloid. It also favors the synthesis of T_3 relative to that of T_4 and alters the composition of thyroglobulin.[21] TSH secretion is increased in response to threatened hypothyroidism. Thus, conditions leading to decreased levels of circulating thyroid hormone, such as iodine deficiency, will prompt TSH release and its attendant effects on the thyroid. In addition, iodine deficiency promotes a greatly increased synthesis of T_3 relative to that of T_4. This is presumably because iodine lack leads to more molecules of MIT precursor than of DIT. It is a useful response, since T_3 has several times the hormonal effect of a similar amount of T_4 and uses only three quarters as much iodine.

Cellular Actions of Thyroid Hormone

A unifying single mechanism of action for the thyroid hormones has not yet been found and, indeed, may not exist. In experimental studies, thyroxine has been shown to (1) increase levels of a number of enzymes, particularly those involved with cell respiration, (2) stimulate mitochondrial respiration, (3) increase synthesis of ribosomal RNA, and (4) promote protein synthesis in general. There is major controversy over which of these, if any, is the primary effect of thyroid hormone. Exhaustive reviews are available and summarize current progress in this difficult field.[22,23]

Clinical Effects of Thyroid Hormone

These are directed chiefly at metabolic rate and growth. Adult humans who become hypothyroid will typically show torpor, bradycardia, hypothermia, delayed reflex relaxation, and dry rough skin on routine examination. More detailed investigation will show a decline in metabolic rate and a general decrease in the rate of many biochemical processes. Hypothyroid children will show, in addition, retarded mental and somatic growth. If the hypothyroidism was present during the first 2 years of life, the mental retardation is likely to be irreversible.[24]

DIETARY SOURCES OF IODINE AND OPTIMAL LEVELS

The most important natural source of iodine for man is ingested food. The iodine content of water is a useful index of the overall iodine supply for particular geographical areas, but water itself does not provide much iodine. Air has approximately 10% the iodine content of nearby water, and iodine inhalation of clinical significance occurs only with air pollution and industrial exposures.[25] The iodine content of foods varies with their source. Plants or livestock raised on iodine-poor land are low in iodine content, while marine fish and plants reflect the high iodine content of sea water.

Table 1 gives the iodine contents of certain foods reported in three studies. The data from Greece of Koutras et al.[26] are representative of a country with a lower total iodine intake than the U.S. and a greater dependence on natural foods. In Athens, poultry, eggs, and seafood were the most important sources of iodine, while bread contributed little. Thessalia, an area of endemic goiter, showed a much lower content of iodine in all food categories, and seafood was not generally available.

The data from Kidd et al.[27] were collected in 1972 from selected areas of four states (Michigan, Kentucky, Texas, and Georgia) as part of the Ten-State Nutrition Study. The ranges of iodine content among samples of the same type were quite wide, reflecting considerable variations in food processing which would not usually be apparent to the consumer. There are obvious risks in comparing these data with those from Maryland in 1964 (Table 1)[28] since differences in methodology and reporting inevitably are present. With that caveat in mind, it appears that meat and vegetables had more iodine in the earlier study, while milk and eggs had less. Most striking are the huge increases in the iodine content of bread, which parallels the introduction of iodate as a dough conditioner in the 1960's.[29] Table 2, also from the study of Kidd et al.[27] relates urinary iodine levels to diet. It shows that the types and amounts of bread and milk consumed are more closely correlated with body iodine levels than is the use of iodized salt. From these data, we might estimate that iodized salt contributed perhaps 35 μg/day to the diet. This is enough to correct most iodine deficiency but represents a fairly small part of the total iodine intake, in striking contrast to its importance in areas with endemic goiter.

In addition to iodized salt and iodate in bread, other artificial sources of iodine include food coloring,[30] agricultural disinfectants, medicines, and radioopaque dyes. Use of these is fairly infrequent in most areas of the world in which iodine deficiency is found, but they may lead to gross iodine excess and iodine goiter in more developed communities. A combination of these and other exogenous sources has increased the urinary iodine excretion in the United States to a median of approximately 250 μg/g creatinine, five times the minimum for goiter prevention.[31]

An iodine intake of 150 to 300 μg/day has been accepted as optimal by the Food and Nutrition Board of the National Research Council[32] and by the Fourth PAHO Technical Group on Endemic Goiter.[33] A nearly identical recommendation (100 to 300 μg) has been made by the 1970 Conference of the National Academy of Sciences for optimal intake, with "safe" levels set between 50 μg and 1000 μg.[34] Factors supporting the minimal levels include balance studies showing an obligatory daily iodine excretion of 57

μg,[18] the absence of endemic goiter (or at least that correctable by iodine administration) when ingestion exceeded approximately 50 μg/day, and analysis of iodine uptake by the thyroid and calculation of the amounts of hormonal iodine metabolized daily in euthyroidism. Maximal levels are less certain, but there is widespread concern that large amounts of iodine may lead to goiter[35] and hyperthryroidism[36] and are perhaps related to clinical manifestations of Hashimoto's thyroiditis and Graves' disease.[37]

Table 1
IODINE CONTENT OF FOOD SAMPLES

μg I/kg

	Greece[a,d]		U.S., NIH (1964)[b,e]	U.S., 4-State (1972)[c,e]	μg Iodine average serving
	Athens	Thessalia (goiter endemia)			
1. Bread	16	5	105		
Continuous mix[f]				3800	190
Conventional mix[f]				1400	70
2. Milk	41[g]	25[g]	139	210	52
3. Seafood	288		540		
Cod				1020	102
Shrimp				360	27
Perch				210	14
Canned tuna				180	12
4. Meat	26	12	175		
Ground beef				90	9
5. Poultry	523	99			
6. Eggs	268	38	145	360	18
7. Vegetables	10	7	280		
Spinach, frozen				60	6
Broccoli, fresh				50	5
Potatoes				50	5
Beans, dried				70	7
8. Fruits			18		
9. Drinking water	5[g]	2[g]			

[a] Calculated from data of Koutras et al.[26]
[b] Data of Vought and London[28]
[c] Data of Kidd et al.[27]
[d] Mean values
[e] Median values
[f] Both processes use iodate as dough conditioner
[g] μgI/l

Table 2
RELATING URINARY IODINE TO
DIETARY HISTORY

Dietary history	Urinary iodine $\mu g/g$ creatinine $\pm$ SD
Iodate in bread[a]	
Continuous mix	503 ± 226
Conventional mix	453 ± 185
Neither	372 ± 215
Milk consumption[a]	
High	501 ± 208
Moderate	460 ± 235
Low	329 ± 202
Salt[b]	
Iodized	470 ± 219
Noniodized	434 ± 231

[a] Difference among three categories significant at $p <$ 0.005 by analysis of variance
[b] Difference between two categories significant at $p <$ 0.05 by t test

From Kidd, P. S., Trowbridge, F. L., Goldsby, J. B., and Nichaman, M. Z., *J. Am. Diet. Assoc.*, 65, 420–422, 1974. With permission.

EFFECTS OF IODINE DEFICIENCY

The major consequence of iodine deficiency is the threat of hypothyroidism. If compensatory mechanisms, particularly TSH secretion, are adequate, euthyroidism can be maintained, although there will be associated thyroidal enlargement. This is the clinical picture of *endemic goiter*. In areas of severe iodine deficiency, there will be, in addition, a small fraction of subjects (up to 10% of the total population) with gross mental retardation and other developmental anomalies. This condition is called *endemic cretinism*. A discussion of these and possible intermediate conditions follows.

Endemic Goiter
Definition and Clinical Description

Goiter has been defined by Perez et al.[38] as "a thyroid gland whose lateral lobes have a volume greater than the terminal phalanx of the thumb of the person being examined." For endemic goiter, the Fourth PAHO Technical Group recommended that "an area is arbitrarily defined as endemic with respect to goiter if more than 10% of its population is found to be goitrous on appropriate survey. The figure of 10% is chosen because a higher prevalence usually implicates an environmental factor, while a prevalence of several percent is common even when all known environmental factors are controlled."[39]

The term "goiter" means enlarged thyroid. For comparison, it implies an accepted size for the normal thyroid, but this has been difficult to establish. In the U.S., a thyroid weight of 20 g has been taken as normal for adults. Even in the absence of iodine deficiency, it varies considerably with the iodine content of the diet. Examples are a figure of 34 to 41 g in central Europe in 1910, 25 to 32 g in Colorado in 1955, 20 to 25 g in central Africa in 1960, and 12 to 14 g in Iceland.[40] This uncertainty poses practical problems for goiter surveys. For example, Perez et al.[38] defined the category of "persons without goiter" as including "persons whose thyroid glands are less than four to five times enlarged," in clear contradiction to their own definition of goiter. This is not merely a semantic quibble because areas harboring many thyroids of four to five times normal size are likely to be iodine deficient and appropriate surveys should expose this fact.

The most frequent finding on physical examination of iodine deficient subjects is goiter (Figures 1 and 2). The degree of enlargement may vary from slight to huge, up to a kilogram or more. During the subject's early years, the gland is diffusely and symmetrically enlarged, but palpable nodules become increasingly frequent with aging. The size of the gland and the degree of nodularity correlate with the severity of iodine deficiency.

Pathogenesis[40]

Marine originally proposed a repetitive sequence of hyperplasia and involution to explain the goiter of iodine deficiency. Initially, there is epithelial hyperplasia, usually generalized, with increased vascularity, papillary infolding, and gradual disappearance of colloid from the thyroid follicles. Eventually, this hyperplastic stage halts. This is perhaps because it has succeeded in producing adequate hormones or there is now adequate iodine. This is followed by involution of the hyperplastic epithelial cells and accumulation of colloid. With recurrence of iodine deficiency, hyperplasia begins anew, but is likely to be more focal, involving only selected follicles. With repetition of this cycle of hyperplasia and involution, the gland becomes increasingly heterogeneous with wide variations in follicle size, colloid accumulation, and cellular activity. An early feature in this process is the development of thyroid nodules. These are markedly heterogeneous in histological appearance and functional activity, and probably represent focal areas of hyperplasia and involution. Severe and continued iodine deficiency may lead to thyroidal atrophy.

These changes are thought to be mediated by increased TSH secretion. Iodine deficiency leads to inadequate synthesis of thyroid hormones, whose decreased

FIGURE 1.   Five women of highland New Guinea with large goiters. (From Garruto, R. M., Gajdusek, D. C., and ten Brink, J., *Human Biology, 46,* 311–319, 1974. Reprinted through courtesy of authors and publishers).

FIGURE 2. Endemic goiter and cretinism in Bolivia. Mother, on left, is goitrous but otherwise normal. Daughter is goitrous, mentally retarded, and a deaf mute, but of normal stature and clinically euthyroid.

circulating levels are recognized by the pituitary. It responds with increased TSH secretion, leading to many changes in the thyroid, as already described in the section titled Utilization of Iodine by Humans. Their net effect is to improve the amount and potency of secreted thyroid hormone with the concomitant development of thyroidal hyperplasia. In the initial response to iodine deficiency, thyroidal enlargement is valuable because it consists of hyperplastic tissue producing more hormone. However, once maximum uptake of iodine has been obtained, further increase in number of thyroid cells does not improve hormone production, and the continued growth of the goiter has been regarded as an unwanted side effect of continued TSH secretion.[41,42]

This sequence of events is supported by studies in experimental animals.[41,43] It is probably operative in human endemic goiter as well, but proof is hampered by methodological limitations. Several studies[44-46] have shown generally increased circulating TSH levels in areas of endemic goiter when compared with nonendemic area, but could not correlate goiter size with TSH levels. In another study[47] goitrous patients from an area of mild iodine deficiency had both higher resting TSH levels than nongoitrous ones and more exaggerated responses to administered TRH (used as an index of pituitary TSH secretory capacity). These same patients had normal T_4 and T_3 levels, suggesting that euthyroidism was maintained by the increased TSH secretion. Their plasma TSH levels returned to near-normal after 6 months of iodide administration.

Laboratory Assessment

Urinary iodide is the simplest method for detecting iodine deficiency. Since 24-hr collections are impractical in surveys, a spot sample with analysis for both iodine and creatinine is usually employed.[48] Values of greater than 50 μg iodine per gram creatinine suggest adequate iodine intake.

The uptake of ^{125}I or ^{131}I by the thyroid is another sensitive measure of iodine deficiency. With moderate levels of dietary iodine (100 to 300 μg/day), from 20 to 50% of the administered isotope can be detected in the thyroid after 24 hr. These values were once accepted as normal for euthyroid patients in the U.S. but have shifted downward (to approximately 5 to 25%) with increasing dietary iodine.[29] In areas of iodine deficiency, the uptake will be much more rapid and approach a peak of 100%, reflecting the thyroid's greatly increased metabolic activity and its avidity for iodine.

Tests of circulating thyroid hormones in areas of endemic goiter have usually been within the normal range. However, when compared with subjects in iodine sufficient areas, the mean serum T_4 has been less (e.g., 6.5 μg/100 ml vs. 8.4 for normals in the U.S.[46]) and the mean serum T_3 level greater (e.g., 161 ng/100 ml vs. 126 for U.S. normals).[46] These values are supported by kinetic studies of iodine deficiency in man[49,50] and in experimental animals.[41] They lead to the conclusion that a shift to production of more T_3 and less T_4 is one of the adaptations to iodine deficiency. Conceptually, this shift is not surprising since a molecule of T_3 provides more hormone activity than one of T_4 with less expenditure of iodine. Also this shift has been identified as a feature of TSH stimulation[21,51] which appears increased in endemic goiter.

Geographical Distribution of Endemic Goiter

This was exhaustively surveyed in 1960 by Kelly and Snedden.[52] A more recent tabulation for Latin America has also been published.[53] Historically, the major areas have included high mountainous regions, notably the Alps, Himalayas, and Andes, but there have also been many other inland areas such as the Great Lakes region of the U.S. and the Amazon basin. The presence and severity of endemic goiter are closely associated with the degree of iodine deficiency in the soil. This, in turn, is attributed to loss of soil iodine by glaciation, flooding, or other erosion.[43]

Extrapolation from the 1960 survey suggested that 200 million persons worldwide had endemic goiter.[52] Several factors severely limit the accuracy of this estimate. For one, much of the survey data is unrepresentative and inaccurate. Examples are the use of the incidence in army recruits (young men always have a lower incidence of goiter than other groups), or in subjects from nonrural or coastal areas of a country (remote mountainous regions routinely have much more goiter). Another problem is the different techniques of survey. As already discussed, the WHO classification will record thyroids up to four to five times normal size as nongoitrous, thus giving an erroneously low estimate of goiter incidence. Finally, there have been changes in worldwide distribution since 1960. Endemic goiter has essentially disappeared from the U.S. and has been greatly reduced in Alpine countries. Progress has been made in prophylaxis programs in some Latin American and African countries, while in others, the total numbers of goitrous subjects have probably been augmented by an increasing birth rate in the face of continuing iodine deficiency.

Complications of Endemic Goiter

These include thyroid nodules, thyroid cancer, hypothyroidism, and hyperthyroidism. Cretinism is a more indirect complication and is discussed separately below.

The enormous nodular thyroids of some goiters compress the trachea, blood vessels, and surrounding neck tissues. Clinical manifestations may include stridor, respiratory occlusion, tracheomalacia, thoracic outlet syndromes, and upper limb palsies. In areas of endemic goiter, the high prevalence of thyroid nodules makes the early detection of thyroid cancer difficult and has been blamed for its higher mortality in such areas.[54]

Evidence regarding the incidence of thyroid cancer in goiter endemias is conflicting. In the Tyrol[54] there was no increase over that reported for goiter-free regions, but there were more sarcomatous and undifferentiated lesions with a higher mortality. In Colombia, the incidence was five times that of Sweden or New York City.[55] One survey[56] noted that some areas of endemic goiter (Cali and Israel) had high incidences of thyroid cancer, and some with low goiter incidence had low cancer rates, but there were also other areas with low goiter prevalence and a high thyroid cancer rate (Iceland and Hawaii). Thus epidemiological evidence does not clarify the relationship.

In making conclusions about the carcinogenesis of iodine deficiency, we should emphasize the following points: (1) Thyroid cancer is often a controversial histological diagnosis and its reported incidence will depend on the criteria used and the extent to which sections are taken for examination. In Minnesota, careful thyroid sectioning showed occult cancer in 5.7% of routine autopsies[57] and in Japan, similar studies have shown incidences of 14%[58] to 28%.[59] (2) Thyroid cancer grows slowly and has a long latent period. Most goiter endemias are in remote areas where many factors shorten longevity, medical care is suboptimal, and histological examination of thyroids is virtually nonexistent. All of these will underestimate the malignant potential of iodine deficiency. (3) There is a large body of evidence from experimental animals which shows that iodine deficiency, as well as many other manipulations leading to increased TSH secretion and goiter, will result in benign and malignant neoplastic lesions of the thyroid. With this background, it is reasonable to conclude that iodine deficiency and endemic goiter in humans may carry a substantial threat of carcinogenesis.

Hyperthyroidism can be a late complication of endemic goiter. It usually occurs in older patients with longstanding nodular goiters and follows an increase in iodine consumption. Presumably, a thyroid which has devoted much of its life to maximal utilization of iodine cannot slow down when more iodine becomes available, and overproduction of hormone results. Some increase in this type of hyperthyroidism, so-called Jod-Basedow, can be anticipated in the several years following effective correction of iodine deficiency.[36,54,60] The incidence is usually low and the manifestations mild.

Endemic Cretinism
Definition and Clinical Description

Endemic cretinism has been defined in many ways. The following was endorsed by the Fourth PAHO Technical Group:[39] "The condition of endemic cretinism is defined by three major features:

1. *Epidemiology.* It is associated with endemic goiter and severe iodine deficiency.

2. *Clinical manifestations.* These comprise mental deficiency, together with either: (a) A predominant neurological syndrome consisting of defects of hearing and speech, and with characteristic disorders of stance and gait of varying degree; or (b) Predominant hypothyroidism and stunted growth. Although in some regions one of the two types may predominate, in other areas, a mixture of the two syndromes will occur.

3. *Prevention.* In areas where adequate correction of iodine deficiency has been achieved, endemic cretinism has been prevented."

The *sine qua non* of cretinism is mental deficiency. This is readily apparent on cursory examination, and the family or community have usually recognized a delay in reaching developmental milestones early in infancy or childhood. It is useful for descriptive purposes to subdivide endemic cretinism into two clinical syndromes, the nervous and myxedematous types.[61] The former will show, in addition to mental retardation, variable degrees of nerve deafness and neurological defects, principally spastic diplegia of probably cerebral origin.[62,63] The myxedematous type shows dwarfism, epiphyseal dysgenesis, and clinical hypothyroidism. The syndrome of nervous cretinism has been seen in purest form in New Guinea[64] and that of hypothyroidism in central Africa.[65] In other areas, there has been a predominance of the nervous type but frequently with additional myxedematous features, including reduced stature and clinical hypothyroidism. The severe endemia of the Himalayas[66] shows most of the features of both syndromes. Several examples are shown in Figures 2 to 4.

Pathogenesis

This is largely unknown. It is clearly associated with severe iodine deficiency and disappears from a community with adequate iodization. It is generally believed to result from hypothyroidism at a critical period during the development of the central nervous system, but this sequence has been difficult to prove. Some inferences may be suggested from extensive studies in experimental animals.[67-69] Rats made hypothyroid at the time of birth show retarded differentiation and proliferation of cells and decreased protein synthesis. In the developing central nervous system, thyroid hormones may have their major impact at the level of mitochondrial protein synthesis. Since myelin proteins are probably synthesized by mitochondria and are among the major proteins affected by thyroid hormones, a defect in their synthesis may account for some of the features of neonatal hypothyroidism in rats. These appear independent of the changes seen in malnutrition without iodine deficiency.[67,68]

In humans, cretinism can usually be recognized within several months after birth, corresponding to the period of most active myelinization of the central nervous system. It was impressively demonstrated by Thilly et al.[70] that newborns of iodine deficient mothers had elevated serum TSH levels and low serum thyroxines, and both measures were similar to those of myxedematous subjects. Neonates from iodine-sufficient mothers had normal values. This study established that clinical hypothyroidism is widespread among neonates in areas of endemic cretinism. Also, it supports extrapolation of conclusions from animal data, that neonatal hypothyroidism is a major cause of endemic cretinism. More direct evidence is not available.

Neuroanatomical studies of cretins have been meager. The described findings are

FIGURE 3. Adult cretin in Bolivia. Severe mental retardation, deaf mutism, and short stature.

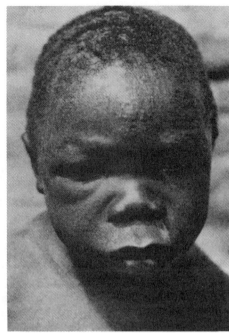

FIGURE 4.   Myxedematous cretin in Zaire, age 15 years. Note short stature, lumbar lordosis, broad flat nose, puffy facies, sparse hair, and dry skin. (From Delange, F., Endemic Goitre and Thyroid Function in Central Africa, S. Karger, Basel, 1974. With permission).

somewhat similar to those of experimental hypothyroidism in rats, but are not specific.[68] They have included nerve cell loss, small brain size, and delayed myelinization.

While it is likely that iodine deficiency produces cretinism by first producing hypothyroidism, no one has yet excluded the possibility that iodine may have developmental effects independent of the thyroid hormone. In some lower organisms without thyroids (e.g., jellyfish and algae), iodine is necessary for development, and in mammals, the choroid plexus has an active iodine concentrating mechanism. These several points keep open the possibility that iodine might have a direct role in maturation of the central nervous system, but this idea still lacks any experimental verification.

Pathogenesis of the deafness of cretinism is very poorly understood. The few pathological studies of the ear in endemic cretinism have been confusing.[71] The major abnormalities have been in the middle ear, and include hypertrophy and deformation of bone and thickening of the mucous membranes. Despite this, clinical testing has consistently shown the deafness to be perceptive rather than conductive. In developing chick embryos,[72] hypothyroidism induced by propylthiouracil was associated with specific morphological lesions in the spiral ganglion of the cochlea and in the sensory hair cells. These could be avoided by simultaneous administration of thyroxine. Such lesions have apparently not been looked for in cretins but are of interest in suggesting a relationship between developmental hypothyroidism and deafness.

Sporadic cretinism occurs in areas of iodine sufficiency and is usually traced to thyroid aplasia (athyreotic cretins) or to one of several biochemical defects in hormone synthesis or utilization.[11] The common result is inadequate hormone effect on peripheral tissues. Clinical features shared with endemic cretinism include mental and developmental retardation, short stature, hypothyroidism, and goiter. Spastic diplegia is unusual. Deafness occurs characteristically only in Pendred's syndrome (familial goiter, defective iodination, and nerve deafness), which usually is not associated with other features of cretinism, although there are well-documented exceptions.[73,74] These considerations suggest that many features of endemic cretinism may be reasonably attributable to neonatal hypothyroidism, but unidentified additional factors may contribute to the hearing and neurological defects.

The clinical features of myxedematous cretinism are easier to integrate than those of nervous cretinism, since all are typical of profound hypothyroidism at an early age, as described in nonendemic areas.

Other Complications of Iodine Deficiency

In addition to goiter and cretinism, iodine deficiency may be responsible for varying degrees of hypothyroidism, mental retardation, and increased risks associated with pregnancy. These were not regarded as significant problems until recently, but improved methods of diagnosis have shown them to be widespread in iodine-deficient populations.

Mild hypothyroidism can be difficult to establish clinically under the best of circumstances, and a great deal of it has probably been missed in areas of endemic goiter. For example, in one village in Nepal with a 90% incidence of goiter and 8% of cretinism, about half of the population had hypothyroidism as determined by elevated TSH levels, low serum thyroxines, and physical examination.[75] This would certainly have been missed by usual survey techniques. In an example from central Java, the noncretinous portion of the population had significantly lower PBI's and slower mean Achilles tendon reflex relaxation times than did those of a control group from a nearby iodine-sufficient area, and about 15% showed clinical hypothyroidism when carefully examined.[76]

The generally elevated TSH levels in areas of endemic goiter support the likelihood of widespread unrecognized hypothyroidism. A semantic argument can be waged as to whether an elevated serum TSH means hypothyroidism, but most would agree that it does indicate a need for more thyroid hormone. The suggestion of widespread unrecognized hypothyroidism has great public health significance, since it puts the *majority* of an iodine deficient population at risk for inadequate hormone production, with its consequences of mental and physical torpor, increased susceptibility to a variety of diseases, and impaired childbearing.

Maternal hypothyroidism is reported to decrease fertility, double the risk of stillbirth, and increase the incidence of congenital anomalies and mental retardation in surviving offspring.[77] Since women with iodine deficiency have increased TSH levels and low values for T_4 and T_3,[78] their offspring would appear to risk all the complications of neonatal hypothyroidism. Data on fertility and stillbirths are difficult to gather in endemic areas, and they will reflect not only iodine deficiency but also other factors such as poor nutrition and inadequate prenatal care. The available information for iodine deficient animals and for hypothyroid humans makes increased fetal loss a likely consequence of iodine deficiency.

The surviving offspring of iodine-deficient mothers also run risks. One is that of cretinism, but in addition there may be milder degrees of developmental retardation which occur with a much higher incidence. Whereas it was once believed that within endemias, subjects were either cretins or intellectually normal, current evidence favors a continuum from the florid cretin leading a vegetative existence through milder degrees of retardation up to an apparent normal.[76,79] This point, like that of hypothyroidism, means that the *entire* population — not just the cretins — carries the risk of some mental impairment from iodine deficiency. In a continuing study of two villages in Andean Ecuador,[79] iodine sufficiency from the moment of conception appeared to improve the mean IQ of school children relative to that of iodine deficient controls, but introduction of iodine sufficiency after the fourth month of gestation did not. In a similar study in Peru,[80] children protected from iodine deficiency from conception onwards had consistently higher IQ's than unprotected controls although statistical significance was not demonstrated. Whether intelligence improves when iodine sufficiency is introduced later in childhood is uncertain.[81]

Other Factors in Endemic Goiter

Iodine deficiency is unquestionably the major feature in endemic goiter, but other factors may contribute. One is protein-calorie malnutrition, which frequently coexists with iodine deficiency. In animal investigations, protein deprivation blunted the thyroid's response to goitrogenic stimuli.[82] Experimental starvation in humans altered the peripheral metabolism of T_4, shifting it from the active hormone T_3 to the inactive isomer "reverse T_3" (3,3'5'-triiodothyronine), and producing a net decrease in hormone effect.[17] Declines in metabolic rate and in hypoxic ventilatory response in semistarved humans have also been described and speculatively related to this mechanism.[83] In many studies, perinatal malnutrition has led to retardation in physical and mental development.[84] Undernutrition in rats may cause alterations in maturation of the central nervous system which are qualitatively different from those of hypothyroidism.[67] From these considerations it seems possible, at least theoretically, that protein-calorie malnutrition occasionally potentiates the threat of hypothyroidism already present in iodine deficiency, and thus contributes to the pathogenesis of endemic goiter and cretinism.

Another feature of endemic goiter is the variation found among individuals in their response to iodine deficiency.[85] Populations have been described in Venezuela,[86] New Guinea,[64] and Zaire[87] with severe iodine deficiency but virtually no goiter or detectable hypothyroidism. Within the same population in an endemic goiter area of Greece, balance studies suggested more iodine deficiency among goitrous than nongoitrous subjects.[85] Goiters tend to be familial in both endemic and nonendemic areas. There is considerable variation among "normal" thyroids in iodine metabolism and chemical structure,[13] and it is reasonable to suggest that some thyroids will be better than others in coping with iodine deficiency or other stress.

Dietary goitrogens are an additional factor which may be operative in endemic goiter. Possible agents include thiocyanates, isothiocyanates, goitrin (1,5,-vinyl-2-thiooxazolidine), lithium, fluoride, calcium, and many other more poorly identified substances.[88] Most are not potent enough to induce goiter unless eaten in gargantuan amounts. In the Cauca valley of Colombia, extensive studies have implicated a sulfate-containing hydrocarbon of uncertain structure as an etiological factor in persistence of mild endemic goiter despite adequate iodine intake.[89] Thiocyanate-containing compounds have been suggested as contributing factors in endemias in central Africa[65] and in Czechoslovakia.[90] Mild endemic goiter in an iodine-sufficient area of Virginia was tentatively associated with water pollution.[91] While such findings are of considerable scientific interest, we emphasize that it is only with iodine deficiency that large goiters and cretinism are found.

TREATMENT AND PREVENTION

Iodization methods

Iodized salt is the best means of correcting iodine deficiency in most areas of endemic goiter. Potassium iodide or potassium iodate are the common additives, the latter being preferred for its greater stability in humid climates. Fine, free-running crystalline salt can be simply iodinated by dry mixing[92] and this is the method generally used by industrialized countries. For most parts of the world, including those with endemic goiter, refined crystalline salt is too expensive, and coarser particulate salt is used instead. This is best iodinated by solution spray of potassium iodate. Details of this operation are given by Hunnikin.[93] The cost of iodization is estimated at from 1/4 to 2¢ per capita per year, assuming a salt consumption of 10 g per day per person. The machinery is easy to build and maintain and has been successfully operated in large areas of the Far East.

Different national programs have used ratios of iodine to salt ranging from 1:100,000 to 1:10,000, delivering from 100 to 1000 μg of iodine per day if a 10 g daily intake of salt is assumed.[94] A WHO study group recommended 1:100,000 in 1952, but long-term results from Switzerland[95] and Yugoslavia[96] suggest this figure to be too low for effective goiter prevention. The Fourth PAHO Technical Group recommended fortification at iodine to salt ratios ranging from 1:30,000 to 1:10,000, with most members favoring 1:30,000.[97] Matovinovic[43] has suggested 1:50,000 which would provide 200 μg q.d. Choice of the level of iodization will be influenced by salt consumption, by the form of iodine used and its stability (most countries including the U.S. still use potassium iodide), and by the severity of iodine deficiency. Constant attention is needed to avoid excessive iodine, particularly in countries such as the U.S. and Japan where large amounts are frequently ingested in food products. The effort here should be toward reducing and standardizing the iodine content of these other sources rather than eliminating iodized salt, which provides a constant and reasonable supply. In the U.S., iodization at a lower level than the unnecessarily high current one of 1:10,000 should be considered.

Additions of iodine to food other than salt have occasionally been undertaken. Vehicles have included drinking water[94] and bread.[98] Unique advantages of salt include a universal requirement for it and the relative ease of controlling its distribution. For these reasons, iodization by other means is unlikely to replace iodized salt.

Iodized oil administration can also be used for correction of iodine deficiency. The dose and frequency of administration have varied among different programs.[65,99] iodine/ml for ages 1 to 45 years, and 0.5 ml for the first 12 months, with reinjection every 3 or 4 years. A program in Argentina found that iodized oil given orally at 1.4 times the recommended intramuscular dose gave lower but quite adequate levels of iodine.[101] In preliminary results from a more severe endemia in Bolivia, oral iodized oil has been considerably less satisfactory when given at the same dose as recommended parenterally.[102]

The PAHO Group estimated in 1974 that the cost of an iodized oil injection program was 9¢ per person per year (27¢ per injection, repeated in 3 years) when existing public health personnel are used, or 16¢ per person per year if special injection teams are required.[100]

Current recommendations by the Fourth PAHO Technical Group[100] call for a single intramuscular injection of 1 ml of 37% iodized oil (Ethiodol®) containing 475 mg

Results of Iodization Programs

It has been repeatedly demonstrated during the past 50 years that iodization can eliminate endemic cretinism and almost eradicate endemic goiter. The pockets of goiter remaining after iodization are those few instances where a goitrogen has been implicated.[65,89-91] As already discussed, the goiters in these instances are not large, there is no cretinism, and the public health significance is minor compared to the ravages of iodine deficiency. Most successful programs have used iodized salt, but impressive early results have been obtained with iodized oil as well. In the U.S., endemic goiter has disappeared since the introduction of iodized salt. Other countries with impressive results include the Alpine countries,[52] Guatemala,[103] Colombia,[104] Uruguay,[105] and in several parts of Asia. In these areas, endemic cretinism has also disappeared. Many other countries have implemented programs within the past 2 decades and are showing impressive progress. More noteworthy is the fact that despite adequate demonstration of the effectiveness of iodization, many countries continue to be iodine deficient. The obstacles have variously been political, geographical, economic, and social. A catalogue of progress by country in Latin America has recently been published[53] and causes for failure considered.[106]

Organization of Iodization Programs

In considering iodization for a particular area of endemic goiter, it is important to know the severity of iodine deficiency and thus the urgency for its correction. The Fourth PAHO Technical Group recommended the following classification of endemias by severity, using the urinary iodine as yardstick:[39]

Grade I: Goiter endemias with an average urinary iodine excretion of more than 50 μg/g creatinine. At this level, thyroid hormone supply is adequate, and normal mental and physical development can be anticipated.

Grade II: Goiter endemias with an average urinary iodine excretion of between 25 and 50 μg/g creatinine. In these circumstances, adequate thyroid hormone formation may be impaired. This group is at risk for hypothyroidism but not for overt cretinism.

Grade III: Goiter endemias with an average urinary iodine excretion below 25 μg/g creatinine. Endemic cretinism is a serious risk in such a population.

Advice on survey technique is included in the same report. In Grade I endemias, adequate iodine is important to decrease complications of the goiter itself. If urinary iodide levels are well above 50 μg/mg creatinine, the possibility of a goitrogen should be considered. Grade II is more serious because of hypothyroidism. Grade III, however, is by far the most urgent because of the associated cretinism. Therapy with iodized oil is particularly appropriate in Grade III endemias, where immediate correction is mandatory to avoid further cretinism and other irreparable damage. Grade III endemias tend to be in isolated areas which may require many years for effective introduction of iodized salt. Iodized oil offers an important temporary form of therapy in such instances, as has been shown dramatically in New Guinea,[99] Peru,[78] Ecuador,[79] and central Africa.[65] However, salt should be considered the ultimate vehicle of iodization for nearly all endemias, because it is cheaper, provides fairly constant intake, and is a simpler program to continue once in operation.

The administrative structure of iodization programs is crucial to their success. Initially, the most important feature is a strong law specifying mandatory iodization, vesting authority for implementation in a responsible government agency, and providing resources for surveillance and penalties for noncompliance. Voluntary use of iodized salt is an acceptable alternative only in those rare communities with mild deficiency and with the sophistication to recognize the importance of its correction. The cost of iodization, which is usually quite minimal, must not appreciably change the purchase price of salt for the consumer. The law should specify that all salt for human and animal consumption be iodized. Domestic animals suffer from iodine deficiency as do humans[107] and iodine has been used in feeds for over 50 years. Its most impressive result is in increased fetal salvage. In some places, humans consume salt sold for animal use because it is cheaper, and this is another reason for iodinating it.

In setting up iodization programs, care should be taken that no group is economically penalized. Salt producers, especially the smaller ones, may feel particularly threatened by government regulations and by the demand for additional machinery. A rational program must meet these concerns with aid or realistic assurance against financial loss. No government is so poor that it cannot aid in an iodization program if it accords it the priority it deserves. International agencies have also helped. It is highly likely, although unproven, that the gain in health of citizens and animals would advance economic productivity to levels far in excess of the cost of iodization.

Once a program is operating, continuing surveillance is necessary.[104,108,109] The iodine content of salt should be monitored both at the factory and community store, and periodic surveys should be made of goiter incidence and urinary iodine levels. There may be an increase in the incidence of hyperthyroidism (toxic nodular goiter) during the several years after effective iodization.[54,60,98] This so-called "Jod-Basedow" occurs on the introduction of iodine sufficiency to long-standing nodular glands adapted to maximal iodine utilization. The hyperthyroidism is usually mild and effectively treated. It is worth alerting physicians to this complication of iodization, but it is of negligible importance when balanced against the benefits of correcting iodine deficiency, and is not a rational basis for delaying iodization programs.

Different countries have tailored their iodization programs to differing conditions and needs. Excellent examples are Uruguay,[105] Colombia,[104] and Guatemala,[103] all of which have been quite effective. The most important feature in each has been the realization by persons in power that elimination of iodine deficiency deserves high priority among national goals. Once that conviction exists, a rational program can be developed and executed with every promise of success.

REFERENCES

1. **Dunn, J. T. and Medeiros-Neto, G. A., Eds.,** *Endemic Goiter and Cretinism: Continuing Threats to World Health,* Report of the Fourth Meeting of the PAHO Technical Group on Endemic Goiter, PAHO/WHO, Scientific Publication No. 292, Washington, D.C., 1974.
2. **Stanbury, J. B., Ed.,** *Endemic Goiter* PAHO/WHO, Scientific Publication No. 193, Washington, D.C., 1969.
3. World Health Organization, *Endemic Goitre, WHO Monogr. Ser.* No. 44, Geneva, 1960.
4. **Stanbury, J. B. and Kroc, R. L., Eds.,** *Human Development and the Thyroid Gland: Relation to Endemic Cretinism,* Advances in Experimental Medicine and Biology, Vol. 30, Plenum Press, New York, 1972.
5. **Hetzel, B.S. and Pharoah, P. O. D., Eds.,** *Endemic Cretinism,* Institute of Human Biology, Papua, New Guinea, 1971.
6. International Symposium on Endemic Goiter, Innsbruck, Austria, Oct. 5-7, 1972, *Acta Endocrinol.,* Suppl. 179, 7–120, 1973.
7. **Greer, M. A. and Solomon, D. H., Eds.,** *Handbook of Physiology,* Section 7, Vol. 3, *Thyroid,* American Physiological Society, Washington, D.C., 1974.
8. **De Groot, L. J. and Stanbury, J. B.,** *The Thyroid and Its Diseases,* 4th ed., John Wiley & Sons, New York, 1975.
9. **Werner, S. C. and Ingbar, S. H., Eds.,** *The Thyroid,* 3rd ed., Harper & Row, New York, 1971.
10. **Bastomsky, C. H.,** Thyroid iodide transport, in *Handbook of Physiology,* Section 7, Vol. 3, Thyroid, Greer, M. A. and Solomon, D. H., Eds., American Physiological Society, Washington, D.C., 1974, 81–99.
11. **Stanbury, J. B.,** Familial goiter, in *Metabolic Basis of Inherited Disease,* 3rd ed., Stanbury, J. B., Wyngaarden, J. B., and Frederickson, D. S., Eds., McGraw-Hill, 1972, 223–265.
12. **Ui, N.,** Synthesis and chemistry of iodoproteins, in *Handbook of Physiology,* Section 7, Vol. 3, Thyroid, Greer, M. A. and Solomon, D. H., Eds., American Physiological Society, Washington, D.C., 1974, 55–80.
13. **Dunn, J. T. and Ray, S. C.,** The heterogeneity of thyroglobulin from normal and goitrous human thyroids, *Clin. Res.,* 24, 43A, 1976.
14. **Dunn, J. T.,** Thyroglobulin and other factors in the utilization of iodine by the thyroid, in *Endemic Goiter and Cretinism: Continuing Threats to World Health,* Report of the Fourth Meeting of the PAHO Technical Group on Endemic Goiter, Dunn, J. T. and Medeiros-Neto, G. A., Eds., PAHO/WHO, Scientific Publication No. 292, Washington, D.C., 1974, 17–24.
15. **Oppenheimer, J. H. and Surks, M. I.,** Quantitative aspects of hormone production, distribution, metabolism and activity, in *Handbook of Physiology,* Section 7, Vol. 3, Thyroid, Greer, M. A. and Solomon, D. H., Eds., American Physiological Society, Washington, D.C., 1974, 197–214.
16. **Chopra, I. J. and Smith, S. R.,** Circulating thyroid hormones and thyrotropin in adult patients with protein-calorie malnutrition, *J. Clin. Endocrinol. Metab.,* 40, 221–227, 1975.
17. **Vagenakis, A. G., Burger, A., Portnay, G. I., et al.,** Diversion of peripheral thyroxine metabolism from activating to inactivating pathways during complete fasting, *J. Clin. Endocrinol. Metab.,* 41, 191–194, 1975.
18. **Vought, R. L. and London, W. T.,** Iodine intake, excretion and thyroidal accumulation in healthy subjects, *J. Clin. Endocrinol. Metab.,* 27, 913–919, 1967.
19. **Van Middlesworth, L.,** Metabolism and excretion of thyroid hormones, in *Handbook of Physiology,* Section 7, Vol. 3, Thyroid, Greer, M. A. and Solomon, D. H., Eds., American Physiological Society, Washington, D.C., 1974, 215–231.

20. Tong, W., Actions of thyroid-stimulating hormone, in *Handbook of Physiology,* Section 7, Vol. 3, Thyroid, Greer, M. A. and Solomon, D. H., Eds., American Physiological Society, Washington, D.C., 1974, 255–283.

21. **Dunn, J. T. and Ray, S. C.,** Changes in the structure of thyroglobulin following the administration of thyroid-stimulating hormone, *J. Biol. Chem.,* 250, 5801–5807, 1975.

22. **Hoch, F. L.,** Metabolic effects of thyroid hormones, in *Handbook of Physiology,* Section 7, Vol. 3, Thyroid, Greer, M. A. and Solomon, D. H., Eds., American Physiological Society, Washington, D.C., 1974, 391–411.

23. **Tate, J. R.,** Growth and developmental action of thyroid hormones at the cellular level, in *Handbook of Physiology,* Section 7, Vol. 3, Thyroid, Greer, M. A. and Solomon, D. H., Eds., American Physiological Society, Washington, D.C., 1974, 469–478.

24. **Smith, D. W., Blizzard, R. M., and Wilkins, L.,** The mental prognosis in hypothyroidism of infancy and childhood, *Pediatrics,* 19, 1011–1022, 1957.

25. **Vought, R. L., London, W. T., and Brown, F. A.,** A note on atmospheric iodine and its absorption by man, *J. Clin. Endocrinol. Metab.,* 24, 414–416, 1964.

26. **Koutras, D. A., Papapetrou, P. D., Yataganas, X., and Malamos, B.,** Dietary sources of iodine in areas with and without iodine-deficiency goiter, *Am. J. Clin. Nutr.,* 23, 870–874, 1970.

27. **Kidd, P. S., Trowbridge, F. L., Goldsby, J. B., and Nichaman, M. Z.,** Sources of dietary iodine, *J. Am. Diet. Assoc.,* 65, 420–422, 1974.

28. **Vought, R. L. and London, W. T.,** Dietary sources of iodine, *Am. J. Clin. Nutr.,* 14, 186–192, 1964.

29. **Pittman, J. A., Dailey, G. E., and Beschi, R. J.,** Changing normal values for thyroidal radioiodine uptake, *N. Engl. J. Med.,* 280, 1431–1434, 1969.

30. **Vought, R. L., Brown, F. A., and Wolff, J.,** Erythrosine: An adventitious source of iodine, *J. Clin. Endocrinol. Metab.,* 34, 747–752, 1972.

31. **Trowbridge, F. L., Hand, K. A., and Nichaman, M. Z.,** Findings relating to goiter and iodine in the ten-state nutrition survey, *Am. J. Clin. Nutr.,* 28, 712–716, 1975.

32. National Research Council, *Food and Nutrition Board: Recommended Dietary Allowances,* Washington, D.C., 1948, 1–31.

33. **Robinson, W., Chopra, J., de Leon, R., et. al.,** Iodization of salt, in *Endemic Goiter and Cretinism: Continuing Threats to World Health,* Report of the Fourth Meeting of the PAHO Technical Group on Endemic Goiter, Dunn, J. T. and Medeiros-Neto, G. A., Eds., PAHO/WHO, Scientific Publication No. 292, Washington, D.C., 1974, 276–277.

34. Food and Nutrition Board, National Academy of Science, *Iodine Nutriture in the United States,* National Research Council, Washington, D.C., 1970, 51.

35. **Wolff, J.,** Iodide goiter and the pharmacologic effects of excess iodide, *Am. J. Med.,* 47, 101–124, 1969.

36. **Stewart, J. C. and Vidor, G. I.,** Thyrotoxicosis induced by iodine contamination of food – A common unrecognized condition? *Br. Med. J.,* 1, 372–375, 1976.

37. **Nagataki, S.,** Effect of excess quantities of iodide, in *Handbook of Physiology,* Section 7, Vol. 3, Thyroid, Greer, M. A. and Solomon, D. H., Eds., American Physiological Society, Washington, D.C., 1974, 329–344.

38. **Perez, C., Scrimshaw, N. S., and Munoz, J. A.,** Technique of endemic goitre surveys, in *Endemic Goitre, WHO Monogr. Ser.* No. 44, World Health Organization, Geneva, 1960, 369–383.

39. **Querido, A., Delange, F., Dunn, J. T., et. al.,** Definition of endemic goiter and cretinism, classification of goiter size and severity of endemias and survey techniques, in *Endemic Goiter and Cretinism: Continuing Threats to World Health,* Report of the Fourth Meeting of the PAHO Technical Group on Endemic Goiter, Dunn, J. T. and Medeiros-Neto, G. A., Eds., PAHO/WHO, Scientific Publication No. 292, Washington, D.C., 1974, 267–272.

40. **De Smet, M. P.,** Pathological anatomy of endemic goitre, in *Endemic Goitre, WHO Monogr. Ser.* No. 44, World Health Organization, Geneva, 1960, 315–349.

41. **Studer, H., Kohler, H., and Burgi, H.,** Effect of excess quantities of iodide, in *Handbook of Physiology,* Section 7, Vol. 3, Thyroid, Greer, M. A. and Solomon, D. H., Eds., American Physiological Society, Washington, D.C., 1974, 303–328.

42. **Ermans, A. M.,** Intrathyroid iodine metabolism in goiter, in *Endemic Goiter* PAHO/WHO, Stanbury, J. B., Ed., Scientific Publication No. 193, Washington, D.C., 1969, 1–13.

43. **Matovinovic, J., Child, M. A., Nichaman, M. Z., and Trowbridge, F. L.,** Iodine and endemic goiter, in *Endemic Goiter and Cretinism: Continuing Threats to World Health,* Report of the Fourth Meeting of the PAHO Technical Group on Endemic Goiter, Dunn, J. T. and Medeiros-Neto, G. A., Eds., PAHO/WHO, Scientific Publication No. 292, Washington, D.C., 1974, 67–94.

44. **Buttfield, I. H., Hetzel, B.S., and Odell, W. D.,** Effect of iodized oil on serum TSH determined by immunoassay in endemic goiter subjects, *J. Clin. Endocrinol. Metab.,* 28, 1664–1666, 1968.

45. **Delange, F., Hershman, J. M., and Ermans, A. M.,** Relationship between the serum thyrotropin level, the prevalence of goiter and the pattern of iodine metabolism in Idjwi Island, *J. Clin. Endócrinol. Metab.,* 33, 261–268, 1971.

46. **Chopra, I. J., Hershman, J. M., and Hornabrook, R. W.,** Serum thyroid hormone and thyrotropin levels in subjects from endemic goiter regions of New Guinea, *J. Clin. Endocrinol. Metab.,* 40, 326–333, 1975.

47. **Medeiros-Neto, G. A., Penna, M., Monteiro, K., Kataoka, K., Imai, Y., and Hollander, C.,** The effect of iodized oil on the TSH response to TRH in endemic goiter patients, *J. Clin. Endocrinol. Metab.,* 41, 504–510, 1975.

48. **Jolin, T. and Escobar del Rey, F.,** Evaluation of iodine/creatinine ratios of casual samples as indices of daily urinary iodine output during field studies, *J. Clin. Endocrinol. Metab.,* 25, 540–542, 1965.

49. **Koutras, D. A., Berman, M., Sfontouris, J., et al.,** Endemic goiter in Greece: Thyroid hormone kinetics, *J. Clin. Endocrinol. Metab.,* 30, 479–487, 1970.

50. **Silva, E., Pineda, G., and Stevenson, C.,** The importance of triiodothyronine as a thyroid hormone, in *Endemic Goiter and Cretinism: Continuing Threats to World Health,* Report of the Fourth Meeting of the PAHO Technical Group on Endemic Goiter, Dunn, J. T. and Medeiros-Neto, G. A., Eds., PAHO/WHO, Scientific Publication No. 292, Washington, D.C., 1974, 52–63.

51. **Greer, M. A. and Rockie, C.,** Effect of thyrotropin and the iodine content of the thyroid on the triiodothyronine: Thyroxine ratio of newly synthesized iodothyronines, *Endocrinology,* 85, 244–250, 1969.

52. **Kelly, F. C. and Snedden, W. W.,** Prevalence and geographical distribution of endemic goitre, in *Endemic Goitre, WHO Monogr. Ser.* No. 44, World Health Organization, Geneva, 1960, 27–233.

53. **Schaefer, A. E.,** Status of salt iodization in PAHO member countries, in *Endemic Goiter and Cretinism: Continuing Threats to World Health* (Report of the Fourth Meeting of the PAHO Technical Group on Endemic Goiter), Dunn, J. T. and Medeiros-Neto, G. A., Eds., PAHO/WHO, Scientific Publication No. 292, Washington, D.C., 1974, 242–250.

54. **Riccabona, G.,** Hyperthyroidism and thyroid cancer in an endemic goiter area, in *Endemic Goiter and Cretinism: Continuing Threats to World Health* (Report of the Fourth Meeting of the PAHO Technical Group on Endemic Goiter), Dunn, J. T. and Medeiros-Neto, G. A., Eds., PAHO/WHO, Scientific Publication No. 292, Washington, D.C., 1974, 156–165.

55. **Gaitan, E.,** Discussion: Hyperthyroidism and thyroid cancer in endemic goiter area, in *Endemic Goiter and Cretinism: Continuing Threats to World Health,* Report of the Fourth Meeting of the PAHO Technical Group on Endemic Goiter, Dunn, J. T. and Medeiros-Neto, G. A., Eds., PAHO/WHO, Scientific Publication No. 292, Washington, D.C., 1974, 165.

56. **Correa, P., Cuella, C., and Eisenberg, H.,** Epidemiology of different types of thyroid cancer, in *Thyroid Cancer,* Hedinger, C. E., Ed., Springer-Verlag, Berlin, 1969, 81–93.

57. **Sampson, R. J., Woolner, L. B., Bahn, R. C., and Kurland, L. T.,** Occult thyroid carcinoma in Olmsted County, Minnesota: Prevalence at autopsy compared with that in Hiroshima and Nagasaki, Japan, *Cancer,* 34, 2072–2076, 1974.

58. **Sasaki, J., Seta, K., Yagawa, K., et al.,** Clinicopathological studies on latent and occult carcinoma of the thyroid, *Excerpta Med.,* Int. Congr. Series No. 361, pp. 99–100, 1975.

59. **Sampson, R. J., Key, C. R., Buncher, C. R., and Iijima, S.,** Thyroid carcinoma in Hiroshima and Nagasaki. I. Prevalence of thyroid carcinoma at autopsy, *J.A.M.A.,* 209, 65–70, 1969.

60. **Perinetti, H., Staneloni, L. N., Nacif Nora, J., Sanchez-Tejeda, J., and Perinetti, H. A.,** Results of salt iodization in Mendoza, Argentina, in *Endemic Goiter and Cretinism: Continuing Threats to World Health,* Report of the Fourth Meeting of the PAHO Technical Group on Endemic Goiter, Dunn, J. T. and Medeiros-Neto, G. A., Eds., PAHO/WHO, Scientific Publication No. 292, Washington, D.C., 1974, 217–226.

61. **Delange, F., Costa, A., Ermans, A. M., Ibbertson, H. K., Querido, A., and Stanbury, J. B.,** A survey of the clinical and metabolic patterns of endemic cretinism, in *Human Development and the Thyroid Gland: Relation to Endemic Cretinism,* (Advances in Experimental Medicine and Biology, Vol. 30) Stanbury, J. B. and Kroc, R. L., Eds., Plenum Press, New York, 1972, 175–187.

62. **Hornabrook, R. W.,** Neurological aspects of endemic cretinism in eastern New Guinea, in *Endemic Cretinism,* Hetzel, B. S. and Pharoah, P. O. D., Eds., Institute of Human Biology, Papua, New Guinea, 1971, 105–107.

63. **Dodge, P. R., Ramirez, I., and Fierro-Benitez, R.,** Neurological aspects of endemic cretinism, in *Endemic Goiter* PAHO/WHO, Scientific Publication No. 913, Stanbury, J. B., Ed., Washington, D.C., 1969, 373–377.

64. Choufoer, J. C., van Rhijn, M., and Querido, A., Endemic goiter in Western New Guinea. II. Clinical picture, incidence and pathogenesis of endemic cretinism, *J. Clin. Endocrinol. Metab.*, 25, 385–402, 1965.

65. Delange, F., *Endemic Goitre and Thyroid Function in Central Africa*, S. Karger, Basel, 1974.

66. Ibbertson, H. K., Pearl, M., McKinnon, J., Tait, J. M., Lim, T., and Gill, M. B., Endemic cretinism in Nepal, in *Endemic Cretinism*, Hetzel, B. S. and Pharoah, P.O.D., Eds., Institute of Human Biology, Papua, New Guinea, 1971, 71–88.

67. Balazs, R., Effects of hormones and nutrition on brain development, in *Human Development and the Thyroid Gland: Relation to Endemic Cretinism* (Advances in Experimental Medicine and Biology, Vol. 30), Stanbury, J. B. and Kroc, R. L., Eds., Plenum Press, New York, 1972, 385–415.

68. Rosman, N. P., The neuropathology of congenital hypothyroidism, in *Human Development and the Thyroid Gland: Relation to Endemic Cretinism* (Advances in Experimental Medicine and Biology, Vol. 30), Stanbury, J. B. and Kroc, R. L., Eds., Plenum Press, New York, 1972, 337–366.

69. Dunn, J. T., The effects of thyroid hormone on protein synthesis in the central nervous system of developing animals, in *Human Development and the Thyroid Gland: Relation to Endemic Cretinism* (Advances in Experimental Medicine and Biology, Vol. 30), Stanbury, J. B. and Kroc, R. L., Eds., Plenum Press, New York, 1972, 367–383.

70. Thilly, C. H., Delange, F., Camus, M., Berquist, H., and Ermans, A. M., Fetal hypothyroidism in endemic goiter: the probably pathogenic mechanism of endemic cretinism, in *Endemic Goiter and Cretinism: Continuing Threats to World Health*, Report of the Fourth Meeting of the PAHO Technical Group on Endemic Goiter, Dunn, J. T. and Medeiros-Neto, G. A., Eds., PAHO/WHO, Scientific Publication No. 292, Washington, D. C., 1974, 121–128.

71. Koenig, M. P. and Neiger, M., The pathology of the ear in endemic cretinism, in *Human Development and the Thyroid Gland: Relation to Endemic Cretinism* (Advances in Experimental Medicine and Biology, Vol. 30), Stanbury, J. B. and Kroc, R. L., Eds., Plenum Press, New York, 1972, 325–333.

72. Bargman, G. J. and Gardner, L. I., Experimental production of otic lesions with antithyroid drugs, in *Human Development and the Thyroid Gland: Relation to Endemic Cretinism* (Advances in Experimental Medicine and Biology, Vol. 30), Stanbury, J. B. and Kroc, R. L., Eds., Plenum Press, New York, 1972, 305–323.

73. Almeida, F., Temporal, A., Calvalcanti, N., Lins Neto, S., Albuquerque, R., and Cristina, T., Pendred's Syndrome in an area of endemic goiter in Brazil: Genetic and metabolic studies, in *Endemic Goiter and Cretinism: Continuing Threats to World Health*, Report of the Fourth Meeting of the PAHO Technical Group on Endemic Goiter, Dunn, J. T. and Medeiros-Neto, G. A., Eds., PAHO/WHO, Scientific Publication No. 292, Washington, D.C., 1974, 167–171.

74. Cave, W. T., Jr. and Dunn, J. T., Studies on the thyroidal defect in an atypical form of Pendred's Syndrome, *J. Clin. Endocrinol. Metab.*, 41, 590–599, 1975.

75. Ibbertson, H. K., Gluckman, P. D., Croxson, M. S., and Strang, L. J. W., Goiter and cretinism in the Himalayas: A reassessment, in *Endemic Goiter and Cretinism: Continuing Threats to World Health*, Report of the Fourth meeting of the PAHO Technical Group on Endemic Goiter, Dunn, J. T. and Medeiros-Neto, G. A., Eds., PAHO/WHO, Scientific Publication No. 292, Washington, D.C., 1974, 129–134.

76. Querido, A., Djokomoeljanto, R., and van Hardeveld, C., The consequences of iodine deficiency for health, in *Endemic Goiter and Cretinism: Continuing Threats to World Health*, Report of the Fourth Meeting of the PAHO Technical Group on Endemic Goiter, Dunn, J. T. and Medeiros-Neto, G. A., Eds., PAHO/WHO, Scientific Publication No. 292, Washington, D.C., 1974, 8–14.

77. Burrow, G. N., in *Medical Complications During Pregnancy*, Burrow, G. N. and Ferris, T. F., Eds., W. B. Saunders, Philadelphia, 1975, 196–241.

78. Pretell, E. A., Palacios, P., Tello, L., Wan, M., Utiger, R., and Stanbury, J. B., Iodine deficiency and the maternal-fetal relationship, in *Endemic Goiter and Cretinism: Continuing Threats to World Health*, Report of the Fourth Meeting of the PAHO Technical Group on Endemic Goiter, Dunn, J. T. and Medeiros-Neto, G. A., Eds., PAHO/WHO, Scientific Publication No. 292, Washington, D.C., 1974, 143–155.

79. Fierro-Benitez, R., Ramirez, I., Estrella, E., and Stanbury, J. B., The role of iodine in intellectual development in an area of endemic goiter, in *Endemic Goiter and Cretinism: Continuing Threats to World Health*, Report of the Fourth Meeting of the PAHO Technical Group on Endemic Goiter, Dunn, J. T. and Medeiros-Neto, G. A., Eds., PAHO/WHO, Scientific Publication No. 292, Washington, D.C., 1974, 135–142.

80. **Pretell, E. A., Torres, T., Zenteno, V., and Cornejo, J.,** Prophylaxis of endemic goiter with iodized oil in rural Peru, in *Human Development and the Thyroid Gland: Relation to Endemic Cretinism* (Advances in Experimental Medicine and Biology, Vol. 30), Stanbury, J. B. and Kroc, R. L., Eds., Plenum Press, New York, 1972, 249–265.

81. **Dodge, P. R., Palkes, H., Fierro-Benitez, R., and Ramirez, I.,** Effect on intelligence of iodine in oil administered to young Andean children – A preliminary report, in *Endemic Goiter* PAHO/WHO, Scientific Publication No. 193, Stanbury, J. B., Ed., Washington, D.C., 1969, 378–380.

82. **Ramalingaswami, V., Vickery, A. L., Stanbury, J. B. and Hegsted, D. M.,** Some effects of protein deficiency on the rat thyroid, *Endocrinology,* 77, 87–95, 1965.

83. **Doebel, R. C., Jr., Zwillich, C. W., and Scoggin, C. H., et al.,** Clinical semistarvation depression of hypoxic ventilatory response, *N. Engl. J. Med.,* 295, 358–361, 1976.

84. **Chase, H. P. and Martin, H. P.,** Undernutrition and child development, *N. Engl. J. Med.,* 282, 933–939, 1970.

85. **Koutras, D.,** Variation in incidence of goiter within iodine-deficient populations, in *Endemic Goiter and Cretinism: Continuing Threats to World Health,* Report of the Fourth Meeting of the PAHO Technical Group on Endemic Goiter, Dunn, J. T. and Medeiros-Neto, G. A., Eds., PAHO/WHO, Scientific Publication No. 292, Washington, D.C., 1974, 95–101.

86. **Roche, M.,** Elevated thyroidal I 131 uptake in the absence of goiter in isolated Venezuelan Indians, *J. Clin. Endocrinol. Metab.,* 19, 1440–1445, 1959.

87. **Delange, F., Thilly, C., and Ermans, A. M.,** Iodine deficiency, a permissive condition in the development of endemic goiter, *J. Clin. Endocrinol. Metab.,* 28, 114–116, 1968.

88. **Van Etten, C.,** in *Toxic Constituents of Plant Foodstuffs,* Leiner, I. E., Ed., Academic Press, New York, 1969, 103–142.

89. **Gaitan, E., Meyer, J. D., and Merino, H.,** Environmental goitrogens in Colombia, in *Endemic Goiter and Cretinism: Continuing Threats to World Health,* Report of the Fourth Meeting of the PAHO Technical Group on Endemic Goiter, Dunn, J. T. and Medeiros-Neto, G. A., Eds., PAHO/WHO, Scientific Publication No. 292, Washington, D.C., 1974, 107–116.

90. **Podoba, J., Michajlovskij, M., and Stukovsky, R.,** Iodine deficiency, environmental goitrogens and genetic factors in the etiology of endemic goiter, *Acta Endocrinol.,* Suppl. 179, 36–37, 1973.

91. **Vought, R. L., London, W. T., and Stebbing, G. E.,** Endemic goiter in Northern Virginia, *J. Clin. Endocrinol. Metab.,* 27, 1381–1389, 1967.

92. **Holman, J. C. M. and McCartney, W.,** Iodized salt, in *Endemic Goitre, WHO Monogr. Ser.* No. 44, World Health Organization, Geneva, 1960.

93. **Hunnikin, C.,** The spray method of iodating salt, in *Endemic Goiter and Cretinism: Continuing Threats to World Health,* Report of the Fourth Meeting of the PAHO Technical Group on Endemic Goiter, Dunn, J. T. and Medeiros-Neto, G. A., Eds., PAHO/WHO, Scientific Publication No. 292, Washington, D.C., 1974, 184–190.

94. **Matovinovic, J. and Ramalingaswami, V.,** Therapy and prophylaxis of endemic goitre, in *Endemic Goitre, WHO Monogr. Ser.,* No. 44, World Health Organization, Geneva, 1960.

95. **Steck, A., Steck, B., Koenig, M. P., and Studer, H.,** Aus Wirkungen Einer Verbesserten Jodprophylaxe auf Kropfendemie und Jodstoffwechsel, *Schweiz. Med. Wochenschr.,* 102, 829–837, 1972.

96. **Buzina, R.,** Ten years of goiter prophylaxis in Croatia, Yugoslavia, *Am. J. Clin. Nutr.,* 23, 1085–1089, 1970.

97. **Robinson, W., Chopra, J., de Leon, R., Hunnikin, C., Matovinovic, J., and A. Pardo, S.,** Iodization of salt, in *Endemic Goiter and Cretinism: Continuing Threats to World Health,* Report of the Fourth Meeting of the PAHO Technical Group on Endemic Goiter, Dunn, J. T. and Medeiros-Neto, G. A., Eds., PAHO/WHO, Scientific Publication No. 292, Washington, D.C., 1974, 276–277.

98. **Connolly, R. J., Vidor, G. I., and Stewart, J. C.,** Increase in thyrotoxicosis in endemic goitre area after iodation of bread, *Lancet,* 1, 500–502, 1970.

99. **Buttfield, I. H. and Hetzel, B. S.,** Endemic goiter in New Guinea and the prophylactic program with iodinated poppyseed oil, in *Endemic Goiter* PAHO/WHO, Scientific Publication No. 193, Stanbury, J. B., Ed., Washington, D.C., 1969, 132–145.

100. **Pretell, E., Degrossi, O., Riccabona, G., Stanbury, J., and Thilly, C.,** The use of iodized oil, in *Endemic Goiter and Cretinism: Continuing Threats to World Health,* Report of the Fourth Meeting of the PAHO Technical Group on Endemic Goiter, Dunn, J. T. and Medeiros-Neto, G. A., Eds., PAHO/WHO, Scientific Publication No. 292, Washington, D.C., 1974, 278–281.

101. Watanabe, T., Moran, D., El Tamer, E., et al., Iodized oil in the prophylaxis of endemic goiter in Argentina, in *Endemic Goiter and Cretinism: Continuing Threats to World Health,* Report of the Fourth Meeting of the PAHO Technical Group on Endemic Goiter, Dunn, J. T. and Medeiros-Neto, G. A., Eds., PAHO/WHO, Scientific Publication No. 292, Washington, D.C., 1974, 231–241.

102. Bautista, A., Barker, P. B., and Dunn, J. T., The effects of oral iodized oil on the somatic growth, thyroid status, and mental performance of children from an area of severe iodine deficiency, to be published.

103. De Leon, J. Romeo and Retana, O. G., Eradication of endemic goiter as a public health problem in Guatemala, in *Endemic Goiter and Cretinism: Continuing Threats to World Health,* Report of the Fourth Meeting of the PAHO Technical Group on Endemic Goiter, Dunn, J. T. and Medeiros-Neto, G. A., Eds., PAHO/WHO, Scientific Publication No. 292, Washington, D.C., 1974, 227–230.

104. Mora, J. O., Pardo, F., and Rueda-Williamson, R., Surveillance of salt iodization in Columbia, in *Endemic Goiter and Cretinism: Continuing Threats to World Health,* Report of the Fourth Meeting of the PAHO Technical Group on Endemic Goiter, Dunn, J. T. and Medeiros-Neto, G. A., Eds., PAHO/WHO, Scientific Publication No. 292, Washington, D.C., 1974, 209–213.

105. Salveraglio, F. J., Gaining public acceptance of prophylaxis: Experience from the campaign against endemic goiter in Uruguay, in *Endemic Goiter and Cretinism: Continuing Threats to World Health,* Report of the Fourth Meeting of the PAHO Technical Group on Endemic Goiter, Dunn, J. T. and Medeiros-Neto, G. A., Eds., PAHO/WHO, Scientific Publication No. 292, Washington, D.C., 1974, 198–204.

106. Rueda-Williamson, R., Baeta, L. B., Merino, H., Salveraglio, F., and Ticas, J., Overall planning of goiter prevention programs, in *Endemic Goiter and Cretinism: Continuing Threats to World Health,* Report of the Fourth Meeting of the PAHO Technical Group on Endemic Goiter, Dunn, J. T. and Medeiros-Neto, G. A., Eds., PAHO/WHO, Scientific Publication No. 292, Washington, D.C., 1974, 284–287.

107. Robertson, W. B., Technical and economic aspects of salt-iodination, in *Endemic Goiter and Cretinism: Continuing Threats to World Health* Report of the Fourth Meeting of the PAHO Technical Group on Endemic Goiter, Dunn, J. T. and Medeiros-Neto, G. A., Eds., PAHO/WHO, Scientific Publication No. 292, Washington, D.C., 1974.

108. Salvaneschi, J. P. and Degrossi, O. J., The program for surveillance of salt iodization in Argentina, in *Endemic Goiter and Cretinism: Continuing Threats to World Health,* Report of the Fourth Meeting of the PAHO Technical Group on Endemic Goiter, Dunn, J. T. and Medeiros-Neto, G. A., Eds., PAHO/WHO, Scientific Publication No. 292, Washington, D.C., 1974, 205–208.

109. Schaefer, A., Salvaneschi, J., Simmons, J., et al., Surveillance of iodine prophylaxis, in *Endemic Goiter and Cretinism: Continuing Threats to World Health,* Report of the Fourth Meeting of the PAHO Technical Group on Endemic Goiter, Dunn, J. T. and Medeiros-Neto, G. A., Eds., PAHO/WHO, Scientific Publication No. 292, Washington, D.C., 1974, 282–283.

MANGANESE DEFICIENCY — MAN*

R. M. Leach, Jr.

There is only one reported occurrence of manganese deficiency in humans.[1] In Doisey's study of vitamin K requirements, a patient failed to respond to vitamin K supplementation when fed a semipurified diet. Symptoms included inability to elevate depressed clotting proteins in response to vitamin K, hypocholesterolemia, slowed growth of hair and nails, weight loss, and reddening of hair and beard. Symptoms were alleviated when the patient was returned to the house diet. Retrospective analysis indicated that manganese had been omitted from the trace mineral mixture during the study. The manganese intake was calculated to be 0.34 mg/day. There were also substantial decreases in blood and stool manganese content.

As a follow-up to this study, it was reported[1] that a moderate manganese deficiency in the young chick also reduced the clotting response to supplemental vitamin K. However, studies in another laboratory[2] failed to demonstrate an impairment in the response to vitamin K with manganese deficiency.

There have been a number of dietary studies on manganese intakes in humans. These have been recently summarized by Waslien.[3] Infant intakes range from 0.006 to 0.34 mg/day while the intakes of children have been reported to be between 0.24 and 9.5 mg per day. Dietary intakes from 0.4 to 10.7 mg/day have been reported for adults. The majority of the studies indicate intakes in excess of 2 mg/day for most adults. This is very close to the 2 to 5 mg/day intake reported by Schroeder et al.[4] Such intakes would represent 10 to 25% of the total body pool of manganese which is about 20 mg. Cereals, nuts, and vegetables are foods relatively rich in manganese while meats and dairy products are relatively low in this essential element.

Because the deficiency symptoms are well defined in laboratory and some farm animals, several investigations have attempted to determine if impaired manganese metabolism could be associated with the occurrence of several human diseases. Since manganese deficiency greatly impedes skeletal growth and development, Schor and associates[5] determined the manganese content of costal cartilage in two patients with dwarfism due to impaired cartilage development. The cartilage manganese content was found to be normal in the dwarf individuals. However, Cotzias et al.[6] have reported that seven out of eight patients with rheumatoid arthritis showed a decrease in manganese turnover. A patient with hydralazine disease also showed a similar decrease in manganese turnover.

Limited studies have also been conducted on the manganese concentration in blood and its possible relationship to disease. Versieck et al.[7] have reported that normal human blood had an average manganese concentration of 0.057 mg/100 ml. No changes in serum manganese concentrations were observed following myocardial infarction, although there was a significant increase in serum copper and a decrease in serum zinc associated with this condition.[8] Hepatitis was also studied.[9] During the active phase, manganese concentrations were usually increased. In addition, there was a highly significant positive correlation between serum manganese concentration and serum aminotransferase activity in subjects with acute or chronic hepatitis or postnecrotic cirrhosis. These changes probably reflect liver cell damage.

Comment: Although manganese deficiency has been described for a large number of animal species, the chick is the only species in which a deficiency is easily produced with natural diets. For many mammals, prolonged depletion on diets containing less than 1 mg/kg diet is necessary to produce symptoms of manganese deficiency. Thus, man is not greatly different from many of these species of mammals.

* Authorized as Paper No. 5260 in the Journal Article Series of The Pennsylvania Agricultural Experiment Station, University Park, Pennsylvania, 16802, on March 3, 1977.

REFERENCES

1. **Doisey, E. A., Jr.,** Effects of deficiency in manganese upon plasma levels of clotting proteins and cholesterol in man, in *Trace Element Metabolism in Animals,* Vol. 2, Hoekstra, W. G., Suttie, J. W., Ganther, H. E., and Mertz, W., Eds., University Park Press, Baltimore, 1974, 668—670.
2. **Leach, R. M., Jr.,** Metabolism and function of manganese in *Trace Elements in Human Health and Diseases,* Vol. 2, Prasad, A. S., Ed., Academic Press, New York, 1976, 235—247.
3. **Waslien, C. I.,** Human intake of trace elements, in *Trace Elements in Human Health and Disease,* Vol. 2, Prasad, A. S., Ed., Academic Press, New York, 1976, 347—370.
4. **Schroeder, H. A., Balassa, J. J., and Tipton, I. H.,** Essential trace metals in man: Manganese, *J. Chronic Dis.,* 19, 545—571, 1966.
5. **Schor, R. A., Prussin, S. G., Jewett, D. L., Ludowieg, J. J., and Bhatnagar, R. S.,** Trace levels of manganese, copper, and zinc in rib cartilage as related to age in humans and animals, both normal and dwarfed, *Clin. Orthop. Relat. Res.,* 93, 346—355, 1973.
6. **Cotzias, G. C., Papavasiliou, P. S., Hughes, E. R., Tang, L., and Borg, D. G.,** Slow turnover of manganese in active rheumatoid arthritis accelerated by prednisone, *J. Clin. Invest.,* 47, 992—1001, 1968.
7. **Versieck, J., Barbier, F., Speeke, A., and Hoste, J.,** Normal manganese concentrations in human serum, *Acta Endocrinol.,* 76, 783—788, 1974.
8. **Versieck, J., Barbier, F., Speeke, A., and Hoste, J.,** Influence of myocardial infarction on serum manganese, copper and zinc concentrations, *Clin. Chem.,* 21, 578—581, 1975.
9. **Versieck, J., Barbier, F., Speeke, A., and Hoste, J.,** Manganese, copper, and zinc concentrations in serum and packed blood cells during acute hepatitis, chronic hepatitis, and posthepatic cirrhosis, *Clin. Chem.,* 20, 1141—1145, 1974.

NUTRITIONAL DEFICIENCIES IN MAN: ZINC

A. S. Prasad

INTRODUCTION AND HISTORICAL BACKGROUND

The essentiality of zinc for the growth of microorganisms has been known for the past hundred years.[1] In 1934, zinc was shown to be necessary for the growth and well-being of the rat.[2] Clinical manifestations in zinc-deficient animals include growth retardation, testicular atrophy, skin changes, and poor appetite.

In man, deficiency of zinc was suspected to have occurred for the first time in 1961 in Iran.[3] The clinical symptoms in Iranian subjects consisted of dwarfism, anemia, hypogonadism, hepatosplenomegaly, rough, dry skin, mental lethargy, and geophagia. Dietary history revealed that they ate only bread made of wheat flour, and their intake of animal protein was negligible. They consumed nearly 1 lb of clay daily; the practice of geophagia is not uncommon in the villages around Shiraz, Iran. Hematological studies established the presence of severe iron deficiency. There was no evidence of blood loss, and parasitic infestations, such as hookworm and schistosomiasis, were not commonly seen in that part of Iran. The following factors were considered to be responsible for the iron deficiency: 1. the total amount of available iron in the diet was insufficient; 2. excessive sunburn and sweating probably caused greater iron loss from the skin than would occur in a temperate climate; and 3. geophagy may have further decreased iron absorption. In every case, the anemia was completely corrected by oral administration of iron.

Because of the clinical similarities observed in these patients with those seen in several species with zinc deficiency, it was considered reasonable to attribute the dwarfism and hypogonadism in these subjects to a deficiency of zinc. Inasmuch as the food content of iron and of zinc are known to parallel each other and factors responsible for the availability of iron may also affect that of zinc, it was considered possible that one may encounter deficiency of both iron and zinc in a given population.

Subsequently, studies in Egypt established that such patients were zinc deficient.[5,6] The clinical symptoms of Egyptian patients were remarkably similar to those of the Iranian subjects. Their dietary intake of animal protein was negligible, and they consumed mainly bread and beans.

The zinc concentrations in plasma, red cells, and hair were decreased. Radioactive[65] Zn studies revealed that: the plasma zinc turnover rates in these patients was greater than in the control subjects; the 24-hr exchangeable pool was smaller than in the controls; and the excretion of [65]Zn in stool and urine was less than that of control subjects.

Further studies in Egypt showed that growth rate was greater in patients who received supplemental zinc compared to those who received iron instead or only an animal protein diet, which consisted of bread, beans, lamb meat, chicken, eggs, and vegetables daily.[7] Pubic hair appeared in all cases within 7 to 12 weeks after zinc supplementation was started, genitalia became normal, and secondary sexual characteristics developed within 12 to 24 weeks in all patients receiving zinc. On the other hand, no such changes were observed in a comparable length of time in the iron-supplemented group nor in the group on an animal protein diet alone. Thus, these studies established for the first time that growth retardation and gonadal hypofunction in these subjects were related to a deficiency of zinc (Figure 1).

Clinical features similar to those reported by us in zinc-deficient dwarfs have been observed in many countries, such as Turkey, Portugal, and Morocco.[8] Also, zinc deficiency should be prevalent in other countries where primarily cereal proteins are consumed by the population. Clinically, it is not difficult to recognize extreme examples

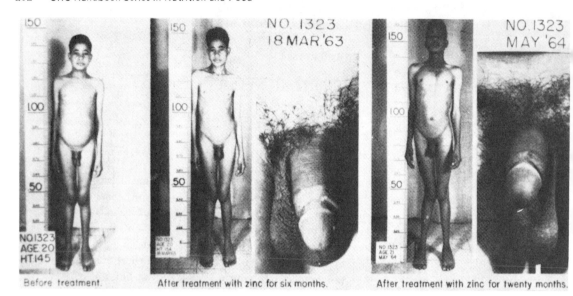

FIGURE 1.   Zinc-deficient dwarf before and after zinc supplementation. (From Prasad, A. S., in *Zinc Metabolism*, Prasad A. S., Ed., Charles C Thomas, Springfield, Ill., 1966, chap. 15.)

of zinc-deficient dwarfs in a given population, but the marginally deficient subjects are likely to present diagnostic problems. It is now becoming clear that not only nutritional deficiency, but also conditioned deficiency, of zinc may complicate many disease states. Recent reports indicate that marginal deficiency of zinc in man is probably not uncommon in the U.S.[9]

ETIOLOGICAL FACTORS RESPONSIBLE FOR ZINC DEFICIENCY IN MAN

Nutritional Factors

Several dietary factors influence zinc nutrition by making zinc unavailable for absorption. (See Table 1.) The formation of insoluble complexes with calcium and phytate in the alkaline intestinal environment has been shown to markedly decrease the availability of zinc for intestinal absorption by experimental animals. Dietary fiber has also been shown to decrease the availability of zinc for intestinal absorption by man.[10] It seems probable that both dietary phytate and fiber may contribute to the occurrence of zinc deficiency in populations which subsist largely on bread and other foods rich in fiber and phytate. There may be other factors in the diet which affect the absorption of dietary zinc.

Alcohol intake is known to induce hyperzincuria in man.[11] The mechanism by which hyperzincuria due to alcohol is induced is unknown. Excessive intake of alcohol may thus play a role in producing deficiency of zinc.

Liver Disease

Hypozincemia and hyperzincuria in patients with cirrhosis of the liver have been reported.[12] It has been suggested that zinc deficiency in the alcoholic cirrhotic patient may be a conditioned deficiency somehow related to hyperzincuria.

Malabsorption

Of the variety of mechanisms proposed for zinc deficiency in patients with

Table 1
CAUSES OF ZINC DEFICIENCY IN HUMAN SUBJECTS

Dietary	Excessive intake of phytate, fiber, polyphosphates, clay and laundry starch, and alcohol
Malabsorption	Pancreatic insufficiency Steatorrhea Gastrectomy Intestinal mucosal disease
Cirrhosis of the liver	
Renal disease	Nephrotic syndrome Renal tubular disease
Renal dialysis	
Chronically debilitating diseases	Neoplastic infections
Psoriasis (skin loss)	
Burns	
Parasitic infections (chronic blood loss)	
Iatrogenic	Penicillamine therapy Total parenteral nutrition Surgical trauma
Genetic	Sickle cell disease Acrodermatitis enteropathica
Pregnancy	

gastrointestinal disease, steatorrhea may be the most common. Exudation of large amounts of zinc-protein complexes into the intestinal lumen may also contribute to the decrease in plasma zinc concentrations which occur in patients with inflammatory disease of the bowel and protein-losing enteropathy. Another potential cause of a negative zinc balance is a massive loss of intestinal secretions.

Renal Diseases

The potential causes of conditioned deficiency of zinc in patients with renal disease include proteinuria and failure of tubular reabsorption. In the former instance, the loss of zinc-protein complexes across the glomerulus is the mechanism. In the latter, an impairment in the metabolic machinery of tubular reabsorption due to a genetic abnormality or toxic substances results in zinc loss.

Burns

The causes of zinc deficiency in patients with burns include losses in exudates. Starvation of patients with burns is a well-recognized cause of morbidity and mortality. The contribution of conditioned zinc deficiency to the morbidity of burned patients is not defined. Limited studies indicate that epithelization of burns may be improved by treatment with zinc.[13]

Pregnancy

The plasma concentration of zinc decreases during human pregnancy. Similar changes have also been noted in patients receiving oral contraceptive agents. During pregnancy, presumably the decrease in plasma zinc reflects in part the uptake of zinc by the fetus and the other products of conception. It has been estimated that the pregnant women must retain approximately 750 μg of zinc per day for growth of the products of conception during the last two thirds of pregnancy. Whether or not decreased levels of plasma zinc in these subjects truly represent a deficient state with respect to zinc has not yet been settled.

GENETIC DISORDERS

Sickle Cell Anemia

Recently deficiency of zinc in sickle cell disease has been recognized.[14,15] Certain clinical features are commonly found in some sickle cell anemia patients and zinc-deficient dwarfs, data for the latter coming from the Middle East. Inasmuch as zinc is an important constituent of erythrocytes, it appears possible that prolonged hemolysis in patients with sickle cell anemia might lead to a zinc-deficient state, which would account for some of the clinical manifestations observed. Zinc concentrations in plasma, erythrocytes, and hair were decreased, and urinary zinc excretion was increased in sickle cell anemia patients as compared to controls. Erythrocyte zinc and daily urinary zinc excretion were inversely correlated in the anemia patients, suggesting that hyperzincuria may have caused zinc deficiency in these patients. Carbonic anhydrase, a zinc metalloenzyme, correlated significantly with erythrocyte zinc.

Recent studies have demonstrated a potentially beneficial effect of zinc on the sickling process in vitro, mediated by its effect on the oxygen dissociation curve and the erythrocyte membrane.[16] In limited uncontrolled studies, zinc appeared to have been effective in decreasing the symptoms and the crisis of sickle cell anemia patients.

Acrodermatitis Enteropathica

Acrodermatitis enteropathica is a lethal, autosomal, recessive trait which usually occurs in infants of Italian, Armenian, or Iranian lineage. The disease develops in the early months of life, soon after weaning from breast feeding. Dermatological manifestations include progressive bullous-pustular dermatitis of the extremities and the oral, anal, and genital areas, paronychia, and generalized alopecia. Other clinical signs include blepharitis, conjunctivitis, chronic diarrhea, malabsorption, steatorrhea, irritability, and emotional disorders. The patients generally are retarded in growth and have hypogonadism. Plasma zinc is profoundly reduced. Zinc supplementation to these patients lead to complete clearance of skin lesions and restoration of normal bowel function, both of these symptoms having been found resistant to various dietary and drug regimens.[17] The underlying mechanism of the zinc deficiency in these patients is most likely due to malabsorption of zinc; however, the cause of poor zinc absorption is obscure.

Iatrogenic Causes of Zinc Deficiency

Zinc deficiency occurring in patients following penicillamine therapy for Wilson's disease has been reported recently.[18] The manifestations consisted of parakeratosis, "dead" hair and alopecia, keratitis, and centrocecal scotoma. Following supplementation with zinc several clinical manifestations were reversed.

Similar clinical features have been reported to occur in patients receiving total parenteral nutrition.[19] The clinical manifestations consisted of diarrhea, mental apathy and depression, moist eczematoid dermatitis, most severe in the peri-oral area, and alopecia. The response to intravenous zinc therapy was very striking. Hyperzincuria in such patients has been related to excessive levels of aminoaciduria.[20] Inasmuch as zinc is bound to several amino acids, it is likely that excessive excretion of amino acids in the urine may enhance excretion of zinc in the urine of patients on total parenteral nutrition.[21]

Miscellaneous Disorders

A low level of plasma zinc has been reported to occur in many conditions.[22,23] Most patients who are chronically debilitated exhibit hypozincemia. This is mostly related to poor appetite and lack of intake of zinc-containing foods. Starvation, which often leads to a depletion of zinc from the body store, is known to cause hyperzincuria.

CLINICAL ASPECTS OF ZINC DEFICIENCY

Acute or Severe Deficiency

Symptoms of acute deficiency of zinc may be seen in patients receiving penicillamine therapy, total parenteral nutrition, and acrodermatitis enteropathica. (See Table 2.) The clinical features in all these conditions are quite similar. Skin changes seem to predominate. They may consist of parakeratosis or moist eczematoid dermatitis, most severe in the peri-oral, -anal and -orbital areas, and alopecia. In acrodermatitis enteropathica, dermatological manifestations also include progressive bullous-pustular dermatitis of the extremities and the oral, anal, and genital areas, paronychia, and generalized alopecia. Loss of hair is fairly common. In addition, diarrhea, malabsorption, steatorrhea, and lactose intolerance have been reported in acrodermatitis enteropathica. Mental apathy and depression as well as retardation of growth and hypogonadism in adolescents have been reported particularly in patients with acrodermatitis enteropathica. Ophthalmic signs, such as centrocecal scotoma and photophobia, may be related to zinc deficiency, inasmuch as there is a high concentration of zinc in the retina, and deficiency of zinc may affect opthalmic functions.

Chronic Deficiency

Our studies in the Middle East revealed that growth retardation and hypogonadism were the most consistent clinical features in adolescents with zinc deficiency. Other manifestations observed in Iran included rough, dry skin, mental lethargy, and poor appetite. The liver and spleen were palpable; however, it is not known if this can be related to deficiency of zinc.

Endocrinological studies in subjects from the Middle East failed to reveal any evidence of hypothyroidism or hypoadrenalism. In animals, zinc deficiency does not affect pituitary function; its major effect is on the testes with impairment of testosterone synthesis.[24] In human adolescent males, lack of facial, body, and pubic hair is a common clinical feature of zinc deficiency. In all age groups and both sexes, deficiency of zinc may affect weight adversely. Height may be affected due to deficiency of zinc in both sexes during their growth period.

Table 2
CLINICAL AND LABORATORY FEATURES OF ZINC DEFICIENCY IN MAN

Clinical characteristics

  Chronic — Growth retardation in adolescents, decreased weight, hypogonadism in males, skin changes (rough and dry), mental lethargy, poor appetite, impaired wound healing, and susceptibility to infections

  Acute (Severe) — Skin and nail changes (parakeratosis, moist eczematoid dermatitis, most severe in peri-oral, -anal and -orbital areas, bullous-pustular dermatitis, paronychia and alopecia), diarrhea, malabsorption, steatorrhea, mental apathy, depression, and ophthalmic signs (centrocecal scotoma and photophobia)

Laboratory findings

  Decreased level of zinc in plasma, red cells, hair, and urine; zinc deficiency may be associated wth hyperzincuria in liver disease, hemolytic anemia (sickle cell anemia), certain renal diseases, following injury and surgical trauma, burns, acute starvation, and as a result of total parenteral nutrition

  ^{65}Zn studies show increased plasma zinc turnover rate and decreased 24-hour exchangeable pool; zinc balance study reveal positive retention of zinc

  Activity of ribonuclease in plasma may be increased; following supplementation with zinc, activity of alkaline phosphatase in plasma increases

Table 3
ZINC IN PLASMA, ERYTHROCYTES,
HAIR, AND URINE

Type of sample	Normal values mean ± SD
Plasma	112 ± 13.6 μg percent
Erythrocytes	40 ± 4 μg/g hemoglobin
Hair	190 ± 17 μg/g
Urine	633 ± 158 μg/day

Impaired wound healing has been noted in chronically debilitated subjects as a result of zinc deficiency.[13] Susceptibility to infection is another manifestation of zinc deficiency; however, the mechanism by which zinc affects immunity is not well understood at this time.

The effects of zinc deficiency on pregnancy and the fetus in human subjects have not been investigated in depth. In animals, however, if the deficiency is severe, the fetus exhibits many abnormalities.[25] If the deficiency of zinc is mild in pregnant rats, the offspring exhibit impaired behavioral function.[26] In a recent report, lowered serum zinc concentration during early pregnancy in human subjects correlated with inefficient labor, atonic bleeding, and immaturity and congenital malformations in infants.[27]

Diagnosis of Zinc Deficiency in Man

The laboratory criteria for the diagnosis of zinc deficiency are not well-established. The response to therapy with zinc is probably the most reliable index for making a diagnosis of the zinc-deficient state in man.

Low levels of plasma zinc have been observed in many illnesses. It is not certain, at present, if a low level of plasma zinc is indicative of "zinc deficiency" in all of these conditions. It seems likely that a low plasma zinc concentration reflects impaired zinc nutrition in many, while in others, it may reflect a shift of zinc from the plasma to another body pool. The concentration of zinc in hair and red cells probably reflects the status of chronic zinc nutrition (Table 3).

Excretion of zinc in urine decreases as the duration of zinc deficiency progresses. This test, however, requires a complete collection of urine on a 24-hour basis. Although there are certain exceptions, most cases of zinc deficiency in man are associated with hypozincuria. In cirrhosis of the liver and sickle cell disease, when deficiency of zinc is present, hyperzincuria is observed. Hyperzincuria has also been observed in certain renal diseases and infections and following injury, burns, and acute starvation.

Zinc balance study, although difficult, may provide a good basis for assessment of zinc status; a positive retention of zinc by human subjects would be indicative of zinc deficiency. By use of [65]Zn, it was shown that in zinc deficiency the plasma zinc turnover was increased, the 24-hour exchangeable pool was decreased, and cumulative excretion of [65]Zn in urine and stool was low.[5] Unfortunately, [65]Zn is not available for common use.

A good correlation between zinc status and the activity of ribonuclease has been observed.[14] The activity of ribonuclease in the plasma is increased during the zinc-deficient state. Also observed was an increase in the activity of plasma alkaline phosphatase following zinc supplementation to subjects with zinc deficiency.

METABOLIC AND BIOCHEMICAL ASPECTS OF ZINC IN MAN

An adult human body contains approximately 2 g of zinc. Liver, kidney, bone, retina, prostate, and muscle appear to be rich in zinc. In man, the zinc content of testes and skin have not been determined accurately, although it appears that clinically these tissues are

Table 4
A PARTIAL LIST OF
ZINC-DEPENDENT ENZYMES

Alcohol dehydrogenase
Alkaline phosphatase
Carboxy peptidase
Carbonic anhydrase
D-Lactate cytochrome reductase
Δ-Aminolevulinic acid dehydratase
Glyceraldehyde-phosphate dehydrogenase
Leucine aminopeptidase
Nucleic acid enzymes
    DNA polymerase
    RNA polymerase
    Reverse transcriptase
    Thymidine kinase
    Ribonuclease (inhibited by Zn)

sensitive to zinc depletion. In plasma, most zinc is protein bound.[21] Although serum albumin binds most zinc, other proteins, such as α_2-macroglobulin, transferrin, and ceruloplasmin, also have great affinity for binding zinc. A small fraction of plasma zinc is bound to amino acids and is, thus, ultrafiltrable. Histidine, glutamine, threonine, cystine, and lysine showed the most marked effects with respect to zinc binding. It is suggested that the amino acids, bound to zinc may have an important role in the biological transport of this element.

Approximately 20 to 30% of ingested dietary zinc is absorbed. Data on both the site(s) of absorption in man and on the mechanism(s) of absorption, whether it be active, passive or facultative transport, are meager. Zinc absorption is variable in extent and is highly dependent upon a variety of factors. For one thing, zinc is more available for absorption from animal proteins. Among other factors that might affect zinc absorption are body size, the level of zinc in the diet, and the presence in the diet of other potentially interfering substances, such as calcium, phytate, fiber, and other chelating agents.[28]

Normal zinc intake in a well-balanced American diet with animal protein is approximately 12 to 15 mg/day. Urinary zinc loss is approximately 0.5 mg/day. Loss of zinc by sweat may be considerable under certain climatic conditions; under normal conditions, approximately 0.5 mg of zinc may be lost daily by sweating. Endogenous zinc loss in the gastrointestinal tract may amount to 1 to 2 mg/day.

Many enzymes need zinc for their function. During the past 15 years, at least 30 enzymes have been identified requiring zinc for their activity. If related enzymes from different species are included, then, over 70 zinc metalloenzymes would be on record (Table 4). Zinc is present in several dehydrogenases, aldolases, peptidases, and phosphatases. In these metalloenzymes, the metal is located in the active site and participates in the actual catalytic process. Carbonic anhydrase, alkaline phosphatase, and alcohol and lactic dehydrogenase are some examples of zinc metalloenzymes.[29]

Zinc has been found in both DNA and RNA polymerases. The recent demonstration that RNA-dependent DNA polymerase in the reverse transcriptase of avian myeloblastosis and other viruses is also a zinc metalloenzyme indicates a direct relationship between zinc metabolism and malignancy and opens up a new area for investigation. Recent studies suggest that thymidine kinase is a zinc-dependent enzyme and is very sensitive to a lack of zinc.[30] The activity of RNAase is inhibited by the presence of zinc; thus, zinc seems to play a very significant role in RNA and DNA synthesis and catabolism of RNA. Growth retardation, so commonly seen as a result of zinc deficiency, is most likely due to its effect on nucleic acid metabolism and decreased protein synthesis.

Table 5
ENZYME CHANGES IN ZINC DEFICIENCY

Enzyme	Changes in zinc deficiency
Alkaline phosphatase	Decreased activity in plasma, bone, kidney, stomach, intestine, pancreas, thymus, and testis of experimental animals; activity increases in plasma of zinc-deficient dwarfs following Zn supplementation
Carboxypeptidase	Decreased activity in pancreas of experimental animals
Carbonic anhydrase	Decreased activity in blood and stomach of experimental animals; decreased activity in red cells of sickle cell anemia patients (conditioned deficiency of zinc)
Alcohol dehydrogenase	Decreased activity in liver, bone, kidney, esophagus, and testis of experimental animals
RNA polymerase	Decreased activity in liver nuclei in experimental animals
RNAase	Increased activity in testis, kidney, bone, and thymus of experimental animals; increased activity in plasma of sickle cell anemia patients (conditioned deficiency of zinc)
Thymidine kinase	Decreased activity in connective tissue harvested following sponge implantation and in regenerating liver of experimental animals

It has now been shown that the activity of various zinc-dependent enzymes is affected adversely by zinc deficiency (Table 5). Changes resulting from zinc deficiency in alkaline phosphatase, alcohol dehydrogenase, carboxypeptidase, and carbonic anhydrase have been reported in only the sensitive tissues of experimental animals, since the probability of detecting these biochemical changes is highest in those tissues which respond rather sensitively to a lack of available zinc. It is also believed that those metalloenzymes which bind zinc with a very high affinity are still fully active, even in the extreme stages of zinc deficiency.

TREATMENT

An oral supplement of 15 mg of zinc (90 mg of zinc sulfate) was sufficient to reverse the clinical manifestations of zinc deficiency in patients from Egypt. It should be emphasized, however, that they also received a well-balanced diet containing adequate animal protein; thus, the availability of zinc from dietary sources was optimal in their cases. It is reasonable to expect that in subjects subsisting mainly on cereal proteins with high phytate and fiber content, 15 mg of zinc as an oral supplement may not be enough to correct deficiency symptoms. It is clear that in order to arrive at a reasonable level of oral zinc-supplement dose, one must consider availability of zinc from dietary sources as an important factor.

For the treatment of wound-healing, 660 mg of zinc sulfate in three divided doses have been given to human subjects for several weeks without any side effects. In sickle cell anemia patients, 110 mg of zinc acetate every 4 hr has been used for the prevention of a pain crisis. In view of the fact that zinc is known to compete with copper at the cellular level, one should anticipate copper deficiency in subjects receiving high doses of zinc for a prolonged period.

In comparison with other trace elements, such as lead, cadmium, arsenic, and antimony, zinc is relatively nontoxic. Zinc is noncumulative, and the proportion absorbed is believed to be inversely related to the amount ingested. Vomiting, a protective mechanism, occurs after ingestion of large quantities of zinc; in fact, 2 g of zinc sulfate is recommended as an emetic. However, in one instance, toxicity was observed in a 16-year-old male who ingested 12 g of zinc sulfate over a 2-day period. It was characterized by drowsiness, lethargy, and increased serum lipase and amylase levels.

The use of zinc sulfate as an oral supplement may produce minor gastric irritation, which is believed to be due to the sulfate radical itself. Zinc acetate is perhaps a better choice for an oral supplement, inasmuch as patients with sickle cell anemia seem to tolerate it much better.

REFERENCES

1. Raulin, J., Études cliniques sur la végétation, *Ann. Sci. Nat. Bot. Biol. Veg.*, 11, 93, 1869.
2. Todd, W. R., Elvehjem, C. A., and Hart, E. B., Zinc in the nutrition of the rat *Am. J. Physiol.*, 107, 146–156, 1934.
3. Prasad, A. S., Halsted, J. A., and Nadimi, M., Syndrome of iron deficiency anemia, hepatosplenomegaly, hypogonadism, dwarfism and geophagia, *Am. J. Med.*, 31, 532–546, 1961.
4. Minnich, V., Okevogla, A., Tarcon, Y., Arcasoy, A., Yorukoglu, O., Renda, F., and Demirag, B., The effect of clay on iron absorption as a possible cause for anemia of Turkish subjects with pica, *Am. J. Clin. Nutr.*, 21, 78–86, 1968.
5. Prasad, A. S., Miale, A., Jr., Farid, Z., Schulert, A., and Sandstead, H. H., Zinc metabolism in patients with the syndrome of iron deficiency anemia, hypogonadism and dwarfism, *J. Lab. Clin. Med.*, 61, 537–549, 1963.
6. Prasad, A. S., Miale, A., Jr., Farid, Z., Sandstead, H. H., and Darby, W. J., Biochemical studies on dwarfism, hypogonadism and anemia, *A.M.A. Arch. Internal Med.*, 111, 407–428, 1936.
7. Sandstead, H. H., Prasad, A. S., Schulert, A. R., Farid, Z., Miale, A., Jr., Bassilly, S., and Darby, W. J., Human zinc deficiency, endocrine manifestations and response to treatment, *Am. J. Clin. Nutr.*, 20, 422–442, 1967.
8. Halsted, J. A., Smith, J. C., Jr., and Irwin, M. I., A conspectus of research on zinc requirements of man, *J. Nutr.*, 104, 345–378, 1974.
9. Hambidge, K. M., Hambidge, C., Jacobs, M., and Baum, J. D., Low levels of zinc in hair, anorexia, poor growth, and hypogeusia in children, *Pediatr. Res.*, 6, 868–874, 1972.
10. Reinhold, J. G., Faradji, B., Abadi, P., and Ismal-Beigi, F., Binding of zinc to fiber and other solids of wholemeal bread; with a preliminary examination of the effects of cellulose consumption upon the metabolism of zinc, calcium and phosphorus in man, in *Trace Elements in Human Health and Disease*, Vol. 1, Prasad, A. S., Ed., Academic Press, New York, 1976, 163–179.
11. Gudbjarnason, S. and Prasad, A. S., Cardiac metabolism in experimental alcoholism, in *Biochemical and Clinical Aspects of Alcohol Metabolism*, Sardesai, V. M., Ed., Charles C Thomas, Springfield, Ill., 1969, 266–272.
12. Vallee, B. L., Wacker, W. E. C., Bartholomay, A. F., and Hoch, F. L., Zinc metabolism in hepatic dysfunction, *N. Engl. J. Med.*, 257, 1055–1065, 1957.
13. Pories, W. J., Mansouri, E. G., Plecha, F. R., Flynn, A., and Strain, W. H., Metabolic factors affecting zinc metabolism in the surgical patient, in *Trace Elements in Human Health and Disease*, Vol. 1, Prasad, A. S., Ed., Academic Press, New York, 1976, 115–136.
14. Prasad, A. S., Schoomaker, E. B., Ortega, J., Brewer, G. J., Oberleas, D., and Oelshlegel, F. J., Zinc deficiency in sickle cell disease, *Clin. Chem.*, Winston-Salem, N.C., 21, 582–587, 1975.
15. Prasad, A. S., Abbasi, A., Oberleas, D., Rabbani, P., Fernandez-Madrid, F., and Ryan, J., Experimental production of zinc deficiency in man, *Clin. Res.*, 24, 486A, 1976, abstr.
16. Brewer, G. J., Oelshlegel, F. J., Jr., and Prasad, A. S., Zinc in sickle cell anemia, in *Erythrocyte Structure and Function*, Vol. 1, Brewer, G. J., Ed., Alan R. Liss, New York, 1975, 417–435.
17. Barnes, P. M. and Moynahan, E. J., Zinc deficiency in acrodermatitis enteropathica: multiple dietary intolerance treated with synthetic diet, *Proc. R. Soc. Med.*, 66, 327–329, 1973.
18. Klingberg, W. G., Prasad, A. S., and Oberleas, D., Zinc deficiency following penicillamine therapy, in *Trace Elements in Human Health and Disease*, Vol. 1, Prasad, A. S., Ed., Academic Press, New York, 1976, 51–65.

19.   **Kay, R. G. and Tasman-Jones, C.,** Acute zinc deficiency in man during intravenous alimentation, *Aust. N. Z. J. Surg.,* 45, 325–330, 1975.
20.   **Van Rij, A. M., McKenzie, J. M., and Dunckley, J. V.,** Excessive urinary zinc losses and aminoaciduria during intravenous alimentation, *Proc. Univ. Otago Med. Sch.,* 53, 77–78, 1975.
21.   **Prasad, A. S. and Oberleas, D.,** Binding of zinc to amino acids and serum proteins in vitro, *J. Lab. Clin. Med.,* 76, 416–425, 1970.
22.   **Prasad, A. S.,** Metabolism of zinc and its deficiency in human subjects, in *Zinc Metabolism,* Prasad, A. S., Ed., Charles C Thomas, Springfield, Ill., 1966, 250–303.
23.   **Prasad, A. S.,** Deficiency of zinc in man and its toxicity, in *Trace Elements in Human Health and Disease,* Vol. 1, Prasad, A. S., Ed., Academic Press, New York, 1976, 1–17.
24.   **Lei, K. Y., Abbasi, A., and Prasad, A. S.,** Function of pituitary-gonadal axis in zinc-deficient rats, *Am. J. Physiol.,* 230, 1730–1732, 1976.
25.   **Hurley, L. S.,** Perinatal effects of trace element deficiencies, in *Trace Elements in Human Health and Disease,* Vol. 2, Prasad, A. S., Ed., Academic Press, New York, 1976, 301–310.
26.   **Caldwell, D. F., Oberleas, D., Clancy, J. J., and Prasad, A. S.,** Behavorial impairment in adult rats following acute zinc deficiency, *Proc. Soc. Exp. Biol. Med.,* 133, 1417–1421, 1970.
27.   **Jameson, S.,** Effects of zinc deficiency in human reproduction, *Acta Med. Scand.,* Suppl. 593, 1–89, 1976.
28.   **Spencer, H., Osis, D., Kramer, L., and Norris, C.,** Intake, excretion, and retention of zinc in man, in *Trace Elements in Human Health and Disease,* Vol. 1, Prasad, A. S., Ed., Academic Press, New York, 1976, 345–359.
29.   **Vallee, B. L.,** Biochemistry, physiology and pathology of zinc, *Physiol. Rev.,* 39, 443–490, 1959.
30.   **Prasad, A. S. and Oberleas, D.,** Thymidine kinase activity and incorporation of thymidine into DNA in zinc deficient tissue, *J. Lab. Clin. Med.,* 83, 634–639, 1974.

EFFECT OF NUTRIENT DEFICIENCIES IN MAN: SELENIUM

H. Mitchell Perry, Jr.

There is currently no convincing evidence of selenium deficiency in man. However, the possibility of such a deficiency exists since selenium is essential to at least one human enzyme system, glutathione peroxidase; selenium deficiency diseases of herbivorous mammals living in areas with selenium poor soil are well known. Selenium metabolism is complex, it is closely interrelated with vitamin E metabolism, and it is not completely understood. A deficiency has been suspected or suggested as contributing to the human disease states listed in Table 1.

Selenium has a strong tendency to complex heavy metals. Availability of selenium could therefore alter human toxicity to environmentally significant metals such as mercury and cadmium. The evidence for this is by analogy with animal studies. In animals, selenium provides a marked protective effect against the acute toxicity from injected mercury[1] and against the chronic toxicity from feeding methyl mercury.[2] In like manner, selenium protects animals against toxicity from both injected[3] and fed[4] cadmium. Other elements, particularly arsenic and copper, interact with selenium and protect against toxic amounts of selenium.[5] Silver and manganese also interact with selenium; silver potentiates selenium deficiency, presumably by binding what selenium is available,[6] and manganese deficiency results in lowered tissue levels of both manganese and selenium.[7] **Present Status:** Despite the lack of confirmatory human data, low selenium states seem likely to increase toxic effects of mercury or cadmium in subjects exposed to excesses of these metals; however, the likelihood of selenium status influencing toxicity to other metals is much more speculative.

An inverse correlation between selenium content of the soil and cardiovascular death rate, particularly coronary and cerebrovascular death rates, has been reported.[8] Despite lack of confirmatory data or a postulated mechanism, this correlation is mentioned because the extreme public health importance of the diseases involved warrants consideration of any possible explanation of their currently unknown causation and any practical ways of reducing their present very high inci-

Table 1
POSSIBLE INVOLVEMENT OF SELENIUM
DEFICIENCY IN HUMAN DISEASE

Disease conditions[a]	Ref.
Enhancement of chronic cadmium or mercury toxicity	2,4
Partial cause of excess cardiovascular deaths in low selenium (possibly soft water) areas	8,13
Contributing factor in kwashiorkor	14,16,17
Partial cause of excess cancer deaths in low-selenium areas	19
Partial cause of muscular dystrophy	22,23
Contributing factor in sudden infant death (crib death)	25,26

[a]   The table lists six suggested effects of selenium on human disease; however, there is presently no convincing evidence that selenium deficiency has any significant effect on human health. A low selenium state might potentiate chronic toxicity to excess cadmium or mercury exposure, but there is currently little reason to postulate a significant role for selenium deficiency in other human disease states.

dence. The suggestion that low-selenium soil is correlated with cardiovascular mortality recalls the unexplained, but probably real, inverse correlation which has been reported between cardiovascular death rates and hardness of drinking water.[9,10] Moreover, it poses the question of whether selenium might be a protective factor present in hard water.[8] A possible but currently completely unsubstantiated mechanism for such an effect could involve the detoxification of cadmium by selenium. The proposed argument goes as follows: First, chronically feeding rats low doses of cadmium induces a mild to moderate degree of hypertension[11] which is blocked by feeding selenium.[4] Second, human beings exposed to cadmium in the environment accumulate renal cadmium concentrations comparable to those found in the cadmium-exposed hypertensive rats.[12] Third, although there is no evidence that cadmium induces hypertension in man, it could be one of various pressor influences, in which case selenium excess might antagonize it and selenium deficiency potentiate it. Somewhat against the suggestion that selenium deficiency is related to excess cardiovascular disease is the failure of preliminary findings from a WHO/IAEA joint research program involving tissue metal content in human cardiovascular disease to show any abnormalities in tissue selenium; however, the available data were far too meager to be conclusive.[13] **Present Status:** There is no evidence that selenium deficiency is related to excess cardiovascular disease in man; a relationship cannot be excluded, but the possibility is not seriously considered by most investigators.

There is suggestive evidence that selenium deficiency occurs in children with protein-calorie malnutrition, but it seems unlikely that selenium deficiency plays any specific or special role in kwashiorkor. Decreased plasma and whole blood selenium levels have been observed in protein-calorie malnutrition in Guatemala[14] and Thailand.[15] In five Guatemalan children with kwashiorkor, the initially low selenium levels in blood gradually rose to normal during 6 to 12 months of treatment without selenium supplementation.[14] The available data favoring a specific selenium effect include inadequately controlled reports of weight gain in two malnourished children[16] and a reticulocyte response in three of five anemic infants[17] following selenium administration. **Present Status:** Although a low selenium state may occur in kwashiorkor, it probably does not represent a specific deficiency; it seems more likely that selenoproteins are depleted as part of the general protein depletion.

A relationship of selenium to cancer has been postulated on the basis of a limited statistical correlation; it is mentioned only because of the great importance and unknown causation of the disease. Selenium has been reported to both cause cancer and protect against it[18] in animals. In man, an unexplained inverse relationship between selenium content of soil and cancer has been reported.[19] In addition, a low blood selenium level has been reported in cancer patients,[2] and an association between serum selenium and the status of the cancer has been reported.[21] However, there is very little to suggest any direct causative relationship. **Present Status:** There is no evidence that selenium deficiency causes or aggravates cancer in man.

White-muscle disease of sheep and cattle is a widespread, naturally occurring form of muscular dystrophy which involves both cardiac and skeletal muscle.[22,23] It is caused by selenium deficiency, and it is cured by selenium supplementation.[22] Because this entity resembles human muscular dystrophy, it is tempting to implicate selenium in the pathogenesis of the human disease, which is currently of an unknown etiology. However, selenium treatment has been of little value in human muscular dystrophy,[24] and there is little reason to assume that selenium is involved in human disease. **Present Status:** There is no evidence that selenium deficiency is involved in human muscular dystrophy despite the apparent clinical similarity of the muscular dystrophy of selenium deficiency in animals.

Because of some similarities between sudden death in piglets with selenium and vitamin E deficiencies and sudden infant death, selenium-vitamin E deficiency has been proposed as a cause of sudden infant death ("crib death"). An initial report of low postmortem blood levels of selenium in affected infants[25] was followed by a report of low vitamin E, but normal selenium, levels in fatal cases of the syndrome.[26] **Present Status:** There is no evidence that selenium deficiency is related to sudden infant death.

REFERENCES

1. **Parizek, J. and Ostadalova, I.,**The protective effect of small amounts of selenite in sublimate intoxication,*Experientia,*23, 142, 1967.
2. **Ganther, H.E., Gaudic, C., Sunde, M.L., Kopecky, M.J., Wagner, P., Oh, S.H., and Hoekstra, W.G.,**Selenium: relation to decreased toxicity of methylmercury added to diets containing tuna,*Science,*175, 1122, 1972.
3. **Chen, R.W., Wagner, P.A., Hoekstra, W.G., and Ganther, H.E.,**Affinity labeling studies with 109 Cd and Cd induced testicular injury in rats,*J. Reprod. Fertil.,*38, 293, 1974.
4. **Perry, H.M., Jr., Perry, E.F., and Erlanger, M.W.,**Reversal of cadmium-induced hypertension by selenium or hard water, trace substances in environmental health, Vol. 8, University of Missouri, Columbia, Mo., 1974, 51.
5. **Hill, C.H.,**Interrelationships of selenium with other trace elements,*Fed. Proc. Fed. Am. Soc. Exp. Biol.,*34, 2096, 1975.
6. **Ganther, H.E., Wagner, P.A., Sunde, M.L., and Hoekstra, W.G.,**Protective Effects of Selenium Against Heavy Metal Toxicities, Trace Substances in Environmental Health, Hemphill, D.D., Ed., Vol. 6, University of Missouri, Columbia, Mo., 1972, 247.
7. **Burch, R.E., Williams, R.V., Hahn, H.K.J., Jetton, M.M., and Sullivan, J.F.,**Tissue trace element and enzyme content in pigs fed a low manganese diet. I. A relationship between manganese and selenium,*J. Lab. Clin. Med.,*86, 132, 1975.
8. **Shamberger, R.J., Tutko, S.A., and Willis, C.E.,**Selenium and Heart Disease. Trace Substances in Environmental Health, Hemphill, D.D., Ed., Vol. 9, University of Missouri, Columbia, Mo., 1975, 15.
9. **Schroeder, H.A.,**Municipal drinking water and cardiovascular death rates,*JAMA,*195, 81, 1966.
10. **Morris, J.N., Crawford, M.D., and Heady, J.A.,**Hardness of local water supplies and mortality from cardiovascular disease,*Lancet,*1, 860, 1961.
11. **Perry, H.M., Jr., Erlanger, M.W., and Perry, E.F.,** Elevated systolic pressures following chronic low-level cadmium feeding,*Am. J. Physiol.,*232, H114, 1977.
12. **Perry, H.M., Jr., Tipton, I.H., Schroeder, H.A., Steines, R.L., and Cook, M.J.,**Variation in the concentrations of cadmium in human kidneys as a function of age and geographic origin,*J. Chron. Dis.,* 14, 259, 1961.
13. **Masironi, R. and Parr, R.,**Selenium and Cardiovascular Disease. Preliminary Results of the WHO/IAEA Joint Research Program. Proc. Industrial Health Found. Symp. on Selenium and Tellurium in the Environment, Frost, D., Ed., University of Notre Dame, Notre Dame, Ind., 1976.
14. **Burk, R.F., Pearson, W.N., Word, R.P., and Viteri, F.,**Blood selenium levels and *in vitro* red blood cell uptake of ^{75}Se in kwashiorkor,*Am. J. Clin. Nutr.,*20, 723, 1967.
15. **Levine, R.J. and Olson, R.E.,**Blood selenium in Thai children with protein calorie malnutrition,*Proc. Soc. Exp. Biol. Med.,* 134, 1030, 1970.
16. **Schwarz, K.,**Development and status of experimental work on factor 3 — selenium,*Fed. Proc. Fed. Am. Soc. Exp. Biol.,*20, 666, 1961.
17. **Hopkins, L.L. and Majaj, A.S.,**Selenium in human nutrition, in Selenium in Biomedicine, Muth, O.H., Ed., AVI Publishing, Westport, Conn., 1967, 203.
18. **Shapiro, J.R.,**Selenium and carcinogenesis: a review, *Ann. N.Y. Acad. Sci.,* 192:215, 1972.
19. **Shambergh, R.J. and Willis, C.E.,**Selenium distribution and human cancer mortality,*CRC Crit. Rev. Clin. Lab. Sci.,*2, 211, 1971.

20. **Shambergh, R.J., et al.,**Anti-oxidents and cancer. I. Selenium in the blood of normals and cancer patients, *J. Nat. Cancer Inst.,*50, 863, 1973.
21. **Broghamer, W.L., Jr., McConnell, K.P., and Blotcky, A.L.,**Relationship between serum selenium levels and patients with carcinoma, *Cancer,*37, 1384, 1976.
22. **Hartley, W.J. and Grant, A.B.,**A review of selenium-responsive diseases of New Zealand livestock, *Fed. Proc. Fed. Am. Soc. Exp. Biol.,*20, 679, 1961.
23. **Muth, A.H., et al.,**Effects of selenium and vitamin E on white muscle disease, *Science,*128, 1090, 1958.
24. **Levander, O.A.,** Selenium and chromium in human nutrition, *J. Am. Diet. Assoc.,* 66, 338, 1975.
25. **Money, D.F.L.,**Vitamin E and selenium deficiencies and their possible etiological role in the sudden deaths in infants syndrome, *N.Z. Med. J.,*71, 32, 1970.
26. **Rhead, W.J., Cary, E.E., Alloway, W.H., Saltzstein, S.L., and Schrauzer, G.N.,**The vitamin E and selnium status of infants and the sudden infant death syndrome, *Bioinorg. Chem.,*1, 289, 1972.

NUTRITIONAL DEFICIENCIES IN MAN — CHROMIUM

K. M. Hambidge

INTRODUCTION

The nutritional importance of chromium has received only recent recognition,[1,2] and data are limited on human deficiency states. However, there is suggestive evidence that inadequate chromium nutrition may occur frequently in man. The most notable and earliest detectable effect of mild chromium deficiency in animals is an impairment of glucose utilization, attributable to a diminished response of peripheral tissues to insulin.[2] Therefore, studies in man have been directed primarily towards those conditions that are associated with an impairment of glucose tolerance.

GLUCOSE TOLERANCE FACTOR

The biologically active form of chromium is a naturally occurring, low molecular weight, organic complex containing Cr^{+++} that has been termed "glucose tolerance factor" (GTF). GTF is probably a nicotinic acid-chromium complex.[3] In contrast to GTF, inorganic Cr^{+++} exhibits very little biological activity and is absorbed very poorly in the intestinal tract (<1% absorption).[2] Currently, calculations of chromium nutritional requirements and assessment of chromium nutritional status are complicated by difficulties in measuring the GTF fraction of the total chromium in food items, tissue samples, blood, and urine.

CHROMIUM IN TISSUES AND BODY FLUIDS

The total body content of chromium has been calculated to be approximately 6 mg. Concentrations in different tissues have been found to vary from less than 1 to several thousand parts per billion (ppb) on a wet weight basis.[2] However, the chromium content of most tissues is less than 100 ppb.

Autopsy Tissues

Considerable geographical variation has been reported[4-7] in the chromium content of autopsy samples with relatively low levels for adults in the U.S. (Table 1). The concentration of chromium in most tissues in the U.S. has been found to decline with increasing age during childhood[7-9] (Table 2). This decline is not as notable in many other countries and has led to speculation that chromium depletion may be unusually frequent in North American adults. The concentration of chromium in the liver of diabetic subjects has been reported to be lower than normal.[10]

Blood

There is no universal agreement on the normal concentration of chromium in blood plasma or serum. In general, levels reported since 1970 have been considerably lower than earlier literature values; these are thought to reflect improvements in analytical techniques.[9] However, considerable variation still exists between different laboratories, with mean values ranging from approximately 1 to 10 ng/ml (ppb).[11,12] Fasting plasma chromium concentrations have not been considered a valid index of chromium nutritional status,[13] although unusually low levels have been observed during pregnancy[14] and in association with acute infectious illnesses.[11] Normally, an increase in plasma chromium may follow the administration of oral glucose or intravenous insulin.[12,15,16] This

Table 1

CONCENTRATION OF CHROMIUM IN
ADULT HUMAN TISSUES FROM
FOUR AREAS OF THE WORLD[a]

	U.S.	Africa	Near East	Far East
Aorta	1.9	5.5	11.0	15.0
Brain	0.2	0.6	2.4	2.7
Heart	1.6	0.9	4.0	6.3
Kidney	0.8	2.0	5.3	3.3
Liver	0.8	1.3	2.1	2.0
Lung	14.0	16.0	22.0	23.0
Pancreas	1.6	2.5	4.2	6.7
Spleen	0.5	1.3	3.1	3.1
Testis	1.6	4.2	7.0	8.3
Total (μg)[b]	<6,000	7,400	11,800	12,400

[a] Median values for males aged 20 to 59 in parts per million ash.[4]
[b] Estimated total in body of standard man, based on comparable organ and tissue concentrations.

increment is thought to be GTF chromium that may be released into the circulation in response to an increase in circulating insulin.[2,13] However, this increment in plasma chromium has not been detectable in all normal subjects,[17] and there have been reports of a decrease rather than an increase in concentration following administration of insulin or glucose.[11,14] These discrepancies may be attributable to variable loss of volatile GTF chromium with some analytical techniques.[18] Although measurement of the plasma chromium response does not provide a reliable means of assessing chromium nutritional status, differences in this response, which have been observed in some diabetics,[15] during an acute infection,[11] and in pregnancy[14,17] may reflect changes in chromium metabolism or nutritional status.

Urinary Excretion of Chromium

At least 80% of chromium is excreted via the kidneys, and measurement of urine chromium excretion may provide a useful index of chromium nutritional status.[9,21] Variations in reported values for chromium in urine (Table 3) probably reflect not only analytical errors, but also differences in chromium nutritional status[24] and variable loss of a volatile chromium fraction.[25] Rates of excretion may be increased by a glucose load[4] and by a high sucrose diet.[26] In insulin-dependent diabetics, excretion rates have been reported to be either unusually high[17,27] or low.[23]

Chromium in Hair

The hair content of chromium is relatively high. Significant variations have been reported between different populations. The lower levels for the groups included in Table 4 may reflect a body depletion of chromium, although confirmation is lacking on the value of hair chromium as an index of chromium nutritional status.

In summary, no reliable biochemical indices of human chromium nutritional status have been established. However, results of measuring the chromium concentration in blood, urine, hair, and autopsy tissues have indicated a number of circumstances in which there may be an increased risk of chromium deficiency.

Table 2

CHROMIUM IN HUMAN TISSUES BY AGE, U.S. VALUES (μg/g ASH)

	0–45 days	45 days to 10 years	10–20 years	20–30 years	30–40 years	40–50 years	50–60 years	60–70 years	70–80 years	80 + years
Kidney	51.8[a]	57.5	3.7	2.3	2.1	3.9	3.0	2.4	2.0	10.0
%	100(21)[b]	96(25)	79(14)	78(18)	82(38)	91(23)	85(20)	79(11)	100(5)	100(1)
Liver	17.9	16.6	4.6	2.9	1.8	3.1	2.5	1.3	1.1	1.3
%	100(28)	92(25)	79(14)	83(18)	72(39)	91(23)	95(21)	73(11)	100(5)	100(2)
Lung	85.2	10.1	6.8	8.2	15.6	31.8	24.2	21.0	38.0	1.1
%	100(12)	100(6)	100(12)	100(18)	100(38)	100(22)	100(21)	100(10)	100(5)	100(1)
Aorta	19.5	7.0	7.0	4.2	9.1	4.9	4.2	4.4	2.6	0
%	100(4)	100(1)	67(7)	82(11)	78(27)	77(22)	67(15)	80(5)	33(3)	0(1)
Heart	82.4	1.7	2.1	3.3	3.8	6.0	3.7	3.0	2.9	0
%	100(9)	100(6)	90(10)	94(18)	87(38)	95(23)	85(20)	80(10)	100(5)	0(1)
Pancreas	—	2.6	3.4	2.5	6.5	2.5	3.6	3.7	2.9	0.9
%	100(0)	100(2)	83(12)	93(17)	85(39)	100(21)	88(16)	90(10)	100(4)	100(1)
Spleen	27.0	4.1	1.6	1.1	1.7	1.5	1.8	3.0	1.9	0.3
%	100(7)	100(5)	67(12)	61(18)	73(37)	62(22)	75(20)	67(9)	50(4)	100(1)
Testes	—	—	2.1	1.8	3.1	3.2	2.3	3.6	1.1	1.0
%	(0)	(0)	100(5)	100(8)	95(20)	100(12)	91(11)	100(10)	100(5)	100(1)

[a] Mean values when present.
[b] % Occurrence. Numbers in parentheses are numbers of samples.

From Schroeder, H. A., Balassa, J. J., and Tipton, I. H., *J. Chronic Dis.*, 15, 941–964, 1962. With permission.

Table 3
EXCRETION OF CHROMIUM IN URINE

Year	No.	Subjects description	μg/l Mean ± SE	μg/l Range	μ/24 hr Mean ± SE	μ/24 hr Range	Ref.
1963	154	Normal adults	4.0	(1.8—11.0)	—	—	19
1966	—	—	5.0	—	—	—	20
1975	9	Healthy, young women	9.4 ± 0.6	(7.2—11.0)	7.2 ± 0.4	(5.9—10.0)	21
1971	20	Normal young adults	—	—	8.4	(1.6—21.0)	17
1971	18	Children	—	—	5.5	(0.8—11.5)	17
1972	10	Normal adults	5.2	(2.6—10.6)	—	—	22
1968	—	Normal adults	—	(20—40)	—	—	4
1975	17	Normal adults	30.8 ± 4.1	—	—	—	23
1975	5	Normal adults (Turkey)	3.1 ± 0.8	—	2.7 ± 0.7	—	24

Table 4
POPULATION GROUPS WITH LOW HAIR CHROMIUM CONCENTRATIONS

Subjects	Mean hair chromium (μg/g dry wt)	Mean hair chromium of control subjects	Significance of difference	Ref.
Parous women	220	750	p < 0.01	28
Juvenile diabetics	560	850	p < 0.001	29
Premature newborn	325	970	p < 0.005	9
Newborn — intrauterine growth retardation	154	970	p < 0.005	9
Diabetics in Thailand	94	241	p < 0.01	30
Pregnant women at term	360	570	—	31

CHROMIUM SUPPLEMENTATION STUDIES

Studies of human dietary chromium supplementation have been limited in number and scope, but the results of some of these studies have provided good evidence for the occurrence of chromium deficiency in man. Improvements in or correction of impaired glucose tolerance following supplementation with Cr^{+++} has been observed in the conditions listed in Table 5. Glucose tolerance has improved in some diabetics treated with inorganic Cr^{+++} (150 to 1000 μg/day) for periods of several weeks or months, but this has not been a consistent finding[4,34] and no improvement was observed in one double-blind study.[37] Similarly, glucose tolerance of some middle-aged[35] and elderly[16] subjects has improved following oral Cr^{+++} supplementation. In Turkey,[32] Jordan, and Nigeria,[33] but not in Egypt,[38] a single oral dose of 250 μg Cr^{+++} was followed by significant improvement in, or total correction of, the abnormal glucose tolerance in infants suffering from protein-calorie malnutrition. Improvement over the same time interval was not observed in malnourished infants who did not receive chromium. The subsequent weight gain of the Turkish infants who received chromium was more rapid than that of controls.[39]

Pure GTF chromium is not yet available for human supplementation studies. Administration of batches of brewer's yeast that had substantial quantities of GTF-chromium activity has been followed by improved glucose tolerance, a reduction in insulin/glucose ratios after glucose loading, a reduction in exogenous insulin require-

Table 5

STUDIES IN WHICH
CHROMIUM SUPPLEMENTATION
HAS BEEN ASSOCIATED WITH
IMPROVED GLUCOSE TOLERANCE

Condition	Ref.
Protein calorie malnutrition	32, 33
Diabetes mellitus	34
Middle-aged Americans	35
Elderly Americans	16
Prolonged total parenteral nutrition	36

Table 6

CHROMIUM CONTENT OF DIETS

Geographical area	Type of diet	Chromium (μg/day)	Ref.
U.S.	Institutional	78	7
U.S.	Hospital	101	43
U.S.	Institutional	52	16
U.S.	Ad libitum	65	16
U.S.	Self-selected	200–400	48
Rome	Dinner only	64	2
Japan	Self-selected	130–140	49
Egypt	Egypt[a]	58–588	50

[a] Calculated values for family having child with kwashiorkor. The range reflects variation in measurements according to analytical techniques employed.

ments for insulin-dependent diabetics, and a reduction in serum levels of cholesterol and triglycerides.[40] It has been tentatively concluded, although not proven, that these improvements are attributable to the GTF content of the brewer's yeast.

DIETARY CHROMIUM INTAKE AND REQUIREMENTS

Average measured or calculated daily dietary intakes of chromium are given in Table 6. The variation between these figures may be attributed in part to errors in analytical techniques. However, wide variations between individual diets occur as a result of individual food preferences, as illustrated in Table 7. Spices, mushrooms, brewer's yeast, and animal meats, especially liver and kidney, have been reported to have a relatively high chromium content, while seafoods, for example, contain relatively little chromium.[2,7,16,41,42] Cooking acidic foodstuffs in stainless steel ware may leach appreciable quantities of chromium from the steel into the food. Refining processes may remove a large percentage of the chromium from some major food items, e.g., sugar and cereal grains.[43-46] The concentration of chromium in water is generally low[47] and does not contribute significantly to the dietary intake of this element.

The daily requirement for chromium is uncertain and recommended dietary allowances have not been established. The quantity required to maintain chromium balance will depend to a large extent on the type of chromium ingested. While a diet containing 30 μg or less of GTF chromium may be adequate, several hundred micrograms of inorganic Cr^{+++} may be required. Measurement of the chromium present in a 50%

Table 7
COMPOSITION OF ELDERLY SUBJECTS' DIET

Day 1 — 115 µg Cr per day
> Breakfast: orange juice, farina, toast, milk, coffee
> Lunch: cream of tomato soup, egg souffle, tossed salad, applesauce, tea
> Dinner: meat loaf with gravy, mashed potato, beets, cottage cheese, cookies, milk, tea

Day 2 — 47 µg Cr per day
> Breakfast: prune juice, oatmeal, 1 egg, 2 slices toast, 1 cup milk, coffee
> Lunch: chicken noodle soup, roast beef hash, crushed pineapple, 1 cup tea
> Dinner: baked chicken, dressing, gravy, mashed potato, 1 slice bread, peas, fruit gelatin, milk, tea

Day 3 — 5 µg Cr per day
> Breakfast: prune juice, farina, egg, toast, milk, coffee
> Lunch: clam chowder, tuna fish sandwich, white cake, tea
> Dinner: creamed cod fish, mashed potato, peas, apricots, 1 slice bread, milk

From Levine, R. A., Streeten, D. H. P., and Doisy, R. J., *Metab. Clin. Exp.,* 17, 114—125, 1968.

ethanol extract of foods (but not of total chromium) provides a good indication of the GTF biological activity of these foods.[46] Food items found to contain the highest concentrations of ethanol-extractable chromium include brewer's yeast, liver, and beef and, on a dry weight basis, mushrooms and beer. Those with the lowest concentrations include skimmed milk, fish, flour, chicken, fruits, and vegetables. A percentage of the chromium present in food items and other biological samples is very volatile and, consequently, may be lost during sample preparation or analyses. It is believed that this fraction is GTF chromium.[18]

REFERENCES

1. **Schwarz, K. and Mertz, W.,** Chromium (III) and the glucose tolerance factor, *Arch. Biochem. Biophys.,* 85, 292—295, 1959.
2. **Mertz, W.,** Chromium occurrence and function in biological systems, *Physiol. Rev.,* 49, 163—239, 1969.
3. **Mertz, W., Toepfer, E. W., Roginski, E. E., and Polansky, M. M.,** Present knowledge of the role of chromium, *Fed. Proc.,* 33, 2275—2280, 1974.
4. **Schroeder, H.,** The role of chromium in mammalian nutrition, *Am. J. Clin. Nutr.,* 21, 230—244, 1968.
5. **Tipton, I. H. and Cook, M. J.,** Trace elements in human tissue. Part II. Adult subjects from the United States, *Health Phys.,* 9, 103—145, 1963.
6. **Tipton, I. H., Schroeder, H. A., Perry, H. M., Jr., and Cook, M. J.,** Trace elements in human tissue. Part III. Subjects from Africa, the Near and Far East and Europe, *Health Phys.,* 11, 403—451, 1965.
7. **Schroeder, H. A., Balassa, J. J., and Tipton, I. H.,** Abnormal trace metals in man — chromium, *J. Chronic Dis.,* 15, 941—964, 1962.
8. **Schroeder, H. A.,** Cadmium, chromium and cardiovascular disease, *Circulation,* 35, 570—582, 1967.
9. **Hambidge, K. M.,** Chromium nutrition in man, *Am. J. Clin. Nutr.,* 27, 505—514, 1974.
10. **Morgan, J. M.,** Hepatic chromium content in diabetic subjects, *Metab. Clin. Exp.,* 21, 313—316, 1972.
11. **Pekarek, R. S., Hauer, E. C., Rayfield, E. J., Wannemacher, R. W., Jr., and Beisel, W. R.,** Relationship between serum chromium concentrations and glucose utilization in normal and infected subjects, *Diabetes,* 24, 350—353, 1975.

12. **Behne, D. and Diel, F.,** Relations between carbohydrate and trace element metabolisms investigated by neutron activation analysis, in *Nuclear Activation Techniques in the Life Sciences,* IAEA-SM-157/11, International Atomic Energy Agency, Vienna, 1972, 407–413.

13. **Mertz, W. and Roginski, E. E.,** Chromium metabolism: the glucose tolerance factor, in *Newer Trace Elements in Nutrition,* Mertz, W. and Cornatzer, W. E., Eds., Marcel Dekker, New York, 1971, 123–153.

14. **Davidson, I. W. F. and Burt, R. L.,** Physiologic changes in plasma chromium of normal and pregnant women: effect of a glucose load, *Am. J. Obstet. Gynecol.,* 116, 601–608, 1973.

15. **Glinsmann, W. F., Feldman, F. J., and Mertz, W.,** Plasma chromium after glucose administration, *Science,* 152, 114–128, 1966.

16. **Levine, R. A., Streeten, D. H. P., and Doisy, R. J.,** Effect of oral chromium supplementation on the glucose tolerance of elderly subjects, *Metab. Clin. Exp.,* 17, 114–125, 1968.

17. **Hambidge, K. M.,** Chromium nutrition in the mother and the growing child, in *Newer Trace Elements in Nutrition,* Mertz, W. and Cornatzer, W. E., Eds., Marcel Dekker, New York, 1971, 169–194.

18. **Wolf, W., Mertz, W., and Masironi, R.,** Determination of chromium in refined and unrefined sugars by oxygen plasma ashing flameless atomic absorption, *J. Agric. Food Chem.,* 22, 1037–1042, 1974.

19. **Imbus, H. R., Cholak, J., Miller, L. H., and Sterling, T.,** Boron, cadmium, chromium and nickel in blood and urine, *Arch. Environ. Health,* 6, 286–295, 1963.

20. **Pierce, J. R., II and Cholak, J.,** Lead, chromium and molybdenum by atomic absorption, *Arch. Environ. Health,* 13, 208–212, 1966.

21. **Mitman, F. W., Wolf, W. R., Kelsay, J. L., and Prather, E. S.,** Urinary chromium levels of nine young women eating freely chosen diets, *J. Nutr.,* 105, 64–68, 1975.

22. **Davidson, I. W. F. and Secrest, W. L.,** Determination of chromium in biological materials by atomic absorption spectrometry using a graphite furnace atomizer, *Anal. Chem.,* 44, 1808–1813, 1972.

23. **Canfield, W. K. and Doisy, R. J.,** Evidence for an unrecognized metabolic defect in diabetic subjects, *Diabetes,* 24, 56, 1975.

24. **Gurson, C. T., Saner, G., Mertz, W., Wolf, W. R., and Sokucu, S.,** Nutritional significance of chromium in different chronological age groups and in populations differing in nutritional backgrounds, *Nutr. Rep. Int.,* 12, 9–17, 1975.

25. **Wolf, W. R., Greene, F. E., and Mitman, F. W.,** Determination of urinary chromium by low temperature ashing – flameless atomic absorption, *Fed. Proc. Abstr.,* 33, 659, 1974.

26. **Mitman, F. W., Wolf, W. R., Kelsay, J. L., and Prather, E. S.,** The influence of dietary sucrose on urinary excretion of chromium, *Fed. Proc. Abstr.,* 33, 659, 1974.

27. **Doisy, R. J., Streeten, D. H. P., Souma, M. L., Kalafer, M. E., Rekant, S. E., and Dalakos, T. G.,** Metabolism of ^{51}chromium in human subjects – normal, elderly and diabetic subjects, in *Newer Trace Elements in Nutrition,* Mertz, W. and Cornatzer, W. E., Eds., Marcel Dekker, New York, 1971, 155–168.

28. **Hambidge, K. M. and Rodgerson, D. O.,** Comparison of hair chromium levels of nulliparous and parous women, *Am. J. Obstet. Gynecol.,* 103, 320–321, 1969.

29. **Hambidge, K. M., Rodgerson, D. O., and O'Brien, D.,** Concentrations of chromium in the hair of normal children and children with juvenile diabetes mellitus, *Diabetes,* 17, 517–519, 1968.

30. **Benjanuvatra, N. K. and Bennion, M.,** Hair chromium concentration of Thai subjects with and without diabetes mellitus, *Nutr. Rep. Int.,* 12, 325–330, 1975.

31. **Burt, R. L. and Davidson, I. W. F.,** Carbohydrate metabolism in pregnancy: a possible role of chromium, *Acta Diabetol. Lat.,* 10, 770–778, 1973.

32. **Gurson, C. T. and Saner, G.,** Effect of chromium on glucose utilization in marasmic protein-calorie malnutrition, *Am. J. Clin. Nutr.,* 24, 1313–1319, 1971.

33. **Hopkins, L. L., Jr., Ransome-Kuti, O., and Majaj, A. S.,** Improvement of impaired carbohydrate metabolism by chromium (III) in malnourished infants, *Am. J. Clin. Nutr.,* 21, 203–211, 1968.

34. **Glinsmann, W. H. and Mertz, W.,** Effect of trivalent chromium on glucose tolerance, *Metabol. Clin. Exp.,* 15, 510–520, 1966.

35. **Hopkins, L. L., Jr. and Price, M. G.,** Effectiveness of chromium (III) in improving the impaired glucose tolerance of middle-aged Americans, *Western Hemisphere Nutrition Congr., Puerto Rico,* Vol. 2 (Abstr.), 40–41, 1968.

36. **Jeejeebhoy, K. N., Chu, R., Marliss, E. B., Greenberg, C. R., and Bruce-Robertson, A.,** Chromium deficiency diabetes and neuropathy reversed by chromium infusion in a patient on total parenteral nutrition for 3½ years, *Clin. Res.,* 23, 636A, 1975.

37. **Sherman, L., Glennon, J. A., Brech, W. J., Klomberg, G. H., and Gordon, E. S.,** Failure of trivalent chromium to improve hyperglycemia in diabetes mellitus, *Metabol. Clin. Exp.,* 17, 439, 1968.

38. **Carter, J. P., Kattab, A., Abd-El-Hadi, K., Davis, J. T., El Gholmy, A., and Patwardhan, V. N.,** Chromium (III) in hypoglycemia and in impaired glucose utilization in kwashiorkor, *Am. J. Clin. Nutr.,* 21, 195—202, 1968.

39. **Gurson, C. T. and Saner, G.,** Effects of chromium supplementation on growth in marasmic protein-calorie malnutrition, *Am. J. Clin. Nutr.,* 26, 988—991, 1973.

40. **Doisy, R. J., Streeten, D. H. P., Freiberg, J. M., and Schneider, A. J.,** Chromium metabolism in man and biochemical effects, in *Trace Elements in Human Health and Disease,* Prasad, A. S., Ed., Academic Press, New York, 1976, 79—104.

41. **Toepfer, E. W., Mertz, W., Roginski, E. E., and Polansky, M. M.,** Chromium in foods in relation to biological activity, *Agric. Food Chem.,* 21, 69—73, 1973.

42. **Gormican, A.,** Inorganic elements in foods used in hospital menus, *J. Am. Diet. Assoc.,* 56, 397—403, 1970.

43. **Schroeder, H. A., Nason, A. P., and Tipton, I. H.,** Chromium deficiency as a factor in altherosclerosis, *J. Chronic Dis.,* 23, 123—142, 1970.

44. **Czerniejewski, C. P., Shank, C. W., Bechtel, W. G., and Bradley, W. B.,** The minerals of wheat flour and bread, *Cereal Chem.,* 41, 65—72, 1964.

45. **Masironi, R., Wolf, W., and Mertz, W.,** Chromium in refined and unrefined sugars — possible nutritional implications in the etiology of cardiovascular diseases, *Bull. WHO,* 49, 327—432, 1973.

46. **Schroeder, H. A.,** Losses of vitamin and trace minerals resulting from processing and preservation of foods, *Am. J. Clin. Nutr.,* 24, 562—573, 1971.

47. **Committee on Biologic Effects of Atmospheric Pollutants,** *Chromium,* National Academy of Sciences, Washington, D. C., 1974, 28—34.

48. **Tipton, I. H., Stewart, P. L., and Martin, P. G.,** Trace elements in diets and excretion, *Health Phys.,* 12, 1683—1689, 1966.

49. **Murakami, Y., Suzuki, Y., Yamagata, T., and Yamagata, N.,** Cr and Mu in Japanese diet, *J. Radiat. Res.,* 6, 105—110, 1965; *Chem. Abstr.,* 65, 11033, 1966.

50. **Maxia, V., Meloni, S., Rollier, M. A., Brandone, A., Patwardhan, V. N., Waslien, C. I., and Said-El-Shami,** in *Nuclear Activation Techniques in the Life Sciences,* IAEA-SM-157/67, International Atomic Energy Agency, Vienna, 527—550.

Other Nutrients

EFFECT OF NUTRIENT DEFICIENCIES IN ANIMALS AND MAN:

PROTEIN-ENERGY DEFICIENCY

J.C. Edozien

INTRODUCTION

Kwashiorkor is the end result of severe protein deficiency in children. This disease, which is characterized by edema, fatty liver, and hypoproteinemia, occurs when the diet has a low protein content or is based on a staple of low-quality protein such as maize. Other frequently observed but inconsistent signs of the disease include depigmentation of the hair and skin, dermatosis, and anemia. Chronic starvation causes severe deficiencies of energy and protein and produces the clinical condition of marasmus in children as well as adults. Chronic semistarvation in infants and children usually leads to marasmic-kwashiorkor in which the effects of energy restriction are confounded, to a varying degree, with those of concomitant deficiency of protein. This occurs because some of the dietary protein is used to meet energy needs, and the extent to which this utilization of protein for energy takes place varies according to the degree of starvation and the amount of protein in the diet. The adult equivalent of marasmic-kwashiorkor (nutritional edema) often occurs in times of famine or war.

Kohman[1] was probably the first person to demonstrate that a kwashiorkor-like syndrome with edema can be produced experimentally by feeding young rats a low-protein diet with carrots as the only source of protein. Other studies using force-feeding techniques[2-6] have subsequently reported the induction of experimental kwashiorkor in experimental animals. Kirsch et al.[7] reported that young rats of the Wistar strain fed a diet containing 5% mixed protein developed clinical, biochemical, and histological disorders similar to those seen in children suffering from kwashiorkor. Edozien[8] reported that weanling, male Sprague-Dawley rats freely fed a diet containing 0.5% lactalbumin (a high-quality protein) developed experimental kwashiorkor after a variable time period ranging from 2 to 4 months. Rats fed restricted amounts of diets containing 1 to 18% lactalbumin protein developed the experimental equivalent of marasmus.[8,9] These experimental prototypes of kwashiorkor and marasmus in the rat are illustrated in Figure 1.

Less severe degrees of protein-energy deficiency lead to retarded physical and mental growth and development, delayed maturation, reduced fertility, adaptive metabolic changes, and alterations in the chemical composition of the body. Many of these changes are believed to be responses to poorly understood adaptive alterations in endocrine function. Since these developmental, morphological, and biochemical changes have been described in a previous section,[10] this section will summarize the results of recent studies undertaken in the Department of Nutrition, School of Public Health, The University of North Carolina at Chapel Hill, to determine the nature and extent of the hormonal changes and to measure the relative contributions of levels of dietary protein, energy, and fat in bringing about these changes.

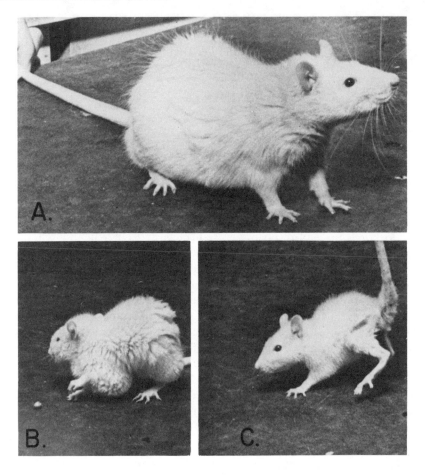

FIGURE 1.   Comparison of three 12-week-old rats fed diets of varying protein level either ad libitum or in restricted amounts for 8 weeks.(A) Fed 15% lactalbumin protein ad libitum;(B) fed 0.5% lactalbumin protein ad libitum; and (C) fed 1.0% lactalbumin protein restricted to 50% of the amount of diet consumed by rats fed 0.5% lactalbumin protein ad libitum. Hence, rats shown in B and C ate the same daily amount of lactalbumin protein. Edema was observed only in rats fed 0.5 to 1% lactalbumin protein diets ad libitum; those fed 2% lactalbumin protein diets ad libitum had fatty liver but no evident edema.

MATERIALS AND METHODS

Male Sprague-Dawley rats* weighing 70 to 80 g were purchased and fed a laboratory stock diet** until attaining a weight of 90 to 110 g. They were then fed isocaloric diets containing 2, 5, 10, 15, 25, or 50% lactalbumin-protein diets given freely or in restricted amounts for 8 weeks. At each protein level, the diets of the ad libitum-fed animals contained 5, 11.9, or 21.1% fat as cottonseed oil. Only the diets containing 11.9% fat were given in restricted amounts. The experimental design is summarized in Table 1 and the composition of the 15% protein diets is shown in Table 2. Rats on restricted intake were fed either 50 or 75% of the amount of diet consumed ad libitum by rats of similar weight irrespective of their age; their ration was usually provided each day between 1500 and 1700 hr, and all of it was generally consumed by the early evening.

*   Charles River Laboratories, Wilmington, Massachusetts.
**   Purina Laboratory Chow, Ralston Purina Company, Checkerboard Square, St. Louis, Missouri.

Table 1
DESIGN OF EXPERIMENTS TO INVESTIGATE THE EFFECTS OF LEVELS OF DIETARY PROTEIN, FAT, AND ENERGY ON THE BODY COMPOSITION OF THE RAT

Protein (%)	Fat (%) 5	11.9			21.1
2	—	√	√	√	—
5	√	√	√	√	√
10	√	√	√	√	√
15	√	√	√	√	√
25	√	√	√	√	√
50	√	√	√	√	√
Energy intake: % of ad libitum	100	100	75	50	100

The rats were housed individually in wire-mesh cages located in a room with automatically controlled 12 hr of light (0700 to 1900 hr) and 12 hr of darkness (1900 to 0700 hr) every day. They were given water ad libitum. Four rats from each dietary group were transferred to a metabolic cage for 2 days of each week for the measurement of food intakes. The food dishes were filled and weighed, and the amount consumed was calculated from the difference in weight 24 hr later. All spilled food was collected and weighed and was taken into account in calculating food intakes. The food consumption of each rat was taken as the average of the 2-day food intake study period. The food consumption data were used to adjust the amount of diet fed to the rats on restricted intakes.

Diets were removed from the rats 10 to 15 hr before they were anesthetized by intraperitoneal injections of phenobarbital* (5 mg/100 g body weight). The abdominal wall and thoracic cage were opened and blood was drawn from the right atrium into a heparinized syringe. All blood samples were drawn between 0800 and 1000 hr. Blood collection was completed within 5 min after injection of the anesthetic in order to minimize any effect of the drug on blood constituents. For the glucagon assay, 0.1 ml of Trasylol** containing 500 KIU was added to a 1-ml aliquot of blood; the sample was centrifuged 1200 × G for 10 min and the plasma was frozen at −70°C until the time of analysis. The balance of the blood sample was placed in a plastic centrifuge tube and kept on ice. Within 1 hr of collection it was centrifuged at 10,000 rpm*** for 10 min at 4°C. A 0.2- to 1.0-ml aliquot of the plasma was placed in a vial containing 25 mg $Na_2S_2O_5$ and was used for the assay of catecholamines. The remaining plasma sample was divided into several aliquots, depending on the number of assays planned in order to minimize thawing and refreezing of samples. All samples were kept frozen at −70°C until they were analyzed.

As soon as blood collection was completed and the blood sample was being processed, the liver was quickly excised and weighed, then placed in a plastic bag and immediately frozen in a large beaker containing alcohol and dry ice. It was kept frozen at −70°C until the time of analysis. Small pieces of some livers were fixed in buffered 10% formalin and later used for histology.

* Nembutal, Abbott Laboratories, North Chicago, Illinois.
** FBA Pharmaceuticals, New York, New York.
***Type JA-21 rotor in a J-21 centrifuge, Beckman Instruments, Inc., Palo Alto, California.

Table 2
COMPOSITION OF 15% LACTALBUMIN
PROTEIN DIETS WITH 5, 11.9, AND
21.1% FAT

Food component[a]	Percentage of fat in diet[b,c]		
	5	11.9	21.1
Lactalbumin[d]	187.50	187.50	187.50
Dextrin, white technical[e]	477.83	459.85	282.34
Sucrose (cane sugar)[e]	238.92	195.65	141.17
Salt mix[f]	50.00	50.00	50.00
Vitamin mixture[f]	5.00	5.00	5.00
Choline chloride (100%)	2.00	2.00	2.00
Cottonseed oil	38.75	100.00	200.03
Fiber[g]	—	—	131.97

[a]   Amounts are in grams per kilogram diet.

[b]   Diets were made up, as specified, by General Biochemicals, Laboratory Park, Chagrin Falls, Ohio 44022.

[c]   All diets provide about 4.0 kcal/g. The metabolizable energy (ME) values used were in kilocalories per grams dextrin, 3.64; sucrose, 4.0; and cottonseed oil, 9.0. The percentages of total energy derived from fat in the 5, 11.9, and 21.1% fat diets were 11.4, 22.0, and 45.0%, respectively. The diets containing 2, 5, 10, 25, and 50% protein; 5, 11.9, or 21.1% fat were made by varying the protein and carbohydrate contents of each of the above three diets. For example, the diet with 2% protein and 11.9% fat contained per kilogram: 25.0 g lactalbumin, 546.30 g dextrin, 271.70 g sucrose, and 100 g cottonseed oil.

[d]   Each 100 g of lactalbumin contains (in grams): protein, 80; ash, 4; moisture, 6; lactose, 3; ether extractables, 6; other, 1.

[e]   The dextrin:sucrose ratio was determined by the optimal amounts for pelletability whenever pellets were desired. Only powdered diets were, however, used in the experiments reported in this paper.

[f]   Rogers and Harper.[11]

[g]   Nonnutritive fiber was added to the diets containing 21.1% fat to make up bulk. The fiber is a white, finely ground product purified from wood by Brown Company, New York, New York. It contains (in grams per 100 g) crude fiber, 64.85; pentosans, hexosans, and galactosans, 26.99; moisture, 7.7; ash, 0.3; and reducing substances, trace.

Total liver lipid was determined by extracting the lipids from a weighed portion of liver with a 2:1 solution of chloroform:methanol. The extract was washed with water by the method of Folch et al.[12] and an aliquot was dried in a hot-air oven. The total lipid content was then determined gravimetrically. DNA was measured by the diphenylamine method of Burton[13] against a standard of calf thymus DNA after separation of the RNA from DNA by the procedure of Munro and Fleck.[14] RNA content of the liver was estimated by the technique of Schmidt and Thannhauser[15] as modified by Fleck and Munro[16] using extinction coefficient $E_{1cm}1\% = 260$ [$32\mu g$/ ml = 1.0 A]. Liver protein concentrations were estimated according to Lowry et al.[17] with bovine serum albumin as the reference.

Hemoglobin was estimated by the cyanmethemoglobin method described by Drabkin and Austin;[18] a commercial system* was used for sample preparation. Hematocrit was determined by the micromethod of Strumia et al.[19] using a microhematocrit centrifuge** and reader.*** Mean corpuscular hemoglobin concentration (MCHC, hemoglobin concentration in grams per 100 ml red blood corpuscles) was calculated from the equation:

$$MCHC = \frac{\text{hemoglobin concentration in g/100ml}}{\text{hematocrit (\%)}} \times 100$$

Total plasma protein was measured by the Buiret method.[20] Albumin was estimated by the direct photometric procedure[21] using rat plasma albumin as the standard.[22] Plasma glucose was estimated by the ultraviolet enzymatic hexokinase method.[23,24] An Autoanalyzer II† was used for the simultaneous, automated, enzymatic determinations of plasma cholesterol and triglycerides. Plasma phospholipid was measured by the method of Baginski et al.[25]

Triiodothyronine (T3) was determined by radioimmunoassay,[26]†† while thyroxine (T4) was estimated by the competitive protein binding technique of Kaplan.[27]†††The free thyroxine index of Clark and Horn[28] referred to as T7, was calculated as the product of the percent of free T3 (expressed as a decimal) and the T4 value (micrograms per 100 ml). Insulin was measured by the radioimmunoassay method of Yalow and Berson[29,30] with modifications by Hale and Randle.[31,32]‡ Glucagon was also determined by radioimmunoassay[33] using [^{125}I] · iodoglucagon‡‡ and a specific pancreatic glucagon antiserum.‡‡‡ Corticosterone was determined by radioimmunoassay[34] using a specific antiserum which had a very low reactivity with cortisol. Cortisol itself was also measured by radioimmunoassay[35] using a highly specific antiserum,⧧ but the results will not be reported here because the difference between zero and the observed mean values for each of the treatment groups was within the analytical error of the method. Plasma norepinephrine (NE) and epinephrine (E) were measured by the double-isotope derivative method[36,37] with slight modifications.

The data were analyzed by multiple-regression analysis.[38] The statistical significance of differences between groups was indicated by a p value which is the probability associated with the F-statistic for testing the differences after adjusting for all other independent variables in the regression. Differences were considered significant for $p < 0.05$.

The magnitude of the differences between groups was shown by calculating the mean values for each group after adjustments for all other factors.[38]

---

* Unopette test for hemoglobin, Catalog No. 5857, Becton-Dickinson and Company, Rutherford, New Jersey.
** Microhematocrit centrifuge, IEC Model 3411, International Equipment Company, Needham Heights, Massachusetts 02194.
***Microhematocrit reader, IEC Model 2201, International Equipment Company, Needham Heights, Massachusetts 02194.
† Technicon Instruments Corporation, Tarrytown, New York 10591. Method No. SE4-0040FC6 was used for plasma cholesterol and Method No. 4-0039PC6 for plasma triglycerides.
†† Available as a kit: T3 (RIA) Diagnostic Kit, Abbott Laboratories/Radiopharmaceuticals, Chicago, Illinois.
†††Available as a kit: Tetrasorb-125, Abbott Laboratories/Radiopharmaceuticals, Chicago, Illinois.
‡ Available as a kit: Amersham/Searle, Arlington Heights, Illinois.
‡‡ Purchased from Nuclear Medical Laboratories, Inc., Dallas, Texas.
‡‡‡Purchased from the laboratory of Roger H. Unger, M.D. Department of Internal Medicine, University of Texas Health Science Center and Veterans Administration Hospital, Dallas, Texas.
⧧ Purchased from Endocrine Sciences, Tarzana, California 91356. Antiserum No. B21-42 was used for corticosterone and F21-53 for cortisol.

RESULTS

Food consumption of ad libitum-fed rats increased with an increase in the dietary protein level up to 15% and then declined (Table 3). Efficiency of protein utilization (PER) was 3.1 when the diet contained 5 to 15% lactalbumin protein, but was reduced when the diet contained 25 to 50% protein (Figure 2). In contrast to the protein utilization results, feed efficiency was maximal when the diet contained 25 or 50% protein and was reduced when the diet contained 15% or less of lactalbumin protein (Figure 3).

The rate of growth was directly related to the dietary levels of protein and energy (Figures 4 and 5). Maximal growth rate occurred when the diet contained about 23% protein (Figure 6). The results of multiple-regression analyses to show the effects of levels of dietary protein, fat, and energy on the body weight of 12-week-old rats after 8 weeks on the experimental diets are displayed in Table 4. Body weight increased with increases in the level of dietary protein up to about 25% and then declined. There was a small progressive increase in body weight with increases in the level of fat in the diet. Rats fed the diets ad libitum were heavier than those fed the diets at 75% of ad libitum intake, and the latter group, in turn, were heavier than those fed the diets at 50% of ad libitum intake. The growth of rats fed diets containing 11.9% fat at 50 or 75% of ad libitum intake was accelerated by increasing the level of dietary protein up to 25% protein (Figures 7 and 8).

The effects of levels of dietary protein, energy, and fat on liver weight were similar to those on body weight except that the livers of rats fed diets containing the intermediate level of fat were heavier and had more DNA than the livers of rats fed diets with low and high fat content (Tables 5 to 7). Feeding of a 2% protein diet as well as restriction of energy intake caused a relative increase in the weight of the liver expressed as a percentage of body weight. The level of dietary fat had no effect on liver weight or liver weight per body weight.

Rats fed a 2% protein diet had the highest concentration of liver DNA, followed by rats fed 5% protein diets; these results indicated that rats fed the low-protein diets (2 and 5%) had smaller relative size of liver cells (grams liver per milligram DNA) than the other rats. Rats fed ad libitum had lower concentration of liver DNA and, hence, larger cells than those fed the same diet in restricted amounts. The level of fat in the diet had no effect on the concentration of DNA and the size of liver cells.

The concentration of protein in the liver as well as the relative amount of protein per cell (milligrams protein per milligram DNA) were directly related to the level of dietary protein (Table 5) and to energy intake (Table 6). Rats fed 2 and 5% protein diets had greatly reduced values whereas differences between rats fed 10% or higher levels of dietary protein were relatively minor. Rats fed diets ad libitum had higher values than those fed the same diets at 75 or 50% of ad libitum intake. Liver protein concentration and relative cellular protein content had a biphasic relationship to dietary fat level. The lowest values were obtained in rats fed diets containing 11.9% fat, and an elevation as well as a decrease in dietary fat level caused an increase in both indices of liver protein content.

The concentration of lipid in the liver of rats fed a 2% protein diet ad libitum was higher than in the others (Table 5). The results, however, showed that the relative cellular lipid content (milligrams lipid per milligram DNA) of rats fed the 2 and 5% protein diets ad libitum was significantly lower than that of rats fed the higher protein diets ad libitum. Therefore, the elevated concentration of lipid in the liver of rats fed a 2% protein diet ad libitum was not due to an increase in the amount of lipid in each cell, but rather to a slightly reduced amount of lipid inside cells of

Table 3
WEEKLY MEAN FOOD INTAKE OF AD
LIBITUM-FED RATS

Lactalbumin in diet[a] (%)	Week			
	1	2	3	4
2	9.6±1.7	11.2±2.3	10.6±1.8	9.9±1.4
5	14.2±2.1	13.2±4.4	12.3±5.2	12.5±3.2
10	14.4±1.2	16.5±3.9	16.7±5.7	16.5±2.3
15	15.2±0.9	14.9±2.0	14.8±1.7	14.7±1.9
25	14.4±1.6	14.9±3.0	14.6±1.6	14.6±1.7
50	13.2±1.1	13.5±5.7	13.5±3.1	13.6±2.7

[a] Mean of four rats per group with standard deviation (given in grams) of diet consumed in 24 hr. All diets contained 11.9% fat.

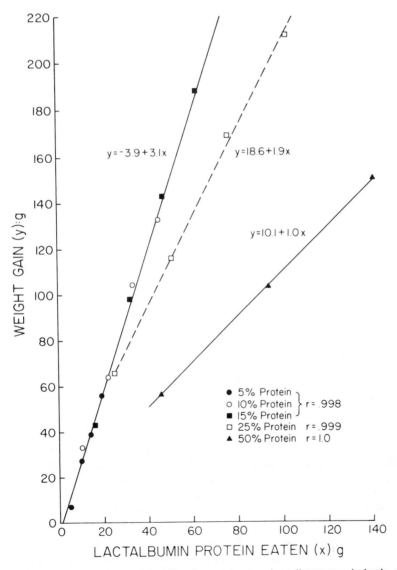

FIGURE 2.   Utilization of lactalbumin protein at various dietary protein levels. The slope of the linear regressions is the PER. Protein utilization was reduced when the diet contained 25 or 50% protein.

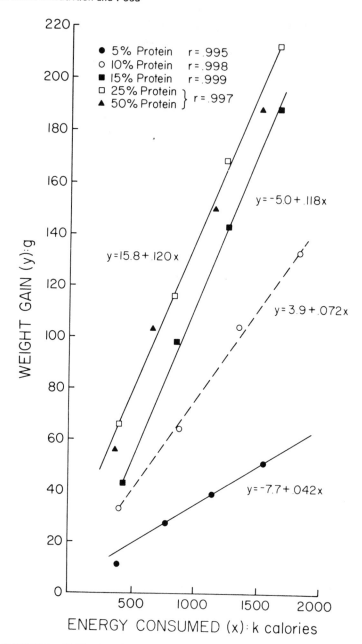

FIGURE 3. Utilization of energy for growth at various dietary protein levels. Efficiency of energy utilization was reduced when the diet contained 15% or less of lactalbumin protein.

greatly reduced size. These relationships of dietary protein level to lipid concentration, cell size, and cellular lipid content are illustrated in Figure 9. Concentration of lipid in the liver as well as relative cellular lipid content were directly related to the level of dietary fat (Table 7) and to the level of energy intake (Table 6).

Rats fed a 2% protein diet had low values for plasma total protein, albumin, and globulin (Table 8). When the dietary protein level was increased, the plasma total protein and albumin rose and reached a peak in rats fed diets containing 10% protein. Thereafter, their values declined with further increases in dietary protein level. The plasma globulin concentration showed a similar trend, but it did not attain a

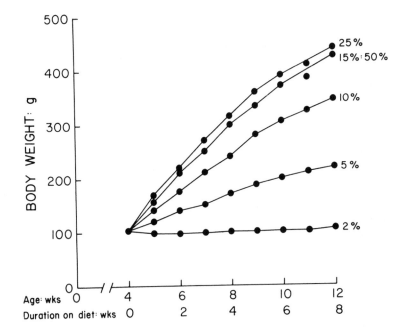

FIGURE 4.   Mean body weights of rats fed ad libitum diets containing 11.9% fat and varying levels of lactalbumin protein. The differences between rats fed 2, 5, 10, 15, and 25% are statistically significant (p < 0.05). Rats fed 15 and 50% protein diets had similar weights and were lighter (p < 0.05) than rats fed the 25% protein diet.

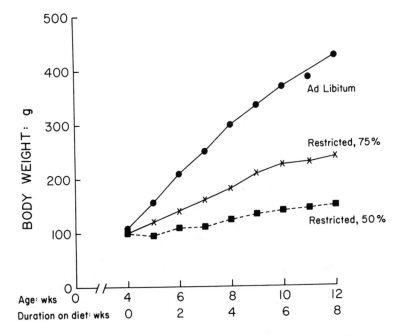

FIGURE 5.   Mean body weights of rats fed either ad libitum or restricted amounts of a diet containing 11.9% fat and 15% lactalbumin protein. The restricted rats were fed 75 or 50% of the ad libitum intake of rats of the same weight.

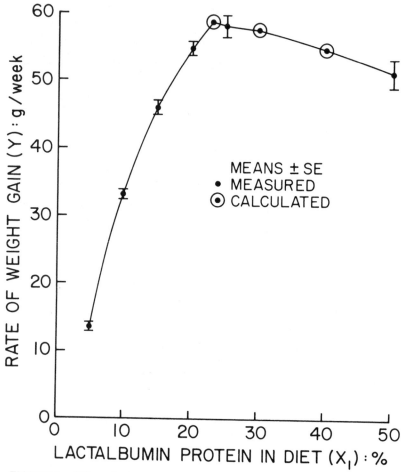

FIGURE 6.   Effect of dietary protein level on the growth of male Sprague-Dawley rats. The calculated body weights were derived from the equation: Body weight = $-54.1 -0.53x_1 -17.4x_2 + 1.66x_3 -0.51x_4$ where: x_1 = dietary protein level (%); x_2 = PER; x_3 = an appetite factor, and x_4 = energy/protein ratio in kilocalories per gram protein. (Adapted from Edozien, J.C. and Switzer, B.R., *J. Nutr.*, in press. With permission.)

peak value until the dietary protein level reached 25%. As a consequence of these changes, the albumin/globulin ratio reached its highest value in rats fed diets containing 5% protein; it declined slowly with increases in dietary protein level. The plasma concentration of total protein, albumin, and globulin was directly related to the level of energy intake. Starvation was associated with an increase in the albumin/globulin ratio (Table 9). The level of dietary fat had no influence on the plasma globulin concentration (Table 10). The plasma albumin concentration and, hence, also the albumin/globulin ratio were, however, lower in rats fed diets containing 11.9% fat than those containing 5 or 21.1% fat, but the differences were not of sufficient magnitude to produce a significant effect on total plasma protein (Table 10).

Blood hemoglobin and hematocrit values increased progressively with increases in the level of dietary protein (Table 8). Rats fed diets containing 15% protein or greater had significantly higher MCHC values than rats fed diets containing 2, 5, or 10% protein. The blood hemoglobin level was unaffected by the plane of energy consumption, but rats fed restricted amounts of diet had lower hematocrit values

Table 4
RESULTS OF MULTIPLE REGRESSION ANALYSES TO SHOW THE EFFECTS OF DIETARY PROTEIN AND FAT, AND LEVEL OF ENERGY INTAKE ON THE WEIGHT OF 12-WEEK-OLD RATS FED EXPERIMENTAL DIETS FOR 8 WEEKS

Level of nutrient in diet (%)	N	Adjusted mean body weight (g)
Protein		
2	96	$51^{a,b}$
5	90	130
10	43	228
15	92	280
25	48	285
50	47	260
Fat		
5	39	$187^{b,c}$
11.9	321	202
21.1	39	215
Energy		
50% of ad libitum intake	49	$116^{b,d}$
75% of ad libitum intake	49	186
Ad libitum	301	301

[a] Values are adjusted for level of dietary fat and of energy intake.
[b] Differences are statistically significant ($p < 0.05$).
[c] Values are adjusted for level of dietary protein and of energy intake.
[d] Values are adjusted for levels of dietary protein and fat.

than rats fed the same diet ad libitum (Table 9); hence, restriction of energy intake was associated with a relative elevation of MCHC values. Rats fed low- (5%) and high- (21.1%) fat diets had lower concentrations of hemoglobin and hematocrit than rats fed diets containing an intermediate (11.9%) level of fat (Table 10). The fat content of the diet had no effect on MCHC values.

An increase in dietary protein level correlated directly with plasma concentration of glucose (Table 8). Restriction of energy intake caused an elevation of plasma glucose (Table 9); alteration of dietary fat level had no influence on plasma glucose concentration (Table 10).

Dietary protein level had no effect on plasma cholesterol concentration, whereas protein deficiency (2 and 5% dietary protein levels) was associated with low values for plasma triglyceride and elevated plasma phospholipid concentration (Table 8). Surprisingly, starvation caused an elevation of plasma cholesterol (Table 9). Plasma triglyceride and phospholipid had their highest values in rats fed diets at 75% of ad libitum intake; both an increase and a decrease in the level of energy consumption caused the plasma triglyceride and phospholipid levels to fall. Dietary fat level had no influence on the plasma concentrations of cholesterol, triglyceride, or phospholipid.

The effects of changes in the dietary levels of protein, energy, and fat on the

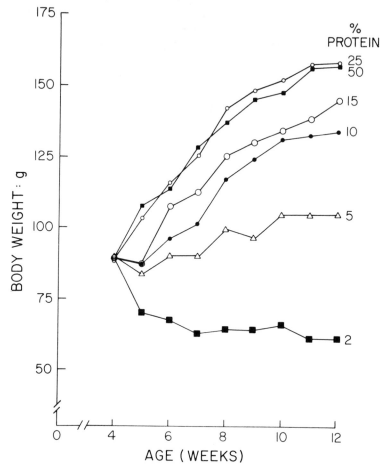

FIGURE 7.   Mean body weight of rats fed isocaloric diets containing 11.9% fat at 50% of ad libitum intakes for 8 weeks starting at 4 weeks of age. The differences between rats fed 2, 5, 10, 15, and 25% are statistically significant (p < 0.05). Rats fed 25 and 50% protein were similar. Growth was accelerated by an increase in dietary protein level up to 25% protein.

concentrations of hormones in rat plasma are reported respectively, in Tables 11, 12, and 13. The plasma level of triiodothyronine declined progressively as the level of protein in the diet increased; protein intake had the opposite effect on plasma level of thyroxine. The net effect was that the T3/T4 ratio declined as dietary protein level increased up to about 15% protein. Rats fed restricted amounts of diet had a reduced plasma T3 level, and hence, a reduction in the T3/T4 ratio. The dietary fat level had no influence on the plasma concentrations of T3 or T4.

Dietary protein level correlated directly with fasting plasma insulin level but was inversely related to the fasting plasma concentration of glucagon. Rats fed a restricted amount of diet had higher values for fasting plasma concentrations of insulin and glucagon than rats fed the same diet ad libitum. The level of dietary fat had no influence on the fasting plasma insulin level, but it was inversely related to the fasting plasma glucagon concentration. Because of these effects of diet on the plasma concentrations of insulin and glucagon, the insulin/glucagon ratio was directly related to the dietary levels of protein and fat but independent of the plane of energy consumption.

The plasma corticosterone concentration showed a biphasic response to changes

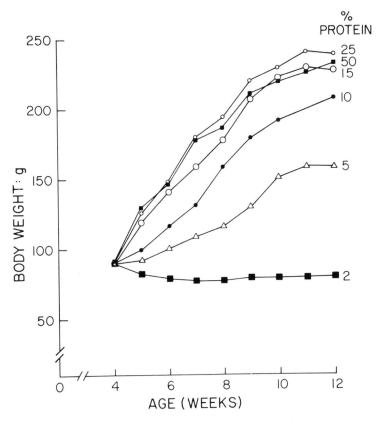

FIGURE 8.   Mean body weight of rats fed isocaloric diets containing 11.9% fat at 75% of ad libitum intakes for 8 weeks starting at 4 weeks of age. The differences between rats fed 2, 5, 10, 15, and 25% are statistically significant (p < 0.05). Rats fed 25 and 50% protein were similar. Growth was accelerated by an increase in dietary protein level to 25% protein.

in dietary protein level between 5 and 50%. The lowest values of fasting plasma corticosterone were obtained in rats consuming 10 to 15% protein diets. An increase of dietary protein level to 25 to 50% as well as a decrease to 5% caused an elevation in the plasma concentration of the hormone; there was, however, a fall when the dietary protein level was further reduced from 5 to 2%. Energy restriction had no effect on the plasma corticosterone level, but rats consuming the intermediate (11.9%) fat diet had a lower plasma corticosterone concentration than rats fed diets with low (5%) and high (21.1%) fat content. Unlike the other hormones investigated, plasma corticosterone concentration was highly related to body weight. If body weight is included as an independent variable in the regression analyses, the biphasic responses disappear, with the net results that plasma corticosterone concentration was actually directly related to the dietary levels of protein, energy, and fat; a steep fall occurred when the dietary protein level was reduced from 5 to 2%.

The level of dietary protein was inversely related to the plasma concentrations of norepinephrine, epinephrine, and total catecholamines, but it did not alter the norepinephrine/epinephrine ratio. The plasma level of epinephrine was directly related to energy intake. The plasma level of norepinephrine showed a trend in the opposite direction, but the changes were not statistically significant; the net effect, however, was that starvation had no influence on the plasma level of total catecholamines, whereas it significantly increased the norepinephrine/epinephrine ratio. The level

Table 5
EFFECTS OF DIETARY PROTEIN LEVEL ON SIZE AND COMPOSITION OF RAT LIVER[a]

			Dietary protein level (%)				Significance of differences
Variable	2	5	10	15	25	50	
N	89	97	58	110	53	58	√[b]
Weight							
g	1.5	3.6	7.5	8.6	8.5	8.5	√
Body weight (%)	3.7	3.0	3.3	3.2	3.4	3.4	√
DNA (g/mg)	0.19	0.24	0.30	0.30	0.30	0.32[c]	√
DNA							
Liver (mg/g)	5.4	4.3	3.4	3.5	3.4	3.2[c]	√
mg/liver	10.6	15.7	24.8	29.0	28.3	26.2[c]	√
RNA							
Liver (mg/g)	7.8	8.2	7.9	7.4	7.3	7.1	√
mg/liver	10.7	29.9	58.4	64.7	62.9	59.4	√
DNA (mg/mg)	1.5	2.0	2.3	2.2	2.2	2.2	√
Protein							
Liver (mg/g)	174	200	212	214	215	226[c]	√
mg/liver	222	708	1606	1845	1876	1934	√
DNA (mg/mg)	33.5	47.7	63.0	62.2	64.3	71.7[c]	√
RNA (mg/mg)	22.6	24.5	27.6	28.8	29.5	31.9[c]	√
Lipid							
Liver (mg/g)	76.9	54.7	52.1	53.2	54.9	49.1[c]	√
DNA (mg/mg)	13.8	13.1	16.1	16.4	17.0	16.5	√
RNA (mg/mg)	9.9	6.6	6.8	7.3	7.6	7.1	√
Protein/lipid ratio	3.0	3.7	4.1	4.0	4.0	4.5[c]	√

[a]   Results are mean values adjusted for level of dietary fat and of energy intake.
[b]   Differences are statistically significant at 0.05 level.
[c]   Significantly different (0.05 level) from rats fed 25% protein ad libitum.

of dietary fat had no effect on the plasma concentrations of norepinephrine or epinephrine.

Most of the results reported in this section are adjusted mean values. For example, the reported mean values of a given variable for rats fed a 2, 5, 10, 25, or 50% protein diet were adjusted statistically for the effects of levels of dietary fat and energy intake so the effects of a change in dietary protein level alone is evident. Therefore, in order that the results of this study can be readily compared to other published data, the mean body weight and the unadjusted mean values of liver, blood, and plasma variables have also been reported in Tables 14 and 15 for rats fed a diet containing 25% protein and 5 or 11.9% fat.

DISCUSSION AND COMMENTS

The results show that rats fed a 2% lactalbumin protein diet ad libitum for 8 weeks developed fatty liver similar to the finding in kwashiorkor (Figure 10). This fatty liver of kwashiorkor has, at various times, been attributed to a specific deficiency of lipotropic factors,[40] an increase in the rate of fat synthesis,[41,42] or an impairment of lipid transport due to a decreased synthesis of the protein part of the lipoprotein molecule.[43-46] However, when our data were analyzed to show the relative amounts of lipid per cell, it became evident that rats fed ad libitum diets con-

Table 6
EFFECTS OF RESTRICTION OF ENERGY INTAKE ON WEIGHT AND COMPOSITION OF RAT LIVER[a]

Variable	Levels of energy consumption[b]			Significance of difference
	50	75	100	
N	84	87	294	√[c]
Weight				
g	4.1	6.2	8.8	√
Body weight (%)	3.8	3.2	2.9	√
DNA (g/mg)	0.27	0.27	0.28	√
DNA				
Liver (mg/g)	4.0	3.9	3.7	√
mg/liver	15.2	21.8	30.3	√
RNA				
Liver (mg/g)	7.3	7.4	8.0	√
mg/liver	29.4	44.6	69.1	√
DNA (mg/mg)	1.9	2.0	2.2	√
Protein				
Liver (mg/g)	205	201	214	√
mg/liver	890	1262	1944	√
DNA (mg/mg)	55	55	61	√
RNA (mg/mg)	28.3	27.2	27.0	√
Lipid				
Liver (mg/g)	43	46	81	√
DNA (mg/mg)	11.6	12.4	22.4	√
RNA (mg/mg)	6.1	6.3	10.3	√
Protein/lipid ratio	4.5	4.2	2.9	√

[a]   Results are mean values adjusted for level of dietary protein and fat.

[b]   Percentages of ad libitum intake.

[c]   Differences are statistically significant at the 0.05 level.

taining 10% protein or more had a nearly constant cellular lipid content, whereas rats fed 2 or 5% protein diets ad libitum had slightly reduced cellular lipid content. Hence, there was, in reality, no accumulation of lipids in the liver cells of rats consuming the 2% lactalbumin protein diets ad libitum; the appearance of fatty liver was due to an increase in the amount of lipid per gram of liver which occurred because there was a greater reduction in liver cell size than in liver cell lipids. For example, there was a 34% difference in cell size between rats fed 2 and 15% protein diets but only a 14% difference in their cellular lipid content. This absence of a true lipid accumulation in the liver cells of protein-deficient rats suggests that there may be no real abnormalities in the synthesis and/or mobilization of fat.

The results also indicate that the fat content of the diet and the level of energy consumption can influence the amount of fat in the liver. Rats fed diets in restricted amounts had a reduced amount of cellular lipid irrespective of the protein content of the diet. Those fed the 2% protein in restricted amounts also had shrunken cells, but they did not develop fatty liver (Figure 10) because there was a relatively greater reduction in liver cell lipids than in liver cell size (Figure 9). The difference between the fatty livers of children suffering from kwashiorkor and the normal appearance of the livers of marasmic children can be explained by similar effects of levels of protein and energy intakes on the lipid content and size of human liver cells. These

Table 7
EFFECTS OF DIETARY FAT LEVEL ON SIZE AND
COMPOSITION OF RAT LIVER[a]

Variable	Dietary fat level (%)			Significance of difference
	5	11.9	21.1	
N	38	387	40	
Weight				
g	6.1	6.4	6.6	x[b]
Body weight (%)	3.3	3.3	3.2	x
DNA (g/mg)	0.28	0.27	0.27	x
DNA				
Liver (mg/g)	3.8	3.8	3.9	x
mg/liver	20.9	23.1	23.3	x
RNA				
Liver (mg/g)	7.8	7.5	7.4	√[c]
mg/liver	47.9	48.0	47.2	x
DNA (mg/mg)	2.2	2.0	2.0	√
Protein				
Liver (mg/g)	216	195	209	√
mg/liver	1378	1285	1432	√
DNA (mg/mg)	60.8	53.7	56.7	√
RNA (mg/mg)	27.8	26.3	28.4	√
Lipid				
Liver (mg/g)	46.2	60.9	63.4	√
DNA (mg/mg)	12.8	16.2	17.4	√
RNA (mg/mg)	5.9	8.1	8.7	√
Protein/lipid ratio	4.3	3.6	3.7	√

[a] Results are mean values adjusted for level of dietary protein and of energy intake.
[b] Differences are not statistically significant at the 0.05 level.
[c] Differences are statistically significant at the 0.05 level.

findings also explain why kwashiorkor does not permanently damage the liver and recovery from the disease usually restores the organ to its normal appearance and structure.

Liver protein content represents the balance between liver protein synthesis and turnover. Liver protein synthesis is influenced by the amount of RNA, the quality of ribosomes, and amino acid supply. The last two variables are reflected in the protein/RNA ratio. The results reported in Table 5 indicate that the amount of RNA per cell (milligrams RNA per milligram DNA) as well as the protein/RNA ratio are reduced in rats fed 2 and 5% protein diets. These changes are reflected in the liver protein content of rats fed these low-protein diets.

The small increase in the protein/RNA ratio in rats whose energy consumption was reduced to 50% of ad libitum intake was probably due to improved amino acid supply since rats fed very low-protein diets ad libitum did not differ from rats fed restricted amounts of the same diets in ribosome quality measured by in vitro protein synthesis or by the percentage of polysomes in the ribosome profile.[9] This earlier study also showed that rats fed restricted amounts of a low-protein diet had a higher concentration of total amino acids than those fed the same amount of protein ad libitum. It may also be relevant that partially starved rats have a higher concentration of fasting plasma insulin than rats fed the same diet ad libitum (Table 12). This enhancement of liver protein synthesis by starvation may become very important when the diet contains very small amounts of protein and probably explains

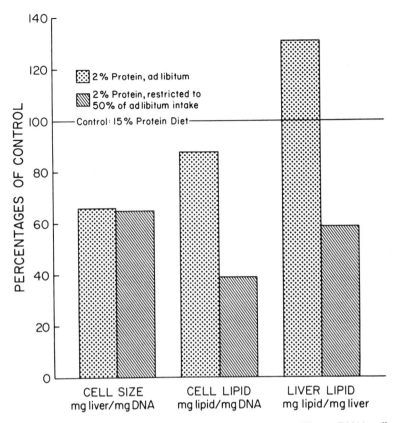

FIGURE 9. Comparisons of cell size (milligrams liver per milligram DNA), cell lipid (milligrams lipid per milligram DNA) and concentration of lipid in liver (milligrams lipid per milligrams liver) in rats fed a 2% lactalbumin protein diet ad libitum and those fed the same diet at 50% of ad libitum intake.

the higher level of plasma albumin in marasmic rats as compared to kwashiorkor rats.[8,9]

Plasma albumin, but not globulin (Table 10), had a biphasic response to dietary fat level similar to that reported in Table 7 for the effects of dietary fat level on liver protein. These effects of dietary fat level on plasma albumin and liver protein can be explained by the known enhancement of liver protein synthesis by glucocorticoids, since fasting plasma corticosterone concentration showed a similar biphasic response to variation in dietary fat level (Table 13).

The data relating to plasma concentrations of total protein and albumin indicate that changes in their values occur only in severe protein-energy deficiency, and, therefore, their measurements are not useful for the detection of mild degrees of deficiency.

The role of hormones in the regulation of blood nutrient concentrations in the immediate postprandial period has been widely recognized. The short-term interactions of diet and hormones which occur, for example, during fasting or acute protein deprivation, have also been extensively investigated. In contrast to these established areas of knowledge, the chronic effects of dietary changes on endocrine function have received relatively little attention. Studies of the hormonal changes associated with human protein calorie malnutrition[47-50] have not provided definitive data because of the complex interactions of multiple nutrient deficiencies, infection, and other environmental variables in patients suffering from kwashiorkor and mar-

Table 8

EFFECTS OF DIETARY PROTEIN LEVEL ON BIOCHEMICALS IN RAT PLASMA[a]

| Variable | Level of dietary protein (%) | | | | | | Significance of difference |
	2	5	10	15	25	50	
Total protein (g/100 ml)	5.1 (37)	5.8 (92)	6.8 (74)	6.8 (85)	6.7 (70)	6.4 (68)	√[b]
Albumin (g/100 ml)	2.9 (37)	3.5 (92)	4.0 (74)	3.9 (85)	3.8 (70)	3.6 (48)	√
Globulin (g/100 ml)	2.2 (37)	2.3 (92)	2.8 (74)	2.9 (85)	2.9 (70)	2.8 (48)	√
A/G ratio	1.3 (37)	1.5 (92)	1.5 (74)	1.4 (85)	1.3 (70)	1.3 (48)	√
Hemoglobin (g/100 ml)	12.0 (80)	14.5 (146)	15.2 (117)	15.6 (152)	15.8 (97)	16.3 (84)	√
Hematocrit (%)	37.1 (80)	45.5 (146)	47.1 (117)	47.3 (152)	47.7 (97)	48.9 (84)	√
MCHC (g/100 ml RBC)	32.2 (80)	32.2 (146)	32.3 (117)	33.3 (152)	33.2 (97)	33.3 (84)	√
Glucose (mg/100 ml)	199 (35)	178 (50)	212 (58)	225 (67)	228 (56)	234 (59)	√
Cholesterol (mg/100 ml)	67 (12)	65 (17)	69 (27)	69 (57)	63 (28)	61 (34)	x[c]
Triglyceride (mg/100 ml)	15 (12)	22 (17)	46 (27)	45 (57)	46 (28)	43 (34)	√
Phospholipid (mg/100 ml)	8.4 (12)	6.6 (17)	5.6 (27)	5.3 (57)	5.3 (28)	4.7 (34)	√

[a]   Results are mean values adjusted for level of dietary fat and of energy intake. Numbers in parenthesis are numbers of rats. Hemoglobin and hematocrit data relate to whole blood while the MCHC relates to the erythrocytes.

[b]   Differences are statistically significant at the 0.05 level.

[c]   Differences are not statistically significant at the 0.05 level.

Table 9
EFFECTS OF ENERGY RESTRICTION ON BIOCHEMICALS IN RAT PLASMA[a]

Variable	Level of energy consumption[b]						Significance of difference
	50		75		100		
Total protein (g/100 ml)	5. 9	(92)	6. 2	(97)	6. 7	(237)	√[c]
Albumin (g/100 ml)	3. 5	(92)	3. 5	(97)	3. 8	(237)	√
Globulin (g/100 ml)	2. 4	(92)	2. 6	(97)	2. 9	(237)	√
A/G ratio	1. 5	(92)	1. 4	(97)	1. 3	(237)	√
Hemoglobin (g/100 ml)	14. 8	(94)	14. 8	(98)	15.	(298)	x[d]
Hematocrit (%)	45	(94)	45	(98)	47	(298)	√
MCHC (g/100 ml RBC)	33. 1	(94)	33.	(98)	31. 8	(298)	√
Glucose (mg/100 ml)	224	(88)	223	(93)	191	(144)	√
Cholesterol (mg/100 ml)	74	(25)	72	(30)	44	(98)	√
Triglycerides (mg/100 ml)	33	(25)	50	(30)	39	(98)	√
Phospholipids (mg/100 ml)	5. 2	(25)	6. 1	(30)	5. 3	(98)	√

[a]   Results are mean values adjusted for levels of dietary protein and fat. Numbers in parenthesis are numbers of rats. Hemoglobin and hematocrit data related to whole blood while the MCHC relates to the erythrocytes.
[b]   Percentages of ad libitum intake.
[c]   Differences are statistically significant at the 0.05 level.
[d]   Differences are not statistically significant at the 0.05 level.

Table 10
EFFECTS OF DIETARY FAT LEVEL ON BIOCHEMICALS IN RAT PLASMA[a]

Variable	Level of dietary fat (%)						Significance of difference
	5		11.9		22.1		
Total protein (g/100 ml)	6. 4	(51)	6. 2	(324)	6. 3	(51)	x[b]
Albumin (g/100 ml)	3. 7	(51)	3. 5	(324)	3. 7	(51)	√[c]
Globulin (g/100 ml)	2. 7	(51)	2. 7	(324)	2. 6	(51)	x
A/G ratio	1. 4	(51)	1. 3	(324)	1. 4	(51)	√
Hemoglobin (g/100 ml)	14. 7	(102)	15. 2	(485)	14. 8	(89)	√
Hematocrit (%)	45. 3	(102)	46. 4	(485)	44. 8	(89)	√
MCHC (g/100 ml RBC)	32. 4	(102)	32. 6	(485)	32. 9	(89)	x
Glucose (mg/100 ml)	213	(32)	210	(263)	214	(30)	x
Cholesterol (mg/100 ml)	62. 3	(26)	62. 9	(103)	64. 6	(24)	x
Triglyceride (mg/100 ml)	38. 8	(26)	43. 1	(103)	40. 0	(24)	x
Phospholipid (mg/100 ml)	5. 7	(26)	5. 4	(103)	5. 4	(24)	x

[a]   Results are mean values adjusted for level of dietary protein and of energy intake. Numbers in parenthesis are numbers of rats. Hemoglobin and hematocrit data relate to whole blood while the MCHC relates to the erythrocytes.
[b]   Differences are statistically not significant at the 0.05 level.
[c]   Differences are statistically significant at the 0.05 level.

asmus. Jackson[51] and Keys et al.[52] have reviewed the earlier literature on the morphology of the endocrine glands in protein-energy deficiency. More recently, Platt et al.[5] reported detailed histological studies of the endocrine glands in experimental protein deficiency in the pig. However, morphological alterations in the endocrine organs often do not accurately reflect functional changes; this is exemplified by the highly variable histological appearance of the adrenals in protein-energy deficiency in children[52-57] in contrast to the consistent reports of elevated plasma cortisol levels in patients suffering from kwashiorkor and marasmus.[56-58]

Table 11
EFFECTS OF LEVEL OF DIETARY PROTEIN ON PLASMA HORMONE CONCENTRATIONS IN THE RAT[a]

Variable	Levels of protein (%)												Significance of difference
	2		5		10		15		25		50		
Triiodothyronine (T3) (ng/100 ml)	281	(12)	209	(33)	123	(31)	51	(19)	45	(23)	26	(16)	√[b]
Thyroxine (T4) (µg/100 ml)	0. 5	(12)	0. 7	(33)	2. 2	(31)	3. 1	(19)	3. 4	(23)	4. 4	(16)	√
T3/T4 (%)	47	(12)	27	(33)	7	(31)	3	(19)	3	(23)	3	(16)	√
T7 (µg/100 ml)	0. 3	(13)	0. 4	(24)	0. 9	(15)	1. 1	(14)	1. 2	(13)	1. 4	(9)	√
Insulin (µU/ml)	21. 0	(13)	25. 7	(51)	34. 7	(72)	39. 3	(56)	47. 8	(48)	65. 7	(24)	√
Glucagon (pg/ml)	256	(13)	209	(51)	174	(72)	159	(56)	165	(48)	152	(24)	√
Insulin/glucagon	0. 05	(13)	0. 11	(51)	0. 21	(72)	0. 26	(56)	0. 31	(48)	0. 53	(24)	√
Corticosterone (µg/100 ml)	22. 1	(89)	30. 1	(130)	24. 5	(102)	24. 8	(102)	29. 5	(85)	29. 9	(67)	√
Norepinephrine (NE) (ng/ml)	19. 8	(18)	19. 9	(37)	16. 1	(49)	15. 8	(44)	14. 7	(51)	13. 1	(54)	√
Epinephrine (E) (ng/ml)	24. 5	(18)	22. 7	(37)	19. 1	(49)	16. 6	(44)	16. 1	(51)	14. 9	(54)	√
NE + E (ng/ml)	44. 4	(18)	42. 6	(37)	35. 2	(49)	32. 4	(44)	30. 8	(51)	27. 7	(54)	√
NE/E ratio	1. 6	(18)	1. 3	(37)	1. 3	(49)	1. 4	(44)	1. 4	(51)	1. 3	(54)	x[c]

[a] Results are mean values adjusted for levels of fat and evergy.
[b] Differences are statistically significant at the 0.05 level.
[c] Differences are not statistically significant at the 0.05 level.

Table 12

EFFECTS OF RESTRICTION OF ENERGY INTAKE ON PLASMA HORMONE LEVELS IN THE RAT[a]

Variable	Levels of energy consumption (%)[b] 50		75		100		Significance of difference
Triiodothyronine (T3) (ng/100 ml)	111	(34)	98	(12)	158	(88)	$\sqrt{}$[c]
Thyroxine (T4) (μg/100 ml)	2. 4	(34)	2. 3	(12)	2. 5	(88)	x[d]
T3/T4 (%)	12	(34)	12	(12)	15	(88)	$\sqrt{}$
T7 (μg/100 ml)	0. 8	(12)	0. 8	(6)	1. 1	(61)	$\sqrt{}$
Insulin (μU/ml)	41. 9	(18)	42. 5	(24)	32. 7	(222)	$\sqrt{}$
Glucagon (pg/ml)	210	(18)	190	(24)	156	(222)	$\sqrt{}$
Insulin/glucagon	0. 23	(18)	0. 26	(24)	0. 25	(222)	x
Corticosterone (μg/100 ml)	24. 3	(51)	26. 6	(58)	28. 4	(501)	$\sqrt{}$
Norepinephrine (NE) (ng/ml)	17. 4	(54)	16. 7	(69)	15. 5	(130)	x
Epinephrine (E) (ng/ml)	16. 0	(54)	18. 6	(69)	22. 4	(130)	$\sqrt{}$
NE + E (mg/ml)	33. 5	(54)	35. 4	(69)	38. 0	(130)	x
NE/E ratio	2. 0	(54)	1. 4	(69)	0. 9	(130)	$\sqrt{}$

[a] Results are mean values adjusted for levels of dietary protein and fat.
[b] Percentages of ad libitum intake.
[c] Differences are statistically significant at the 0.05 level.
[d] Differences are not statistically significant at the 0.05 level.

Table 13

EFFECTS OF DIETARY FAT LEVEL ON PLASMA HORMONE CONCENTRATIONS IN THE RAT[a]

Variable	Level of dietary fat (%) 5		11.9		21.1		Significance of difference
Triiodothyronine (T3) (ng/100 ml)	115	(19)	125	(94)	127	(21)	x[b]
Thyroxine (T4) (μg/100 ml)	2. 3	(19)	2. 6	(94)	2. 3	(21)	x
T3/T4 (%)	14	(19)	15	(94)	16	(21)	x
T7 (μg/100 ml)	0. 9	(19)	0. 9	(94)	0. 9	(21)	x
Insulin (μU/ml)	38. 1	(66)	40. 4	(145)	38. 6	(53)	x
Glucagon (pg/ml)	228	(66)	171	(145)	158	(53)	$\sqrt{}$[c]
Insulin/glucagon	0. 18	(66)	0. 24	(145)	0. 28	(53)	$\sqrt{}$
Corticosterone (μg/100 ml)	25. 2	(101)	23. 7	(421)	30. 7	(88)	$\sqrt{}$
Norepinephrine (NE) (ng/ml)	15. 8	(37)	15. 8	(187)	18. 0	(29)	x
Epinephrine (E) (ng/ml)	16. 0	(37)	20. 3	(187)	20. 0	(29)	x
NE + E (ng/ml)	32. 6	(37)	36. 1	(187)	38. 1	(29)	x
NE/E ratio	1. 6	(37)	1. 3	(187)	1. 4	(29)	x

[a] Results are mean values adjusted for level of dietary protein and of energy intake.
[b] Differences are not statistically significant at the 0.05 level.
[c] Differences are statistically significant at the 0.05 level.

The fasting, resting levels of hormones in plasma do not provide information on the response of the endocrine glands to provocative stimuli, nor do they, by themselves, give any indication of tissue response to the hormones. Nevertheless, they do provide useful information which usually predicts the functional level of the endocrine system. Within these limitations, the results reported here illustrate the profound chronic effects of diet on hormonal function.

The principal role of these hormonal adjustments is probably to control the supply of amino acids and metabolic fuel and to regulate protein synthesis. Insulin regulates protein synthesis through its effects on amino acid transport, on the activity of RNA polymerase, and on the degree of aggregation of ribosomes; hence,

Table 14
MEAN VALUES OF BODY WEIGHT AND SELECTED LIVER VARIABLES IN RATS FED AD LIBITUM DIETS CONTAINING 25% LACTALBUMIN PROTEIN AND 5 OR 11.9% FAT

Variable	25% protein, 5% fat, ad libitum[a]	25% protein, 11.9% fat, ad libitum
Body weight (g)	364 ± 43 (8)[b]	413 ± 38 (12)[c]
Liver variables		
Weight		
g	11.2± 0.8(8)	12.7± 2.3(12)[c]
DNA (mg/mg)	290 ± 30 (8)	295 ± 36 (12)
Body weight (%)	3.0± 0.2(8)	3.0± 0.3(12)
DNA		
Liver (mg/g)	3.5± 0.4(8)	3.4± 0.4(12)
mg/liver	38.9± 4.8(8)	43.6± 8.4(11)
RNA		
Liver (mg/g)	8.0± 0.6(8)	7.8± 0.5(12)
mg/liver	89 ± 10 (8)	100 ± 17 (12)
DNA (mg/mg)	2.3± 0.2(8)	2.3± 0.2(12)
Protein		
Liver (mg/g)	226 ± 22 (8)	232 ± 17 (12)
mg/liver	2533 ± 401 (8)	2969 ± 651 (12)
DNA (mg/mg)	65.6± 11.0(8)	68.9± 7.2(12)
RNA (mg/mg)	28.4± 3.2(8)	29.7± 2.6(12)
Lipid		
Liver (mg/g)	66 ± 9 (8)	79 ± 11 (12)[c]
mg/liver	738 ± 126 (8)	1000 ± 279 (12)[c]
DNA (mg/mg)	19.2± 4.1(8)	23.4± 4.2(12)[c]
RNA (mg/mg)	8.3± 1.7(8)	10.2± 1.8(12)[c]
Protein/lipid ratio	3.5± 0.4(8)	2.9± 0.4(12)[c]

[a] Of all the diets used in the study, this diet has the closest composition to Purina Laboratory Chow, Ralston Purina Company, Checkerboard Square, St. Louis, Missouri.

[b] Data are means ± SD. The number in parenthesis is the number of rats.

[c] Significantly different ($p<0.05$) from rats fed the 25% protein, 5% fat diet.

expectedly, its level in plasma increases progressively with increases in protein intake. Similarly, a high protein intake is associated with a depressed plasma level of T3 in spite of a high concentration of total thyroid hormones, while a low protein diet is associated with an elevated plasma level of T3 and a low concentration of total thyroid hormones. These adaptive differences in T3/T4 ratios may reflect variations in tissue sensitivity to thyroid hormones and/or differences in the basal metabolic rate (BMR), whereas the differences in the concentration of total thyroid hormones may reflect the effects of diet on the secretory activity of the thyroid gland. The elevation of plasma corticosterone when rats were fed a low-protein diet may play a role in the maintenance of liver protein synthesis by making amino acids from muscle catabolism available to the liver. Therefore, the reduction in glucocorticoid levels in severe protein deficiency may be a crucial factor in the resulting breakdown of adaptation which precipitates kwashiorkor.

The physiological and clinical significance of changes in the balance between structurally related hormones which are usually produced in the same organ are also not clearly understood. Dietary changes are associated with alterations in the ratios

Table 15
MEAN VALUES OF SELECTED BLOOD AND PLASMA VARIABLES IN RATS FED AD LIBITUM DIETS CONTAINING 25% LACTALBUMIN PROTEIN AND 5 OR 11.9% FAT

Variable	25% protein, 5% fat, ad libitum[a]	25% protein, 11.9% fat, ad libitum
Blood		
Hemoglobin (g/100 ml)	15.8 ± 1.2 (22)[b]	16.0 ± 1.1 (20)
Hematocrit (%)	49.0 ± 1.7 (22)	49.8 ± 3.3 (20)
MCHC (g/100 ml RBC)	32.0 ± 1.8 (22)	32.1 ± 2.2 (20)
Plasma		
Total protein (g/100 ml)	7.1 ± 0.7 (22)	6.9 ± 0.3 (20)
Albumin (g/100 ml)	4.0 ± 0.3 (22)	3.9 ± 0.3 (20)
Globulin (g/100 ml)	3.1 ± 0.3 (22)	3.0 ± 0.3 (20)
Albumin/globulin ratio	1.3 ± 0.2 (22)	1.3 ± 0.2 (20)
Glucose (mg/100 ml)	209 ± 14 (22)	213 ± 30 (20)
Cholesterol (mg/100 ml)	34.0 ± 2.3 (22)	36.7 ± 2.8 (20)
Triglyceride (mg/100 ml)	47.6 ± 25.5 (22)	48.8 ± 15.8 (20)
Phospholipid (mg/100 ml)	4.7 ± 0.2 (22)	4.7 ± 0.6 (20)
Triiodothyronine (ng/100 ml)	70 ± 29 (10)	89 ± 39 (10)
Thyroxine (μg/100 ml)	3.5 ± 0.1 (10)	3.6 ± 0.1 (10)
T3/T4 (%)	2.0 ± 0.1 (10)	2.5 ± 0.2 (10)
T7 (μg/100 ml)	1.30± 0.01(10)	1.40± 0.16(10)
Insulin (μI/ml)	38.2 ± 7.7 (16)	41.3 ± 10.3 (30)
Glucagon (pg/ml)	178 ± 17 (16)	118 ± 15 (27)[c]
Insulin/glucagon	0.19± 0.02(12)	0.35± 0.05(23)[c]
Corticosterone (μg/100 ml)	26.6 ± 10.5 (22)	29.3 ± 10.4 (30)
Norepinephrine (NE) (ng/ml)	12.2 ± 3.7 (10)	15.6 ± 9.4 (20)
Epinephrine (E) (ng/ml)	15.7 ± 7.9 (10)	21.3 ± 9.2 (20)
NE + E (ng/ml)	27.9 ± 10.5 (10)	36.9 ± 16.4 (20)
NE/E ratio	0.98± 0.54(10)	0.83± 0.61(20)

[a] Of all the diets used in the study, this diet has the closest composition to Purina Laboratory Chow, Ralston Purina Company, Checkerboard Square, St. Louis, Missouri.

[b] Data are means ± SD. The number in parenthesis is the number of rats.

[c] Significantly different from rats fed the 25% protein, 5% fat diet.

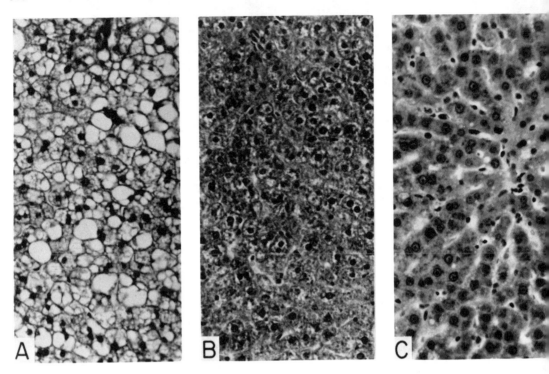

FIGURE 10.   Microscopic appearances of the livers of 12-week-old rats fed diets of varying protein content and/or e█
level for eight weeks. (A) Fed 2% lactalbumin protein ad libitum; (B) fed 2% lactalbumin protein at 50% of ad lib█
intake; and (C) fed 25% lactalbumin protein ad libitum.

of T3/T4, norepinephrine/epinephrine, and possibly cortisol/corticosterone. It was not possible to determine the effects of diet on the cortisol/corticosterone ratio because of the negligible amounts of cortisol in rat plasma. There is evidence that an alteration of the cortisol/corticosterone ratio can be induced in the rabbit by prolonged treatment with ACTH,[59,60] and the reported differences in the urinary excretion of glucocorticoids and 17-ketosteroids between kwashiorkor and marasmus[61] may be due to differences in the relative amounts of the two hormones since they have different turnover rates.[62] The two hormones of each pair differ in their dose-response relationships with target organs, and changes in their relative amounts may provide the fine control of their combined actions. For example, the elevated T3/T4 ratio in protein deficiency represents a shift towards the biologically more active hormone, and this may be a feedback response to a diminished sensitivity of peripheral tissues to thyroid hormones or it may reflect a need for the increase in basal metabolic rate in protein-deficient rats.[63]

Alterations in the plasma concentrations of hormone-binding proteins which occur in severe cases of protein-energy malnutrition may have some influence on plasma hormone levels.[64-66] Leonard and MacWilliam[64] reported that changes in cortisol binding in the serum in kwashiorkor caused an increase in the percentage of free cortisol even in patients who had normal plasma cortisol levels. Ingenbleek et al.[65] attributed their observed high values of serum triiodothyronine in healthy Senegalese children to the associated high levels of thyroxine-binding proteins. Our findings suggest, however, that the relationship between nutritional status and plasma hormone levels is more complex than is indicated by these relatively simple explanations. For example, when the dietary protein level was raised from 15 to 50% there was a marked increase in the plasma T4 concentration (Table 11) and a 50% reduction in the plasma T3 concentration (Table 11), while there was only a very small decrease in the plasma concentrations of albumin and globulin (Table 8) and no change in T3 resin uptake. The changes are adaptive responses which cover a very broad spectrum of nutrient intake levels and are not limited to cases of severe malnutrition.

The results of this study adequately explain the conflicting reports in the literature on the hormonal changes in protein-energy deficiency. First, protein deficiency and starvation rarely occur in their pure forms in man; semistarvation, in which the effects of energy restriction are always confounded, to varying degree, with those of concomitant protein deficiency, is the prevalent type. Second, the nature and/or the extent of the hormonal changes in semistarvation depend on the degree of energy restriction and on the intake of protein. Third, the adaptive mechanisms may fail and the plasma level of a hormone may fall during the transition from subclinical to clinically manifest protein-energy deficiency. This is illustrated by the fall in plasma concentration of corticosterone when the protein level was reduced from 5 to 2% (Table 11). A similar drop has been reported for plasma corticosterone concentration in experimental protein-energy deficiency,[9] for cortisol in kwashiorkor,[67] and for T3 in kwashiorkor.[68] It is not surprising, therefore, that the hormonal changes reported in protein-energy deficiency cover a very wide profile. Hence, in order to understand the endocrine status in each case of malnutrition, it is essential to know its type, degree, and stage of development.

Low values for daily urinary excretion of 17-ketosteroids and 17-ketogenic steroids have been reported for adults in Africa[69,70] and Asia.[71-73] Apparently healthy but stunted Nigerian children have elevated levels of plasma cortisol and growth hormone, depressed plasma level of protein-bound iodine, and low daily urinary excretion of 17-ketosteroids and 17-ketogenic steriods; when these children were provided supplements of skimmed milk powder, their plasma hormone levels and steroid excretion were altered towards values usually considered normal for North American and European children.[67] These reports and the results of this study indicate that some geographic and racial differences in endocrine function are due to environmental factors and, in particular, to variations in nutrient intakes.

Although metabolic adaptation is the primary role of these hormonal changes, they may have other important consequences. For example, changes in the pattern of insulin secretion may predispose to certain forms of diabetes while elevation of plasma levels of neuroendocrine hormones may be involved in the causation of hypertension or in behavior modification. Therefore, further studies of the chronic effects of diet on endocrine function could open up new perspectives in our understanding of the role of diet in metabolic adaptation, reproduction, growth and development, aging, behavior, and the pathogenesis of chronic diseases.

REFERENCES

1. **Kohman, E. A.,** The experimental production of edema as related to protein deficiency, *Am. J. Physiol.,* 51, 378—405, 1922.
2. **Follis, R. H., Jr.,** Kwashiorkor-like syndrome observed in monkeys fed maize, *Proc. Soc. Exp. Biol. Med.,* 96, 523—528, 1957.
3. **Widdowson, E. M. and McCance, R. A.,** Effect of a low protein diet on the chemical composition of the bodies and tissues of young rats, *Br. J. Nutr.,* 11, 198—212, 1957.
4. **Sidransky, H. and Farber, E.,** Chemical pathology of acute amino acid deficiencies. I. Morphologic changes in immature rats fed threonine-, methionine-, or histidine-devoid diets, *Arch. Pathol.,* 66, 119—149, 1958.

5. **Platt, B. S., Heard, C. R. C., and Stewart, R. J. C.,** Experimental protein calorie deficiency, in *Mammalian Protein Metabolism,* Vol. 2, Munro, H. N. and Allison, J. B., Eds., Academic Press, New York, 1964, 445—521.

6. **Deo, M. G., Sood, S. K., and Ramalingaswami, V.,** Experimental protein deficiency, *Arch. Pathol.,* 80, 14—23, 1965.

7. **Kirsch, R. E., Brock, J. F., and Saunders, S. J.,** Experimental protein-calorie malnutrition, *Am. J. Clin. Nutr.,* 21, 820—826, 1968.

8. **Edozien, J. C.,** Experimental kwashiorkor and marasmus, *Nature,* 220, 917—919, 1968.

9. **Anthon, L. E. and Edozien, J. C.,** Experimental protein and energy deficiencies in the rat, *J. Nutr.,* 105, 631—648, 1975.

10. **Nolan, G. A.,** Nutrient deficiencies in animals: energy, in *CRC Handbook Series in Nutrition and Food,* CRC Press, Cleveland, to be published.

11. **Rogers, Q. R. and Harper, A. E.,** Amino acid diets and maximal growth in the rat, *J. Nutr.,* 87, 267—273, 1965.

12. **Folch, J., Lees, M., and Stanley, G. H. S.,** A simple method for the isolation and purification of total lipides from animal tissues, *J. Biol. Chem.,* 226, 497—509, 1957.

13. **Burton, K.,** A study of the conditions and mechanisms of the diphenylamine reaction for the colorimetric estimation of deoxyribonucleic acid, *Biochem. J.,* 62, 315—323, 1956.

14. **Munro, H. N. and Fleck, A.,** Recent developments in the measurement of nucleic acids in biological materials, *Analyst,* 91, 78—88, 1966.

15. **Schmidt, G. and Thannhauser, S. J.,** Method for the determination of deoxyribonucleic acid and phosphoproteins in animal tissues, *J. Biol. Chem.,* 161, 83—89, 1945.

16. **Fleck, A. and Munro, H. N.,** The precision of ultraviolet absorption measurements in the Schmidt-Thannhauser procedure for nucleic acid estimation, *Biochim. Biophys. Acta,* 55, 571—-583, 1962.

17. **Lowry, O. H., Rosebrough, N. J., Farr, A. L., and Randall, R. J.,** Protein measurement with the Folin phenol reagent, *J. Biol. Chem.,* 193, 265—275, 1951.

18. **Drabkin, D. L. and Austin, J. H.,** Spectrophotometric studies. II. Preparations from washed blood cells; nitric oxide, hemoglobin and sulfhemoglobin, *J. Biol. Chem.,* 112, 51—65, 1935.

19. **Strumia, M. M., Sample, A. B., and Hart, E. D.,** An improved microhematocrit method, *Am. J. Clin. Pathol.,* 24, 1016—1024, 1954.

20. **Wolfson, W. Q., Cohn, C., Calvary, F., and Ichiba, F.,** Studies in serum proteins. V. A rapid procedure for the estimation of total protein, true albumin, total globulin, alpha globulin, beta globulin and gamma globulin in 1.0 ml of serum, *Am. J. Clin. Pathol.,* 18, 723—730, 1948.

21. **Rodkey, F. L.,** Direct spectrophotometric determination of albumin in human serum, *Clin. Chem.,* 11, 478—487, 1965.

22. **Doumas, B. T., Watson, W., and Biggs, H. G.,** Albumin standards and the measurement of serum albumin with bromcresol green, *Clin. Chim. Acta,* 31, 87—96, 1971.

23. **Slein, M. W.,** D-Glucose. Determination with hexokinase and glucose-6-phosphate dehydrogenase, in *Methods of Enzymatic Analysis,* Bergmeyer, H.U., Ed., Academic Press, New York, 1963, 117.

24. **Bondar, R. J. L. and Mead, D.,** Evaluation of glucose-6-phosphate dehydrogenase from leuconostoc mesenteroides in the hexokinase method for determining serum glucose, *Clin. Chem.,* 20, 586—590, 1974.

25. **Baginski, E. S., Foa, P. P., and Zak, B.,** Microdetermination of inorganic phosphate, phospholipids and total phosphate in biologic materials, *Clin. Chem.,* 13, 326—332, 1967.

26. **Gharib, H., Ryan, R. J., Mayberry, W. E., and Hockert, T.,** Radioimmunoassay for triiodothyronine (T3). I. Affinity and specificity of the antibody for T3, *J. Clin. Endocrinol. Metab.,* 33, 509—516, 1971.

27. **Kaplan, B. C.,** A Simple Method for the Determination of Serum Thyroxine, presented before the 133rd meeting of the Bioanalysis Section, American Association for the Advancement of Science, 1966.

28. **Clark, F. and Horn, D. B.,** Assessment of thyroid function by the combined use of the serum protein-bound iodine and resin uptake of ^{131}I-triiodothyronine, *J. Clin. Endocrinol.,* 25, 39—45, 1965.

29. **Yalow, R. D. and Berson, S. A.,** Radiobiology-assay of plasma insulin in human subjects by immunological methods, *Nature,* 184, 1648—1649, 1959.

30. **Yalow, R. D. and Berson, S. A.,** Immunoassay of endogenous plasma insulin in man, *J. Clin. Invest.,* 39, 1157—1175, 1960.

31. **Hales, C. N. and Randle, P. J.,** Immunoassay of insulin with insulin antibody precipitate, *Lancet,* 1, 200, 1963.

32. **Hales, C. N. and Randle, P. J.,** Immunoassay of insulin with insulin antibody precipitate, *Biochem. J.,* 88, 137—146, 1963.

33. **Faloona, G. R. and Unger, R. H.**, in *Methods of Hormone Radioimmunoassay*, Jaffe, B. M. and Behrman, H. R., Eds., Academic Press, New York, 1974.

34. **Mayes, D. M.**, Plasma Corticosterone Radioimmunoassay Procedure, Endocrine Sciences, Tarzana, Cal., November 1972.

35. **Mayes, D. M.**, Plasma Cortisol Radioimmunoassay Procedure, Endocrine Sciences, Tarzana, Cal., November 1972.

36. **Engelman, K., Portnoy, B., and Lovenberg, W.**, A sensitive and specific double-isotope derivative method for the determination of catecholamines in biological specimens, *Am. J. Med. Sci.*, 255, 257—268, 1968.

37. **Engelman, K. and Portnoy, B.**, A sensitive double-isotope derivative assay for norepinephrine and epinephrine, *Circ. Res.*, 26, 53—57, 1970.

38. **SAS.** A User's Guide to the Statistical Analysis System, SAS Institute Inc., Raleigh, N. C., 1972, 94—120.

39. **Edozien, J. C. and Switzer, B. R.**, Influence of diet on growth in the rat, *J. Nutr.*, in press.

40. **Waterlow, J. C.**, Fatty liver disease in infants in the British West Indies, *Med. Res. Counc. G. B. Spec. Rep. Ser.*, No. 263.

41. **MacDonald, I.**, Hepatic lipid of malnourished children, *Metabolism*, 9, 838—846, 1960.

42. **MacDonald, I.**, Liver and depot lipids in children on normal and high carbohydrate diets, *Am. J. Clin. Nutr.*, 12, 431—436, 1963.

43. **Flores, H., Pak, N., Maccioni, N., and Monkeberg, F.**, Lipid transport in kwashiorkor, *Br. J. Nutr.*, 24, 1005—1012, 1970.

44. **Seakins, A. and Waterlow, J. C.**, Effect of a low-protein diet on the incorporation of amino acids into rat serum lipoproteins, *Biochem. J.*, 129, 793—795, 1972.

45. **Truswell, A. S., Hansen, J. D. L., Watson, C. E., and Wannerburg, P.**, Relation of serum lipids and lipoproteins to fatty liver in kwashiorkor, *Am. J. Clin. Nutr.*, 22, 569—576, 1969.

46. **Waterlow, J. C.**, Amount and rate of disappearance of liver fat in malnourished infants in Jamaica, *Am. J. Clin. Nutr.*, 28, 1330—1336, 1975.

47. **Gardner, L. I. and Amacher, P.**, Eds., *Endocrine Aspects of Malnutrition*, The Kroc Foundation, Santa Ynez, Cal., 1975.

48. **Hansen, J. D. L.**, in *Protein-Calorie Malnutrition*, Olson, R.E., Ed., Academic Press, New York, 1975, 229—246.

49. **Waterlow, J. C. and Alleyne, G. A. O.**, Protein malnutrition in children: advances in knowledge in the last ten years, *Adv. Protein Chem.*, 25, 117—241, 1971.

50. **Pimstone, B.**, Endocrine function in protein-calorie malnutrition, *Clin. Endocrinol.*, 5, 79—95, 1976.

51. **Jackson, C. M.**, *The Effects of Inanition and Malnutrition Upon Growth and Structure*, Blackiston, Philadelphia, 1925, 417—456.

52. **Keys, A., Brozek, J., Henschel, A., Mickelsen, O., and Taylor, H. L.**, *The Biology of Human Starvation*, Vol. 1, University of Minnesota Press, Minneapolis, 1950, 209—217.

53. **Gillman, J. and Gillman, T.**, *Perspectives in Human Malnutrition*, Grune and Stratton, New York, 1951, 409—414.

54. **Trowell, H. C., Davies, J. N. P., and Dean, R. F. A.**, *Kwashiorkor*, Arnold, London, 1954, 155—156.

55. **Campbell, J. A. H.**, The morbid anatomy of infantile malnutrition in Cape Town, *Arch. Dis. Child.*, 31, 310—314, 1956.

56. **Alleyne, G. A. O. and Young, V. H.**, Adrenocoritical function in children with severe protein-calorie malnutrition, *Clin. Sci.*, 33, 189—200, 1967.

57. **Abassy, A., Mikhail, M., Zeitoun, M., and Ragab, M.**, The suprarenal cortical function as measured by the plasma 17-hydroxycorticosteroid level in malnourished children. I. In wasted children, *J. Trop. Pediatr.*, 13, 87—95, 1967.

57a. **Abassy, A., Mikhail, M., Zeitoun, M., and Ragab, M.**, The suprarenal cortical function as measured by the plasma 17-hydroxycorticosteroid level in malnourished children. II. In kwashiorkor, *J. Trop. Pediatr.*, 13, 154—162, 1967.

58. **Rao, K. S. J., Srikantia, S. G., and Gopalan, C.**, Plasma cortisol levels in protein-calorie malnutrition, *Arch. Dis. Child.*, 43, 365—367, 1968.

59. **Kass, E. H., Hechter, O., Macchi, I. A., and Mau, T. W.**, Changes in patterns of secretion of corticosteroids in rabbits after prolonged treatment with ACTH, *Proc. Soc. Exp. Biol. Med.*, 85, 583—587, 1954.

60. **Ganjam, V. K., Campbell, A. L., and Murphy, B. E. P.**, Changing patters of circulating corticosteroids in rabbits following prolonged treatment with ACTH, *Endocrinology*, 91, 607—611, 1972.

61. **Castellanos, H. and Arroyave, G.**, Role of the adrenal cortical system in the response of children to severe protein malnutrition, *Am. J. Clin. Nutr.*, 9, 186—195, 1961.

62. **Peterson, R. E.,** The miscible pool and turnover rate of adrenocortical steroids in man, *Recent Prog. Horm. Res.,* 15, 231—274, 1959.

63. **Tulp, O., Horton, E. S., Tyzbir, E. D., Danforth, E., Krupp, P. P., and Bollinger, J.,** Thyroid function in experimental protein malnutrition, *Am. J. Clin. Nutr.,* 30 (Abstr.), 621, 1977.

64. **Leonard, P. J. and MacWilliam, K. M.,** Cortisol binding in the serum in kwashiorkor, *J. Endocrinol.,* 29, 273—276, 1964.

65. **Ingenbleek, Y., DeVisscher, M., and DeNayer, P.,** Measurement of prealbumin as index of protein-calorie malnutrition, *Lancet,* 2, 106—108, 1972.

66. **Ingenbleek, Y., DeNayer, P., and DeVisscher, M.,** Thyroxine-binding globulin in infant protein-calorie malnutrition, *J. Clin. Endocrinol.,* 39, 178—180, 1974.

67. **Edozien, J. C., Khan, M. A. R., and Waslien, C. L.,** Human protein deficiency: results of a Nigerian village study, *J. Nutr.,* 106, 312—328, 1976.

68. **Ingenbleek, Y. and Beckers, C.,** Triiodothyronine and thyroid-stimulating hormone in protein-calorie malnutrition in infants, *Lancet,* 2, 845—847, 1975.

69. **Barnicot, N. A. and Wolffson, D.,** Daily urinary 17-ketosteroid output of African negroes, *Lancet,* 1, 893—895, 1952.

70. **Edozien, J. C.,** Biochemical normals in Nigerians: urinary 17-oxosteroids and 17-oxogenic steroids, *Lancet,* 1, 258—259, 1960.

71. **Friedmann, H. C.,** 17-Ketosteroid excretion in Indian males, *Lancet,* 2, 262—268, 1954.

72. **Lugg, J. W. H. and Bowness, J. M.,** Renal excretion of 17-ketosteroids by members of some ethnic groups living in Malaya, *Nature,* 174, 1147—1148, 1954.

73. **Ramachandran, M., Venkatachalam, P. S., and Gopalan, G.,** Urinary excretion of 17-ketosteroids in normal and undernourished subjects, *Indian J. Med. Res.,* 44, 227—230, 1956,

NUTRIENT DEFICIENCIES IN MAN: DIETARY FIBER

H. C. Trowell and D. P. Burkitt

NOMENCLATURE AND DEFINITION

Crude fiber (CF) is the residue of plant food after sequential hydrolysis by dilute acid and dilute alkali (Figure 1). The term has been defined by the Association of Official Agricultural Chemists, who have determined internationally accepted methods of analysis. The term crude fiber (CF) has legal status with regard to animal foodstuffs. It was considered, erroneously, that CF represented most of the indigestible material. An average of 80% of hemicelluloses or pentosans and from 50 to 90% of lignins are removed, and cellulose recovery is only 50 to 80%.[1] Dietary fiber (DF), as will be discussed, contains all these structural polymers; it has recently been redefined as all structural and storage polysaccharides and lignin of plant food that are not digested by the enzymes of man.[2]

However, general agreement on the definition of the basic substances cellulose and hemicelluloses is lacking.[1] In regard to isolation of these substances, it has proved difficult to separate lignin from cellulose and to remove all traces of hemicellulose from cellulose. Of course, the celluloses, hemicelluloses, and lignins from different plant materials differ chemically and biologically. Therefore, there have been many attempts to find better methods of analyzing fiber. The earliest attempts indentified fiber with cellulose and in many nutritional textbooks the two substances were reported to be identical. Early attempts to analyze fiber gave little attention to lignin. Eventually it was realized that the largest constituents of fiber (namely, the hemicelluloses and pectic substances) had been virtually overlooked.

In the United States many procedures have been devised to study the constituents of plant cell walls, especially by those workers studying the feeding of ruminants. In the rumen, bacteria ferment celluloses, hemicelluloses, and pectic substances to contribute much absorbed metabolizable energy to the host. There has been a tendency among those conducting research in animal foodstuffs to concentrate almost exclusively on the constituents of the plant cell wall.[1] Little attention has been paid to storage polysaccharides, present, for example, within the cells of certain leguminous seeds. Ruminants are seldom fed these seeds. However, as these polysaccharides are not hydrolyzed by human alimentary enzymes, they consititute a part of the DF[3] (Figure 2).

A totally different approach to the undigested components of food occurred in England. McCance and Lawrence[4] divided the carbohydrates into two groups. Those hydrolyzed by human alimentary enzymes (such as starch, amylopectin, dextrin, and sugar) were termed available carbohydrates (Figure 2). These were analyzed by direct determination and reported as monosaccharides in food tables.[5] The remainder were

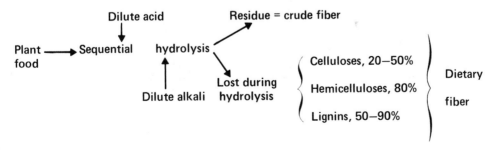

FIGURE 1.   Schematic representation of derivation of crude fiber and dietary fiber from plant food.

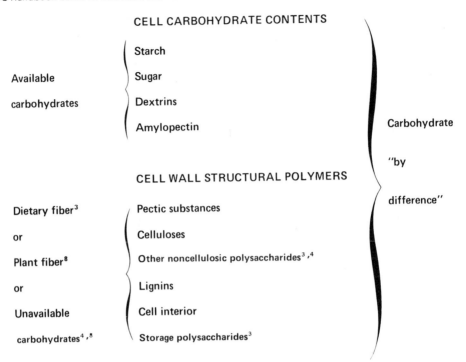

FIGURE 2. Relationship of available carbohydrates and carbohydrate 'by difference' to unavailable carbohydrates, which are almost identical to both dietary fiber and nonpurified plant fiber.[8] Unavailable carbohydrates originally included lignin,[12] and polysaccharides not hydrolyze by human alimentary enzymes.

termed unavailable carbohydrates. These were not estimated by analysis of their constituent parts, but assessed by difference. This group of substances contain lignins, which are not carbohydrates, so that the term "unavailable carbohydrates and lignins" is more appropriate. No attempt was made to limit unavailable carbohydrates to the cell wall. Storage polysaccharides, such as inulin of artichokes, and the galactomannans of leguminous seeds, neither of which are digestible by human alimentary enzymes, were included.

Trowell[2,6,7] and Trowell et al.[3] suggested that a new term, "dietary fiber," be introduced; it was defined as the remnants of plant food unhydrolyzed by human alimentary enzymes. It was thus similar to unavailable carbohydrates and lignin.

Spiller and Amen[8,9] pointed out the need for a better nomenclature on fiber. They suggested that it should be called plant fiber (PF). In its natural state, associated with all associated substances (cutins, trace minerals, glycoprotein, etc.), it should be called nonpurified plant fiber (NPPF). When these polymers have been purified and isolated, as has occurred with cellulose and pectin, they should be called purified plant fibers (PPF). Synthetic enzyme-indigestible material such as cellophane should be called synthetic nonnutritive fiber (NNF). Trowell[10] regarded "synthetic fiber" as a useful new term, but criticized the term "nonnutritive fiber," for some synthetic hemicelluloses would provide energy to the host after bacterial degradation in the colon. He suggested that the term "dietary fiber" (DF) be restricted to the undigested polysaccharides, both structural and storage, and lignins. If it was desirable to note the presence of associated substances such as glycoprotein and trace minerals, then, together, all of these might be called the dietary fiber complex (DFC).[10] The terminology concerning fiber is as tangled as the fibrils of fiber itself.

ANALYSIS

Crude fiber, as reported in food tables, is analyzed by the method approved by the Association of Official Analytical Chemists. A sample of plant food is hydrolyzed by 0.255 N sulfuric acid solution, followed by hydrolysis with 0.313 N sodium hydroxide. The residue is incinerated. Crude fiber is the difference between the weight of the residue before and after incineration. This method reports only a variable and small proportion of the structural and storage undigested polymers, and replacement by modern methods of analysis is essential. Most food tables report only crude fiber. In Britain the food tables[5] report the unavailable carbohydrate of fruit, vegetables, and nuts, but not of cereals. The figures of unavailable carbohydrate are comparable to those of the dietary fiber (see Table 3).

In England, Southgate's[11,12] method of analysis has been developed to analyze the principal constituents of dietary fiber of human foods. A series of extractions with various chemical solutions allows estimates of cellulose, hemicellulose, and lignin to be made, but the associated substances such as cutins, epicuticular waxes, and glycoprotein are not estimated.

In the U.S., Van Soest and Robinson[85] have developed methods to analyze plant foods. At first these methods were employed to analyze foodstuffs of ruminants, but human foods have more recently been analyzed by these procedures. Analysis is based on the use of detergents. The neutral detergent fiber (NDF) of van Soest is possibly the best single analytical estimate of DF at the present time and should replace CF figures in food tables.[9] Table 1 presents data on the NDF and CF of certain foods.[13] The van Soest method of analysis can provide data concerning the cellulose, hemicellulose, and lignin constituents of fiber. These figures compare favorably with those obtained by Southgate's method of analysis.[14] Neither of these methods includes or estimates the pectic substances; these are also excluded from CF estimations. If it is desired to estimate pectic substances, special procedures are employed.[9] Table 2 reports the total pectic substances present in certain fruits and vegetables.

In the Netherlands Hellendoorn and his associates[15] favor the estimation of the indigestible residue (IR) of plant foods employing enzymatic digestion by neutral pancreatin for 1 hr following pepsin digestion for 18 hr. Table 3 compares CF analyzed by two modifications of the Weende procedure: the standard U.S. method that employs hydrolytic treatment (CF-US) and a rapid procedure that employs oxidation treatment as used in the Netherlands[15] and other countries in Europe (CF-NETH). Table 3 also presents data derived from three different methods of measuring plant polysaccharides and lignin (DF) that are resistant to hydrolysis by the digestive enzymes of man. This residue has been assessed in the U.S. as NDF,[9] in the Netherlands as the IR,[15] and in Britain as unavailable carbohydrate (UC).[5]

Those reporting CF should specify the method employed. For example, white bread in the U.S. has been reported to contain CF 1.7%, in the Netherlands, CF 0.4% (Table 3), and in England 0.12%.[16] The standard U.S. method of analyzing CF gives higher figures than those obtained in the Netherlands[15] or in accepted food tables in England.[17] With regard to the undigested portion of plant foods, the data obtained by enzymatic digestion in the Netherlands (IR) should be compared with the UC assessed in England. It should be noted that indigestible soluble pectin and galactooligosaccharides escape recognition in the enzymatic procedures and in all estimations of crude fiber.

As fiber is recognized to be an essential constituent of the diet, international agreement on definition, terminology, and methods of analysis will be achieved in the next few years. Meanwhile, it represents an extremely confused area of nutritional data.

Table 1

COMPARISON OF "NEUTRAL DETERGENT
FIBER" AND "CRUDE FIBER"[a,9,13]

	Dry matter (%)	Neutral detergent fiber (%)	Crude fiber (%)
Cereals			
Bread, whole wheat	64.4	14.9	5.1
Bread, white enriched	64.2	3.3	1.7
Kelloggs All Bran	97.3	34.0	9.2
Kelloggs Corn Flakes	96.3	7.9	1.4
Nabisco Shredded Wheat	93.5	22.4	3.6
Ralston Purina® Wheat Chex®	96.4	17.6	3.5
Post Grape Nuts	94.9	11.9	2.2
Quaker Puffed Wheat	94.3	8.9	3.5
Scotts Porridge Oats	91.1	10.4	2.0
Fruit and vegetables			
Apples	11.7	7.6	3.7
Beans, green	10.2	22.0	10.6
Beet root	12.5	11.8	4.7
Broccoli	11.0	11.3	12.6
Cabbage	7.8	14.2	8.4
Carrots	10.4	9.2	5.7
Cauliflower	8.7	15.1	10.1
Celery	5.0	14.4	13.3
Cucumber, peeled	3.2	12.7	13.4
Lettuce, romaine	4.9	17.3	12.4
Onions	9.1	7.1	9.7
Oranges, peeled	13.6	3.7	2.7
Potatoes, peeled	22.2	4.7	0.5
Radishes	11.9	14.3	11.4

[a] Neutral detergent fiber and crude fiber as percent of dry matter.

Table 2

TOTAL PECTIC SUBSTANCES
OF FRUIT AND VEGETABLES[9]

	Pectic substances (%)
Fruits	
Apples	0.7—0.8
Apricots	0.7—1.3
Bananas	0.6—1.3
Blackberries	0.7—1.2
Cherries	0.2—0.5
Grapes	0.1—0.3
Grapefruit	3.3—4.5
Lemons	2.8—3.0
Oranges	2.3—2.4
Raspberries	1.0
Vegetables	
Beans	0.3—1.1
Carrots	1.2—2.9
Sweet potatoes	0.8

Table 3

COMPARISON OF CRUDE FIBER, U.S. (CF-US),[a] NETHERLANDS
(CF-NETH),[b] NEUTRAL DETERGENT FIBER (NDF),[9]
INDIGESTIBLE RESIDUE (IR),[c] AND UNAVAILABLE
CARBOHYDRATE (UC)[5] OF VARIOUS FOODS

Food	Crude fiber		Dietary fiber		
	CF-US	CF-NETH	NDF	IR	UC
Cereals					
White bread	1.7	0.4	3.3	2.2	
Whole wheat bread	5.1	1.2	14.9	9.3	
Rye bread		1.1		14.0	
Rolled oats		1.5		7.0	
Rice, polished		0.6		1.4	
Wheat bran		9.0		49.0	
Legumes					
Brown beans		2.2		12.0	20
Soya beans		2.2		4.6	
White beans		0.8		4.7	
Marrow fat peas		0.6		5.1	
Peas		0.6		3.3	5.2
Peanuts		2.5		7.7	5.6
Vegetables					
Brussels sprouts		1.4		1.8	3.6
Cabbage	8.4	1.6	14.2	1.7	2.2
Carrots	5.7	1.0	9.2	1.4	3.7
Onions	9.7	1.0	7.1	0.8	1.3
Potatoes	0.5	0.6	4.7	2.0	1.0
Turnips		1.0		1.3	2.2
Fruit					
Apples	3.7	0.6	7.6	1.1	2.2
Pears		1.4		1.1	1.6

[a] CF-US by Weende standard procedure hydrolytic treatment.[9]
[b] CF-NETH by rapid Weende procedure oxidation treatment.[15]
[c] Indigestible residue by enzyme digestion.[15]

PHYSIOCOCHEMICAL PROPERTIES OF FIBER

Water adsorption — Certain polysaccharides swell and form a jelly-like mass. The water is partly bound with hydrogen bonds. It is also stored as immobilized water in the interstices of the swollen polymers. Hemicelluloses and pectic substances can adsorb much water; celluloses have less capacity. Lignins do not adsorb water; they may even aid the formation of dry fecal material. Water adsorption capacity varies considerably: fiber from wheat bran, carrots, cabbage, and apple can hold much water; fiber from other vegetables and fruit has less capacity to do so.[18] The bulk-forming capacity of fiber and its laxative effect are largely due to the water adsorption of various forms of fiber in the diet.

Cation exchangers — Acidic polysaccharides with uronic acid moieties can act as cation exchangers and bind metals. Pectic substances and alginates from seaweed are strong cation exchangers; hemicelluloses have a moderate capacity.

Adsorption — Unconjugated bile acids present in the colon are strongly adsorbed by certain forms of fiber; the conjugated bile acids present in the duodenum and small intestine are only slightly adsorbed.[19,20] Fiber can adsorb certain toxic substances. The deleterious effects of ionic detergents such as Tween[®] 20 and 60 were counteracted by feeding lucerne fiber.[21] Therefore, fiber may offer protection against certain poisonous substances present in the diet or formed during digestion in the gut.

Gel-filtration system — Polysaccharides that form a gel produce a gel-filtration system that has molecular exclusion capacity. There has been little study of this property in the human gastrointestinal tract.

FIBER AND INTESTINAL MICROFLORA

Digestion of Fiber by Colonic Microflora

Fiber is not digested by human alimentary enzymes, but colonic bacteria partially digest fiber. The original studies of Williams and Olmsted[22] in 1936 demostrated that a small amount of cellulose, but a large proportion of hemicelluloses, were digested by the microflora of the large bowel. Recent research[23] has reported that colonic bacteria degrade cellulose 15% and hemicelluloses 70 to 90%. Hemicellulose digestion varies considerably. Lignin is almost entirely undigested by bacteria, but its presence may alter the digestibility of other cell-wall polymers such as cellulose. Pectin also is digested by colonic bacteria. This microflora degrades constituents of the fiber, especially the hemicelluloses, to form volatile fatty acids (acetic, butyric, and proprionic). These are the major anions of the feces.[24] Certain gases such as methane, carbon dioxide, and hydrogen are formed; if excessive, they may cause bloating.

Alteration of Colonic Bacteria by Fiber

There has been little investigation by modern techniques of the microflora at different levels of the colonic contents; this is especially true for the anaerobes. There has been a small amount of investigation of fecal bacteria of the extruded stool. Thus, fecal bacteria of persons living in developing countries, wherein high-fiber diets are consumed, differ from the microflora of persons in Western countries, wherein low-fiber diets are eaten. The microflora of U.S. adult volunteers altered considerably when they were switched from a fiber-free semisynthetic diet to one containing fiber in natural foods.[25] Fiber also increases the unconjugated bile acids in the colon and these acids inhibit the growth of certain anaerobic bacteria.[26]

DIETARY CHANGES DURING ECONOMIC DEVELOPMENT

The World Health Organization report[27] outlined the complex dietary changes that follow modern economic development (Figure 3). The overall pattern consists of a decreased consumption of unrefined, high-fiber, starchy foods which are replaced by an increased consumption of fiber-free fat and sugar. Fiber from cereals decreases considerably. With regard to fiber from vegetables and fruit, rural inhabitants of developing countries usually take a moderate amount. These expensive foods decrease considerably during urbanization. They subsequently increase in western society with increasing affluence.

The changes in CF intakes that rapidly accompany urbanization of South African Bantu are depicted in Figure 4. Rural male South African Blacks eat a daily average of approximately 20 to 25 g CF, almost half from unrefined maize meal, the other half from vegetables. Urban male Blacks eat a daily average of approximately 6 g CF, about 5 g from lightly refined maize meal; vegetables and fruit provide less than 1 g.[28]

The changes in CF intakes that slowly accompanied modern economic development in England from 1770 to 1970 are depicted in Figure 5. Data are derived from official estimates of food supplies from 1860 to 1970 and an assessment of the 1770 foods in order to portray the probable trends in supply.[10] A considerable reduction in starchy foods, cereals, and potatoes, occurred around the turn of the present century. In 1850

ENERGY % FROM VARIOUS FOODS

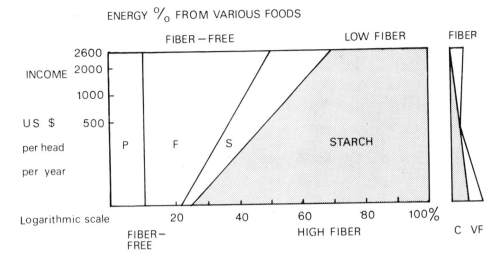

FIGURE 3. Changing dietary patterns with economic development and income (U.S. dollars per head per year). Energy percentages derived from fiber-containing starch decrease and these foods contain less fiber; energy from fiber-free foods increases. (P) protein, (F) fat, and (S) sugar. Fiber intakes have been assessed and represented diagrammatically on the right.[10] Cereal fiber (C) decreases; vegetable and fruit fiber (VF) decreases with urbanization, then increases with affluence. See also Figures 4 and 5. (Modified from *WHO Tech. Rep. Ser.*, No. 522, 19 – 22, 1973. With permission.)

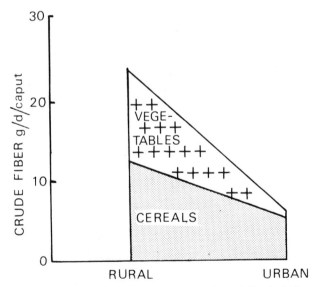

FIGURE 4.   Changes in crude fiber intakes of South African Bantu men following urbanization. (From Trowell, H. C., *Am. J. Clin. Nutr.*, 29, 417–427, 1976. With permission.)

cereal CF intakes per person averaged 1.7 g/day; they fell to 0.5 g/day in 1970. Potato consumption and fiber intakes declined similarly. On the other hand, fruit and vegetable CF has been rising during the present century, but no data exist concerning the supply of these foods in the previous century. Cereal fiber intakes rose during the period of wartime food controls from 1941 to 1954, with decreased death rates from diabetes mellitus and diverticular disease.[10] Apart from the war time changes, a comparable decreased consumption of starchy food and, therefore, of its fiber occurred in the U.S.[29] The

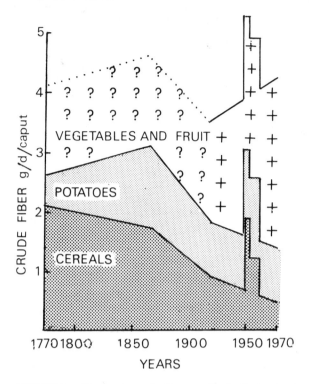

FIGURE 5.   Changes in crude fiber supplies in England and Wales, 1700 to 1970, omitting 1914 to 1918 as there are no data for these years. Increased fiber supplies 1941 to 1954 due to high-fiber National bread. No data on vegetable and fruit supplies prior to 1910; approximate assessments 1770 to 1860. (From Trowell, H. C., *Am. J. Clin. Nutr.*, 29, 417—427, 1976. With permission.)

English daily CF intake per person in 1970 was estimated to be 4.2 g, derived from cereal CF, 0.5 g; potato CF, 0.9 g; and vegetable and fruit CF, 2.7 g.[30] English white flour, 70% extraction, contains CF 0.12 g/100 g.[16]

FECAL BULK, WEIGHT, AND TRANSIT TIMES

The laxative effect of the unabsorbable fiber content of food has long been recognized. Hippocrates observed that there was "a great difference whether the bread be made of fine flour or coarse." Persian physicians in the ninth century AD reported that "chupattis containing more bran came out of the digestive tract quicker."[31] In the U.S. during the 1930s, Cowgill[32] measured the effect of wheat bran on laxation. Cowgill was the first to assess the daily requirements of fiber to prevent constipation at 90 mg CF per kilogram body weight.

Many different methods have been employed to measure intestinal transit times; these vary according to the method employed. Radioopaque plastic pellets have been employed much in recent years. Transit time is usually calculated as the number of hours that elapse between swallowing the "shapes" and the recovery of 20 of 25 pellets in the stools. This involves the examination of several stools. A recent modification allows transit time to be estimated on the examination of a single stool.[33] Different radioopaque "shapes" are swallowed on three successive days with breakfast. The first stool to be passed on the fourth morning, or at any subsequent period, is examined. The number of each variety of the markers present is counted. The single stool transit time (SST) is calculated from the following formula

$$SST = \frac{t_1 S_1 + t_2 S_2}{S_1 + S_2}$$

where t_1 and t_2 are the times in hours from ingestion until defecation of the two markers present in the greatest number in the single stool and S_1 and S_2 are the number of each marker present. This simplified method is used considerably during epidemiological surveys.

Figure 6 shows the relationship between transit times, measured by radioopaque pellets, and stool weight in various groups of persons who ate different amounts of fiber (residue) in their diet.[34] African villagers eating a high-fiber diet had daily stool weights of 400 to 500 g. English boarding school pupils and British naval personnel eating low-fiber (residue) diets had daily stool weights of 50 to 150 g. Those eating an intermediate-type diet, whether African boarding school pupils or British vegetarians, had intermediate stool weights and transit times. As regards transit times, a critical point is attained when daily fecal weights reach 200 g. Above this weight there is little further decrease in transit times. These studies[30] have been confirmed in Rhodesia[35] and Malaysia.[36] There have been numerous studies in Britain of the effect of adding wheat bran (15 to 30 g/day) to the diet; stool weight is always considerably increased; transit times are usually, but not invariably, decreased. Long transit times are shortened to approximately 2 days; short transit times are prolonged to approximately 2 days (Figure 7).[37] Fiber from vegetables and fruit is also laxative, but there is considerable variation. Thus, carrots and apples aid fecal bulk, but many fruits and certain vegetables, in proportion to their weight, have little power to absorb water or aid bowel laxation.[38]

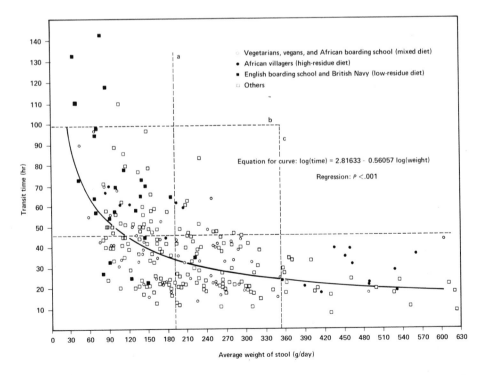

FIGURE 6.   Relationship between transit times, stool weights of four groups of persons, and diet which contained varying amounts of fiber (residue). (From Burkitt, D. P., Walker, A. R. P., and Painter, N. S., *JAMA*, 229, 1068, 1974. With permission.)

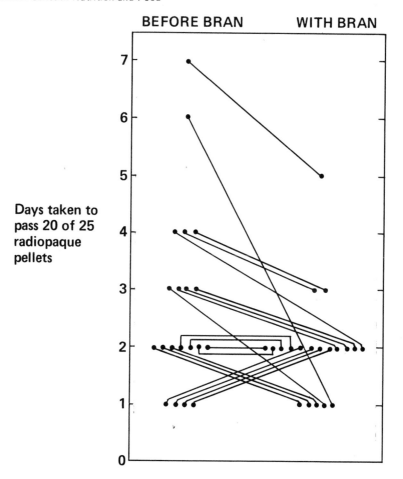

FIGURE 7.   Changes in intestinal transit times in 20 normal subjects before and during the addition of bran (20 g/day). (From Payler, D. K., Pomare, E. W., Heaton, K. W., and Harvey, R. F., *Gut,* 16, 209, 1975. With permission.)

Fecal bulk and consistency affect colonic intraluminar pressures and segmentation.[31] High-fiber diets increase fecal volume within the colon, which is wide in bore and large in volume; colonic intraluminar pressures decrease and segmentation proceeds in a more relaxed manner because the feces are soft. Low-fiber diets are accompanied by a small volume of feces in a small-volume colon; intraluminar pressures are raised and muscular segmentation is forcible. Any dietary change to a high-fiber diet must proceed therefore very slowly. Several weeks may be required before excessive flatus formation and bloating are overcome.

COLONIC DISEASE ASSOCIATED WITH LOW—FIBER INTAKES

Constipation — Gastroenterologists have stated that the adult rectal muscle can accommodate 150 to 200 ml of feces before the defecation reflex is fully operative.[39] In Britain, the adult stool averages little more than 100 g/day.[38] Therefore, constipation is common in western countries. Elderly persons with small appetites take little food and fiber; they are especially prone to constipation and have much diverticular disease. At all ages, many persons strain for several minutes in order to defecate, thereby raising the intraabdominal pressure considerably.[40]

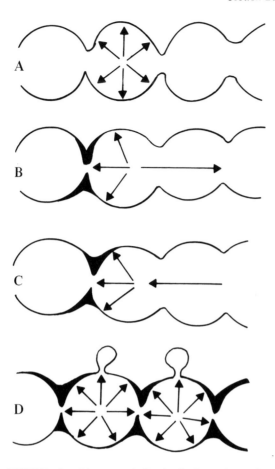

FIGURE 8. Diagram of longitudinal section of colon. (A) High pressure localized to one segment. (B) Pressure released as one contraction relaxes and contents move to right. (C) Contents halted in any movement to left. (D) High pressures due to forcible segmentation forms pouches (diverticula). (From Painter, N. S., *Ann. R. Coll. Surg. Engl.*, 34, 98, 1966. With permission.)

Diverticular disease — There is more evidence relating diverticular disease to the lifelong consumption of low-fiber diets[41] than to any other organic disease of the colon. It is the commonest organic disease of the colon in the Western world.[42] Colonic diverticula are caused by high intraluminar pressures resulting from the forcible segmenting action of the colon acting on inspissated fecal masses (Figure 8). Experimental studies in various animals[42] and the response to wheat bran fiber support the view that the disease is due to a low intake of bulk-forming fiber.[43] Epidemiological data from many countries support the role of fiber as the principal etiological factor.[41,42] The British male and female crude death rates of diverticular disease rose steeply from their first registration in 1923 until 1937; then both remained stationary all years of the high-fiber national flour (1942 to 1953).[41]

Irritable bowel syndrome — Many gastroenterologists consider this to be the most common functional disease of the colon in the Western world. It appears rarely to be diagnosed by gastroenterologists among South African Bantu, who consume cereals containing much fiber: the first two cases have been reported recently in Johannesburg.[44] Gastroenterologists in the U.S. and Europe are treating this disease with wheat bran. Many report success, but emphasize that the bran should be of coarse particle size; this ensures increased water absorption.[45]

Appendicitis — There is considerable evidence that acute appendicitis is still a rare disease in rural communities of Africa and Asia.[46,47] Appendicitis was a rare disease in Europe and U.S. during the last century; pathologists reported a few cases of "suppurative perityphlitis" around the cecum. There is a negative correlation between appendectomy rates and crude fiber intakes among three ethnic groups in South Africa.[48] In many European countries the incidence of appendicitis decreased markedly when less refined cereal flour was eaten during the last war. On the other hand, the acute appendicitis incidence rose in African troops who began to eat more refined cereals in their army rations. At present, the pathological evidence suggests that the initial lesion is obstructive in nature rather than inflammatory. More epidemiological studies and animal experiments are required before a relationship to fiber can be considered to be fully established.

Hemorrhoids — There is considerable evidence that hemorrhoids are rarely encountered in Africans, although they are becoming increasingly diagnosed in the larger towns. Two hypotheses link the prevalence of hemmorrhoids to hard, inspissated feces due to low-fiber diets. First, it has been postulated that straining to pass hard fecal masses raises the intravenous pressure not only in the femoral vein,[40] but also in the hemorrhoidal plexus.[49] Second, it has been suggested that hard, inspissated fecal masses exert direct pressure on the hemorrhoidal anal cushions so that these become congested and distended and the piles become prolapsed.[50]

Carcinoma of the large bowel — No other form of cancer is so closely associated with modern economic development and westernization of the diet and lifestyle. It has been postulated that carcinogens form in the feces due to the action of certain anaerobic bacteria on the bile salts.[51,52] The diets with a low incidence of colonic cancer are often, but not always, low in fat; therefore, high-fat diets have been considered to be a risk factor. However, there appears to be a low incidence of colonic cancer in milk-drinking African tribes such as the Masai, and Eskimos; both consume much fat. If fecal fat is largely endogenous in origin and increased by high-fiber diets,[53] the role of fiber becomes more likely, especially because it always increases fecal bulk and usually decreases transit times. At the present time there is insufficient data to relate the incidence of large bowel cancer to any single environmental factor, dietetic or otherwise.

Hiatus hernia — Epidemiological data, admittedly limited, reports the rarity of this disease in Africa, certain areas of India, Iraq, and South Korea.[54] Numerous radiologists in those countries have reported that the rarity of the disease contrasts with the high incidence in many Western countries. It has been suggested that raised intraabdominal pressures during defecation have produced, over a period of many years, an upward protrusion of the stomach through the esophageal hiatus.[55]

Varicose veins — There is considerable evidence that varicose veins are rare in rural and urban areas of Africa[56] and India.[57] However, there is still inadequate data in terms of prevalence in age and sex groups, and varicose veins are an age-related disease. Raised intraabdominal pressures which accompany straining to defecate[40] are communicated down the femoral vein. It is postulated that over a period of many years the long saphenous vein dilates and renders the valves incompetent sequentially from above downwards. The hydrostatic pressure of the venous column of blood also aids distension of the vein.[40] More studies of blood pressure and blood flow within the venous system of the legs are required before the hypothesis can be considered established. However, the epidemiological data are very striking.

FIBER AND LIPID METABOLISM

Bile Salts and Neutral Sterols

Natural forms of fiber bound certain bile salts, sodium taurocholate and glycocholate, in vitro; synthetic forms of fiber such as cellulose and cellophane had no such effect.[58] Plant fiber from celery, lettuce, potato, and string beans had a similar binding action in vitro.[9] Rats fed bagasse, the fibrous residue obtained in sugar extraction, excreted more bile acids and more neutral sterols than rats fed cellulose or fiber-free diets.[59] Pectic substances and fiber-rich laboratory chow had a similar action.[9]

In man, increased bile acid excretion and moderate reduction of serum cholesterol levels occurred in Indian adults who ate large amounts of fiber-rich leguminous seeds (*Cicer arietinum*).[60] This may be relevant to the decreased incidence of ischemic heart disease in areas of India and other Asian countries wherein these seeds are an important constituent of the diet.

Serum cholesterol levels can be reduced in animals and man by some of the constituents of dietary fiber. Pectin is hypocholesterolemic in animals and man.[61] The artificially high serum cholesterol levels produced in experiments in rats which were fed much cholesterol and fat as well as other hyperlipidemic substances have been reduced by many natural forms of fiber such as rice bran[62] and Black gram (*Phaseolus mungo*).[63] A polysaccharide fraction isolated from rice bran had a marked hypocholesterolemic action in these rats.[64] Fiber-rich rolled oats decreased serum cholesterol levels in rats.[65] This action appeared to be due to the combined action of dietary fiber of the oats with some unsaturated fat extracted from this cereal. The hypocholesterolemic effect of fiber-rich carbohydrate foods fed to experimental animals has been reviewed.[66]

Coronary Heart Disease (CHD)

The Department of Health and Social Security in England[67] reported in 1974 that "populations who eat a diet rich in fibre (especially fibre from cereals and legumes) usually have a lower serum cholesterol concentration and a lower mortality from ischaemic heart disease than those who eat a western type diet relatively low in this kind of fibre." Although many varieties of natural foodstuff rich in fiber have lowered serum cholesterol levels in artificially-fed hyperlipidemic animals, neither wheat bran (30 to 50 g/day) nor wholemeal bread (180 g/day) have reduced serum cholesterol levels in man.[68] Most, but not all, of these subjects were young normolipidemic subjects.

This absence of any hypolipidemic effect on blood lipids led to a restatement of the hypothesis concerning CHD and fiber, "CHD mortality is considerably lower in populations who eat a large proportion of energy derived from starchy carbohydrate foods, cereals, pulses, starchy roots and starchy fruits, than in those who eat a western-type diet. These foods are usually undepleted, or only lightly depleted of fiber. Such diets contain a low proportion of fiber-free foods, fat, sucrose and animal proteins."[10] Large amounts of starchy carbohydrate foods can only be eaten if fats and sucrose contribute only a small proportion of energy; whether any benefit is due to these decreased intakes remains uncertain. If starchy foods rich in fiber are indeed a protective factor in CHD, it is possibly partly through the brisk fibrinolysis that has been reported in many communities in which CHD is uncommon. Therefore, the role of carbohydrate foods rich in fiber in preventing or alleviating CHD remains under investigation, but is undecided.

Diabetes Mellitus

There is more epidemiological and experimental evidence linking diabetes mellitus to the consumption of fiber-depleted starchy food[69,70] than to other metabolic diseases such as CHD, thromboembolic disease, gallstones, and obesity. Diabetes mellitus has not been reported in food-gatherer hunter groups; their women and children (likewise the men, on many occasions) eat large amounts of unprocessed starchy foods. Few sophisticated diabetes prevalence surveys have been conducted in rural groups of developing countries, but the rarity of glycosuria in all hospital patients is well documented in East African Blacks, South African Bantu,[71] Yemenite Jews, and New Guineans.[72] In Western countries, but not in developing countries such as New Guinea,[72] 16% of adults over 50 years of age have a diabetic-type glucose tolerance test. Diabetes death rates in England and Wales declined 55% in men, and 54% in women, from 1941 until 1954 to 1957; these were the years in which high-fiber national flour was compulsory.[73] Diabetes has been produced in small rodents which were fed low-fiber cereal chows, and the condition was reversed by a dietary switch to high-fiber cereal chows. Both diets were fed ad libitum.[69,70] The investigation of the role of fiber in the carbohydrate metabolism of man is beginning. Wheat bran has increased carbohydrate tolerance. Indigestible polysaccharides such as pectin and guar gum have increased carbohydrate tolerance in normal persons[74] and diabetics.[75]

Obesity

Gross obesity is a major liability in terms of evolution in all animals who are either predators or prey. Primitive man was a food-gatherer hunter for a million years, then became a peasant agriculturalist for 10,000 years. Obesity would have been a serious threat to survival. Plant foods were the basis of the diet of primitive man, who lived like the African Bushmen today. These foods were unprocessed or lightly processed, and their staple carbohydrate foods contained much fiber.[76] Widespread severe obesity has appeared in Western societies only during the past 2 or 3 centuries. This has occurred recently in certain parts of East Africa, where it was unknown even 40 years ago. Obesity commonly appears when fiber-free energy foods such as fat, sugar, white flour, and white rice increasingly replace the traditional high-fiber lightly processed starchy foods.[76]

The inherited appetite and satiety centers of the hypothalamus over long periods of time adjust energy intakes to equal energy expenditure. Energy absorption is decreased from 1 to 2% by high-fiber starchy food.[23] Modern research in human subjects has reported that the unavailable carbohydrates of dietary fiber increased the sensation of satiety.[77] More experimental studies are required before the role of fiber-rich starchy foods in preventing obesity can be assessed. Experiments on animals fed cellulose are not an adequate investigation of the problem. Doubtless, obesity in man has many causative factors, but the role of fiber has seldom been considered.

Gallstones

Heaton[78] has suggested that the metabolic defect responsible for gallstone formation is the secretion of bile supersaturated with cholesterol, and that this is caused by eating refined carbohydrates, especially sugar. Overnutrition and obesity are key factors in cholelithiasis; these manifest themselves as high energy intakes, increased cholesterol secretion, and hypertriglyceridemia. Fiber-depleted foods tend to produce overnutrition. They are easy to ingest, very palatable, and have an increased energy/satiety ratio.[79] Several animal species form cholesterol gallstones when fed artificial diets containing fiber-depleted energy foods, especially sugar. Gallstone patients have a small circulating bile acid pool. This suggests suppressed synthesis of bile acids. Experimentally, synthesis and pool size diminish with refined carbohydrate foods. Deoxycholate, a bacterial metabolite absorbed from the colon, suppresses chenodeoxycholate synthesis and renders bile more saturated.[80] Feeding bran reduces deoxycholate adsorption, increases circulating chenodeoxycholate, and renders bile less saturated.[81] Fiber in the colon adsorbs bile acids. Such adsorption probably explains the action of bran, although bran does not increase bile acid turnover. However, this has been reported with massive intake of cellulose and gel-forming fiber preparations.

FIBER REQUIREMENTS

More epidemiological data in terms of disease prevalence in age and sex groups from many countries are required before the numerous hypotheses can be evaluated. More experimental studies both in animals and man are required before any causal relationship can be demonstrated. In the colonic disorders, increasing the crude fiber content of the Western diet by 1 to 2 g/day, especially if this is cereal fiber, will cure constipation, alleviate diverticular disease and hemorrhoids, and relieve the symptoms of many patients suffering from irritable colon. The fiber content of the diet can be increased by eating whole cereals, vegetables, and fruit. Wheat bran (10 to 30 g/day) (increasing the initial dose slowly to avoid excessive flatus) will help those unable or unwilling to make any other dietary change.

For the improvement of the more severe metabolic disorders such as atherosclerosis, coronary heart disease, and diabetes, it has been recommended that unprocessed carbohydrate foods (75% of daily energy) be tried,[10] but this suggestion awaits reported evaluation.

There already exist comprehensive reviews[10,24,82,83] which are justifiably critical of many of the numerous fiber hypotheses[84] set forth in our book.[31,54] It is quite certain that many hypotheses will be proved incorrect and others will be modified, but fiber can no longer be ignored as a possible factor in many diseases characteristic of modern Western culture.

ACKNOWLEDGMENT

The investigation of Dr. Trowell was carried out with the aid of a grant from the British Heart Foundation and the secretarial assistance of Mrs. Priscilla Milton.

REFERENCES

1. Van Soest, P. J. and McQueen, R. W., The chemistry and estimation of fibre, *Proc. Nutr. Soc.*, 32, 123–130, 1973.
2. Trowell, H. C., Crude fibre, dietary fibre and atherosclerosis, *Atherosclerosis*, 16, 138–140, 1972.
3. Trowell, H., Southgate, D. A. T., Wolever, T. M. S., Leeds, A. R., Gassull, M. A., and Jenkins, D. J. A., Dietary fibre redefined, *Lancet*, 1, 967, 1976.
4. McCance, R. A. and Lawrence, R. D., *The Carbohydrate Content of Foods*, Medical Research Council, London, Spec. Rep. Ser. No. 135, Her Majesty's Stationery Office, London, 1929.
5. McCance, R. A. and Widdowson, E. M., *The Composition of Foods*, Medical Research Council, London, Spec. Rep. Ser. No. 297, Her Majesty's Stationery Office, London, 1960.
6. Trowell, H. C., Dietary fibre, ischaemic heart disease and diabetes mellitus, *Proc. Nutr. Soc.*, 32, 150–157, 1973.
7. Trowell, H. C., Definitions of fibre, *Lancet*, 1, 503, 1974.
8. Spiller, G. A. and Amen, R. J., Nomenclature of fiber. *Am. J. Clin. Nutr.*, 28, 675, 1975.
9. Spiller, G. A. and Amen, R. J., Dietary fiber in human nutrition, *CRC Crit. Rev. Food Sci. Nutr.*, 7, 39–70, 1975.
10. Trowell, H. C., Definition of dietary fiber and hypotheses that it is a protective factor in certain diseases, *Am. J. Clin. Nutr.*, 29, 417–427, 1976.
11. Southgate, D. A. T., Determination of carbohydrates in food. I. Available carbohydrates, *J. Sci. Food Agric.*, 20, 326–330, 1969.
12. Southgate, D. A. T., Determination of carbohydrates in foods. 2. Unavailable carbohydrates, *J. Sci. Food Agric.*, 20, 331–335, 1969.

13. Robertson, J. B., personal communication; Spiller, G. A. and Amen, R. J., Dietary fiber in human nutrition, *CRC Crit. Rev. Food Sci. Nutr.*, 7(1), 39–70, 1975.

14. McConnell, A. A. and Eastwood, M. A., Comparison of methods of measuring "fibre" in vegetable material, *J. Sci. Food Agric.*, 25, 1451–1456, 1974.

15. Hellendoorn, E. W., Noordhoff, M. G., and Slagman, J., Enzymatic determination of the indigestible residue (dietary fibre) content of human food, *J. Sci. Food Agric.*, 26, 1461–1468, 1975.

16. Eastwood, M. A., Fisher, N., Greenwood, C. T., and Hutchinson, J. B., Perspectives on the bran hypothesis, *Lancet*, 1, 1029–1033, 1974.

17. Platt, B. S., *Tables of Representative Values of Foods Commonly Used in Tropical Countries*, Medical Research Council, London, Spec. Rep. Ser. No. 302, Her Majesty's Stationery Office, London, 1962.

18. McConnell, A. A., Eastwood, M. A., and Mitchell, W. D., Physical characteristics of vegetable foodstuffs that could influence bowel function, *J. Sci. Food Agric.*, 25, 1457–1464, 1974.

19. Eastwood, M. A., Vegetable fibre: its physical properties, *Proc. Nutr. Soc.*, 32, 137–143, 1973.

20. Kritchevsky, D. and Story, J. A., Binding of bile salts in vitro by non-nutritive fibre, *J. Nutr.*, 104, 458–462, 1974.

21. Ershoff, B. H., Antitoxic effect of plant fiber, *Am. J. Clin. Nutr.*, 27, 1395–1398, 1974.

22. Williams, R. D. and Olmsted, W. H., Effect of cellulose, hemicellulose and lignin on weight of stool, *J. Nutr.*, 11, 433–449, 1936.

23. Southgate, D. A. T. and Durnin, J. V. G. A., Calorie conversion factors. An experimental reassessment of the factors used in the calculation of the energy value of human diets, *Br. J. Nutr.*, 24, 517–535, 1970.

24. Cummings, J. H., Progress report. Dietary fibre, *Gut*, 16, 323–327, 1975.

25. Winitz, M., Seedman, D. A., and Graaff, J., Studies in metabolic nutrition employing chemically defined diets, *Am. J. Clin. Nutr.*, 23, 525–559, 1970.

26. Binder, H. J., Filburn, B., and Floch, M., Bile acid inhibition of intestinal anaerobic arganisms. *Am. J. Clin. Nutr.*, 28, 119–125, 1975.

27. WHO, Energy and protein requirements, *WHO Tech. Rep. Ser.*, No. 522, 19–22, 1973.

28. Lubbe, A. M., Dietary evaluation, *S. Afr. Med. J.*, 45, 1289–1297, 1971.

29. Antar, M. A. and Ohlson, M., Perspectives in nutrition, *Am. J. Clin. Nutr.*, 14, 169–178, 1964.

30. Robertson, J., Changes in the fibre content of the British diet, *Nature*, 238, 290–292, 1972.

31. Painter, N. and Burkitt, D. P., Diverticular disease of the colon, in *Refined Carbohydrate Foods and Disease*, Burkitt, D. P. and Trowell, H. C., Eds., Academic Press, London, 1975, 69–84.

32. Cowgill, G. R. and Sullivan, A. J., Further studies on the use of wheat bran as a laxative, *JAMA*, 100, 795–802, 1933.

33. Cummings, J. H. and Wiggins, H. S., Transit times through the gut measured by analysis of a single stool, *Gut*, 17, 219–223, 1976.

34. Burkitt, D. P., Walker, A. R. P., and Painter, N. S., Dietary fibre and disease, *JAMA*, 229, 1068–1074, 1974.

35. Holmgreen, G. O. R. and Mynors, J. M., The effect of diet on bowel transit times, *S. Afri. Med. J.*, 46, 918–920, 1972.

36. Balasegram, M. and Burkitt, D. P., Stool characteristics and western diseases, *Lancet*, 1, 152, 1976.

37. Payler, D. K., Pomare, E. W., Heaton, K. W., and Harvey, R. F., The effect of wheat bran on intestinal transit, *Gut*, 16, 209–213, 1975.

38. Eastwood, M. A., Kirkpatrick, J. R., Mitchell, W. D., Bone, A., and Hamilton, T., Effects of dietary supplements of wheat bran and cellulose on faeces and bowel function, *Brit. Med. J.*, 4, 392–394, 1973.

39. Jones, F. A. and Godding, E. W., *Management of Constipation*, Blackwell, Oxford, 1972, 31.

40. Martin, A. and Odling-Smee, W., Pressure changes in varicose veins, *Lancet*, 1, 768–770, 1976.

41. Painter, N. S., *Diverticular Disease of the Colon*, Heinemann, London, 1975.

42. Trowell, H., Painter, N., and Durkitt, D., Aspects of epidemiology of diverticular disease and ischaemic heart disease, *Am. J. Digest. Dis.*, 19, 864–873, 1974.

43. Brodribb, A. J. M. and Humphreys, A. D., Diverticular disease: three studies. II. Treatment with bran, *Brit. Med. J.*, 1, 424–430, 1976.

44. Segal, I. and Hunt, J. A., Irritable bowel syndrome in the urban South African, *S. Afr. Med. J.*, 49, 1645–1646, 1975.

45. Weinrich, J., Bran and the irritable bowel, *Lancet*, 1, 810–811, 1976.

46. Burkitt, D. P., Appendicitis, in *Refined Carbohydrate Foods and Disease*, Burkitt, D. P. and Trowell, H.C., Eds., Academic Press, London 1975, 87–98.

47. Burkitt, D. P., The aetiology of appendicitis, *Br. J. Surg.*, 58, 695–699, 1971.

48. Walker, A. R. P., Walker, B. F., Richardson, B. D., and Woolford, A., Appendicitis, fibre intake, and bowel behaviour in ethnic groups in South Africa, *Postgrad. Med. J.*, 49, 187–193, 1973.

49. **Burkitt, D. P. and Graham-Stewart, C. W.,** Haemorrhoids-postulated pathogenesis and proposed prevention, *Postgrad. Med. J.,* 51, 631–636, 1975.
50. **Thomson, W. H. F.,** The nature of haemorrhoids, *Brit. J. Surg.,* 62, 542–552, 1975.
51. **Drasar, B. S. and Hill, M. J.,** Intestinal bacteria and cancer, *Am. J. Clin. Nutr.,* 25, 1399–1404, 1972.
52. **Hill, M. J.,** Steroid nuclear dehydrogenation and colon cancer, *Am. J. Clin. Nutr.,* 27, 1475–1480, 1974.
53. **Walker, A. R. P.,** Affect of high crude fibre intake on transit time and the absorption of nutrients in South African Negro school children. *Am. J. Clin. Nutr.,* 28, 1160–1161, 1975.
54. **Burkitt, D. P.,** Hiatus hernia, in *Refined Carbohydrate Foods and Disease,* Burkitt, D. P. and Trowell, H. C., Eds., Academic Press, London, 1975, 160–169.
55. **Burkitt, D. P., and James, P. A.,** Low-residue diets and hiatus hernia. *Lancet,* 2, 128–130, 1973.
56. **Burkitt, D. P.,** Varicose veins, deep vein thrombosis, and haemorrhoids: epidemiology and suggested aetiology, *Brit. Med. J.,* 2, 556–561, 1972.
57. **Burkitt, D. P., Jansen, H. K., Mategaonker, D. W., Phillips, C., Chuntsog, Y. P., and Sukhnandan, R.,** Varicose veins in India, *Lancet,* 2, 765, 1975.
58. **Kritchevsky, D. and Story, J. A.,** Bindings of bile salts in vitro by non-nutritive fibre, *J. Nutr.,* 104, 458–462, 1974.
59. **Morgan, B., Heald, M., Aitkin, S. D., and Green, J.,** Dietary fibre and sterol metabolism in the rat, *Brit. J. Nutr.,* 32, 447–455, 1974.
60. **Mathur, K. S., Khan, M. A., and Sharma, R. D.,** Hypocholesterolaemic effect of Bengal gram: a long-term study in man, *Brit. Med. J.,* 1, 30–31, 1968.
61. **Palmer, G. H. and Dixon, D. G.,** Effect of pectin dose on serum cholesterol levels, *Am. J. Clin. Nutr.,* 18, 437–442, 1966.
62. **Vijayagopalan, P. and Kurup, P. A.,** Effect of dietary starches on the serum, aorta and hepatic lipid levels in cholesterol-fed rats, *Atherosclerosis,* 11, 257–264, 1970.
63. **Devi, K. S. and Kurup, P. A.,** Hypolipidemic activity of *Chaseolus Mungo* (Blackgram) in rats fed in high-fat high-cholesterol diet, *Atherosclerosis,* 15, 223–230, 1972.
64. **Vijayagopal, P. and Kurup, P. A.,** Hypolipidemic activity of the protein and polysaccharide fraction from *Chaseolus Mungo* (Blackgram) in rats fed a high-fat high-cholesterol diet, *Atherosclerosis,* 18, 379–387, 1973.
65. **De Groot, A. P., Luyken, R., and Pikaar, N. A.,** Cholesterol-lowering effect of rolled oats, *Lancet,* 2, 303–304, 1963.
66. **Trowell, H. C.,** Ischemic heart disease and dietary fiber, *Am. J. Clin. Nutr.,* 25, 926–932, 1972.
67. Department of Health and Social Security, *Diet and Coronary Heart Disease,* Her Majesty's Stationery Office, London, 1974, 18.
68. **Truswell, A. S. and Kay, R.,** Bran and blood lipids, *Lancet,* 1, 367, 1976.
69. **Trowell, H. C.,** Dietary fibre, ischaemic heart disease and diabetes mellitus, *Proc. Nutr. Soc.,* 32, 151–157, 1973.
70. **Trowell, H. C.,** Dietary-fiber hypothesis of the etiology of diabetes mellitus, *Diabetes,* 24, 762–765, 1975.
71. **Cleave, T. L.,** *The Saccharine Disease,* Wright, Bristol, England, 1974, 80 – 96.
72. **Sinnett, P. F. and Whyte, H. M.,** Epidemiological studies in a total population: Tukisenka, New Guinea. Cardiovascular disease and relevant clinical, electrocardiographic, radiological and biochemial findings, *J. Chronic. Dis.,* 26, 265–290, 1973.
73. **Trowell, H. C.,** Diabetes mellitus death-rates in England and Wales 1920–70 and food supplies, *Lancet,* 2, 998–1002, 1974.
74. **Leeds, A., Gassull, M. A., Jenkins, D. J. A., and Alberti, K. G. M. M.,** Metabolic effects of bran, *Br. Med. J.,* 1, 900–901, 1976.
75. **Jenkins, D. J. A.,** personal communication, 1976.
76. **Trowell, H. C.,** Obesity in the western world, *Plant Foods for Man,* 1, 157–168, 1975.
77. **Leeds, A.R., Gassull, M. A., Metz, G. L., and Jenkins, D. J. A.,** Food: influence of form on absorption, *Lancet,* 2, 1213, 1975.
78. **Heaton, K. W.,** *Bile Salts in Health and Disease,* Churchill Livingstone, Edinburgh, 1972, 184–195.
79. **Heaton, K. W.,** Food fibre as an obstacle to energy intake, *Lancet,* 2, 1418, 1973.
80. **Low-Beer, T. S. and Pomare, E. W.,** Can colonic bacterial metabolites predispose to cholesterol gall-stones? *Br. Med. J.,* 1, 438–440, 1975.
81. **Pomare, E. W. and Heaton, K. W.,** Alteration of bile-salt metabolism by dietary fibre (bran), *Br. Med. J.,* 3, 262–264, 1973.
82. **Mendeloff, A. I.,** Dietary fiber, *Nutr. Rev.,* 33, 321–326, 1975.

83. **Hellendoorn, R. W.,** Physiological importance of indigestible carbohydrates in human nutrition, *Voeding,* 43e, 618–636, 1973.

84. **Reilly, R. W. and Kirsner, J. B., Eds.,** *Fiber Deficiency and Colonic Disorders,* Plenum Press, New York, 1975.

85. **Van Soest, P. J. and Robinson, J. B.,** What is fiber in food?, in Marabou Food and Fibre Symp., Sundyberg, Sweden, 1976.

EFFECT OF NUTRIENT DEFICIENCIES IN MAN: SPECIFIC AMINO ACIDS

Marian E. Swendseid and Marian Wang

Dietary nitrogen requirements for maintenance of nitrogen balance in adult subjects include exogenous sources of eight or nine specific amino acids in addition to adequate amounts of nonspecific or total nitrogen.[1] The same specific amino acids are also required in the diet to promote optimum nitrogen retention and growth in infants and children.[1] These amino acids, which either are not synthesized from ordinary dietary constituents by body tissues or are not synthesized in sufficient quantities, have been designated as essential or indispensable. They are: histidine (His), isoleucine, (Ile), leucine (Leu), lysine (Lys), methionine (Met), phenylalanine (Phe), threonine (Thr), and tryptophan (Trp). In investigations of effects of essential amino acid deficiencies, the diets employed have usually contained mixtures of amino acids as the chief nitrogen source. Some diets have utilized a purified protein or protein from a particular food. All diets were designed to be lacking only in the single essential amino acid under study and to be adequate in all other nutrients. Most studies were conducted for relatively short time periods. Prolonged feeding of any of these diets whould be expected to have effects similar to those of protein-deficient states.

Table 1
THE EFFECTS OF SINGLE ESSENTIAL AMINO ACID DEFICIENCIES IN ADULTS (A), INFANTS (I), AND CHILDREN (C)

Deficient amino acid	Urinary constituents	Blood constituents	Metabolic alterations and clinical symptoms
His	↓ His, A[2,3]	↓ Plasma His, A[2,3] ↓ Serum iron, A[2] ↓ Serum albumin, A[2] ↓ Hematocrit, A[2]	↓ Muscle His, A[2] ↓ Oxidation rate of ^{14}C-His, A[2] Skin lesions A and I[2,4] Temporary loss of memory[2]
Ile	—	↓ Plasma Ile, I[5] ↓ Plasma Tyr and Phe, I[5] ↓ Postprandial plasma Ile, Leu and Val, A[6] ↓ Plasma cholesterol, I[5]	Severe malaise, A[1] Cheilosis, I[5] tremors, I[5]
Leu	↑ Riboflavin, C[7]	↓ Plasma Leu, I[8] ↑ Plasma Gly, Ile, Met, Ser, Thr, and Val, I[8] ↓ Postprandial plasma Ile, A[6] ↑ Postprandial plasma Val and Leu, A[6]	—
Lys	↓ Lys, I[9] ↑ Riboflavin, C[10] ↑ N-methyl nicotinamide, C[10]	—	—
Met	↓ Met, A[11] ↓ Sulfur, A[11] ↑ Riboflavin, C[10] ↑ N-methyl nicotinamide, C[10]	↓ Plasma Met, I[12] ↑ Plasma Thr, Ser, Pro, Tyr, and Phe, I[12] ↓ Plasma cholesterol I[12]	—
Phe	—	↓ Plasma Phe, I[13] ↑ Plasma His, I[13] ↑ Postprandial plasma Phe and Tyr, A[6]	—
Thr	—	↓ Fasting and postprandial plasma Thr, A[6]	—
Trp	↓ Trp[1]	↓ Fasting and postprandial plasma Trp, A[14] ↑ Fasting and postprandial plasma Met and Thr, A[14]	—

Table 1 (continued)
THE EFFECTS OF SINGLE ESSENTIAL AMINO ACID DEFICIENCIES IN
ADULTS (A), INFANTS (I), AND CHILDREN (C)

Deficient amino acid	Urinary constituents	Blood constituents	Metabolic alterations and clinical symptoms
Val	—	↓ Plasma Val, A[16] ↓ Postprandial Ile and Leu, A[15]	—

Note: Effects common to all of the essential amino acid deficiencies include: some degree of lassitude anorexia and general malaise, negative N balance (A), and decreased nitrogen (N) retention and growth (I and C). These effects usually occur after 1 or 2 days of consuming a deficient diet.[1] Histidine-deficient diets fed to adults cause negative N balance only after a more prolonged time period.[2] Additional effects noted with specific amino acids are summarized above.

REFERENCES

1. **Irwin, M. I. and Hegsted, D. M.,** A conspectus of research on amino acid requirements, *J. Nutr.,* 101, 539, 1971.
2. **Kopple, J. D. and Swendseid, M. E.,** Evidence that histidine is an essential amino acid in normal and chronically uremic man, *J. Clin. Invest.,* 55, 881, 1975.
3. **Anderson, H. L. and Linkswiler, H.,** Effect of source of dietary nitrogen on plasma concentration and urinary excretion of amino acids of men, *J. Nutr.,* 99, 91, 1969.
4. **Snyderman, S. E., Boyer, A., Roitman, E., Holt, L. E., Jr., and Prose, P. H.,** The histidine requirement of the infant, *Pediatrics,* 31, 786, 1963.
5. **Snyderman, S. E., Boyer, A., Norton, P. M., Roitman, E., and Holt, L. E., Jr.,** The essential amino acid requirements of infants. IX. Isoleucine, *Am. J. Clin. Nutr.,* 15, 313, 1964.
6. **Ozalp, I., Young, V. R., Nagchaudhuri, J., Tontisirin, K., and Scrimshaw, N. S.,** Plasma amino acid response in young men given diets devoid of single essential amino acids, *J. Nutr.,* 102, 1147, 1972.
7. **Nakagawa, I., Takahashi, T., and Suzuki, T.,** Amino acid requirements of children: isoleucine and leucine, *J. Nutr.,* 73, 186, 1961.
8. **Snyderman, S. E., Roitman, E. L., Boyer, A., and Holt, L. E., Jr.,** Essential amino acid requirements of infants: leucine, *Am. J. Dis. Child.,* 102, 157, 1961.
9. **Snyderman, S. E., Norton, P. M., Fowler, D. I. and Holt, L. E., Jr.,** The essential amino acid requirements of infants: lysine, *Am. J. Dis. Child.,* 97, 175, 1959.
10. **Nakagawa, I., Takahashi, T., and Suzuki, T.,** Amino acid requirements of children: minimal needs of lysine and methionine based on nitrogen balance method, *J. Nutr.,* 74, 401, 1961.
11. **Lakshamanen, F. L., Perera, W. D., Scrimshaw, N. S., and Young, V. R.,** Plasma and urinary amino acids and selected sulfur metabolites in young men fed a diet devoid of methionine and cystine, *Fed. Proc.,* 34, 931, 1975.
12. **Snyderman, S. E., Boyer, A., Norton, P. M., Roitman, E., and Holt, L. E., Jr.,** The essential amino acid requirements of infants. IX. Isoleucine, *Am. J. Clin. Nutr.,* 15, 322, 1965.
13. **Snyderman, S. E., Pratt, E. I., Cheung, M. W., Norton, P., and Holt, L. E., Jr.,** The phenylalanine requirement of the normal infant, *J. Nutr.,* 56, 253, 1955.
14. **Young, V. R., Hussein, M. A., Murray, E., and Scrimshaw, N. S.,** Plasma tryptophan response curve and its relation to tryptophan requirements in young adult men, *J. Nutr.,* 101, 45, 1971.
15. **Young, V. R., Tontisirin, K., Ozalp, I., Lakshamanen, F. L., and Scrimshaw, N. S.,** Plasma amino acid response curve and amino acid requirements in young men: valine and lysine, *J. Nutr.,* 102, 1159, 1972.
16. **Swendseid, M. E., Tuttle, S. G., Figueroa, W. S., Mulcare, D., Clark, A. J., and Massey, F. J.,** Plasma amino acid levels of men fed diets differing in protein content; Some observations with valine-deficient diets, *J. Nutr.,* 88, 239, 1966.

ESSENTIAL FATTY ACID DEFICIENCY IN HUMANS*

R. T. Holman

I. INTRODUCTION

The importance of essential fatty acids (EFA) in human nutrition and metabolism has been difficult to document for several reasons. Humane considerations preclude purposeful precipitation of deficiency, and observations have been limited to description of spontaneously or accidentally occurring disease conditions. The induction of EFA deficiency in animals[1,2] requires rigid exclusion of fat from the diet, and even with supposed low-fat diets for humans, the deficiency state is difficult if not impossible to attain. Natural diets, even poor ones, usually contain adequate amounts of EFA; thus, the deficiency is far rarer than those of protein, vitamins, or minerals. Externally recognizable EFA deficiency induced by a fat-free diet requires almost one eighth of a rat's normal lifetime to develop, and rarely have humans been subjected to low-fat diet under observation for a proportionate span of time. These factors and others contributed to the commonly held opinion in the first three decades following the discovery of essential fatty acids that EFA were not really necessary for humans.

Despite these difficulties and the attitude which developed from early failure to induce EFA deficiency in man, the literature on the role of EFA in human nutrition and metabolism has become voluminous, and the point is now clearly made that essential polyunsaturated acids are of real nutritional and medical importance. The purpose of this review is to summarize some of the information now available on the relationship of EFA to man. The reader should also consult several reviews now available on this subject.[3-10]

II. EARLY OBSERVATIONS

Shortly after the discovery of EFA by Burr and Burr,[11] investigations were begun at the University of Minnesota to assess the role of EFA in human metabolism and disease. Eczema had been reported to develop in some infants maintained on a low-fat diet,[12,13] and the iodine value of serum lipids of EFA-deficient rats had been found to be low.[14] Brown and Hansen[15] and Hansen[16,17] observed that serum lipids of eczematous infants had approximately 25% lower iodine values than those of normals and that eczema and iodine value both responded to dietary essential fatty acids[18] (Figures 1 and 2).

The first attempt to induce EFA deficiency in an adult by feeding a low-fat diet for 6 months[19] produced no noticeable harmful effects, although young rats fed the same diet developed typical EFA deficiency. The adult volunteer began the experiment with much depot fat and, in retrospect, overt deficiency could hardly be expected in so short a period of time, for EFA deficiency is very difficult to induce in adult animals.[1] During the experiment, changes in respiratory metabolism parallel to those in EFA-deficient animals and 50% reductions of contents of linoleic and arachidonic acids in serum lipids were observed. Nevertheless, the inability to induce overt skin symptoms of EFA deficiency left the impression that EFA were not really important in human nutrition. This opinion was strengthened by a later report[20] that an adult maintained by long term "fat-free" intraduodenal infusion developed no clinical signs of deficiency.

* Investigations which led to some of the data presented here were supported by grants from the National Institute of Health, AM 04524, Program Project Grant HL 08214, and The Hormel Foundation.

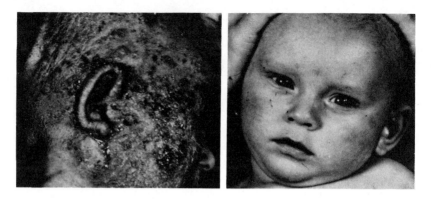

FIGURE 1.   A case of infant eczema treated with 9 to 12 g of lard daily for 1 month. (From Burr, G. O., *Fed. Proc. Fed. Am. Soc. Exp. Biol.*, 1, 224–233, 1942. With permission.)

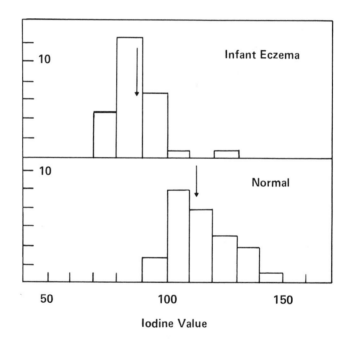

FIGURE 2.   Distribution of iodine values (total unsaturation) of serum total lipids in infants under 2 years of age having eczema and normal infants. Mean values are indicated by arrows. (From Burr, G. O., *Fed. Proc. Fed. Am. Soc. Exp. Biol.*, 1, 224–233, 1942. With permission.)

III. MANIFESTATIONS OF EFA DEFICIENCY

A. Clinical Observations

von Gröer[12] and von Chwalibogowski[21] studied the effects of low-fat diet on infants, but observed no abnormalities of the skin. Abnormality was observed in one of three infants fed low-fat diets by Holt et al.[13] Hansen and Wiese[22] reported that an infant with chylous ascites who had been maintained for an extended period of time on a low-fat diet developed periodic eczematous lesions and a refractive impetigo eruption. Warwick et al.[23] also reported a case who was maintained on a low-fat diet to control chylous ascites and who developed dry and scaly skin, coarse and sparse hair, and extensive eczematous lesions over the abdomen. She was also observed to be prone to upper respiratory infections and pneumonia. Administration of 10 g of ethyl linoleate caused a marked reversal of these signs. Infants who have been sustained by low-fat diets to control steatorrhea have also been found to develop skin eruptions.[24]

Adam et al.[25] and Hansen et al.[26] studied the effects of level of intake of linoleate on infants and described the manifestations of EFA deficiency. Skim milk diet, which is low in fat and linoleic acid, induced clinical manifestations of EFA deficiency. Most infants on this diet developed large and frequent stools, often accompanied by perianal irritation. Within weeks, skin changes were observed in most infants. Dryness of skin developed first, followed by thickening and, later, desquamation and oozing in the intertriginous folds. Histological comparison of normal and EFA-deficient skin is shown in Figure 3. Supplement of saturated fatty acids to the diet did not correct these conditions, whereas trilinolein fed at 2% of calories restored the skin to normal within 1 or 2 weeks. One case of EFA deficiency developed by feeding the sucrose-skim milk formula is shown in Figure 4. This child responded rapidly to administration of EFA.

B. Changes in Fatty Acid Composition of Tissue Lipids
1. Serum Lipids

When the method of alkaline isomerization was developed for measurement of dienoic, trienoic, and tetraenoic acids, it was applied to analysis of human serum lipids in relationship to EFA deficiency and nutritive status.[27,28] The content of dienoic and tetraenoic acids of serum lipids in 18 infants and children who were underweight and malnourished were much lower than in children who were in a good state of nutrition. The relatively deficient group included patients with cystic fibrosis or celiac disease. These data confirmed in greater detail the first observations measured by iodine value.

The changes in fatty acid composition as a consequence of EFA deficiency were first observed in animals and have been studied in most detail in animals.[1] The same abnormalities have been observed in man when inadequate dietary linoleate is provided. The abnormal patterns of polyunsaturated acids, although crudely measured as a group by alkaline isomerization, were easily demonstrable. When gas-liquid chromatographic (GLC) analysis became available, it was found that the decreased dienoic acid in EFA deficiency was largely a decrease in $18:2\omega6$, the increased trienoic acid was largely an increase in $20:3\omega9$, and the decreased tetraenoic acid was principally due to a decrease in $20:4\omega6$. In addition to these major changes, many other fatty acids change in proportion during EFA deficiency. Changes have been observed in the fatty acid patterns of all the gross lipid classes, phospholipids, triglycerides, cholesteryl esters, and free fatty acids, as well as in the total lipid extract.[29] The changes are maximal in the phospholipids and the total lipid and, therefore, are most often analyzed. Analysis of total lipids is simpler, but is a composite of the four major classes which may vary in proportion. Therefore, analysis of the phospholipids is preferred.

Table 1 gives an analysis of serum phospholipids and serum total lipids of one case of severe EFA deficiency,[30] illustrating the abnormal pattern of fatty acids in comparison with average normal values. The content of individual acids in the lipids of normals is subject to sizeable variation, perhaps reflecting dietary differences, so that only major differences of several metabolically related acids can be taken as proof of EFA deficiency. In EFA deficiency, $16:1\omega7$ and $18:1\omega9$ are increased and $18:2\omega6$ and its metabolites ($\omega6$ acids) are decreased. The acid most characteristic of EFA deficiency, $20:3\omega9$, is increased significantly by endogenous synthesis from $18:1\omega9$. Minor acids of the $\omega3$ family are also decreased.

The meaning of the other parameters listed at the bottom of the table will be discussed

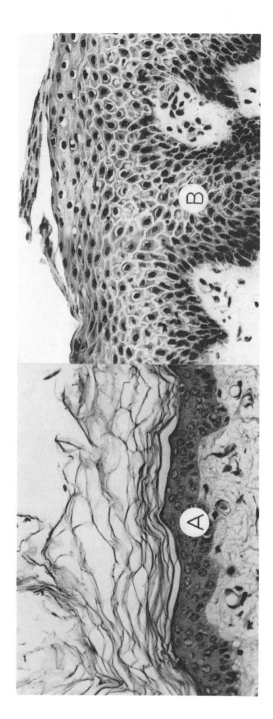

FIGURE 3.   Histological sections of normal human skin (A) and skin from a case of EFA deficiency (B). (Courtesy of A. E. Hansen.)

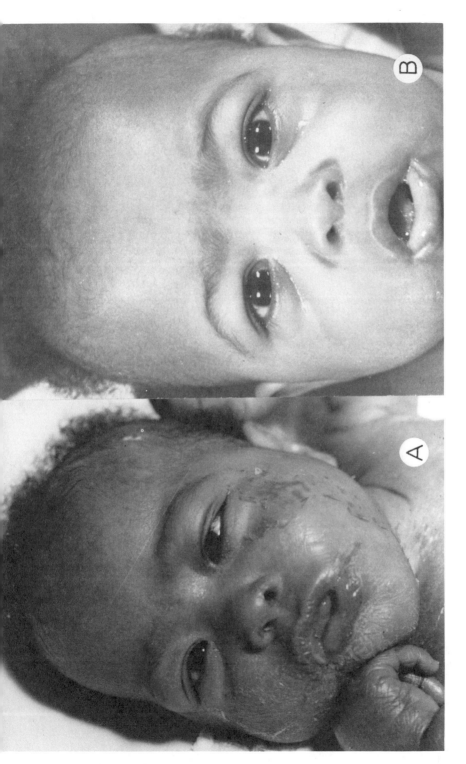

FIGURE 4. Dermatitis of EFA deficiency in an infant fed less than 0.1% of calories for 2.5 months (A). After supplementation with 2% of calories of trilinolein for 2 months, the dermatitis disappeared (B). (Courtesy of A. E. Hansen.)

Table 1
FATTY ACID COMPOSITION OF SERUM LIPIDS FROM A CASE OF ESSENTIAL FATTY ACID DEFICIENCY

Female, Age 82 Days, Maintained Totally on Fat-free Intravenous Feeding

Fatty acids	Serum phospholipids			Serum total lipids		
	%	Normal %	SV[a]	%	Normal %	SV
12:0	0.0	0.0	0.1	0.0	0.1	0.1
14:0	0.0	0.2	0.2	0.7	1.3	0.7
14:1	0.0	0.1	0.2	0.0	0.5	0.3
16:0	23.2	26.1	4.6	31.4	21.0	3.2
16:1ω7	5.1	1.2	0.8	11.8	3.4	1.6
16:2	0.0	0.1	0.2	0.0	0.1	0.1
18:0	15.1	14.0	2.4	8.2	7.8	1.4
18:1ω9	29.6	12.8	2.4	32.7	21.7	5.2
18:2ω6	1.9	18.9	4.6	3.1	25.3	5.5
18:3ω6	0.0	0.4	0.4	0.1	0.5	0.3
18:3ω3	0.0	0.2	0.4	0.8	0.5	0.3
20:2ω9	0.5	0.2	0.3	0.1	0.5	0.4
20:2ω6	0.9	0.3	0.3	0.6	0.3	0.2
20:3ω9	19.4	1.7	0.9	9.9	1.0	0.7
20:3ω6	0.0	3.7	1.4	0.1	1.9	0.7
20:4ω6	1.6	12.7	2.9	0.8	8.9	3.0
20:4ω3	0.0	0.2	0.5	0.0	0.2	0.2
20:5ω3	0.8	1.4	0.7	0.0	0.8	0.6
22:4ω6	1.4	1.9	0.9	0.0	0.5	0.6
22:4ω3	0.0	0.7	1.3	0.0	0.5	0.6
22:5ω6	0.0	0.5	0.7	0.0	1.0	1.0
22:5ω3	0.0	0.6	0.5	0.0	0.5	0.4
22:6ω3	0.0	2.0	1.3	0.0	1.8	1.4
Total	99.5			100.3		
R index	0.504			0.499		
20:3ω9/20:4ω6	12.12	0.132	0.071	12.37	0.120	0.07?
18:2ω6 + 20:4ω6 − 20:3ω9	−15.9	29.9	4.2	−6.0	33.1	6.0
ω6 Metabolites	3.9	19.5	4.0	1.6	13.1	3.9
Total ω6 acids	5.8	38.4	3.8	4.7	38.3	6.5
ω3 Metabolites	0.8	4.8	2.4	0.0	3.9	1.8
Total ω3 acids	0.8	5.0	2.5	0.8	4.3	1.8
ω9 Metabolites	19.9	1.9	1.0	10.0	1.5	1.0
Total ω9 acids	49.5	14.7	2.8	42.7	23.2	4.8
Saturated acid	38.3	40.3	4.2	40.3	30.2	3.6
Monoene acids	34.7	14.2	2.8	44.5	25.6	6.1
Double bond index	1.116	1.518	0.160	0.705	1.487	0.15?
18:2ω6/20:4ω6	1.187	1.670	0.621	3.875	3.548	1.24?
18:1ω9/18:2ω6	15.58	0.800	0.280	10.55	0.932	0.39?

[a] SV, standard variance.

From Palsrud, J. R., Pensler, L., Whitten, C. F., Stewart, S., and Holman, R. T., *Am. J. Clin. Nutr.*, 25, 897–904, 1972. With permission.

in later sections, and it shall suffice here to emphasize that all change significantly in EFA deficiency except the total saturated fatty acid content. This is true for total serum lipids or for the phospholipid fraction.

Phospholipids of Tissues

Fatty acid patterns of phosphatidyl choline (PC) and phosphatidyl ethanolamine (PE) from three tissues of an EFA deficient infant are given in Table 2. This infant had undergone an intestinal resection and had been maintained by intravenous feeding with a fat-free solution of nutrients for approximately 4.5 months.[30] Samples of tissues were taken at autopsy, lipids were extracted and separated by thin layer chromatograph (TLC), and fatty acid composition of each was determined by GLC. This case was severely deficient in EFA, and these and other tissues all showed patterns of fatty acid similar to those seen in EFA-deficient animals. Abnormalities were also observed in the total lipid extract, triglycerides, and cholesteryl esters. Although no comparable analyses have been made on normal human tissues and comparisons cannot be made, these data may be taken as examples of the fatty acid compositions of EFA-deficient tissue.

A comparison of the parameters calculated as indices of EFA status for the three tissues and for serum phospholipids (PL) of the same patient at an earlier time reveals that, fortunately, the serum is an amplified indicator of EFA status. The ratio of $20:3\omega9/20:4\omega6$ in serum PL was already 12.1 at 82 days of deficiency, whereas in the tissues at death (143 days), the ratio was considerably lower. The ratio was lowest in spinal cord, indicating its resistance to change in fatty acid composition.

IV. QUANTITATIVE ESTIMATES OF LINOLEATE REQUIREMENT

A. Estimates Based on Clinical Observations

Hansen and his colleagues[31] conducted a study of 428 infants over a period of 4 years for a total of over 4000 patient months of observation, during which five different infant formulas were evaluated, ranging from 0.04 to 7.3% of calories of linoleate. Evidence of linoleic acid deficiency was observed in young infants who were fed either a skim milk-sucrose diet almost devoid of fat and containing 0.04% of calories of linoleate or an artificial milk containing 42% of calories of fat containing less than 0.07% of calories of linoleic acid. Manifestations of the deficiency[26] disappeared promptly when linoleate was given at 1% or more of calories, either as the ethyl ester or triglyceride. Ectodermal changes indicated EFA deficiency when infants were fed less than 0.1% of calories of linoleate, but no changes were observed when 1.0% or more of calories was fed. The characteristic feature of deficiency is dry and scaly skin.[26] One subject which developed EFA deficiency on the skim milk-sugar diet is shown in Figure 4. Infants receiving diets low in linoleate reacted severely to *Staphylococcal* infections. Growth rate was unsatisfactory for many infants on the low linoleate intake, but was satisfactory in almost all receiving 1.3 to 7.3% of calories of linoleate.

Dienoic acid contents of the total fatty acids of serum lipids were 5.6 ± 1.8% for deficient infants, compared with 12.9 ± 2.6% for normals. On the basis of the clinical observations and the levels of dienoic, trienoic, and tetraenoic acids in serum lipids, their results suggested one or more percent of calories as the minimum required level. Inasmuch as human milk contains 4 to 5% of calories of linoleic acid and optimal caloric efficiency occurred at this level of intake, Adam et al.[32] regarded 4 to 5% of calories of linoleate as optimum.

Hansen and his colleagues analyzed total serum lipids for their fatty acid patterns using alkaline isomerization. They found that as the content of linoleate in the diet diminished from 7.3 to 0.04% of calories, the dienoic acids decreased from 36.5 to 12.5% of fatty

Table 2

FATTY ACID COMPOSITION OF TISSUE PHOSPHATIDYL CHOLINE (PC)
AND PHOSPHATIDYL ETHANOLAMINE (PE) FROM RIGHT VENTRICLE,
LIVER AND SPINAL CORD OF AN ESSENTIAL FATTY ACID-DEFICIENT
CHILD, FEMALE, AGE 4.5 MONTHS

Values are Given as Percent of Total Fatty Acids

Fatty acids	Right ventricle		Liver		Spinal cord	
	PC	PE	PC	PE	PC	PE
12:0	0.0	0.0	0.0	0.0	0.2	0.0
14:0	0.2	0.1	0.7	0.9	0.9	2.6
14:1	0.5	0.7	0.2	1.0	0.2	3.4
16:0	29.4	9.2	32.3	15.0	32.3	7.0
16:1ω7	2.3	5.4	6.9	3.7	5.7	3.6
16:2	0.3	0.0	0.1	0.0	0.3	0.0
18:0	4.8	21.9	10.7	18.6	12.0	12.9
18:1ω9	45.1	9.7	32.6	17.3	41.8	25.1
18:2ω6	0.3	0.2	1.1	0.9	0.6	0.1
18:3ω6	0.0	0.0	0.0	0.0	0.0	0.0
18:3ω3	0.0	0.0	0.0	0.0	0.2	0.4
20:2ω9	0.5	0.4	0.3	0.5	1.5	4.1
20:2ω6	0.3	0.1	0.4	0.2	0.7	2.0
20:3ω9	7.8	23.2	9.9	22.4	1.1	4.5
20:3ω6	0.0	0.0	0.0	0.0	0.1	0.0
20:4ω6	3.1	18.2	2.0	10.7	1.0	4.3
20:4ω3	0.0	0.0	0.0	0.0	0.0	0.0
20:5ω3	0.0	1.5	0.1	0.0	0.0	0.4
22:4ω6	0.1	0.8	0.1	0.5	0.1	4.0
22:4ω3	0.0	0.0	0.0	0.0	0.0	0.0
22:5ω6	0.3	1.4	0.7	1.2	0.1	1.8
22:5ω3	0.0	0.0	0.0	0.0	0.0	0.0
22:6ω3	0.0	2.4	0.6	3.6	0.0	2.2
Other	5.0	4.8	1.3	3.5	1.2	21.6
20:3ω9/20:4ω6	2.516	1.275	4.950	2.093	1.100	1.047
18:2ω6 + 20:4ω6 – 20:3ω9	–4.4	–4.8	–6.8	–10.8	0.5	–0.1
ω6 Metabolites	3.8	20.5	3.2	12.6	2.0	12.1
Total ω6 acids	4.1	20.7	4.3	13.5	2.6	12.2
ω3 Metabolites	0.0	3.9	0.7	3.6	0.0	2.6
Total ω3 acids	0.0	3.9	0.7	3.6	0.2	3.0
ω9 Metabolites	8.3	23.6	10.2	22.9	2.6	8.6
Total ω9 acids	53.4	33.3	42.8	40.2	44.4	33.7
Saturated acid	34.4	31.2	43.7	34.5	45.4	22.5
Monoene acids	47.9	15.8	39.7	22.0	47.7	32.1
Double bond index	0.677	1.991	0.814	1.651	0.656	1.606
18:2ω6/20:4ω6	0.097	0.011	0.550	0.084	0.600	0.023
18:1ω9/18:2ω6	150.3	48.50	29.64	19.22	69.67	251.0

From Palsrud, J. R., Pensler, L., Whitten, C. F., Stewart, S., and Holman, R. T., *Am. J. Clin. Nutr.*, 25, 897–904, 1972. With permission.

acids of serum lipids, the trienoic acids increased from 1.8 to 3.1%, and the tetraenoic acids decreased from 11.2 to 6.5%. Hansen et al. considered that their analyses confirmed their clinical observations that the requirement was at or above 1% of calories of linoleate.

B. Minimum Nutrient Requirement Estimated from Dose-Response Curves

When the method for setting EFA requirements of rats based on dose-response curves was developed,[33] it was desirable to use it to more precisely establish the requirement for humans. The only body of data relating dose of linoleic acid to fatty acid composition of serum lipids in humans was the large study of infants conducted by Hansen et al.[31] From that study, fatty acid compositions of serum lipids were available for 234 infants under 4 months of age. Plots of dienoic, trienoic, and tetraenoic acids as function of linoleate intake are shown in Figure 5, with the estimates of minimum nutrient requirement (MNR) derived from each of these biochemical parameters of EFA deficiency.[35] The MNR is that value of nutrient intake which induces a biological response equal to 70% of maximum change. The MNR derived for the trienoic acid curve is 1.38% of calories and for the tetraenoic acid curve 1.44% of calories. For the dienoic acid deposition, the requirement was estimated to be 3.3% of calories, probably reflecting the higher level of which linoleate is stored in adipose tissues.

C. EFA Requirement Estimated from Triene/Tetraene Ratio

Estimate of linoleate requirement based upon triene/tetraene ratio takes into consideration the two major biochemical responses to change in dietary linoleate.[33] The device was a convenience — an attempt to express, in one number, two biochemical changes which are probably interdependent. The trienoic and tetraenoic acids ($20:3\omega9$ and $20:4\omega6$) are found largely in the 2-positions of phospholipids, and as animals become deficient in EFA, the $20:4\omega6$ is replaced by endogenous $20:3\omega9$.[36] Therefore, the ratio of these two substances in phospholipids or total lipids is an index of the nutritive status and it has been used to establish requirement. The sharp break in the triene/tetraene

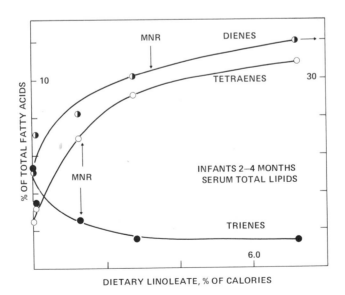

FIGURE 5.   Effect of dietary intake of linoleic acid upon dienes, trienes, and tetraenes in total fatty acids of serum of infants. (From Holman, R. T., Caster, W. O., and Wiese, H. F., *Am. J. Clin. Nutr.,* 14, 70—75, 1964. With permission.)

curve occurs in the vicinity of 1% of calories[33] for a large variety of species and tissues.[1] This was also found to be the case for fatty acids from serum lipids of infants receiving known levels of dietary linoleate[34] (Figure 6). For hospitalized infants 2 to 4 months of age fed solely on formula diets, the variation of each level of linoleate was less than for infants 11 to 12 months of age for whom some solid food may have been taken in addition to the formula diet. In both groups, however, the triene/tetraene ratio was 0.3 to 0.4 at intakes of linoleate above 1% of calories. These estimates of linoleate requirement are in good agreement with the MNR values derived from the trienoic acid contents alone or from the tetraenoic acid alone.

Cuthbertson[37] suggested that the actual infant requirment for linoleate is less than 0.5% of calories and that 0.6% of calories should provide an ample margin of safety. He based his conclusion on the modern knowledge that the alkaline isomerization used by Hansen and Wiese did not measure true linoleic acid content of butterfat, that Hansen did not control the content of tocopherol in the diet, and that there is a low incidence of detectable EFA deficiency in infant populations fed cow's milk. These criticisms may be valid, but they raise very complex and debatable questions. Indeed, alkaline isomerization is a crude method in comparison to GLC, but, with respect to dose-response phenomena (which are probably more sensitive criteria than are clinical observations of dermatitis, which vary with humidity and climate), the same conclusions were reached via alkaline isomerization or GLC analyses.[38] The criticism that alkaline isomerization does not permit measure of true linoleic acid is valid, but neither does GLC, as usually practiced, separate the isomers of 18:2. On the other hand, in EFA deficiency the major change in 18:2 is due to the linoleic acid 9,12-18:2[2] and both methods of analysis reflect this change. Although tocopherol requirement may be elevated by a highly polyunsaturated diet,[39] low tocopherol is not known to affect intensity of EFA deficiency. The incidence of EFA deficiency cannot be measured by dermal symptoms alone, for its dermal symptoms are notoriously variable and affected by such factors as growth rate and humidity.[1] In fact, the phenomenon of EFA deficiency was at first contested by an investigator whose laboratory was in a mild, damp, coastal location where the dermal symptoms could not be observed. It is much easier to induce the dermatitis in dry

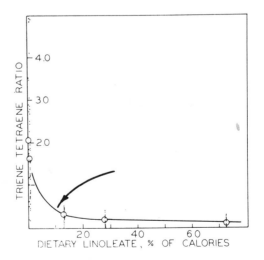

FIGURE 6.   Relationship of triene/tetraene ratio of infant serum to dietary intake of linoleate. Arrow indicates point at which curve reaches 0.4 for the ratio. (From Holman, R. T., Caster, W. O., and Wiese, H. F., *Am. J. Clin. Nutr.*, 14, 70—75, 1964. With permission.)

climates. Therefore, the absence of dermal symptoms has little meaning at marginal intakes. A much more precise criterion is the fatty acid composition; by this measure, cow's milk formula is not adequate as compared to mother's milk. The marginal adequacy of cow's milk in meeting EFA requirement is also shown by a comparison of one serum parameter of EFA sufficiency as a function of time during the feeding of either cow's or mother's milk. The sum (dienoic − trienoic + tetraenoic acids) is a good measure of EFA status (see Section V.B). When infants were fed cow's milk, this measure of EFA intake decreased with time, but when infants were fed mother's milk, the natural tendency was an increase in this parameter.[40] These observations are illustrated in Figure 7.

The setting of recommended allowances downward on the basis of the arguments above would seem hazardous, especially because the requirement for linoleic acid to support serum dienoic acid levels may be greater than 1% of calories set from consideration of trienoic and tetraenoic acid levels (see Figure 5), and it would seem wise not to ignore this biochemical requirement. It would be wiser to set the recommended allowance upward, closer to the optimum level suggested by Adam et al.[32] from measurement of caloric efficiency and the level provided by human milk.

V. QUANTIFIED INDICES OF NUTRITIVE STATUS

One difficulty is immediately apparent in presenting methods of assessing EFA intake of humans. The serum lipid analyses of Hansen et al.[31] on all their cases were performed by alkaline isomerization, an obsolete technique now supplanted by GLC. The first dose-response studies[38] performed with single dietary fatty acids and rats revealed that triene/tetraene ratios (alkaline isomerization) were almost identical to the ratios of $20:3\omega9/20:4\omega6$ (GLC) and, therefore, similar conclusions may be reached on EFA phenomena by either method. Nevertheless, mathematical relationships developed via alkaline isomerization cannot be used with GLC data to make estimates because the modes of expression of data are different and alkaline isomerization measures groups of acids whereas GLC measures individual acids.

Thus far, the only sizeable body of data relating dose of linoleate to response of tissue fatty acids in humans is the study by Hansen et al.,[31] using the method of alkaline isomerization. For moral reasons, administration of a fat-free diet now known to induce

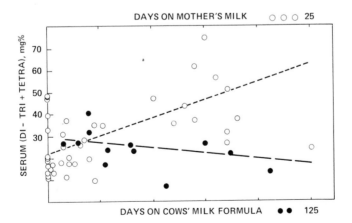

FIGURE 7. Change in the parameter (dienoic − trienoic + tetraenoic acids) as a function of time of feeding infants with human or cow's milk formula. (o) Human milk; (•) cow's milk formula. (From Holman, R. T., Hayes, H. W., Rinne, A., and Söderhjelm, L., *Acta Paediatr. Scand.*, 54, 573–577, 1965. With permission.)

EFA deficiency in humans cannot be repeated merely for the sake of obtaining more precise data. Thus, it seems unlikely that GLC data to derive equations for diagnostic use will be forthcoming. It appears that, at least for the present, some imprecision must be accepted, standards of EFA efficacy be derived by other means, and a way to approximate the constants for estimation equations using GLC data be found.

In developing means for quantifying EFA status, it is immediately apparent that many individual fatty acids of tissue lipids vary with EFA status. In the example of phospholipids and total lipids from serum of one EFA-deficient infant given in Table 1, many of the individual acids are drastically different from normal. EFA deficiency caused a tenfold decrease of $18:2\omega6$ in serum PL, the absence of $20:3\omega6$, and an eightfold decrease of $20:4\omega6$. In the deficiency $16:1\omega7$ increased 4-fold, $18:1\omega9$ increased 2.3-fold, and $20:3\omega9$ increased 11-fold. Any of these could be used directly as indices of EFA deficiency, but this author considers it wiser to use combinations of them to avoid confusing biological variations in individual acids. Had GLC been used when Hansen's controlled nutritional experiment was done and estimates of EFA intake and EFA status were derived, it is probable that the six acids listed above would have been chosen as the terms of a multiple regression equation. Such an equation should have considerable precision.

In the following treatment, several parameters or combinations of data are explored and offered as useful indices of EFA status or severity of EFA deficiency.

A. Triene/Tetraene Ratio as a Measure of EFA Status

In addition to its use in establishing the requirement for linoleate,[33] the triene/tetraene ratio has been used to assess EFA status. From the studies with rats using alkaline isomerization,[33] a value of 0.4 for the ratio was suggested as an upper limit of the range of normalcy (see Figures 5 and 6 in Reference 1), because this value corresponded to approximately 1% of calories of linoleate in most of the species and tissues studied. When the data of Hansen et al.[31] were evaluated from this point of view, it produced the curve shown in Figure 6, in which the triene/tetraene ratio was again approximately 0.4 at 1.0% of calories of linoleate.[34] Subsequently, this limit has frequently been used as a criterion of EFA deficiency or sufficiency.[10,30,41-46]

Recent studies in this author's laboratory on fatty acid composition of serum lipids of a hospitalized population indicate that the value of 0.4 is too high for humans. The upper limit for normal values should be approximately 0.2 for infants. This subject will be considered in greater detail in Section VI.

An example of change in ratio of $20:3\omega9/20:4\omega6$ in response to administering EFA to a deficient man is shown in Figure 8. In this case, EFA deficiency developed during fat-free intravenous feeding, and the deficiency was at least partially relieved by administering a fat emulsion containing EFA. The ratio of $20:3\omega9/20:4\omega6$ dropped from an initial value of approximately 0.75 to a variable level near 0.3, indicating restoration to near-normal value by daily administration of one unit of fat emulsion daily.[29]

B. Multiple Regression Equations

The development of multiple regression equations for estimate of EFA intake is outlined in this Handbook.[1] The evolution from five-term linear equations to three-term logarithmic equations made possible the derivation of estimation equations for data from humans. Equations having the form

$$\log \text{dietary linoleate} = a(\text{diene} - \text{triene} + \text{tetraene}) + b$$

proved to be as accurate in measuring linoleate intake as more complex forms; and the dienoic, trienoic and tetraenoic acids, fortunately, were the same values which had been

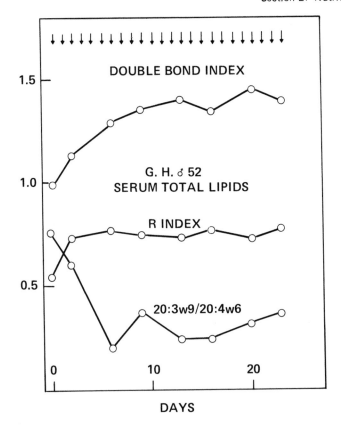

FIGURE 8. The effects of intravenous fat infusions in an EFA-deficient man on double bond index, R index, and ratio of 20:3ω9/20:4ω6 of serum total lipids. Infusions are indicated by arrows. (From Holman, R. T. and Bissen, S., unpublished data.)

measured in the studies of Hansen and his colleagues.[31] This form of equation has the added advantage that it can be easily derived graphically.

The data of Hansen et al.,[31] measured by alkaline isomerization of infant serum total fatty acids and plotted in this form, are shown in Figure 9. The least squares straight line derived from these data is described by the equation

Infant serum total fatty acids, alkaline isomerization

$$\log_{10} (\text{dietary linoleate}, \% \text{ of cal}) = 0.0432(\text{dienoic} - \text{trienoic} + \text{tetraenoic}) - 1.087$$

in which the fatty acids are expressed as percent of total fatty acids of serum. The correlation coefficient between actual dietary linoleate and calculated dietary linoleate is $r = 0.89$. Error of estimation may be on the order of ±25% of true value, but this degree of precision permits assessment of nutritive status with respect to EFA. It often is sufficient to know whether an intake is 4 ± 1% of calories or 1 ± 0.25% of calories.

Similar estimation equations were derived for total fatty acids of serum and triglyceride fatty acids of serum in young men,[47] based on alkaline isomerization analyses.

Adult serum total fatty acids, alkaline isomerization:

$$\log_{10} \text{lin} = 0.024(\text{dienoic} - \text{trienoic} + \text{tetraenoic}) - 0.296$$

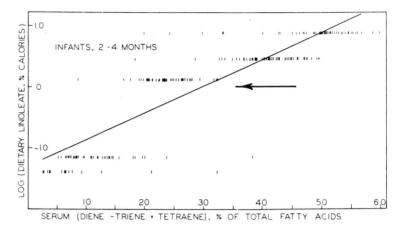

FIGURE 9.   Relationship between $\log_{10}$ dietary intake of linoleate and serum (diene – triene + tetraene) in a group of infants. The arrow indicates intake of linoleate of 1% of calories. (From Holman, R. T., Caster, W. O., and Wiese, H. F., *Am. J. Clin. Nutr.*, 14, 70–75, 1964. With permission.)

Adult serum triglyceride fatty acids, by alkaline isomerization

$$\log_{10}\text{lin} = 0.024(\text{dienoic} - \text{trienoic} + \text{tetraenoic}) + 0.179$$

The latter equation has been applied without change to the calculation of linoleate intake based on GLC analyses.[48] It found application in the assessment of patient adherence to prescribed diets. Apparently, even though the constants in the equation should be different for the two kinds of data, they must be similar enough to permit one equation to make plausible estimates using data from the other.

To make the transition from an obsolete method to the modern, more precise method, it is highly desirable to develop equations for estimate of linoleate intake from GLC analyses of serum lipid fatty acids. Therefore, a search was made for GLC data for serum lipid fatty acids from individuals fed a diet of known linoleate content for an extended period of time. Olegård and Svennerholm[49] published analyses of serum phospholipids from ten infants fed mother's milk and eight fed cow's milk formula fortified with vegetable oil. In our own laboratory, we had made analyses of serum phospholipids of seven infants who were severly EFA deficient as a consequence of long-term intravenous feeding with a fat-free preparation.[30] The average values of these three groups were used to derive graphically (Figure 10) the estimation equation

Infant serum phospholipid fatty acid, GLC

$$\log_{10}\text{lin} = 0.079(18{:}2\omega6 + 20{:}4\omega6 - 20{:}3\omega9) - 1.9$$

Using this equation, the intake of linoleate of normal infants was calculated from the normal value of $18{:}2\omega6 + 20{:}4\omega6 - 20{:}3\omega9$ given in Table 1. The value found (5.1% of calories) is reasonable, judging from the composition of breast milk.[26] Until similar equations can be derived from a larger base of data from infants fed known levels of linoleate, this equation is offered to permit a reasonable estimate of dietary linoleate from analysis of serum phospholipid fatty acids.

In the absence of multiple regression equations derived from prolonged and controlled experiments involving several known intakes of linoleate and using modern GLC analysis, another approximation may be useful. In Table 1, the value for the parameter $18{:}2\omega + 20{:}4\omega6 - 20{:}3\omega9$ for one overtly EFA-deficient infant is 15.9% of serum phospholipids.

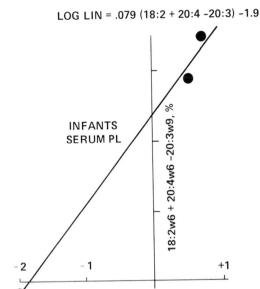

LOG LIN = .079 (18:2 + 20:4 -20:3) -1.9

INFANTS
SERUM PL

18:2w6 + 20:4w6 -20:3w9, %

-2   -1   +1

LOG (Dietary 18:2, % Cal)

FIGURE 10.   Relationship of dietary intake of linoleate
and the sum $(18:2\omega6 + 20:4\omega6 - 20:3\omega9)$ in serum PL.

This value was typical of several measured on EFA-deficient infants. In Table 3, the comparable value for infants hospitalized for nonmetabolic diseases was 34.5 for males and 29.9 for females. These values may be used to judge whether an infant is grossly deficient or essentially normal. Comparable normal values for fatty acids of total serum lipids can also be found in the tables. The response of this parameter to intravenous feeding of fat emulsion to a male adult is shown in Figure 11.

C. Index of Relationship (R) to Normal Fatty Acid Profile

In comparisons of cases against normal values, if variations from normal are minimal, decision may be difficult. This is especially true for fatty acid pattern, because an individual acid may vary only in a minor way, but many of the fatty acids undergo change. The composite changes of a score or more fatty acid concentrations may have significance although change in a single one may not be significant. To assess how greatly two bodies of data differ in pattern (i.e., how much two GLC patterns differ) a formula was developed to calculate the Index of Relationship. This formula is given in an article on EFA deficiency in animals,[1] and the relationship has been used to quantify interspecies relationships for taxonomic purposes.[50] The value ranges from zero, for no coincidence of components, to one for complete identity. The calculation is the sum of terms for each component in the two samples; for each component, the term is the average content of that component in the two samples multiplied by the minor ratio of its concentrations in the two samples.

For the EFA-deficient child described in Table 1, the R index for fatty acid composition of serum phospholipids compared to normals of same age and sex is 0.504. This value indicates only a 50% identity of serum phospholipid pattern with normals for this individual. Thus, the composition of serum phospholipid is drastically altered in EFA deficiency. In an EFA-deficient man whose deficiency was partially corrected by intravenous fat emulsion, the R index rose during treatment from 0.55 to approximately 0.75 (see Figure 7).

Table 3
FATTY ACID COMPOSITION OF SERUM PHOSPHOLIPIDS FROM NORMAL HUMANS FROM AGE 0 TO 90 YEARS

Normal Values May be Calculated at Any Age by
Substituting Appropriate Values in the Equation

Fatty acid(s)	Male[a]			Female[a]		
	Slope A	Intercept B	SV[b]	Slope A	Intercept B	SV
12:0	−0.0029	0.20	0.48	0.0004	0.02	0.10
14:0	0.0039	0.20	0.47	0.0009	0.20	0.24
14:1	0.0006	0.12	0.21	0.0012	0.11	0.16
16:0	0.0884	23.25	4.26	0.0123	26.10	4.61
16:1ω7	0.0044	1.29	1.16	0.0046	1.22	0.78
16:2	−0.0012	0.14	0.18	−0.0000	0.13	0.19
18:0	−0.0409	15.24	2.35	−0.0187	14.00	2.36
18:1ω9	0.0479	10.89	3.02	0.0076	12.84	2.43
18:2ω6	−0.0856	23.53	4.35	−0.0043	18.91	4.57
18:3ω6	−0.0001	0.16	0.24	−0.0024	0.36	0.43
18:3ω3	0.0006	0.21	0.23	0.0021	0.23	0.36
20:2ω9	0.0044	0.14	0.29	0.0048	0.17	0.31
20:2ω6	−0.0094	0.69	0.85	−0.0002	0.28	0.34
20:3ω9	0.0091	0.88	0.92	−0.0094	1.71	0.92
20:3ω6	−0.0103	3.71	1.16	−0.0037	3.69	1.44
20:4ω6	0.0008	11.86	2.91	−0.0096	12.72	2.88
20:4ω3	−0.0039	0.38	0.34	0.0017	0.24	0.46
20:5ω3	0.0051	0.77	0.82	−0.0057	1.36	0.68
22:4ω6	0.0019	1.63	1.06	−0.0055	1.95	0.85
22:4ω3	−0.0112	1.48	1.21	0.0082	0.67	1.31
22:5ω6	−0.0072	0.77	0.59	0.0028	0.53	0.74
22:5ω3	−0.0006	0.71	0.61	0.0022	0.55	0.52
22:6ω3	0.0047	1.86	1.29	0.0093	1.96	1.33
18:2ω6 + 20:4ω6 − 20:3ω9	−0.0939	34.51	4.35	−0.0045	29.92	4.19
Double bond index	−0.0020	1.80	0.16	−0.0003	1.52	0.16
20:3ω9/20:4ω6	0.0008	0.08	0.08	−0.0005	0.13	0.07
18:1ω9/18:2ω6	0.0042	0.52	0.26	−0.0011	0.80	0.28
18:2ω6/20:4ω6	−0.0095	2.19	0.65	−0.0008	1.67	0.62
Total ω6 acids	−0.1099	42.36	4.83	−0.0228	38.44	3.85
Total ω6 − 18:2	−0.0242	18.83	4.00	−0.0185	19.53	4.02
Total ω3 acids	−0.0053	5.41	2.10	0.0179	5.01	2.46
Total ω3 − 18:3	−0.0059	5.20	2.03	0.0158	4.79	2.39
Total ω9 acids	0.0615	11.91	3.16	0.0030	14.72	2.75
Total ω9 − 18:1	0.0136	1.02	1.06	−0.0046	1.88	1.01
Monoene acids	0.0530	12.30	3.50	0.0134	14.17	2.79
Saturated acids	0.0484	38.89	3.68	−0.0051	40.32	4.24

[a] Fatty acid = A (age) + B.
[b] SV, standard variance.

From Holman, R. T. and Bissen, S., unpublished data.

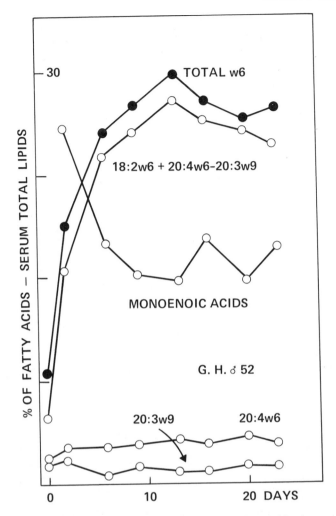

FIGURE 11.   The effect of daily intravenous fat emulsion in an EFA-deficient man on several parameters measured in serum total lipids. (From Holman, R. T. and Bissen, S., unpublished data.)

D. Double Bond Index (DBI)

The average number of double bonds per fatty acid in a sample (DBI) is another useful composite parameter.[51] For normal human serum phospholipids the value is 1.8 ± 0.16 for males and 1.52 ± 0.16 for females at age 0 (Table 3). During EFA deficiency, the DBI dropped to 1.11 for the case described in Table 1. In Figure 8, the DBI for serum total lipids of a 52-year-old male rose from 1.0 in overt deficiency to 1.4 after intravenous fat emulsion was administered. Calculated normal value for a 52-year-old man is 1.42 ± 0.16. The DBI has significance because physical and physiological properties of lipids and lipid-containing membranes are related to the degree of unsaturation of their component fatty acids. The DBI of tissue lipids in EFA deficiency changes toward more saturation and less fluidity of the lipids.[51]

E. Ratio of 18:1/18:2

The ratio of 18:1 to 18:2 was proposed by Reid et al.[52] as having an advantage over the triene/tetraene ratio in estimating EFA status because 20:3 was known to consist of more than one isomer. This criticism is not really valid because 18:1, 18:2, 20:3, and 20:4 all exist as isomers in natural samples. Although most of the isomers of 20:3 and 20:4 in animal and human lipids are separable with GLC as currently practiced, the isomers of 18:1 and 18:2 are not. A ratio of less than 1.5 is considered to indicate that the requirement for linoleate has been met.

The ratio has merit in measuring the ratio of the metabolic precursors of $20:3\omega9$ and $20:4\omega6$. If the available $18:2\omega6$ is low, the synthesis of other $\omega6$ acids should be diminished. However, abnormality in EFA metabolism would not be discovered through the ratio of the endogenous and exogenous substrates as much as through the ratio of the products. For example, in one case of *Achrodermatitis enteropathica*[53] (see Section VII.E), although linoleic acid was present in amounts higher than normal (22 to 24%), the arachidonic and oleic acids were lower than normal. In this case, the 18:1/18:2 ratio was 0.51, 0.63, 0.60, and 1.52 for successive periods of exacerbation, improvement, exacerbation, and improvement, indicating that the requirement for linoleate had been met at all times except under the final regimen which restored the child to near normalcy. Clearly, this parameter was not able to detect the abnormality in EFA metabolism resulting in less than one sixth the normal content of $20:4\omega6$ in serum total fatty acids although linoleate content was somewhat above normal. In this author's opinion, the 18:1/18:2 ratio has limited use, and parameters developed from content of metabolic products rather than from precursors more reliably describe the EFA status.

F. Ratio of $18:2\omega6$ to $20:4\omega6$

The precursor/product ratio should be an indicator of abnormalities existing in the metabolic sequence. Thus, if a defect in metabolism of $18:2\omega6$ to $20:4\omega6$ occurs, the ratio of $18:2\omega6$ to $20:4\omega6$ should detect it. Of course, this precursor/product ratio is affected by competitive suppression due to the presence of high proportions of weak inhibitors such as $18:1\omega9$ or moderate proportions of effective inhibitors such as $18:3\omega3$.[54] In the case of *Achrodermatitis enteropathica* mentioned above, an abnormality of metabolism of $18:2\omega6 \rightarrow 20:4\omega6$ was indicated by a ratio of $18:2\omega6/20:4\omega6$ equal to 12.8, as compared with 3.6 for normals.[53] During the final period of improvement the ratio became 5.8.

For this ratio to be useful in detecting metabolic abnormalities, simple nutritional EFA deficiency should not affect it drastically. Although this ratio is 1.67 in normal infant females at age 0 and a severely EFA-deficient infant had a ratio of 1.19, by comparison, the case of *Achrodermatitis enteropathica* exhibited a ratio of 12.8 during exacerbation. This parameter is not especially useful detecting nutritional deficiency, but may be more useful in detecting errors in metabolism.

G. Total $\omega6$ Acids

Inasmuch as EFA activity is largely, if not exclusively, due to $\omega6$ acids,[1,51] the total of these acids should be a good measure of EFA status. Indeed, the log of total $\omega6$ acids in liver lipids was found to be proportional to the log of dietary linoleate in rats.[1] This parameter varies over a wide range in response to EFA status: from 5.8 in a deficient infant to 38.4 in normals of the same age and sex (Table 1). Recovery of a deficient adult male was plotted by total $\omega6$ acids in Figure 11. Much of the response is due to linoleate itself, and if linoleate is subtracted, the range is 3.9 to 19.5 for the metabolites of $18:2\omega6$. Because conditions are known which affect principally $18:2\omega6$ in lipids and in which its metabolism is modified, it seems wisest to use total $\omega6$ acids to detect both types of abnormality.

The total $\omega6$ acids were found to be a good measure of dietary linoleate,[1] as $\log_{10}$ total $\omega6$ acids of tissue is proportional to $\log_{10}$ dietary linoleate. Using this same general relationship and data from fat deficient infants[30] plus infants fed known linoleate levels,[4,9] an equation based on GLC analysis of serum phospholipids was derived. For infants the relationship is

Serum PL, GLC data

$$\log_{10} \text{ dietary linoleate (cal \%)} = 5.8 \ (\log \omega^6) - 8.5$$

By this equation, the linoleate intake of newborn infants (Table 3) was calculated to be 4.8% of calories, a value in close agreement with that estimated via the parameter $18:2\omega6 + 20:4\omega6 - 20:3\omega9$ (Section V.B).

H. Total ω3 Acids

Although $\omega3$ acids are not directly concerned in EFA deficiency, they do inhibit metabolism of $\omega6$ acids.[54] Moreover, they are widely found in animal and human tissues and are included in the tables for these reasons. Although as a group they are minor acids in normal serum phospholipids, they are diminished during EFA deficiency, which is usually encountered in humans as the result of a fat-free alimentation.

I. Total ω9 Acids

In EFA deficiency, $\omega6$ acids are diminished in lipids because their dietary precursor is absent from the diet and, in substitution, $\omega9$ acids are synthesized endogenously. The common endogenous acid oleic acid is elongated and desaturated in abnormally high amounts. The product $20:2\omega9$ is measurable in many tissue lipids, and the product $20:3\omega9$ becomes a prominent component. These are partially substituted for arachidonic acid in tissue lipids, but obviously do not function as well as does arachidonic acid. Oleic acid itself is increased in amount in phospholipids as a substitute for the linoleic acid, which is not available. Thus, in deficiency, $\omega9$ acids increase dramatically in the same order of magnitude as the decrease of total $\omega6$ acids.

J. Monoenoic and Saturated Acids

Monoenoic acids include primarily oleic ($18:1\omega9$) and palmitoleic acids ($16:1\omega7$), both of which increase in EFA deficiency as the result of increased endogenous synthesis of unsaturated acids (Figure 11). Values for this category of acids and for saturated acids were included in the study of normal and deficient humans primarily for descriptive reasons rather than as potential measurements of EFA deficiency. The monoenoic acids have some value as a measure of severity of EFA deficiency, but better measures are given above. The total saturated acids were found to remain relatively constant in normalcy or deficiency and, thus, have no utility in detecting or measuring EFA deficiency.

VI. NORMAL VALUES FOR FATTY ACID PATTERNS OF SERUM LIPIDS

Several groups of investigators have assembled limited fatty acid analyses for series of normal humans. These have not been sufficiently large, nor have they adequately covered the range of age necessary for our purposes. We have recently analyzed total lipids of serum and serum phospholipids, cholesteryl esters, triglycerides, and free fatty acids for their component fatty acids by GLC as previously described.[1] Samples were analyzed from patients hospitalized for nonmetabolic diseases. At least ten of each sex from every decade from 0 to 90 years of age were included in the study. Tables 3 and 4 give the results for serum phospholipids and serum total lipids, respectively. All values for one sex and for one fatty acid were plotted against age by computer and the best-fitting straight line was derived from the data by computer. The slopes and intercepts for each parameter measured or calculated are given. From these and the age, a normal value can be

Table 4
FATTY ACID COMPOSITION OF SERUM TOTAL LIPIDS FROM NORMAL HUMANS FROM AGE 0 TO 90 YEARS

Normal Values may be Calculated at Any Age by
Substituting Appropriate Values in the Equation

Fatty acid(s)	Male[a]			Female[a]		
	Slope A	Intercept B	SV[b]	Slope A	Intercept B	SV
12:0	−0.0006	0.10	0.13	−0.0003	0.11	0.22
14:0	−0.0059	1.28	0.66	0.0024	0.84	0.71
14:1	−0.0046	0.48	0.32	−0.0011	0.34	0.38
16:0	0.0243	21.04	3.20	0.0062	21.39	3.52
16:1ω7	0.0023	3.45	1.59	0.0019	3.56	1.24
16:2	−0.0010	0.11	0.12	−0.0004	0.10	0.16
18:0	−0.0263	7.81	1.35	−0.0142	7.38	1.43
18:1ω9	0.0416	21.72	5.24	0.0283	22.10	3.79
18:2ω6	−0.0045	25.25	5.54	0.0159	24.25	5.03
18:3ω6	−0.0012	0.46	0.33	−0.0002	0.45	0.57
18:3ω3	0.0003	0.46	0.30	−0.0000	0.52	0.31
20:2ω9	0.0059	0.45	0.44	0.0045	0.52	0.50
20:2ω6	−0.0032	0.29	0.24	−0.0025	0.26	0.20
20:3ω9	−0.0028	1.05	0.68	−0.0008	0.88	0.72
20:3ω6	−0.0020	1.86	0.70	−0.0124	2.46	0.97
20:4ω6	−0.0114	8.92	2.97	−0.0206	9.71	2.91
20:4ω3	−0.0011	0.15	0.20	0.0002	0.10	0.26
20:5ω3	−0.0007	0.81	0.61	−0.0002	0.66	0.47
22:4ω6	0.0028	0.53	0.60	0.0024	0.50	0.52
22:4ω3	−0.0031	0.53	0.57	−0.0001	0.44	0.62
22:5ω6	−0.0038	1.03	1.00	−0.0048	1.10	0.97
22:5ω3	−0.0030	0.53	0.40	0.0022	0.32	0.42
22:6ω3	−0.0033	1.84	1.38	−0.0066	2.06	1.24
18:2ω6 + 20:4ω6 − 20:3ω9	−0.0130	33.13	5.96	−0.0038	33.07	4.68
Double bond index	−0.0013	1.49	0.16	−0.0012	1.24	0.16
20:3ω9/20:4ω6	−0.0002	0.12	0.08	0.0002	0.09	0.08
18:1ω9/18:2ω6	0.0017	0.93	0.39	−0.0003	1.00	0.32
18:2ω6/20:4ω6	−0.0053	3.55	1.25	0.0008	3.14	1.21
Total ω6 acids	−0.0233	38.35	6.47	−0.0221	38.72	4.90
Total ω6 − 18:2	−0.0188	13.09	3.88	−0.0380	14.48	4.13
Total ω3 acids	−0.0109	4.32	1.77	−0.0045	4.11	1.72
Total ω3 − 18:3	−0.0112	3.86	1.76	−0.0045	3.59	1.76
Total ω9 acids	0.0446	23.22	4.84	0.0320	23.51	3.36
Total ω9 − 18:1	0.0030	1.50	1.01	0.0037	1.40	1.12
Monoene acids	0.0405	25.58	6.06	0.0291	26.00	4.45
Saturated acids	−0.0084	30.23	3.59	−0.0060	29.72	3.17

[a] Fatty acid = A (age) + B.
[b] SV, standard variance.

From Holman, R. T. and Bissen, S., unpublished data.

calculated at any age. Standard variance from the best-fitting straight line is also given. The tables present values for 23 measurable fatty acids and 14 parameters calculated from them. The latter include several which are offered above as useful indices of EFA status. The reader should note that the normal values for the ratio 20:3ω9/20:4ω6 are lower than previously thought.

VII. INVOLVEMENT OF EFA IN DISEASE AND THERAPY

A. EFA Deficiency Induced by Fat-free Total Parenteral Nutrition

In 1971 Collins et al.[55] described a case of small bowel resection maintained by fat-free total parenteral nutrition (TPN) for 100 days. The changes observed in fatty acid composition of serum lipids and the dermatitis which developed clearly indicated that the patient was deficient in EFA. The following year, Paulsrud et al.[30] reported very similar findings on seven infants maintained by prolonged TPN. One of these cases was studied in detail and is perhaps the most severe case of EFA deficiency described to date. Since these reports, many additional studies have appeared confirming and extending these observations.[10,41-46,56] There is general agreement that dermatitis appears after several weeks of fat-free TPN and that the fatty acid patterns of serum lipids become that of an EFA deficiency induced in animals by a fat-free diet. The principal changes have been given in detail in Section III.B.

The appearance of these symptoms is surprisingly rapid. Figure 12 shows the onset of

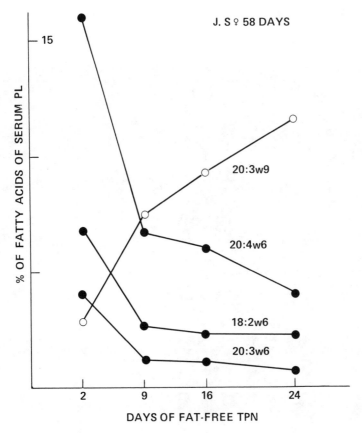

FIGURE 12.   The effect of fat-free TPN on content of 20:3ω9, 18:2ω6, 20:3ω6, and 20:4ω6 of serum PL. Note that most of the changes occurred within 9 days. (From Paulsrud, J. R., Pensler, L., Whitten, C. F., Stewart, S., and Holman, R. T., *Am. J. Clin. Nutr.*, 25, 897–904, 1972. With permission.)

changes in certain fatty acids of serum phospholipids of an infant.[30] The content of ω6 acids drops within a week and the content of 20:3ω9 rises abruptly. The same changes take place in adults fed via fat-free TPN, and the dermatitis of EFA deficiency also appears. In one 60-year-old woman the skin became dry and scaly (Figure 13A) within 46 days, and when intravenous fat emulsion was administered, the skin returned to normal within 14 days (Figure 13B).[43] The ratio of 20:3ω9/20:4ω6, which was found to be 3.7 before emulsion therapy, decreased to 0.3 within 1 week.

The mechanism operating in the induction of EFA deficiency so rapidly by intravenous feeding has recently been explained by Connor.[57] In normal nutrition and metabolism, the individual has two sources of essential fatty acid (linoleate) available for synthesis of membranes and circulating lipoprotein. The dietary supply is intermittent, depending on frequency of meals. In the intervals between meals, adipose tissue, which normally holds several hundred grams of linoleate in human, releases a supply of linoleate by hydrolysis of triglycerides. In fat-free intravenous alimentation containing glucose, both of these supplies are cut off. The high glucose content of most intravenous preparations inhibits the hydrolysis of adipose triglycerides. That this is the primary cause of the rapid onset of EFA deficiency was confirmed by Stegink et al.[58] who observed that during intravenous feeding with solutions containing amino acids but no glucose, the biochemical symptoms of EFA deficiency do not occur. Thus, the availability of adequate stores of linoleate in adipose tissues is blocked by constant infusion of glucose. Perhaps an intermittent infusion of the usual glucose-amino preparation would also accomplish release of reserves of linoleate.

Recovery from EFA deficiency through oral feeding — This restored the fatty acid

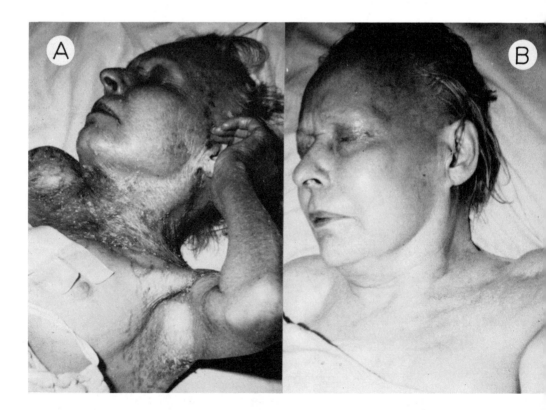

FIGURE 13.   Skin lesions associated with EFA deficiency induced by long-term fat-free TPN (A) and the effect of intravenous fat emulsion on the condition (B). (From Riella, M. C., Broviac, J. W., Wells, M., and Scribner, B. H., *Ann. Intern. Med.*, 83, 786–789, 1975. With permission.)

pattern toward normal within 20 days in all infants followed in the study of Paulsrud et al.[30] Oral administration of safflower oil to EFA-deficient adults also restored serum fatty acid patterns to normal within that time.[46]

Blood or serum transfusions — were ineffective in relieving EFA deficiency[44,57,58] because they do not contain sufficient amounts of EFA.

Intravenous fat emulsion — This treatment relieves EFA deficiency rapidly.[9,10,43] A case of a 52-year-old male is illustrated in Figures 7 and 10. In this instance, most of the parameters were close to normal within 10 days of daily administration of fat emulsion.

Cutaneous administration of oils containing EFA — This treatment has been found by Press et al.,[61] Böhles et al.,[62] and Friedman et al.[45] to correct EFA deficiency. In these cases, restoration of serum fatty acid patterns toward normal indicated efficacy of this route of administration. However, this is not always effective. In the study by Hunt et al.,[63] neither dermatitis of EFA deficiency nor patterns of fatty acids were corrected by cutaneous application of sunflower seed oil. Figure 14 shows one of the cases which failed to respond.

B. Wound Healing

Caldwell et al.[60] reported a case of midgut volvulus requiring resection of 50% of the small bowel on the tenth day of life. Multiple attempts to anastomose the small bowel to the duodenum failed, a duodenostomy and ileostomy were performed, and an enterocutaneous fistula developed. Parenteral feeding was begun on the 18th day of life and continued half a year. During this time, several attempts were made to close the fistulate, but they failed because of inadequate healing. At 22 weeks, skin lesions and plasma fatty acid patterns confirmed an EFA deficiency. At 25 weeks an attempt was made to correct the EFA deficiency by infusion of a fat emulsion which changed the serum fatty acid pattern toward normal, healed the skin lesions, and corrected the thrombocytopenia. At 28 weeks the duodenoileostomy and closure of the enterocutaneous fistula were performed successfully, with complete healing of all surgical wounds. The patient was later able to accept normal food and was discharged. This patient is shown in Figure 15 before and after intravenous alimentation with fat emulsion.

In our unpublished studies of the efficacy of intravenous fat emulsion, we have encountered another case of failure of surgical wounds to heal during EFA deficiency induced by fat-free intravenous feeding; they healed normally after EFA was provided via intravenous fat emulsion.

C. Burn Healing

Victims of extensive burns are presented with two metabolic problems related to lipids. Much new tissue containing essential fatty acids must be synthesized. Indeed, there is biochemical evidence of EFA deficiency in severe cases of burns. Wilmore et al.[62] and Helmkamp et al.[65] found a decrease in the polyunsaturated acids of erythrocyte phospholipids in severe burn cases. The ratio of $20:3\omega9/20:4\omega6$ for one patient was normal, but reached 0.5 in another.

Another effect of the burn is excessive loss of heat by the unprotected burned area.[66] Jelenko et al.[67] isolated from normal skin a lipid which is capable of restricting the loss of body water through burns. Application of this extract to collagen films diminished water loss through the films. The "water holding" lipid was isolated by chromatography and found by mass spectrometry to be primarily an ester of an unsaturated fatty acid with molecular weight 308. Although infrared spectra were not identical, the substance was concluded to be ethyl linoleate. "Water holding" activity and chromatographic behavior of the isolated substance and authentic ethyl linoleate were identical.[67]

Topical application by aerosol of ethyl linoleate to burn wounds of rabbits and mice was found to be an effective means of cutting evaporative loss of water, controlling hypermetabolism, and hastening healing.[68]

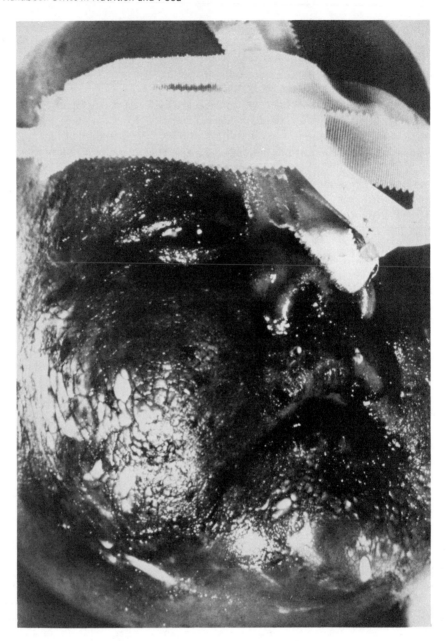

FIGURE 14.   A case of EFA deficiency which failed to respond to the cutaneous application of sunflower seed oil. (From Hunt, C. E., Engel, R. R., Modler, S., Hamilton, W., Bissen, S., and Holman, R. T., *J. Pediatr.,* in press.)

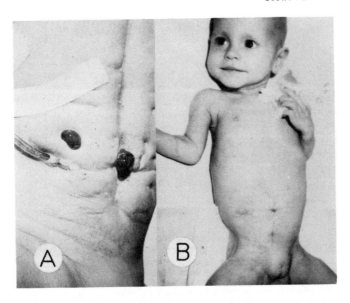

FIGURE 15. A case of multiple surgery on the small bowel maintained by fat-free TPN and in which surgical wounds failed to heal. (A). (B) The same case after intravenous fat emulsion had been administered for 3 weeks. (From Caldwell, M. D., Jonsson, H. T., and Othersen, H. B., *J. Pediatr.*, 81, 894–898, 1972. With permission.)

Topical application of ethyl linoleate to cases of severe burn has had very promising results in humans. One case, involving an 88% burn treated with ethyl linoleate 6 hr after a burn by explosion, is shown in Figure 16A. At the time of discharge 46 days later, the patient had full range of motion and even his natural pigmentation was beginning to return (Figure 16B).

D. Infection

Although controlled studies have not been conducted on the role of EFA in prevention of infection, theoretically, it should be beneficial. Destruction of tissue by organisms requires replacement of structures, including EFA. EFA deficiency may therefore be expected to impede defenses against infection. Although there has been no well-controlled study of the effect of EFA deficiency on infection, a number of observations have been made which suggest that infections increase in the deficiency state. Hansen and Wiese[69] and Hansen et al.[70] reported a high incidence of bacterial infections of the skin and ears and intestinal infections in EFA-deficient dogs, although dogs in the same colony fed EFA had very low incidences of infection. In the study by Hansen et al.[31] of dietary linoleate in a large number of infants, hospitalizations for infections were more frequent for infants given the two formulas of lowest linoleate content. One case of chylous ascites who was maintained for several years on a low-fat diet and developed dermal and biochemical signs of EFA deficiency was observed to be very susceptible to infections.[24] Long-term intravenous feeding with fat-free preparations is now known to induce EFA deficiency (see Section VII.A), and is well known to be accompanied by frequent and severe systemic infections related to deep-dwelling catheters.[71] Administration of fat emulsions by peripheral veins is accompanied by a much lower rate of infection. At least some of the lower susceptibility to infection may be due to the supply of EFA the emulsion provides.

Love et al.[72] have suggested that an abnormal pattern of serum lipid fatty acids resembling a marginal EFA deficiency may be a general phenomenon common to illness.

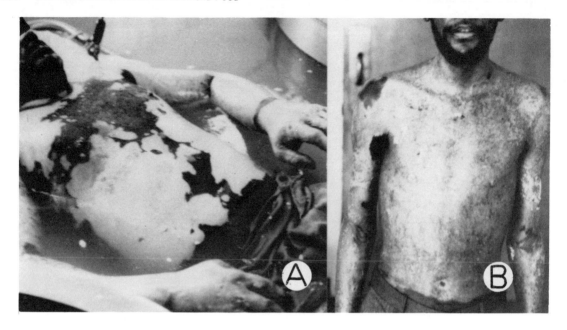

FIGURE 16.   A case of severe burn (A) which responded to aerosol application of ethyl linoleate to wound surface and healed within 46 days (B). (Courtesy of C. Jelenko III.)

If this is true, infections and metabolic abnormalities may increase the requirement for EFA.

E. Achrodermatitis enteropathica

One case of *Achrodermatitis enteropathica* (AE) in which fatty acid composition of serum lipids was studied in detail indicated that an error in metabolism of EFA occurred. Cash and Berger[53] employed intravenous fat emulsion in the diagnosis of AE in a 3-month-old male admitted to the hospital for failure to thrive and a persistent rash. Fatty acid metabolism was studied, beginning with a control period during exacerbation (Figure 17A). The fatty acid pattern of serum total lipids revealed a normal level of $18:2\omega6$, but a very low level of $20:4\omega6$. The level of unusual unidentified fatty acids was high. Oral administration of arachidonic acid (1 g per day for 6 days) diminished the content of $18:2\omega6$ and increased the content of $20:4\omega6$, but elicited no clinical improvement. Intravenous administration of cottonseed oil emulsion for 4 days caused a dramatic clinical improvement within 48 hr (Figure 17B); the level of $18:2\omega6$ serum lipids was again elevated and $20:4\omega6$ continued at the same moderate level. When the supply of intravenous fat emulsion was exhausted, within a period of 6 days the condition had exacerbated and the serum level of $20:4\omega6$ had decreased (Figure 17C). Thereafter, human breast milk was given rather than cow's milk formula, and beginning 3 days later Diodoquin® was administered daily. Ten days later the clinical recovery was complete (Figure 17D), the $18:2\omega6$ level was normal, and the $20:4\omega6$ level remained at a moderate level somewhat below normal. This case proved to be one in which an abnormality in lipid metabolism was clearly demonstrated. Another case of AE did not react similarly, and the serum lipids were closer to the normal pattern.[73]

F. Multiple Sclerosis

Multiple sclerosis (MS) has been postulated to be a deficiency of $\omega3$ acids during maturation of the nervous system[74,75] and a faulty myelination. The content of $22:6\omega3$ in white matter in MS is below half the normal value.[76] The $18:3\omega6$ content of serum

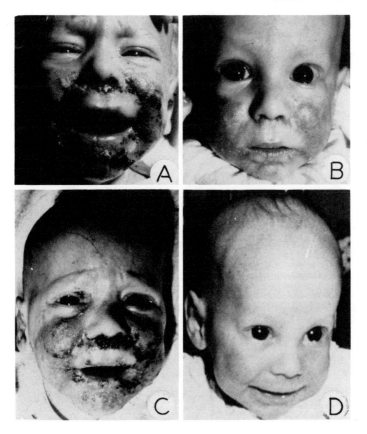

FIGURE 17.   A case of *Achrodermatitis enteropathica* at admission (A), after intravenous fat emulsion, 4 days (B), during exascerbation after 6 days of no therapy (C), and after 10 days of feeding with human milk and Diodoquin® therapy (D). (From Cash, R. and Berger, C. K., *J. Pediatr.*, 74, 717–729, 1969. With permission.)

lipids has also been reported to be low,[77] and the greatest decrease was found to be in the cholesteryl ester fraction.[78] In a study comparing MS with other diseases, Love et al.[72] found that MS definitely had a pattern of fatty acid composition of serum lipids resembling that of EFA deficiency, but this pattern was not limited to MS. Patients with acute nonneurological illnesses were found to have similar patterns, and the pattern of reduced $18:2\omega6$ and elevated $18:1$, $16:1$, and $16:0$ may be a general phenomenon in illness. Millar et al.,[79] studying a series of 75 patients with MS, found that in patients given supplements of sunflower oil emulsion (17.2 g $18:2\omega6$ per day), relapses tended to be less frequent and less severe than in patients given olive oil emulsion supplement (0.4 g $18:2\omega6$ per day). The overall rate of deterioration appeared unaffected.

G. Diabetes

Experimental diabetes superimposed on EFA deficiency accelerates the onset of EFA symptoms,[80] and dermal symptoms of deficiency increase with the severity of diabetes as measured by serum glucose.[81] Desaturation of $18:0$ and $18:2\omega6$ in the diabetic rat is decreased.[82,83] The content of $18:1\omega9$ in adipose tissue of diabetic men is higher than normal.[84] The content of $20:4\omega6$ in serum lipids has been found to be lower than normal.[85] These observations suggest that EFA metabolism may be abnormal in diabetes, and that diabetes may precipitate partial EFA deficiency.

H. Cystic Fibrosis

Underwood et al.[86] measured fatty acid patterns in lipids of erythrocytes and found that, although total polyunsaturated acids were not abnormal in cystic fibrosis (CF) patients, the $18:2\omega6$ was lowered. Campbell et al.[87] studied the ratios of several fatty acids in blood lipids and found that as palmitate increased, oleate and linoleate remained relatively constant in CF, whereas oleate increased and linoleate decreased in normal children. The patterns of fatty acids of serum lipids were clearly changed in CF, but the data published do not permit consideration of certain crucial polyunsaturated acids, precluding judgment as to whether CF modifies EFA metabolism. Results of supplementation of CF patients with EFA have been equivocal.[88]

I. Atherosclerosis and Cholesterol Transport

Despite 25 years of studies attempting to relate polyunsaturated fatty acids to atherosclerosis and an extremely voluminous and controversial literature on the role of polyunsaturated acids in this disease, there is little evidence or agreement as to whether EFA deficiency plays a causative role in the development of atherosclerosis. No attempt can be made here to be comprehensive, and only a few comments will be made.

The early reports that essential fatty acids may play a role in this disease were reviewed by Kinsell,[89] who had first attempted to link EFA to the phenomenon. Polyunsaturated oils have been demonstrated repeatedly to lower serum cholesterol transport levels, at least temporarily, and purified ethyl linoleate has also been demonstrated to have this effect. Although the development of atherosclerosis has been shown to have many causative factors and many physiological conditions and substances influence transport level of cholesterol, it is reasonable to include polyunsaturated acids as one factor which tends to lower cholesterol.

The link of EFA to cholesterol transport is indicated by the observations that dietary EFA and other polyunsaturated acids lower cholesterol content of serum and that diabetes, hypothyroidism, dietary cholesterol, dietary saturated fat, dietary mono-unsaturated fatty acids, and dietary hydrocarbons increase the requirement for EFA.[51,81] The role of EFA in transport of cholesterol may be summarized as follows.

Essential and polyunsaturated acids occur as normal components of cholesteryl esters and phospholipids of all serum lipoproteins. For normal transport of incoming loads of dietary fats, these lipoproteins must remain within the viable range of lipid and fatty acid compositions, probably because physical properties of the lipids must stay within the range of composition which permits fluidity or liquid crystalline structure. For example, if a load of saturated fat must be transported, unsaturated fatty acids must be mobilized to synthesize additional cholesteryl esters and phospholipids to accompany the increased load of triglyceride to be circulated. Moreover, unsaturated acids will be needed to become part of the reconstituted triglycerides which are synthesized in the intestinal wall in order to insure that the triglyceride to be transported is in the suitable physiological range of unsaturation. Polyunsaturated acids are stored only to a minor degree in adipose tissue and must be mobilized from structural complex lipids. If this mobilization is repeated or prolonged, partial or marginal EFA deficiency could result. This phenomenon could operate whether the transport load is of dietary or metabolic origin or the component to be transported in excess is fat, fatty acid, cholesterol, or even hydrocarbon.

J. Abetalipoproteinemia

In the genetic condition in which β-lipoproteins in serum are deficient, chylomicrons are not formed and a major route of fat absorption and transport is blocked. If this is prolonged, a relative essential fatty acid deficiency may develop. Indeed, the content of linoleic acid in cholesteryl ester and phosphatidyl choline is less than normal in abetalipoproteinemia.[90,91] The content of linoleic acid in cholesteryl ester has been suggested as a very sensitive way to monitor EFA deficiency.[92] Therefore, abetalipoproteinemia is a disease in which an error in metabolism results in faulty absorption and transport of fat and a consequent deficiency in EFA.

VIII. DIETARY SOURCES OF ESSENTIAL FATTY ACIDS

The principal dietary essential fatty acid is linoleic acid, 9,12-octadecadienoic acid, precursor of the entire $\omega6$ family of polyunsaturated fatty acids.[93] All members of the $\omega6$ family

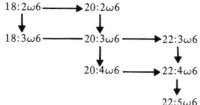

are active as essential fatty acid, and many have been shown to be more active than linoleic acid, their precursor.[51] Linoleic acid and its dehydrogenation product 18:3ω6 are found in highest abundance in plants, but more unsaturated and longer chain members of this family are found principally in animals. Notable exceptions to these generalities are the occurrence of arachidonic acid and other higher members of the group in primitive plants.[94,95] Linoleic acid is the most abundant member of the family and the principal source of EFA in human diet. Meats are the principal source of other ω6 acids, and organ meats are the richest sources.[96] Because polyunsaturated acids of the ω6 family are nearly ubiquitous in plants and animals and in most natural food sources their content is above the minimum nutrient requirement, even random selection of foods is not likely to induce EFA deficiency.

The average content of fat in American diets is 40% of calories. Even if all dietary fat were of one source, it is unlikely that EFA content of the diet would be inadequate. In Figure 18 the levels of linoleate provided by single dietary fats fed at 40% of calories are indicated above the triene/tetraene curve. Beef and mutton tallow, coconut oil, and butterfat would provide almost 1% of calories, and all other common dietary fats would provide more. It is unlikely that voluntary selection of dietary fats would provide less than the requirement for EFA. Any effort to balance sources of dietary fat or to achieve the balanced diet should provide more than the minimum requirement.

A voluminous literature is available indicating the fatty acid content of foods. Analyses of a variety of foods by modern methods have recently been published in a series covering dairy products,[97] beef products,[98] egg products,[99] nuts, peanuts and soups,[95] unhydrogenated fats and oils,[101] cereal products,[102] pork products,[103] and fish.[104] These publications provide contents of linoleic acid and other ω6 acids in a variety of food products and also contain references to a wider literature.

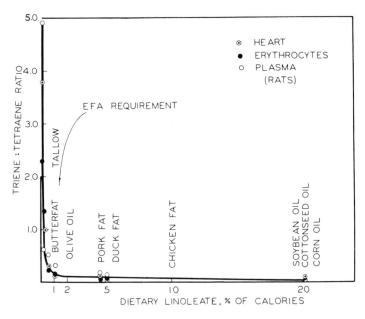

FIGURE 18.   Contribution of dietary linoleate by several common dietary fats when fed at 40% of calories. Values are indicated above the triene/tetraene curve which was used to set EFA requirement.

REFERENCES

1. **Holman, R. T.,** Essential fatty acid deficiency in animals, in *CRC Handbook Series in Nutrition and Food,* Rechcigl, M., Jr., Ed., 1977.
2. **Holman, R. T.,** Essential fatty acid deficiency, in *Progress in the Chemistry of Fats and Other Lipids,* Vol. 9, Holman, R. T., Ed., Pergamon Press, Oxford, 1971, 275–348.
3. **Söderhjelm, L., Wiese, H. F., and Holman, R. T.,** The role of polyunsaturated acids in human nutrition and metabolism, in *Progress in the Chemistry of Fats and Other Lipids,* Vol. 9, Holman, R. T., Ed., Pergamon Press, Oxford, 1971, 555–585.
4. **Holman, R. T.,** Essential fatty acids in human nutrition, in *Advances in Experimental Medicine,* Bazan, N. G., Ed., Plenum Press, New York, in press.
5. **Holman, R. T.,** The deficiency of essential fatty acids, in *Polyunsaturated Fatty Acids,* Kunau, W. and Holman, R. T., Eds., Am. Oil Chem. Soc. Monogr. No. 4, American Oil Chemists' Society, Champaign, Ill., 1977, 163–182.
6. **Holman, R. T.,** Essential fatty acid deficiency in humans, in *Dietary Lipids and Postnatal Development,* Galli, C., Jacini, G., and Pecile, A., Eds., Raven Press, New York, 1973, 127–143.
7. **Holman, R. T.,** Significance of essential fatty acids in human nutrition, in *Lipids,* Vol. 1, Paoletti, R., Porcellati, G., and Jacini, G., Eds., Raven Press, New York, 1976, 215–226.
8. **Holman, R. T.,** Deficiency of essential fatty acids in humans, in *The Essential Fatty Acids: Proceedings of the Miles Symposium 1975,* Hawkins, W. W., Ed., Nutrition Society of Canada, Winnipeg, 1976, 45–58.
9. **Holman, R. T.,** Function and biologic activities of essential fatty acids in man, in *Fat Emulsions in Parenteral Nutrition,* Meng, H. C. and Wilmore, D. W., Eds., American Medical Association, Chicago, 1976, 5–14.
10. **Caldwell, M. D.,** Human essential fatty acid deficiency: a review, in *Fat Emulsions in Parenteral Nutrition,* Meng, H. C. and Wilmore, D. W., Eds., American Medical Association, Chicago, 1976, 24–28.
11. **Burr, G. O. and Burr, M. M.,** A new deficiency disease produced by the rigid exclusion of fat from the diet, *J. Biol. Chem.,* 82, 345–367, 1929.

12. von Gröer, F., Zur Frage der praktischen Bedeutung des Nähr Wertbegriffes nebst einigen Bemerkungen über das Fett minimum des menschlichen Säulings, *Biochem. Z.,* 93, 311–329, 1919.

13. Holt, L. E., Jr., Tidwell, H. C., Kirk, C. M., Cross, D. M., and Neal, S., Studies in fat metabolism: fat absorption in normal infants, *J. Pediatr.,* 6, 427–480, 1935.

14. Hansen, A. E. and Burr, G. O., Iodine numbers of serum lipids in rats fed on fat-free diets, *Proc. Soc. Exp. Biol. Med.,* 30, 1201–1203, 1933.

15. Brown, W. R. and Hansen, A. E., Arachidonic and linoleic acid of the serum in normal and eczematous human subjects, *Proc. Soc. Exp. Biol. Med.,* 36, 113–117, 1937.

16. Hansen, A. E., Study of iodine number of serum fatty acids in infantile eczema, *Proc. Soc. Exp. Biol. Med.,* 30, 1198–1199, 1933.

17. Hansen, A. E., Serum lipids in eczema and in other pathologic conditions, *Am. J. Dis. Child.,* 53, 933–946, 1937.

18. Burr, G. O., Significance of the essential fatty acids, *Fed. Proc. Fed. Am. Soc. Exp. Biol.,* 1, 224–233, 1942.

19. Brown, W. R., Hansen, A. E., Burr, G. O., and McQuarrie, I., Effects of prolonged use of extremely low fat diet on an adult human subject, *J. Nutr.,* 16, 511–524, 1938.

20. Bryant, H. H., Griffiths, J. J., and Smith, D. W., Prolonged drip feeding, *Jackson Mem. Hosp. Bull.* (Miami), 6, 19–24, 1952.

21. von Chwalibogowski, A., Experimentaluntersuchungen über kalorisch ausreichende, qualitativ einseitige Ernährung des Säuglings, *Acta Paediatr.,* 22, 110–123, 1937.

22. Hansen, A. E. and Wiese, H. F., Clinical and blood lipid studies in a child with chylous ascites, *Am. J. Dis. Child.,* 68, 351–352, 1944.

23. Warwick, W. J., Holman, R. T., Quie, P. G., and Good, R. A., Chylous ascites and lymphedema, *Am. J. Dis. Child.,* 98, 317–329, 1959.

24. Luzzatti, L. and Hansen, A. E., Serum lipids in celiac syndrome, *J. Pediatr.,* 24, 417–435, 1944.

25. Adam, D. J. D., Hansen, A. E., and Wiese, H. F., Essential fatty acids in infant nutrition. II. Effect of linoleic acid on caloric intake, *J. Nutr.,* 66, 555–564, 1958.

26. Hansen, A. E., Haggard, M. E., Boelsche, A. N., Adam, D. J. D., and Wiese, H. F., Essential fatty acids in infant nutrition. III. Clinical manifestations of linoleic acid deficiency, *J. Nutr.,* 66, 565–576, 1958.

27. Hansen, A. E. and Wiese, H. F., Unsaturated fatty acid levels in human serum, *Fed. Proc. Fed. Am. Soc. Exp. Biol.,* 11, 445–446, 1952.

28. Hansen, A. E. and Wiese, H. F., Essential fatty acids in human beings, in *The Vitamins,* Vol. 2, Sebrell, W. H. and Harris, R. S., Eds., Academic Press, New York, 1954, 300–311.

29. Holman, R. T. and Bissen, S., unpublished data.

30. Paulsrud, J. R., Pensler, L., Whitten, C. F., Stewart, S., and Holman, R. T., Essential fatty acid deficiency in infants induced by fat-free intravenous feeding, *Am. J. Clin. Nutr.,* 25, 897–904, 1972.

31. Hansen, A. E., Wiese, H. F., Boelsche, A. N., Haggard, M. E., Adam, D. J. D., and Davis, H., Role of linoleic acid in infant nutrition, *Pediatrics,* 31(Suppl. 1), 171–192, 1963.

32. Adam, D. J. D., Hansen, A. E., and Wiese, H. F., Essential fatty acids in infant nutrition. II. Effect of linoleic acid on caloric intake, *J. Nutr.,* 66, 555–564, 1958.

33. Holman, R. T., The ratio of trienoic:tetraenoic acids in tissue lipids as a measure of essential fatty acid requirement, *J. Nutr.,* 70, 405–410, 1960.

34. Holman, R. T., Caster, W. O., and Wiese, H. F., The essential fatty acid requirements of infants and the assessment of their dietary intake of linoleate by serum fatty acid analysis, *Am. J. Clin. Nutr.,* 14, 70–75, 1964.

35. Caster, W. O., Ahn, P., Hill, E. G., Mohrhauer, H., and Holman, R. T., Determination of linoleate requirement and swine by a new method of estimating nutritional requirement, *J. Nutr.,* 78, 147–154, 1962.

36. Pudelkewicz, C. and Holman, R. T., Positional distribution of fatty acids in liver lecithin of rats as a function of dietary linoleate or linolenate, *Biochim. Biophys. Acta,* 152, 340–345, 1968.

37. Cuthbertson, W. F. J., Essential fatty acid requirements in infancy, *Am. J. Clin. Nutr.,* 29, 559–568, 1976.

38. Mohrhauer, H. and Holman, R. T., The effect of dose level of essential fatty acids upon fatty acid composition of the rat liver, *J. Lipid Res.,* 4, 151–159, 1963.

39. Witting, L. A., The interrelationship of polyunsaturated fatty acids and antioxidants *in vivo,* in *Progress in The Chemistry of Fats and Other Lipids,* Vol. 9, Holman, R. T., Ed., Pergamon Press, Oxford, 1971, 517–553.

40. **Holman, R. T., Hayes, H. W., Rinne, A., and Söderhjelm, L.,** Polyunsaturated fatty acids in serum of infants fed breast milk or cow's milk, *Acta Paediatr. Scand.,* 54, 573–577, 1965.

41. **White, H. B., Turner, M. D., Turner, A. C., Miller, R. C., and Rawson, J. E.,** Essential fatty acid deficiency in brain lipids of three premature infants receiving intravenous fat-free alimentation, in *Fat Emulsions in Parenteral Nutrition,* Meng, H. C. and Wilmore, D. W., Eds., American Medical Association, Chicago, 1976, 31–35.

42. **Bozian, R. C. and MacGee, J.,** Total parenteral nutrition and essential fatty acid deficiency – a seven year study in short bowel syndrome, in *Fat Emulsions in Parenteral Nutrition,* Meng, H. C. and Wilmore, D. W., Eds., American Medical Association, Chicago, 1976, 36–37.

43. **Riella, M. C., Broviac, J. W., Wells, M., and Scribner, B. H.,** Essential fatty acid deficiency in human adults, *Ann. Intern. Med.,* 83, 786–789, 1975.

44. **Friedman, Z., Danon, A., Stahlman, M. T., and Oates, J. A.,** Rapid onset of essential fatty acid deficiency in the newborn, *Pediatrics,* 58, 640–649, 1976.

45. **Friedman, Z., Shochat, S. J., Maisels, J., Marks, K. H., and Lamberth, E. L., Jr.,** Correction of essential fatty acid deficiency in newborn infants by cutaneous application of sunflower-seed oil, *Pediatrics,* 58, 650–654, 1976.

46. **Richardson, T. L. and Sgoutas, D.,** Essential fatty acid deficiency in four adult patients during total parenteral nutrition, *Am. J. Clin. Nutr.,* 28, 258–263, 1975.

47. **Holman, R. T., Caster, W. O., and Wiese, H. F.,** Estimation of linoleate intake of men from serum lipid analysis, *Am. J. Clin. Nutr.,* 14, 193–196, 1964.

48. **Fleischman, A. I., Hayton, T., and Bierenbaum, M. L.,** Objective biochemical determination of dietary adherence in the young coronary male, *Am. J. Clin. Nutr.,* 20, 333–337, 1967.

49. **Olegård, R. and Svennerholm, L.,** Effects of diet on fatty acid composition of plasma and red cell phosphoglycerides in three-month-old infants, *Acta Paediatr. Scand.,* 60, 505–511, 1971.

50. **Holman, R. T.,** Quantitative chemical taxonomy based upon composition of lipids, in *Progress in The Chemistry of Fats and Other Lipids,* Vol. 16, Holman, R. T., Ed., Pergamon Press, Oxford, in press.

51. **Holman, R. T.,** Biological activities of and requirements for polyunsaturated fatty acids, in *Progress in The Chemistry of Fats and Other Lipids,* Vol. 9, Holman, R. T., Ed., Pergamon Press, Oxford, 1971, 607–682.

52. **Reid, M. E., Bieri, J. G., Plack, P. A., and Andrews, E. L.,** Nutritional studies with the guinea pig. II. Determination of the linoleic acid requirement, *J. Nutr.,* 82, 401–408, 1964.

53. **Cash, R. and Berger, C. K.,** *Acrodermatitis enteropathica:* defective metabolism of unsaturated fatty acids, *J. Pediatr.,* 74, 717–729, 1969.

54. **Holman, R. T.,** Nutritional and metabolic interrelationships between fatty acids, *Fed. Proc. Fed. Am. Soc. Exp. Biol.,* 23, 1062–1067, 1964.

55. **Collins, F. D., Sinclair, A. J., Royle, J. P., Coats, D. A., Maynard, A. T., and Leonard, R. F.,** Plasma lipids in human linoleic acid deficiency, *Nutr. Metab.,* 13, 150–167, 1971.

56. **Wene, J. D., Connor, W. E., and DenBesten, L.,** The development of essential fatty acid deficiency in healthy men fed fat-free diets intravenously and orally, *J. Clin. Invest.,* 56, 127–134, 1975.

57. **Connor, W. E.,** Pathogenesis and frequency of essential fatty acid deficiency during total parenteral nutrition, *Ann. Int. Med.,* 83, 895–896, 1975.

58. **Stegink, L. D., Freeman, J. B., Wispe, J., and Connor, W. E.,** Absence of the biochemical symptoms of essential fatty acid deficiency in surgical patients undergoing protein sparing therapy, *Am. J. Clin. Nutr.,* in press.

59. **White, H. B., Jr., Turner, M. D., Turner, A. C., and Miller, R. C.,** Blood lipid alterations in infants receiving intravenous fat-free alimentation, *J. Pediatr.,* 83, 305–313, 1973.

60. **Caldwell, M. D., Jonsson, H. T., and Othersen, H. B.,** Essential fatty acid deficiency in an infant receiving prolonged parenteral alimentation, *J. Pediatr.,* 81, 894–898, 1972.

61. **Press, M., Hartop, P. J., and Prottey, C.,** Correction of essential fatty-acid deficiency in man by the cutaneous application of sunflower-seed oil, *Lancet,* 1, 597–598, 1974.

62. **Böhles, H., Bieber, M. A., and Heird, W. C.,** Reversal of experimental essential fatty acid deficiency by cutaneous administration of safflower oil, *Am. J. Clin. Nutr.,* 29, 398–401, 1976.

63. **Hunt, C. E., Engel, R. R., Modler, S., Hamilton, W., Bissen, S., and Holman, R. T.,** Essential fatty acid (EFA) deficiency in neonates: inability to reverse deficiency by topical applications of EFA-rich oil, *J. Pediatr.,* in press.

64. **Wilmore, D. W., Helmkamp, G., Moylan, J. A., and Pruitt, B. A.,** Essential fatty acid deficiency following thermal injury, in *Fat Emulsions in Parenteral Nutrition,* Meng, H. C. and Wilmore, D. W., Eds., American Medical Association, Chicago, 1976, 29–31.

65. **Helmkamp, G. M., Wilmore, D. W., Johnson, A. A., and Pruitt, B. A.,** Essential fatty acid deficiency in red cells after thermal injury: correction with intravenous fat therapy, *Am. J. Clin. Nutr.,* 26, 1331–1338, 1973.

66. **Jelenko, C., III,** Studies in burns. VII. The water-retaining lipids of eschar: their presence and potential usefulness, *Am. Surg.,* 35, 709–718, 1969.

67. **Jelenko, C., III, Wheeler, M. L., and Scott, T. H.,** Studies in burns. X. Ethyl linoleate: the water holding lipid of skin. A. The evidence, *J. Trauma,* 12, 968–973, 1972.

68. **Jelenko, C., III, Wheeler, M. L., Anderson, A. P., Callaway, D., and Scott, R. A., Jr.,** Studies in burns. XIII. Effects of a topical lipid on burned subjects and their wounds, *Am. Surg.,* 41, 466–482, 1975.

69. **Hansen, A. E. and Wiese, H. F.,** Fat in the diet in relation to nutrition of the dog. I. Characteristic appearance and gross changes of animals fed diets with and without fat, *Tex. Rep. Biol. Med.,* 9, 491–515, 1951.

70. **Hansen, A. E., Beck, O., and Wiese, H. F.,** Susceptibility to infection manifested by dogs on a low fat diet, *Fed. Proc. Fed. Am. Soc. Exp. Biol.* (abstr.), 7, 289, 1948.

71. **Curry, C. R. and Quie, P. G.,** Fungal septicemia in patients receiving parenteral hyperalimentation, *N. Engl. J. Med.,* 285, 1221–1225, 1971.

72. **Love, W. C., Cashell, A., Reynolds, M., and Callaghan, N.,** Linoleate and fatty-acid patterns of serum lipids in multiple sclerosis and other diseases, *Br. Med. J.,* 3, 18–21, 1974.

73. **Krivit, W. and Holman, R. T.,** unpublished data.

74. **Bernsohn, J.,** Fatty acids and multiple sclerosis, *Lancet,* 2, 1259, 1967.

75. **Bernsohn, J. and Stephanides, L. M.,** Aetiology of multiple sclerosis, *Nature,* 215, 821–823, 1967.

76. **Kishimoto, Y., Radin, N. S., Tourtellotte, W. C., Parker, J. A., and Itabashi, H. H.,** Gangliosides and glycerophospholipids in multiple sclerosis white matter, *Arch. Neurol.* (Chicago), 16, 44–54, 1967.

77. **Baker, R. W. R., Thompson, R. H. S., and Zilkha, K. J.,** Serum fatty acids in multiple sclerosis, *J. Neurol. Neurosurg. Psychiatry,* 27, 408–414, 1964.

78. **Baker, R. W. R., Saunders, H., Thompson, R. H. S., and Zilkha, K. J.,** Serum cholesterol linoleate levels in multiple sclerosis, *J. Neurol. Neurosurg. Psychiatry,* 28, 212–21711965.

79. **Millar, J. H. D., Zilkha, K. J., Langman, M. J. S., Wright, H. P., Smith, A. D., Belin, J., and Thompson, R. H. S.,** Double-blind trial of linoleate supplementation of the diet in multiple sclerosis, *Br. Med. J.,* 2, 765–768, 1973.

80. **Peifer, J. J. and Holman, R. T.,** Essential fatty acids, diabetes, and cholesterol, *Arch. Biochem. Biophys.,* 57, 520–521, 1955.

81. **Holman, R. T.,** The lipids in relation to atherosclerosis, *Am. J. Clin. Nutr.,* 8, 95–103, 1960.

82. **Peluffo, R. O., de Gomes Dumm, N. T., deAlaniz, M. J. T., and Brenner, R. R.,** Effect of protein and insulin on linoleic acid desaturation of normal and diabetic rats, *J. Nutr.,* 101, 1075–1083, 1971.

83. **deTomas, M. E., Peluffo, R. O., and Mercuri, O.,** Effect of dietary glycerol on the $(1-^{14}C)$stearic acid and $(1-^{14}C)$linoleic acid desaturation of normal and diabetic rats, *Biochim. Biophys. Acta,* 306, 149–155, 1973.

84. **Antonini, F. M., Bucalossi, A., Petruzzi, E., Simoni, R., Morini, P. L., and D'Allessandro, A.,** Fatty acid composition of adipose tissue in normal, atherosclerotic and diabetic subjects, *Atherosclerosis,* 11, 279–289, 1970.

85. **Tuna, N., Frankhauser, S., and Goetz, F. C.,** Total serum fatty acids in diabetes: relative and absolute concentrations of individual fatty acids, *Am. J. Med. Sci.,* 255, 120–131, 1968.

86. **Underwood, B. A., Denning, C. R., and Navab, M.,** Polyunsaturated fatty acids and tocopherol levels in patients with cystic fibrosis, *Ann. N.Y. Acad. Sci.,* 203, 237–247, 1972.

87. **Campbell, I. M., Crozier, D. N., and Coton, R. B.,** Abnormal fatty acid composition and impaired oxygen supply in cystic fibrosis patients, *Pediatrics,* 57, 480–486, 1976.

88. **Elliott, R. B.,** Therapeutic trial of fatty acid supplementation in cystic fibrosis, *Pediatrics,* 57, 474–479, 1976.

89. **Kinsell, L. W.,** Relationship of dietary fats to atherosclerosis, in *Progress in The Chemistry of Fats and Other Lipids,* Vol. 6, Holman, R. T., Ed., Pergamon Press, Oxford, 1963, 114–135.

90. **Barnard, G., Fosbrooke, A. S., and Lloyd, J. K.,** Neutral lipids of plasma and adipose tissue in abetalipoproteinemia, *Clin. Chim. Acta,* 28, 417–422, 1970.

91. **Jones, J. W. and Ways, P.,** Abnormalities of high density lipoproteins in abetalipoproteinemia, *J. Clin. Invest.,* 46, 1151–1161, 1967.

92. **Conner, W. E.,** personal communication.

93. **Mead, J. F.,** The metabolism of the polyunsaturated fatty acids, in *Progress in The Chemistry of Fats and Other Lipids,* Vol. 9, Holman, R. T., Ed., Pergamon Press, Oxford, 1971, 159–192.

94. **Schlenk, H. and Gellerman, J. L.,** Arachidonic 5,11,14,17-eicosatetraenoic and related acids in plants – identification of unsaturated fatty acids, *J. Am. Oil Chem. Soc.,* 42, 504–511, 1965.

95. **Gellerman, J. L., Anderson, W. H., Richardson, D. G., and Schlenk, H.,** Distribution of arachidonic and eicosapentaenoic acids in the lipids of mosses, *Biochim. Biophys. Acta,* 388, 277–290, 1975.

96. **Holman, R. T. and Greenberg, S. I.,** Highly unsaturated fatty acids. I. Survey of possible animal sources, *J. Am. Oil Chem. Soc.,* 30, 600–601, 1953.

97. **Posati, L. P., Kinsella, J. E., and Watt, B. K.,** Comprehensive evaluation of fatty acids in foods. I. Dairy products, *J. Am. Diet. Assoc.,* 66, 482–488, 1975.

98. **Anderson, B. A., Kinsella, J. E., and Watt, B. K.,** Comprehensive evaluation of fatty acids in foods. II. Beef products, *J. Am. Diet. Assoc.,* 67, 35–41, 1975.

99. **Posati, L. P., Kinsella, J. E., and Watt, B. K.,** Comprehensive evaluation of fatty acids in foods. III. Eggs and egg products, *J. Am. Diet. Assoc.,* 67, 111–115, 1975.

100. **Fristrom, G. A., Stewart, B. C., Weihrauch, J. L., and Posati, L. P.,** Comprehensive evaluation of fatty acids in foods. IV. Nuts, peanuts, and soups, *J. Am. Diet. Assoc.,* 67, 351–355, 1975.

101. **Brignoli, C. A., Kinsella, J. E., and Weihrauch, J. L.,** Comprehensive evaluation of fatty acids in foods. V. Unhydrogenated fats and oils, *J. Am. Diet. Assoc.,* 68, 224–229, 1976.

102. **Weihrauch, J. L., Kinsella, J. E., and Watt, B. K.,** Comprehensive evaluation of fatty acids in foods. VI. Cereal products, *J. Am. Diet. Assoc.,* 68, 335–340, 1976.

103. **Anderson, B. A.,** Comprehensive evaluation of fatty acids in foods. VII. Pork products, *J. Am. Diet. Assoc.,* 68, 44–49, 1976.

104. **Exler, J., Kinsella, J. E., and Watt, B. K.,** Lipids and fatty acids of important finfish: new data for nutrient tables, *J. Am. Oil Chem. Soc.,* 52, 154–159, 1975.

Index

INDEX

I

N

Z

CRC PUBLICATIONS OF RELATED INTEREST

CRC HANDBOOKS:

CRC FENAROLI'S HANDBOOK OF FLAVOR INGREDIENTS 2nd Edition
Edited, translated, and revised by: **Thomas E. Furia** and **Nicolo Bellanca,** Dynapol, Palo Alto, California.
This two-volume Handbook is an update of the 1st Edition and includes comprehensive review chapters by recognized experts in the field. It is extensively indexed for easy use.

CRC HANDBOOK OF FOOD ADDITIVES, 2nd Edition
Edited by **Thomas E. Furia,** Dynapol, Palo Alto, California.
Nearly 1000 pages of pertinent food additive information is offered, reflecting the important changes that have occurred in recent years in the area of food additives.

CRC UNISCIENCE PUBLICATIONS:

FOOD ANALYSIS: Analytical Quality Control Methods for the Manufacturer and Buyer, 3rd. Edition
By **R. Lees M.R.S.H., A.I.F.S.T.,** food industry consultant. This book brings together methods of analysis which are of most value to the factor control chemist.

LOW CALORIE AND DIETETIC FOODS
Edited by **Basant K. Dwivedi, Ph.D.,** Estee Candy Company, Inc.
This book includes all aspects of low calorie and dietetic foods including discussions on fructose, applications and commercial potential of sweetening agents, and the present status of food products for people with special dietary requirements.

MAN, FOOD AND NUTRITION
Edited by **Miloslav Rechcigl, Jr., M.N.S., Ph.D., F.A.A.A.S., F.A.I.C.,F.W.A.S.,** Agency for International Development, U.S. Department of State.
This interdisciplinary treatise offers a comprehensive and integrated critical review of the nature and the scope of the world food problem. It presents strategies and discusses various technological approaches to overcoming world hunger and malnutrition.

TOXICITY OF PURE FOODS
By **Eldon M. Boyd, Ph.D.** (deceased), Queen's University, Kingston, Ontario. Edited by **Carl E. Boyd, M. D.,** Health and Welfare, Canada.
A systematic study of the toxicity of pure foods is described in this volume.

WORLD FOOD PROBLEM
Edited by.,**Miloslav Rechcigl, Jr., M.N.S., Ph.D., F.A.A.A.S., F.A.I.C.,F.W.A.S.,**Agency for International Development, U.S. Department of State.
This is a comprehensive and up-to-date bibliography on all important facets of the world food problem, encompassing such areas as the availability of natural resources and the present future sources of energy.

CRC MONOTOPIC REPRINTS:

FLEXIBLE PACKAGING OF FOODS
By **Aaron L. Brody, S. B., M.B.A., Ph.D.,** Arthur D. Little Inc.
The aim of this book is to describe food products employing flexible packaging, the requirements dictating the flexible packaging being used and the current state of flexible packaging in the food industry.

FREEZE-DRYING FOODS

By **C. Judson King, B.E., E.M., Sc.D.,** University of California.
This review concentrates on several papers published in recent years, relates them to the rest of the field, gives an evaluation of their findings, and the conclusions they draw.

SOYBEANS AS A FOOD SOURCE, 2nd Edition

By **W. J. Wolf, B.S., Ph.D.,** and **J. C. Cowan, A.B., Ph.D.,** Northern Marketing and Nutrition Research Division, Peoria, Illinois.
The conversion of soybeans to food is summarized in this book. Emphasis is given to the protein content of soybeans because of the current high level of interest.

STORAGE, PROCESSING, AND NUTRITIONAL QUALITY OF FRUITS AND VEGETABLES

By **D.K. Salunkhe, B.S., M.S., Ph.D.,** Utah State University.
This book reviews the nutritional value and quality of fruits and vegetables as influenced by chemical treatments, storage and processing conditions.

THE USE OF FUNGI AS FOOD AND IN FOOD PROCESSING, Parts 1 and 2

By **William D. Gray, A.B., Ph.D.,** Northern Illinois University. Edited by Thomas E. Furia, Dynapol, Palo Alto, California.
This two-volume work describes how filamentous fungi have been used in the past and present as a food source in various areas of the world.

CRC CRITICAL REVIEW JOURNALS:

CRC CRITICAL REVIEWS IN FOOD SCIENCE AND NUTRITION

Edited by **Thomas E. Furia,** Dynapol, Palo Alto, California.